Emerging and Epizootic Fungal Infections in Animals

Seyedmojtaba Seyedmousavi •
G. Sybren de Hoog •
Jacques Guillot •
Paul E. Verweij
Editors

Emerging and Epizootic Fungal Infections in Animals

Springer

Editors
Seyedmojtaba Seyedmousavi
Laboratory of Clinical Immunology and Microbiology (LCIM)
National Institute of Allergy and Infectious Diseases (NIAID)
National Institutes of Health (NIH)
Bethesda, MD
USA

G. Sybren de Hoog
Center of Expertise in Mycology
RadboudUMC/CWZ
Nijmegen, The Netherlands

Westerdijk Fungal Biodiversity Institute
Utrecht, The Netherlands

Jacques Guillot
Department of Parasitology, Mycology and Dermatology
EnvA, Ecole nationale vétérinaire d'Alfort, UPEC
Maisons-Alfort, France

Dynamyc Research Group
EnvA, Ecole nationale vétérinaire d'Alfort, UPEC
Maisons-Alfort, France

Paul E. Verweij
Department of Medical Microbiology
Radboud University Medical Center
Nijmegen, The Netherlands

Center of Expertise in Mycology
RadboudUMC/CWZ
Nijmegen, The Netherlands

ISBN 978-3-030-10155-8 ISBN 978-3-319-72093-7 (eBook)
https://doi.org/10.1007/978-3-319-72093-7

Softcover re-print of the Hardcover 1st edition 2018

Printed on acid-free paper

This Springer imprint is published by the registered company Springer International Publishing AG part of Springer Nature.
The registered company address is: Gewerbestrasse 11, 6330 Cham, Switzerland

Foreword

Four internationally renowned experts in fungal pathogens have edited this commendable multi-authored book on fungal infections in animals and emerging zoopathogenic fungi. The seriousness of fungal diseases in humans has long been underestimated compared to viral and bacterial diseases for various reasons, but the fungal diseases in animals have been neglected even more despite recent warning on how fungal diseases can wreak havoc on certain animals such as amphibians worldwide. It is a timely publication providing an in-depth discernment on epidemics and emerging fungal diseases in domestic as well as various wild animals. The first comprehensive book on fungal diseases in animals was published in 1955 by G.C. Ainsworth and P.K.C. Austwick. A couple more have followed but this book is long overdue in order to bring the subject up to date in this ever changing environment.

This book is organized in four parts consisting of 16 chapters: Part I has two chapters. The first chapter provides basic information on categories of pathogenic species and species that cause epidemics in the Kingdom Fungi. The second chapter details epidemiological definitions, terminology and classifications of fungal infections in animal. Part II describes seven mycoses and the etiologic agents that cause infections both in man and animals. Part II starts with dermatophytoses, a superficial fungal disease and six systemic infections all caused by fungi belonging to *Ascomycota*. Part III contains six chapters that focus on the emerging animal diseases although some of them such as sporotrichosis, cryptococcosis and aspergillosis are also important diseases in man. Part III is particularly important since it contains diseases of lower animals such as chytridiomycosis in amphibians caused by *Batrachochytrium* spp. and other important fungal diseases that can cause serious damage to the ecosystem. Part IV contains a single chapter describing antifungal therapy for animal mycoses and how resistance to antifungal drugs has been emerging.

It is noteworthy that many chapters in the book present very useful simple diagrams that chronologically depict the milestone discoveries in the progress of our understanding of the diseases and their etiologic agents.

I believe this book addresses a wide range of audiences beyond pathogenic mycologists, veterinary mycologists, infectious disease specialists and epidemiologists. I hope this book will also be received with enthusiasm by those who care about animal welfare and the health of our ecosystem.

Molecular Microbiology Section
Laboratory of Clinical Infectious Diseases
National Institute of Allergy and Infectious Diseases
Bethesda, MD
USA

Kyung J. Kwon-Chung

Preface

We usually think that rainy cities like Amsterdam do not provide optimal conditions for exotic animals. However, the trees in the centre are littered with flocks of loud green parrots, and in a bistro at the central station a giant cockatoo walks around freely. In March 2015, we (Amir, Sybren, Jacques) drank a good Dutch beer and apparently got inspired by this unexpected "wildlife". We noticed that there is not much recent literature on fungal infections in animals, and after some more beers we thought that an overview of emerging and epizootic infections might be appreciated. Paul joined us as a co-editor a few months later and a long process of writing and editing started, and soon we were happy to be supported by an excellent panel of authors of chapters.

Fungi are relatively uncommon causes of disease in animals: the spectrum of fungal infections is much smaller than in the single host *Homo sapiens*. Yet animal hosts are much more exposed than humans to infectious propagules, both in husbandry and in natural ecosystems. For reasons that could not always be explained, an increasing number of recalcitrant fungal diseases in animals have emerged during the last decades, originating from a wide range of opportunistic and pathogenic fungi. With this book we aim to provide more insight into major epizootic and emerging mycoses in various animal groups, with speculations on fungal life cycles, epidemiology and evolution. Information on treatment options, antifungal use in veterinary practice, and emerging resistance is also included. The different chapters have been written by experts in the field, most of them being members of the Veterinary Mycology working group of ISHAM (International Society of Human and Animal Mycology). Inevitably the production of a multi-authored book takes a significant amount of time, and we were unable to cover all relevant fungi, and to include some recent developments in epidemiology and host resistance—underlining that natural processes of host-pathogen interaction may change and develop at a surprising speed.

The book addresses medical and veterinary mycologists, microbiologists, veterinarians, infectious disease specialists, epidemiologists, ecologists, public health scientists from academia and industry as well as graduate students, PhD students and postdocs in the field.

Bethesda, MD, USA — Amir (Seyedmojtaba) Seyedmousavi
Utrecht, The Netherlands — G. Sybren de Hoog
Maisons-Alfort, France — Jacques Guillot
Nijmegen, The Netherlands — Paul E. Verweij

Contents

Part I

Pathogenic Fungi, Definitions, Terminology, Methods of Classification

1 Distribution of Pathogens and Outbreak Fungi in the Fungal Kingdom

G. Sybren de Hoog, Sarah A. Ahmed, Patrizia Danesi, Jacques Guillot, and Yvonne Gräser

Abstract

Over 625 fungal species have been reported to cause infection in vertebrates. The fungal kingdom contains 167 orders, of which 40 (24%) were repeatedly cited in the medical literature. Recurrence indicates that these species have a certain predisposition to cause infection. In the present chapter, the different categories of pathogens and outbreak fungi are presented and discussed. Most emerging fungi concern infections that are non-transmissible; their frequency may show moderate increase due to changes of host conditions. Outbreaks may concern multiple infections from a common environmental source, known as sapronoses.

G. S. de Hoog (✉)
Center of Expertise in Mycology RadboudUMC/CWZ, Nijmegen, The Netherlands

Westerdijk Fungal Biodiversity Institute, Utrecht, The Netherlands
e-mail: s.hoog@westerdijkinstitute.nl

S. A. Ahmed
Westerdijk Fungal Biodiversity Institute, Utrecht, The Netherlands
e-mail: sara3707@gmail.com

P. Danesi
Istituto Zooprofilattico Sperimentale delle Venezie, Legnaro, Italy
e-mail: pdanesi@izsvenezie.it

J. Guillot
Department of Parasitology, Mycology and Dermatology, EnvA, Ecole nationale vétérinaire d'Alfort, UPEC, Maisons-Alfort, France

Dynamyc Research Group, EnvA, Ecole nationale vétérinaire d'Alfort, UPEC, Maisons-Alfort, France
e-mail: jacques.guillot@vet-alfort.fr

Y. Gräser
Nationales Konsiliarlabor für Dermatophyten, Institut für Mikrobiologie und Hygiene, Universitätsmedizin Berlin – Charité, Berlin, Germany
e-mail: yvonne.graeser@charite.de

S. Seyedmousavi et al. (eds.), *Emerging and Epizootic Fungal Infections in Animals*,
https://doi.org/10.1007/978-3-319-72093-7_1

When their life cycle has an invasive phase with adaptations to reside inside host tissue, the fungi are referred to as environmental pathogens. Host-to-host transmission occurs in zoophilic pathogens, which have no environmental phase. This kind of fungi is responsible for mycoses which may occur with changes in host factors and when naive host populations are exposed to novel fungal genotypes. The most dramatic transmissible mycoses are expected with a combination of host and fungal changes. Successful outbreak fungi are recognizable by low genetic diversity.

1.1 Opportunistic and Pathogenic Fungi

The animal body can be regarded as an extreme environment for fungal growth; only a small fraction of the species known in the fungal kingdom is able to cause infection. When a fungus occupies an environmental niche but is able to survive in animal host tissue—because some of the factors needed in its environmental niche coincidentally enhance survival in the host—the infection is referred to as **opportunistic**. However, the infection is **non-transmissible** and the etiologic agent will finally die with the animal host at the end of its life span (Fig. 1.1). Infection is thus

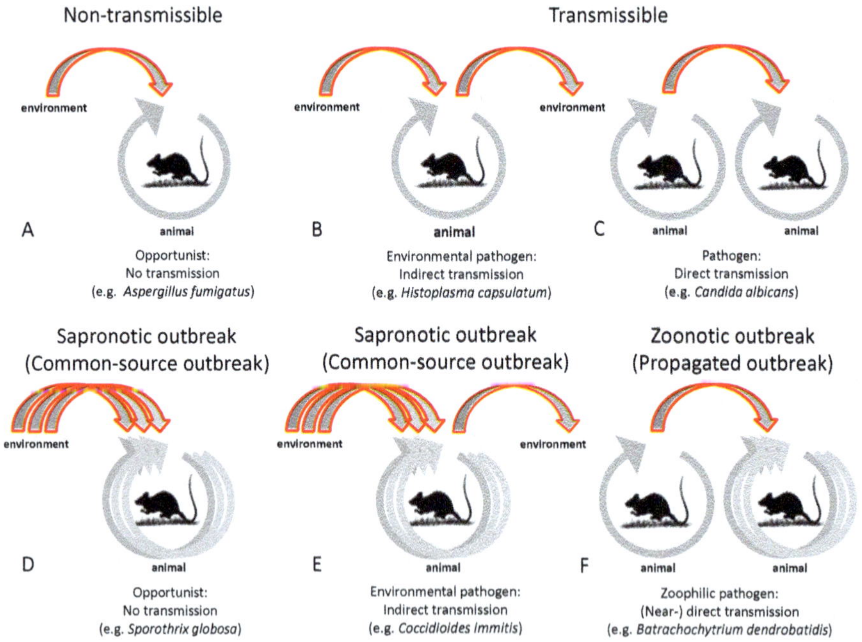

Fig. 1.1 Diagram of possible life cycles. (**a**) Non-transmissible opportunist; (**b**) transmissible environmental pathogen, with infectious propagule distribution via the environment; (**c**) transmissible zoopathogen via contagious hosts; (**d**) sapronosis by opportunist, non-transmissible; (**e**) sapronosis by environmental pathogen, with infectious propagule distribution via the environment; (**f**) zoonosis by transmissible zoophilic pathogen via contagious hosts

not profitable for the fungus: the infectious individuals will be lost for the fungal population and thus have a detrimental effect on the fitness of the species. For example, fungi belonging to *Mucorales* (*Rhizopus*, *Lichtheimia*, *Saksenaea*) are responsible for diseases with a severe morbidity, but do not propagate and are even difficult to culture from the host; their preferred habitat remains dead material such as foodstuffs, on which their sporulation is massive. Another common example is *Aspergillus fumigatus*, which is thermotolerant because of the conditions in its natural habitat of self-heated compost, and this factor also enables survival in immunocompromised patients or in birds whose internal body temperature (up to 43 °C) is much higher than in mammals. In general, decreased host immunity allows extended fungal growth and increases severity of the infection.

A small fraction of the extant fungal species is adapted to the animal host, i.e., have an increased fitness when a living mammal host is used in their life cycle. The infection is **transmissible**, so that mutations acquired during infection can be passed on to next generations. In a **zoophilic pathogen**, this transmission takes place from host-to-host; the infected animal is contagious. For most transmissible fungi, the animal host may be used as a vehicle for fungal dispersal, but the main reservoir of the fungus remains environmental. Such fungi have a double life cycle: in the environment versus in the animal; the host is infected via propagules that were produced in the environment, and thus infected individuals are not contagious. We call such fungi **environmental pathogens**.

Of the estimated 1.5 million extant fungal species, only about 625 have been reported to cause infection in vertebrates. The fungal kingdom presently contains 167 orders, of which 40 (24%) are listed in the *Atlas of Clinical Fungi* (de Hoog et al. 2017) containing members that were repeatedly cited in the medical literature. Recurrence indicates that these species have a certain predisposition to cause infection. If we count the number of fungal genera within the 40 orders, the distribution of fungi that are predisposed to infection is even more skewed: of the approximately 18,000 described genera (Crous et al. 2014), only 206 (0.01%) are listed in the *Atlas*. Considering that only 45 genera, dispersed over 10 orders, belong to the relatively common infectious agents, it is obvious that vertebrate pathogenicity is an extremely rare ability of fungal species. These figures (i.e., the number of species involved) have not really changed with the event of hospitalized patient populations with severe immune disorders; just the frequency of infectious species has increased and some have become quite common. Thus, the popular statement of the "immunocompromised host as a living Petri dish," suggesting that such patients can be infected by "any" fungus, is far from correct.

Nevertheless, the number of known infectious agents is growing. This is largely due to developments in awareness and diagnostics and particularly in molecular taxonomic approaches which enable species distinction with much larger resolution. Also next-generation sequencing improves our knowledge of commensals on the skin and the intestinal microbiome, with concomitant increase of the number of known human- or animal-associated fungi.

1.2 Epidemic and Epizootic Expansions

A small number of opportunistic and pathogenic species or species groups tend to occur in the form of outbreaks, epidemics (in humans), or epizootics (in animals) (Fisher et al. 2012). Fungal epidemics have already been reported in the early twentieth century. Beurmann and Gougerot (1912) described an expansion of sporotrichosis in France, caused by *Sporothrix schenckii*. Zhang et al. (2015) demonstrated that each closely related species of *Sporothrix* has a consistent pattern of outbreaks. Common-source outbreaks are usually reported, which die out when the environmental conditions no longer support growth of the fungus. An epidemic of sporotrichosis by *S. schenckii* among miners in South Africa involved more than 3000 cases and subsequently disappeared when the wood that was used in the mines was treated with preservatives (Govender et al. 2015). When nonliving biological material is the source of infection, the epidemic is referred to as a **sapronosis**. Infected vertebrates do not spread the fungus, and thus sapronoses are caused by non-transmissible opportunistic fungi (Fig. 1.1). Sapronotic outbreaks are linked to a common source of infection, and prevention thus requires physical removal of the infective material.

The epidemiological pattern is fundamentally different when fungal pathogens are transmissible between animals including humans. The agent of chytridiomycosis causing global frog decline, *Batrachochytrium dendrobatidis*, efficiently utilizes the host for its dispersal (Chap. 14). The fungus produces zoosporangia in frog skin, which release massive amounts of zoospores into the environment and contaminate new susceptible hosts. The time of transmission is short, and the environmental phase does not require growth of the thallus but just zoospore dispersal. Transmission can thus be nearly direct, from host-to-host, maximally with a short intermittent phase of motile spores in water. Contagious individuals potentially infect multiple individuals, and hence direct transmission often leads to exponentially expanding epizootics (Fig. 1.1). The fungus completes its life cycle on the host and is thus considered as a pathogen. This kind of mycosis is transmitted directly between living hosts. According to the official definition from the World Health Organization, zoonoses are diseases and infections that are naturally transmitted between vertebrate animals and humans and vice versa. Among transmissible fungal pathogens, a few species should be considered as zoonotic: *Sporothrix brasiliensis* (from cats) (Chap. 10) and some species of dermatophytes, e.g., *Microsporum canis* (from cats), *Trichophyton verrucosum* (from cattle), and *T. benhamiae* (from guinea pigs) (Chap. 3).

Environmental pathogens have a **double life cycle**, combining characteristics of both groups above. Part of their life cycle is completed in the environment, while they also have a reservoir in the vertebrate host. Infection takes place by propagules from the environment, and expansions are thus classified as sapronoses. After infection and completion of a pathogenic life cycle, fungal cells should be able to escape from the dead animal body to return to the environmental habitat—although this has rarely been proven. When sapronoses are caused by environmental pathogens, where the environmental habitat is part of the fungus' natural life cycle, the source will be more difficult to eradicate.

1.3 Expansion Due to Changing Host Factors

In the literature, many fungi are attributed as "emerging," which refers to a prevalence of the infection increasing significantly above the baseline. In most cases, this is triggered by opportunity. Several immunologically naive patient populations have emerged. Host changes in human populations concern, for example, novel medical technologies allowing patients with low immunity, socioeconomic changes, and emerging immune and metabolic diseases. Transplant recipients, patients with chronic diabetes, and those with various long-term chemotherapy are such novelties. These are potentially infected by a gamut of infectious opportunists, such as members of *Mucorales*, non-*albicans Candida*, and various *Aspergillus* and *Fusarium* species. Most of these fungi—with the exception of *Fusarium* which simply seems to have been neglected (Al-Hatmi et al. 2016)—were already known as agents of disease since the nineteenth century, but their incidence has increased due to the expanding populations of susceptible hosts (Fig. 1.2a). Environmental pathogens respond to host changes in a similar way. As endemic or enzootic fungi with a narrowly defined environmental reservoir, their global frequency is low, and their increase, e.g., taking advantage of the AIDS pandemic in humans or FIV/FeLV infection in cats, remains moderate (Fig. 1.2b).

1.4 Expansion Due to Fungal Novelties

In addition to changes in opportunity, also fungal populations change perpetually in their genetic makeup. For instance, virulent genotypes may be novel, or at least the hosts are confronted with fungal genotypes to which they lacked resistance (Fig. 1.2c). Low genetic diversity was at the basis of the Vancouver outbreak of *Cryptococcus gattii* (Kidd et al. 2004), and the Oregon outbreak concerned a single highly virulent genotype (Byrnes et al. 2010) (Chap. 12). Emerging fungal infections in humans and other vertebrates often take us by surprise. Bat white-nose disease suddenly killed thousands of bats in the USA and was caused by a species that was known in Europe and to which European bats were resistant (Chap. 13). The fungus appeared extremely successful in susceptible American bat populations, likely owing to physiological and behavioral differences between new- and old-world bats. The causative agent, *Pseudogymnoascus destructans*, was described only in 2009 (Gargas et al. 2009) and the fungus seemed to appear out of the blue. Several of the species that cause large epidemics or epizootics have only recently been described; first cases were often recorded only a few decades ago. One of the examples is *Aspergillus felis* causing infections in cats in Australia (Barrs et al. 2007) (Chap. 15). One might suppose that this is a fallacy due to insufficient awareness and inadequate diagnostics in the past, but in large collections of, for example, archived Sporothrix material, the recent outbreak fungus *Sporothrix brasiliensis* was not detected (Rodrigues et al. 2014). The frog disease agent *Batrachochytrium dendrobatidis* was described only in 1999 (Longcore et al. 1999) with its newly introduced counterpart *B. salamandrivorans* killing

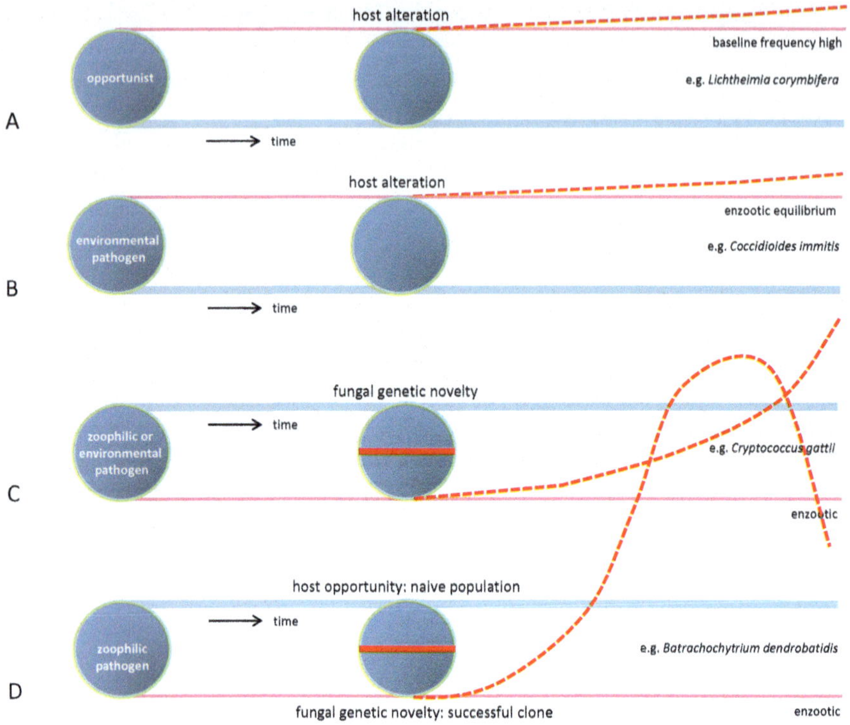

Fig. 1.2 Predicted emergence of species. (**a**) Opportunist with rather high frequency, increase relative to growth of susceptible populations; (**b**) environmental pathogen, relatively frequent in endemic area, in equilibrium with its habitat, increase relative to growth of susceptible populations; (**c**) pathogen developing novelty by which it increases the number of susceptible hosts; (**d**) zoophilic pathogen developing novelty, confronted with naive host populations which leads to epidemic expansion, expected decrease when susceptible hosts are no longer available

salamanders in 2013 (Martel et al. 2013) (Chap. 14). These pathogens have likely been present in low abundance and were undetected until the encounter of a virulent genotype and susceptible host found a match, which was thought to have occurred due to trade-facilitated intercontinental movements. Both species have low genetic diversity suggesting clonal expansion (Morehouse et al. 2003; Fisher et al. 2012). Over a larger sampling area, there is genetic diversity, but local frog mortality is caused by single clones (Morgan et al. 2007). Fastest population expansion is predicted to occur when virulent genotypes of pathogens match with a novel window of opportunity (Fig. 1.2). An extra window of opportunity for frog disease has been suggested to be climate change, weakening frog populations (Clare et al. 2016). The combined changes in fungus and host allow asymptotic expansion of the pathogen, finally leading to decline because susceptible hosts have become rare (Fig. 1.2d). This may account for the wavelike pattern of lethargic crab disease along the Brazilian coast (Ávila et al. 2012) (Chap. 11).

The emergence of sporotrichosis is still largely unexplained. Cat-transmitted *Sporothrix brasiliensis* shows epizootic and epidemic expansion in Southeast Brazil since 1990 (Ortiz Sanchotene et al. 2015) (Chap. 10). The species is one of two near-clonal entities that have evolved from *Sporothrix schenckii* (Rodrigues et al. 2014; Moussa et al. 2017). In the ancestral species, both cat and plant transmission are documented. The clones have specialized in either one of these modes of transmission: both are traumatically inoculated into the skin, but *S. brasiliensis* enters via cat scratches, whereas *S. globosa* infection stems from trauma caused by sharp plant debris or thorns. *Sporothrix globosa* infection is seasonal (Yu et al. 2013), whereas the more virulent *S. brasiliensis* replicates asymptotically to become overabundant, outcompeting the ancestral species, *S. schenckii* (de Araujo et al. 2015). The remarkable difference in transmission mode, with large consequences for evolutionary success and public health is observed at a very small phylogenetic distance (Moussa et al. 2017).

1.5 Recent Outbreaks and Epizootics

Candida albicans, commonly causing mucocutaneous infections in humans, is transmitted directly, from patient to patient, e.g., from mother to child. The host is niche as well as the reservoir of the fungus. The species has an advantage of infecting the vertebrate host and can thus be interpreted as a zoophilic pathogen (Fig. 1.1). Carriage by the healthy host is mostly non-symptomatic, the fungus then maintaining as a commensal (Achkar and Fries 2010) and every new host may be contagious. With only slight impairment of innate or acquired immune systems of the host, *C. albicans* infections become symptomatic. With host-to-host transmission, there is no common source of contamination, as opposed to sapronoses. In the susceptible population of hospitalized patients, other *Saccharomycetales* are emerging, such as the fluconazole-resistant *Candida auris* (Chowdhary et al. 2016). The species has now been reported globally from bloodstream infections in severely compromised patients. It often occurs in the form of local outbreaks (Schelenz et al. 2016). Its mode of infection is as yet unknown, but host-to-host transmission seems an option.

Emergomyces africanus is a novel human pathogen involved in an outbreak of disseminated disease in HIV-positive individuals in South Africa (Kenyon et al. 2014; Dukik et al. 2017). Its origin is unknown; probably the fungus is an environmental pathogen and might occur in animal populations. The genus *Emergomyces* contains a number of species which since recently were repeatedly involved in human infection, while related species reside fairly commonly in the lungs of small rodents (e.g., *Emmonsia crescens*) (Borman et al. 2009, Hubálek et al. 1998) (Chap. 7), infection of humans being coincidental (Dot et al. 2009). Screening of wild animals is required to understand, and to possibly predict, host jumps to humans. Humans are likely non-optimal hosts in most species, and relations between animal hosts may be quite complicated (Vilela et al. 2016). Jiang et al. (2018) noted that small temperature differences of dimorphic switch in *Emergomyces* and *Blastomyces* species may explain predilections for different host animals and thus may be a driver of new directions in the evolution of these fungi. Ecological fitting (Araujo et al. 2015) via the sloppy fitness space of the not-yet-suitable host might be the optimal model to describe the sympatric evolutionary processes in these fungi.

Most *Emergomyces* species and other members of *Ajellomycetaceae* such as *Paracoccidioides* (Chap. 6) are environmental pathogens with a double life cycle (Fig. 1.3), i.e., living permanently in the environment but having a certain advantage (increased fitness) if an animal host is used in any stage of the life cycle. The host may be regarded as a vehicle, which enhances dispersal, optimizing the distribution of the fungus in the environmental niche. For example, *Histoplasma capsulatum* is found on sheltered animal droppings, e.g., bat feces in caves or bird roosting sites (Rocha-Silva et al. 2014). Often the resident animals are asymptomatically colonized (Naiff et al. 1996), return to the cave, and ultimately die there (Chap. 5). Thus, there is no efficient vehicle of dispersal, but susceptible visitors of the cave after contamination and infection are likely to die elsewhere, enabling further dispersal of the pathogen. The fungus thus has a strategy to reach optimal fitness. A common-source outbreak is concerned with the number of infections corresponding with the number of susceptible individuals entering the contaminated site; thus usually no epidemic or

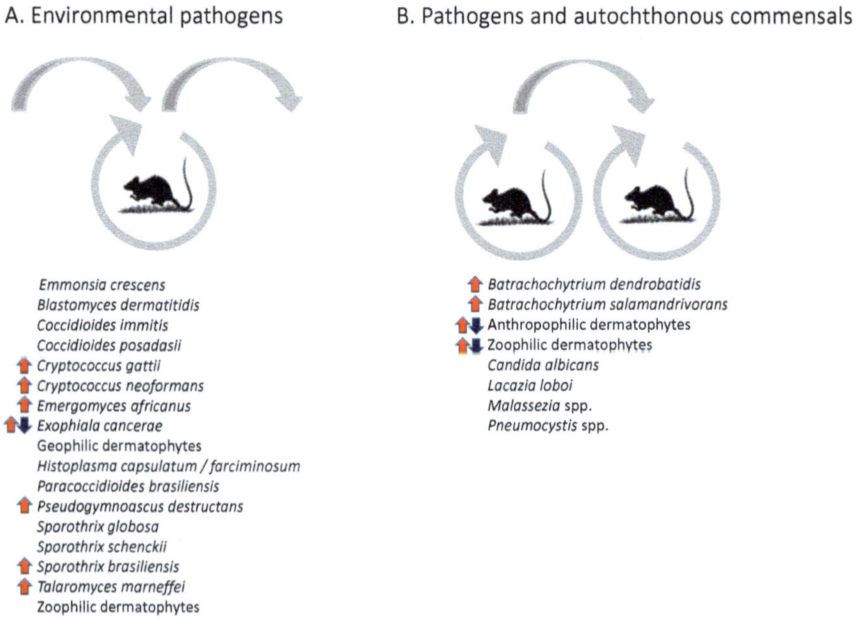

Fig. 1.3 Current approximate changes in frequency in (**a**) environmental pathogens and (**b**) zoophilic pathogens. Large *Exophiala cancerae* zoonosis has disappeared; dermatophyte floras on humans are changing due to socioeconomic changes

epizootic expansion is noted. Hosts are not contagious, but are infected by environmental propagules; the infections are sapronoses.

In the main common-source, sapronotic outbreaks of environmental pathogens such as *Histoplasma*, the role of the susceptible host in fungal expansion is rather insignificant. The size of the inoculum is proportional to the presence of the fungus in its environmental niche. Outbreaks of *Coccidioides immitis* are related to weather conditions and subsequent dust storms that lead to massive inhalation of propagules (Valdivia et al. 2006) (Chap. 4). Over time, the fungal prevalence remains more or less the same, though with fluctuations. *Coccidioides* is hypothesized to form hyphae from cells located in host tissue after its death, to colonize a new environmental site (Lewis et al. 2015). The infection is controlled primarily by acquired cellular immunity and therefore tends to expand with an increasing number of AIDS patients (Rempe et al. 2007).

Talaromyces marneffei is an unrelated environmental pathogen. The species occurs in soil and has a reservoir in large rodents in Southeast Asia but found a susceptible new host due to the human AIDS pandemic. Dormancy inside living pulmonary tissue of the animal requires resistance to innate immunity and control by acquired cellular immunity (Cooper and Vanittanakom 2008), and hence impairment of T-cell function provides an obvious portal for dissemination. Also, this species is overwhelmingly clonal (Fisher et al. 2005) despite local sexuality. Populations are geographically substructured with likely adaptation to different host species (Henk et al. 2012). The environmental habitat of the agent of bat white-nose syndrome (WNS), *Pseudogymnoascus destructans* causing bat decline in northeastern USA (Blehert et al. 2009), is still unknown (Chap. 13) (Rajkumar et al. 2011). The virulent clone causing the bat near extinction presumably already pre-existed in Europe, where bats had found a balance toward infection without causing significant problems. However, when reaching susceptible bat populations in the USA, the fungus was able to expand exponentially (Leopardi et al. 2015).

Recent outbreaks have also been caused by dermatophytes, which are keratinophilic fungi with a gradual adaptation to living tissues (Chap. 3). Anthropophilic dermatophytes are highly specialized species with niche and reservoir on human skin. Transmission takes place from human-to-human, or via propagules which can survive in the environment but do not form assimilative thalli, comparable to *Batrachochytrium*. Zoophilic dermatophytes are adapted to animal skin, with different level of host-specificity. *Trichophyton equinum* is isolated from horses only, whereas *Microsporum canis* can be detected in a wide range of animals (and sometimes in humans). Zoophilic dermatophytes are mostly carried in animal fur, which is often in close contact with soil and plant material of the host's burrow. Furred animal thus may be infected directly from their nest-mates but also from their environment. Carriage in the fur is often asymptomatic, infection only taking place in susceptible nest-mates, e.g., juveniles, or when a non-suitable host is coincidentally infected, such as a predator or a human. Geophilic dermatophytes have a double life cycle with elaborate sexual phases that are produced in soil. Sexuality in anthropophilic dermatophytes is not known to exist or concerns somatic cell fusion at most. A gradual loss of sexuality in dermatophytes with advancing adaptation to mammalian hosts has been observed (de Hoog et al. 2016). Clonal reproduction is often prevalent (Gräser et al. 2006), which is underlined by the fact that potential mating partners having either high mobility group (HMG) or α-box transcription factors can be phenotypically different (Symoens et al. 2013).

1.6 Location of Outbreak Fungi in the Fungal Kingdom

Opportunism and pathogenicity are polyphyletic in the fungal kingdom (Fig. 1.4). Pathogenicity was defined in an ecological sense above as having advantage of the use of a vertebrate host. This is the case in species of *Batrachochytrium* (order *Chytridiales*, *Chytridiomycota*), *Cryptococcus* (*Tremellales*, *Basidiomycota*), dermatophytes and some systemic dimorphic fungi (*Onygenales*, *Ascomycota*), *Talaromyces* and *Pseudogymnoascus* (*Eurotiales*, *Ascomycota*), and perhaps *Sporothrix* (*Ophiostomatales*, *Ascomycota*). If we compare this with criteria of transmission (host-to-host versus host-environment-host), we observe that *Batrachochytrium*, *Candida*, *Pneumocystis*, and anthropophilic (and some zoophilic) dermatophytes lack assimilative thalli in the environment and thus hosts are principally contagious, while *Cryptococcus*, geophilic dermatophytes, and systemic dimorphic fungi are environmental pathogens. Judging from PCR data, *Talaromyces* seems to occur in the environment (Pryce-Miller et al. 2008), suggesting that *T. marneffei* is also an environmental pathogen that is amplified by infection in bamboo rat hosts. *Sporothrix* species are somewhat outside these categories: *S. brasiliensis* is transmitted by cats and the fungus is able to produce a large amount of infective material in feline tissues (which is rarely the case in other mammalian hosts).

Currently, the largest outbreaks, i.e., with the largest degree of acceleration, are *Batrachochytrium*, *Cryptococcus*, *Pseudogymnoascus*, and *Sporothrix*. Given the expansion of *Cryptococcus* in Southern Africa, it might be expected that *Emergomyces africanus*, which also occurs in HIV-positive individuals, might expand in the near future. In contrast, species like *Candida albicans* or *Pneumocystis* spp. are intimately associated with specific hosts, responding to transient host susceptibilities, but on average remain with comparable frequencies in animal or human populations. *Talaromyces marneffei* initially stabilized due to the control of the HIV pandemic as a result of HAART therapy but now emerges in patients with other comorbidities (Chan et al. 2016) and in otherwise healthy individuals (Ye et al. 2015). The emergence of anthropophilic dermatophytes is strongly associated with human socioeconomic changes. In preindustrial societies in developmental transition, a shift is observed from anthropophilic to zoophilic dermatophytes, resulting from increasing hygiene to the growing habit in urban settings to live closely with

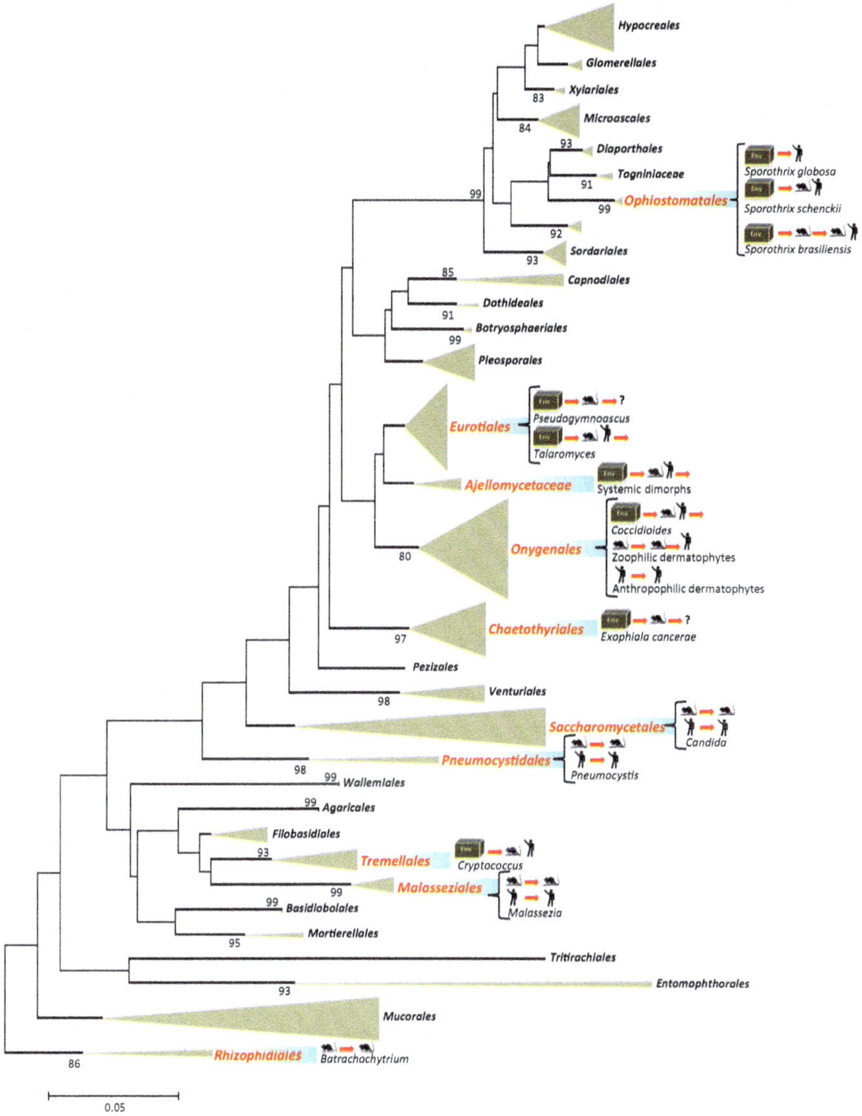

Fig. 1.4 Distribution of zoonotic/epizootic/epidemic fungi over the fungal kingdom. Orders containing pathogens or zoonotic fungi are highlighted. Env → = environmental pathogen. → = zoophilic pathogen

companion animals. This process was observed in Europe half a century ago (Mantovani and Morganti 1977) and currently takes place in China (Zhan et al. 2015). To a certain extent, the dermatophyte species reflect the popularity of pets, including dogs, cats, and more frequently rabbits and small rodents (Nenoff et al. 2014).

Acknowledgments The authors are indebted to Matthew Fisher for constructive discussions and comments on the text.

References

Achkar JM, Fries BC (2010) *Candida* infections of the genitourinary tract. Clin Microbiol Rev 23:253–273

Al-Hatmi AMS, Meis JF, de Hoog GS (2016) *Fusarium*: molecular diversity and intrinsic drug resistance. PLoS Pathog 12(4):e1005464. https://doi.org/10.1371/journal.ppat.1005464

de Araujo ML, Rodrigues AM, Fernandes GF, de Camargo ZP, de Hoog GS (2015) Human sporotrichosis beyond the epidemic front reveals classical transmission types in Espírito Santo, Brazil. Mycoses 58(8):485–590

Araujo SBL, Braga MP, Brooks DR, Agosta SJ, Hoberg EP, von Hartenthal FW, Boeger WA (2015) Understanding host-switching by ecological fitting. PLoS One 10:e0139225

Ávila RA, Mancera PFA, Estevac L, Pied MR, Ferreira CP (2012) Traveling waves in the lethargic crab disease. Appl Math Comput 218:9898–9910

Barrs VR, Beatty JA, Lingard AE, Malik R, Krockenberger MB, Martin P, O'Brien C, Angles JM, Dowden M, Halliday C (2007) Feline sino-orbital aspergillosis: an emerging clinical syndrome? Aust Vet J 85:N23

Beurmann L, Gougerot H (1912) Les Sporotrichose. Librairie Felix Alcan, Paris

Blehert DS, Hicks AC, Behr M, Meteyer CU, Berlowski-Zier BM, Buckles EL, Coleman JTH, Darling SR, Gargas A, Niver R, Okoniewski JC, Rudd RJ, Stone WB (2009) Bat white-nose syndrome: an emerging fungal pathogen? Science 323(5911):227

Borman AM, Simpson VR, Palmer MD, Linton CJ, Johnson EM (2009) Adiaspiromycosis due to *Emmonsia crescens* is widespread in native British mammals. Mycopathologia 68:153–163

Byrnes EJ, Li WJ, Lewit Y, Ma HS, Voelz K, Ren P, Carter DA, Chaturvedi V, Bildfell RJ, May RC, Heitman J (2010) Emergence and pathogenicity of highly virulent *Cryptococcus gattii* genotypes in the Northwest United States. PLoS Pathog 6(4):e1000850. https://doi.org/10.1371/journal.ppat.1000850

Chan JFW, Lau SKP, Yuen KY, Woo PCY (2016) *Talaromyces (Penicillium) marneffei* infection in non-HIV-infected patients. Emerg Microbes Infect 5(3):e19

Clare F, Halder JB, Daniel O, Bielby J et al (2016) Climate forcing of an emerging pathogenic fungus across a montane multihost community. Philos Trans R Soc Lond B Biol Sci 371 (1709):20150454. https://doi.org/10.1098/rstb.2015.0454

Cooper CR, Vanittanakom N (2008) Insights into the pathogenicity of *Penicillium marneffei*. Future Microbiol 3:43–55

Chowdhary A, Voss A, Meis JF (2016) Multidrug-resistant *Candida auris*: 'new kid on the block' in hospital-associated infections? J Hosp Infect 94:209–212

Crous PW, Giraldo A, Hawksworth DL, Robert V, Kirk PM, Guarro J, Robbertse B et al (2014) The genera of fungi: fixing the application of type species of generic names. IMA Fungus 5 (1):141–160

De Hoog GS, Guarro J, Gené J, Figueras MJ (2017) Atlas of clinical fungi. E-edition 4.1.4. Westerdijk Fungal Biodiversity Institute/Universitat Rovira I Virgili, Utrecht/Reus

De Hoog GS, Dukik K, Monod M, Packeu A, Stubbe D, Hendrickx M, Kubsch C, Stielow JB, Göker M, Rezaei-Matehkolaei A, Mirhendi H, Gräser Y (2016) Towards a novel multilocus phylogenetic taxonomy for the dermatophytes. Mycopathologia 82:5–31

Dot J-M, Debourgogne A, Champigneulle J, Salles Y, Brizion M, Puyhardy JM, Collomb J, Plénat F, Machouart M (2009) Molecular diagnosis of disseminated adiaspiromycosis due to *Emmonsia crescens*. J Clin Microbiol 47:1269–1273

Dukik K, Muñoz JF, Jiang Y, Feng P et al (2017) Novel taxa of thermally dimorphic systemic pathogens in the Ajellomycetaceae (Onygenales). Mycoses 60(5):296–309. https://doi.org/10.1111/myc.12601

Fisher MC, Hanage WP, de Hoog GS, Johnson E et al (2005) Low effective dispersal of asexual genotypes in heterogeneous landscapes by the endemic pathogen *Penicillium marneffei*. PLoS Pathog 1(2):e20

Fisher MC, Henk DA, Briggs CJ, Brownstein JS et al (2012) Emerging fungal threats to animal, plant and ecosystem health. Nature 484(7393):186–194

Gargas A, Trest MT, Christense M, Volk TJ, Blehert DS (2009) *Geomyces destructans* sp.nov., associated with bat white-nose syndrome. Mycotaxon 108:147–154

Govender NP, Maphanga TG, Zulu TG, Patel J, Walaza S, Jacobs C, Ebonwu JI, Ntuli S, Naicker SD, Thomas J (2015) An outbreak of lymphocutaneous sporotrichosis among mine-workers in South Africa. PLoS Negl Trop Dis 9(9):e0004096. https://doi.org/10.1371/journal.pntd.0004096

Gräser Y, de Hoog GS, Summerbell RC (2006) Dermatophytes: recognizing species of clonal fungi. Med Mycol 44:199–209

Henk DA, Shahar-Golan R, Ranjana DK et al (2012) Clonality despite sex, the evolution of sexual neighborhoods in the pathogenic fungus *Penicillium marneffei*. PLoS Pathog 8(10):e1002851. https://doi.org/10.1371/journal.ppat.1002851

Hubálek Z, Nesvadbova J, Halouzka J (1998) Emmonsiosis of rodents in an agroecosystem. Med Mycol 36:387–390

Jiang Y, Dukik K, Muñoz J, Sigler L, Schwartz IS et al (2018) Phylogeny, ecology and taxonomy of systemic pathogens in Ajellomycetaceae (Onygenales). Fungal Divers (in press)

Kenyon C, Corcoran C, Govender NP (2014) An *Emmonsia* species causing disseminated infection in South Africa. New Engl J Med 370:283–284

Kidd SE, Hagen F, Tscharke RL, Huynh M, Bartlett KH, Fyfe M, MacDougall L, Boekhout T, Kwon-Chung KF, Meyer W (2004) A rare genotype of *Cryptococcus gattii* caused the cryptococcosis outbreak on Vancouver Island (British Columbia, Canada). Proc Natl Acad Sci U S A 101:17258–17263

Leopardi S, Blake D, Puechmaille SJ (2015) White-nose syndrome fungus introduced from Europe to North America. Curr Biol 25(6):R217–R219. https://doi.org/10.1016/j.cub.2015.01.047

Lewis ERG, Bowers JR, Barker BM (2015) Dust devil: the life and times of the fungus that causes valley fever. PLoS Pathog 11(5):e1004762

Longcore JE, Pessier AP, Nichols DK (1999) *Batrachochytrium dendrobatidis* gen. et sp. nov., a chytrid pathogenic to amphibians. Mycologia 91:219–227

Mantovani A, Morganti L (1977) Dermatophytozoonoses in Italy. Vet Res Comm 1:171–177

Martel A, Spitzen-van der Sluijs A, Blooi M, Bert W, Ducatelle R, Fisher MC, Woeltjes A, Bosman W, Chiers K, Bossuyt F, Pasmans F (2013) *Batrachochytrium salamandrivorans* sp. nov. causes lethal chytridiomycosis in amphibians. Proc Natl Acad Sci U S A 110:15325–15329

Morehouse EA, James TY, Ganley ARD, Vilgalys R, Berger L, Murphy PJ, Longcore JE (2003) Multilocus sequence typing suggests the chytrid pathogen of amphibians is a recently emerged clone. Mol Ecol 12:395–403

Morgan JAT, Vredenburg VT, Rachowicz LJ, Knapp RA et al (2007) Population genetics of the frog-killing fungus *Batrachochytrium dendrobatidis*. Proc Natl Acad Sci U S A 104 (34):13845–13850

Moussa TAA, Kadasa NMS, Al Zahrani HS, Ahmed SA, Feng P, Gerrits van den Ende AHG, Zhang Y, Kano R, Li F, Li S, Yang S, Bilin D, Rossato L, Dolatabadi S, de Hoog GS (2017) Origin and distribution of *Sporothrix globosa* causing sapronoses in Asia. J Med Microbiol 66 (5):560–569

Naiff RD, Barrett TV, Naiff MDF, de Lima Ferreira LC, Arias JR (1996) New records of *Histoplasma capsulatum* from wild animals in the Brazilian Amazon. Rev Inst Med Trop Sao Paulo 38:273–277

Nenoff P, Uhrlaß S, Krüger C, Erhard M, Hipler U-C, Seyfarth F et al (2014) *Trichophyton species* of *Arthroderma benhamiae* – a new infectious agent in dermatology. J Dtsch Dermatol Ges 12 (7):571–581

Ortiz Sanchotene K, Martins Madrid I, Baracy Klafke G, Bergamashi M, Della Terra PP, Rodrigues AM, de Camargo ZP, Orzechowski Xavier M (2015) *Sporothrix brasiliensis* outbreaks and the rapid emergence of feline sporotrichosis. Mycoses 58:652–658

Pryce-Miller E, Aanensen D, Vanittanakom N, Fisher MC (2008) Environmental detection of *Penicillium marneffei* and growth in soil microcosms in competition with *Talaromyces stipitatus*. Fungal Ecol 1:49–56

Rajkumar SS, Li X, Rudd RJ et al (2011) Clonal genotype of *Geomyces destructans* among bats with white nose syndrome, New York, USA. Emerg Infect Dis 17:1273–1276

Rempe S, Sachdev M, Bhakta R, Pineda-Roman M, Vaz A, Carlson R (2007) *Coccidioides immitis* fungemia: clinical features and survival in 33 adult patients. Heart Lung 36:64–71

Rocha-Silva F, Figueiredo SM, Silveira TT, Assunção CB, Campolina SS, Pena-Barbosa JP, Rotondo A, Caligiorne RB (2014) Histoplasmosis outbreak in Tamboril cave-Minas Gerais state, Brazil. Med Mycol Case Rep 4:1–4

Rodrigues AM, de Hoog GS, Zhang Y, de Camargo ZP (2014) Emerging sporotrichosis is driven by clonal and recombinant *Sporothrix* species. Emerg Microbes Infect 3:e32

Schelenz S, Hagen F, Rhodes JL, Abdolrasouli A, Chowdhary A, Hall A, Ryan L, Shackleton J, Trimlett R, Meis JF, Armstrong-James D, Fisher MC (2016) First hospital outbreak of the globally emerging *Candida auris* in a European hospital. Antimicrob Resist Infect Control 5:35. eCollection 2016

Symoens F, Jousson O, Packeu A, Fratti M, Staib P, Mignon B, Monod M (2013) The dermatophyte species *Arthroderma benhamiae*: intraspecies variability and mating behaviour. J Med Microbiol 62:377–385

Valdivia L, Nix D, Wright M, Lindberg E, Fagan T, Lieberman D, Stoffer TP, Ampel NM, Galgiani JN (2006) Coccidioidomycosis as a common cause of community-acquired pneumonia. Emerg Infect Dis 12:958–962

Vilela R, Bossart GD, St. Leger JA, Dalton LM et al (2016) Cutaneous granulomas in dolphins caused by novel uncultivated *Paracoccidioides brasiliensis*. Emerg Infect Dis 22:2063–2069

Yu X, Wan Z, Zhang Z, Li F, Li R, Liu X (2013) Phenotypic and molecular identification of *Sporothrix* isolates of clinical origin in Northeast China. Mycopathologia 176:67–74

Ye F, Luo Q, Zhou Y, Xie J, Zeng Q, Chen G, Su D, Chen R (2015) Disseminated penicilliosis marneffei in immunocompetent patients: a report of two cases. Indian J Med Microbiol 33 (1):161–165. https://doi.org/10.4103/0255-0857.148433

Zhan P, Geng C, Li Z, Jin Y, Jiang Q, Tao L, Luo Y, Xiong L, Wu S, Li D, Liu W, de Hoog GS (2015) Evolution of tinea capitis in the Nanchang area, Southern China: a 50-year survey (1965–2014). Mycoses 58:261–266

Zhang Y, Hagen F, Stielow B, Rodrigues AM, Samerpitak K, Zhou X, Feng P, Yang L, Chen M, Deng S, Li S, Liao W, Li R, Li F, Meis JF, Guarro J, Teixeira M, Al-Zahrani HS, Pires de Camargo Z, Zhang L, de Hoog GS (2015) Phylogeography and evolutionary patterns in *Sporothrix* spanning more than 14,000 human and animal case reports. Persoonia 35:1–20

Epidemiological Definitions, Terminology and Classifications with Reference to Fungal Infections of Animals

2

Matthew C. Fisher

Abstract

Emerging infections caused by fungi have become a widely recognised global phenomenon in animal species and populations worldwide. This chapter details the vocabulary and grammar that is used to discuss such infections. Much of this terminology is specific to the field of mycology, and careful usage is required in the scientific literature and discussion in order to maintain clarity of expression.

2.1 Introduction

Infectious diseases are the class of diseases that are caused by microorganisms, such as bacteria, viruses and eukaryotes, and infectious proteins (prions). Infectious diseases can be **transmissible diseases** (also known as **communicable diseases**) if they are transmitted from an infected to an uninfected individual, directly or indirectly, or they can be **non-transmissible diseases** if an infected host does not generate further infected individuals. Infectious diseases can be **directly transmitted** such as through direct contact or fomites or **indirectly transmitted** by water, air, food or other vectors. Fungi are unusual in that otherwise frequent transmission by insect **vectors**, such as blood-feeding arthropods, is very rare (Rosenberg and Beard 2011). However, vector-borne transmission is not unknown, and the transmission of the microsporidian *Trachipleistophora hominis* into humans with HIV-AIDS by mosquitoes is thought to have occurred, albeit rarely (Mathis et al. 2005). *Histoplasma capsulatum* var. *farciminosum* has been isolated from the alimentary tract of biting flies, suggesting a role in the transmission of the fungus

M. C. Fisher (✉)
Department of Infectious Disease Epidemiology, School of Public Health, Imperial College London, London, UK
e-mail: matthew.fisher@imperial.ac.uk

S. Seyedmousavi et al. (eds.), *Emerging and Epizootic Fungal Infections in Animals*,
https://doi.org/10.1007/978-3-319-72093-7_2

(Gabal and Hennager 1983) and a significant association has been reported between tick infestations and the lesions caused by histoplasmosis in mules (Ameni and Terefe 2004).

2.2 Transmissible Fungal Infections

Transmissible fungal infections include those that are **directly** transmissible between individuals, for instance, some species of dermatophytes. **Indirectly transmissible fungal infections** include examples such as the water-borne chytrids *Batrachochytrium dendrobatidis* and *B. salamandrivorans*. However, most pathogenic fungi are non-transmissible and are acquired from environmental sources. **Exposure** to a transmissible fungal infection can lead to the host becoming a **carrier** that harbours the infection. **Carriage** may lead to **subclinical infection** where no clinical signs of infection are manifested or **clinical infection** whereupon direct or indirect costs of exposure occur (Garner et al. 2009). Much of the terminology used in the wider epidemiological literature refers to the transmissible infections, because transmissible pathogens such as *Plasmodium* spp., *Mycobacterium* spp. and HIV have historically contributed the largest burden of disease and have been the focus of epidemiological modelling and efforts to control disease. Therefore, terms that refer to epidemiological characteristics of transmissible infections are relevant to fungal pathogens that are spread from host to host. Commonly used terms include the following:

Attack Rate This is generally defined as the proportion of individuals who are exposed to an infectious agent and become clinically ill.

Primary/Secondary Cases For transmissible fungal infections such as *Candida* sp., spreading through hospital wards or *Microsporum canis* spreading through a population of cats, the host whom introduces the infection into a susceptible population is called the **primary case**, and the individuals whom are subsequently infected are called **secondary cases**. From these cases, the **reproductive rate** can be defined which describes the potential for the infection to spread through a population. The **basic reproductive rate,** $\boldsymbol{R_0}$, is a critical quantity to understand as it determines the probability that a new infection entering a population will continue to spread. Defined as 'the average number of secondary infections generated from an infected host in an entirely susceptible population', R_0 predicts three possible situations following disease introduction into a population:

- $R_0 < 1 \Rightarrow$ the fungal pathogen will go extinct.
- $R_0 = 1 \Rightarrow$ the fungal pathogen will become endemic.
- $R_0 > 1 \Rightarrow$ an outbreak of the mycosis will occur.

If immunity to a fungal infection builds up within a population over time, then R_0 will decay to less than 1 (for a population with an R_0 of 2, then this will be the time point when roughly half the population have become immune). Therefore, determination of R_0 for transmissible fungi is necessary in order to calculate the proportion of the population that needs to be vaccinated or subjected to treatment, in order to control the onwards spread of infection. As fungal vaccines and other therapies are currently being developed and deployed, this is an increasingly important area of epidemiological interest.

Index Case/Site This is the first case, or region, to be identified during an outbreak and is usually the focus of intense research as it is from the rate that secondary cases are generated and the proportion of cases that develop clinical symptoms that initial epidemiological models are formulated.

Incubation Period and Latent Period The incubation period is the time extending from when the individual is first infected through to the time that they develop symptoms of disease. This can be calculated from clinical observations or within an experimental setting; for example, Doddington et al. (2013) experimentally infected groups of Mallorcan midwife toads with two isolates of the chytrid *B. dendrobatidis* in order to calculate isolate-specific death rates, with which to parameterise epidemiological models aimed at assessing the probability of this species survival on the island following the introduction of the pathogen (Walker et al. 2008).

The latent period is usually defined as the period of time from infection to onset of infectiousness, i.e. the stage at which secondary cases start to be generated. The latent period will vary between individuals and host species depending on the rate of growth of the fungus and time to generation of infectious inocula. However, for mycoses, latent period is often used to describe the period between the time from infection and the time to when clinical symptoms occur, so should really be termed the 'incubation period' as described above. This usage of the term 'latent period' is generally associated with fungal diseases such as human penicilliosis (talaromycosis) or cryptococcal meningitis, where the time from infection to disease can sometimes be measured in decades. For both *Talaromyces* (previously *Penicillium*) *marneffei* and *Cryptococcus neoformans*, pulmonary infection can be quiescent and asymptomatic until the patient becomes immunosuppressed (either through advancing age or other causes such as HIV infection), whereupon clinical signs ensue (Shelburne et al. 2005; Walker et al. 2008).

Infectious Period This is defined as the time from onset of infectiousness to cessation of infectiousness (usually via death or recovery), i.e. the duration of infectiousness. For commensal fungal infections that are probably acquired early in life, such as *Candida albicans* or *Pneumocystis jirovecii*, the infectious period may well be lifelong for a subset of infected individuals.

Incidence and Prevalence Incidence is defined as the number of individuals whom are infected by a fungal infection during *a defined period of time*, often a year, divided by the total population under study. Incidence is often expressed as ×*per 100,000 individuals*. 'Prevalence of infection' relates to the number of individuals who have a specific fungal infection at a *specific time point* divided by the total population. If the average daily incidence I of a fungal disease has a duration of D days, then the average prevalence P will be $P = I \times D$.

In other words, 'prevalence is the product of incidence and duration'. Prevalence of infection is often also used in conjunction with measurements of **intensity of infection**, as this describes the severity of infection and likelihood of severe disease and mortality. For instance, 'Vredenburg's 10,000 zoospore rule' has found wide usage as a term that describes the average zoospore load that an amphibian host can tolerate before succumbing to chytridiomycosis (Vredenburg et al. 2010), although this value is non-absolute and varies amongst and between species (Clare et al. 2016).

Threshold Density This is the minimum population density of hosts at which transmission of an infectious fungus can be sustained with an $R_0 > 1$.

Epidemic/Pandemic and Endemic These terms were coined by the ancient Greek physician Hippocrates in the Hippocratic corpus *Of the Epidemics* written around 400 B.C.E. and relate to the occurrence of diseases that were 'normal' such as coughs and colds and 'abnormal' such as plagues. The modern definition of epidemic has changed somewhat to 'The occurrence in a community or a region of cases of an illness (or an outbreak) with a frequency clearly in excess of normal expectancy' (Heymann 2014). The word 'epidemic' is therefore often linked to the concept of **emerging infectious diseases** as these, by definition, diseases that are in excess of normal expectancy (measured by incidence and/or prevalence of infection). Fungal pathogens that are characteristically associated with a specific region or host species are termed **endemic mycoses**. These mycoses are pathogenic fungi that occupy a specific niche within the environment and therefore are characterised by having spatially constrained distributions. Well-known examples are *Coccidioides immitis*/*C. posadasi* (Fisher et al. 2001, 2002) which are associated with xeric, alkaline soils of the New World and *T. marneffei* which is constrained to the wet tropical region of Southeast Asia (Vanittanakom et al. 2006). Both of these species (alongside several others such as *H. capsulatum* and *Blastomyces dermatitidis*) exhibit **temperature dimorphism** whereupon they can grow as mycelia at environmental temperatures, but can undergo a **dimorphic transition** at 37 °C to grow as yeasts or spherules and cause disseminated, sometimes lethal, infection (Sil and Andrianopoulos 2015).

Epizootic/Panzootic and Enzootic These terms are specific to infectious fungal pathogens that occur in a nonhuman animal population that are either wild (such as the epizootics and panzootics that are described for amphibian chytridiomycosis or

Aspergillus sydowii infections of coral) (Chap. 14) or captive (such as epizootic lymphangitis in horses and mules caused by the dimorphic fungus *H. capsulatum* var. *farciminosum*) (Chap. 5). 'Panzootic' is the nonhuman animal counterpart to the word 'pandemic', and 'enzootic' is the nonhuman animal counterpart to the word 'endemic'. In the context of emerging fungal infections, it is important to recognise that an epizootic mycosis can change to become an established enzootic mycosis over time as host-pathogen dynamics establish a stable equilibrium (Briggs et al. 2010). Such epizootic and enzootic dynamics have been found to occur across varied ecological systems and amphibian communities that have been affected by the introduction of *B. dendrobatidis* (Fisher et al. 2009; Briggs et al. 2010; Retallick et al. 2004).

Zoonoses and Zoonotic Zoonoses are infections of animals that can be transmitted to humans and vice versa. A meta-analysis of 1415 pathogens that are infectious to humans suggested that 307 were caused by species of fungi. For all classes of pathogens, Taylor et al. (2001) showed that the majority were zoonotic comprising 868 (61%); of these, 116 (38%) were considered to be zoonotic fungal infections of humans (Taylor et al. 2001; Fisher et al. 2001). However, closer inspection of the classifications shows that the authors have listed many fungal species that are actually environmental opportunistic fungal infections (or **sapronoses**) (Chap. 1). For instance, the genus *Aspergillus, Alternaria, Acremonium, Bipolaris, Curvularia, Mucor, Paecilomyces* and *Rhizopus* amongst many others are misclassified by Taylor et al. (2001) as 'zoonoses' as they are well understood by experts on fungal diseases to be acquired from their saprophytic environment. From the point of view of a mycologist, the term 'zoonosis' needs to be used in a more specific context to indicate animal-to-human (or human-to-animal) transmission. These 'true' zoonoses are relatively rare and even here there are grey areas. For instance, *Cryptococcus* and *Coccidioides* species are able to grow on nonliving substrates such as lignin (*Cryptococcus*) and/or soil (*Coccidioides*). However, both species are also able to metabolise nitrogen-rich substrates such as pigeon faeces (*C. neoformans*) and small-mammal corpses (*Coccidioides*), and these associations amplify infectious inocula, therefore increasing the risk of human infection. Arguably, these species can therefore be classified as zoonoses owing to their demonstrated association with nonhuman animals as well as human infection. True fungal zoonoses are actually rather rare and information is generally incomplete. For candidate zoonoses such as *T. marneffei* that infects bamboo rats in Southeast Asia, the link to human infection has never been definitively established, although the molecular epidemiology shows that human and bamboo-rat infections are identical on a molecular basis (Gugnani et al. 2004; Henk et al. 2012). Similar arguments have been made suggesting that mosquitoes can vector their microsporidian parasites such as *T. hominis* (Mathis et al. 2005) into immunosuppressed humans and they therefore constitute zoonotic reservoirs of these fungi. The best examples of fungal zoonoses lie within the dermatophytes such as *Microsporum* and *Trichophyton* species, where animal-to-human infection occurs frequently from companion animals (such as dogs, cats, rabbits or small rodents) or cattle (Fehr 2015) (Chap. 3). More recently, human sporotrichosis has emerged as a growing problem in Brazil, where virulent feline sporotrichosis caused by *Sporothrix brasiliensis* is now known to be a true zoonosis that accounts for a growing burden of human disease as infections of *Sporothrix brasiliensis* that have invaded urban feral cat populations spill over to infect humans (Rodrigues et al. 2014; Fehr 2015) (Chap. 10).

Pathogenicity and Virulence Casadevall and Pirofski (2003) argued in a key review that both the host and their infecting microorganism contribute to disease (**pathogenesis**). In their article, they postulate that the amount of damage incurred by the host varies according to the relative input of pathogen and host factors, known as the **'damage-response' framework**. Broadly, **fungal pathogenicity** can be defined as 'the costs accrued through the interaction between an infectious fungus and its host' and will contain both host and pathogen components. For instance, a strong (but damaging) host immunity that is incurred through chronic stimulation of inflammatory cells will incur a cost to the host by causing disease (such as is caused by *Candida albicans*) (Brown et al. 2014). Conversely, direct destruction of host tissue by the fungal infection through lytic processes driven by virulence effectors, such as that caused by proteinase secretion by *B. dendrobatidis* (Joneson et al. 2011) or Candidalysin effector secretion by *C. albicans* (Moyes et al. 2016), will also incur costs to the host and cause disease. The term **virulence** is often (mis)used interchangeably with pathogenicity, when in fact its usage should be more refined and specific. Virulence can be understood as 'the relative capacity of a microorganism to cause damage in a host'. The word 'relative' is used because virulence is often measured in comparison with other variants of the fungal species in question. Virulence is often genetically encoded, so studies commonly focus on defining specific **virulence factors** that are then shown to cause a specific virulence trait. For instance, in *Cryptococcus gattii* infection, **hypervirulence** in humans was shown to occur in isolates that were associated with the well-described 'Vancouver Island' outbreak in the late 1990s (Chap. 12). The hypervirulent isolates were genetically different from a related genotype of *C. gattii* that occurred in the region. These isolates were shown to have well-defined virulence traits, namely, a dramatic ability to tubularise their mitochondria in order to replicate rapidly within macrophages of the mammalian immune system that was linked to distinct genomic differences between lineages of this pathogenic fungus (Ma et al. 2009; Farrer et al. 2015).

Fungal Emerging Infectious Diseases Emerging infectious diseases (EIDs) are those infections that are increasing in either their incidence, geographic and/or host range or virulence (Daszak et al. 2000; Jones et al. 2008) and are found to occur within all groups of living organisms. Until the 1990s, EIDs caused by fungi were generally held to be a feature of plant-infecting fungi, where they have widely challenged food security ever since humans developed agricultural systems (Gurr et al. 2011). However, fungal EIDs are now recognised to occur in animals, and the frequency of their identification appears to be on the increase both spatially, temporally and in terms of their virulence (Fisher et al. 2012). The drivers that underpin this emergence of fungal disease are debated; however, it is clear that worldwide networks of trade contribute to infectious **spillover** into novel naïve hosts as infectious inocula are moved worldwide—increased transcontinental transmission likely accounts for many of the current panzootics of fungal disease that we are witnessing (Fisher and Garner 2007) from endemic centres of infection (Fig. 2.1). Notorious fungal EIDs of animals include amphibian chytridiomycosis caused by *B. dendrobatidis* and *B. salamandrivorans* (Martel et al. 2013) (Chap. 14), cryptococcosis caused by *C. gattii* lineage VGII in the Pacific Northwest of the USA (Chap. 12), bat white nose syndrome caused by *Pseudogymnoascus* (Blehert et al. 2009) (Chap. 13) and cat-transmitted sporotrichosis caused by *Sporothrix brasiliensis* (Rodrigues et al. 2014) (Chap. 10) (Fig. 2.2).

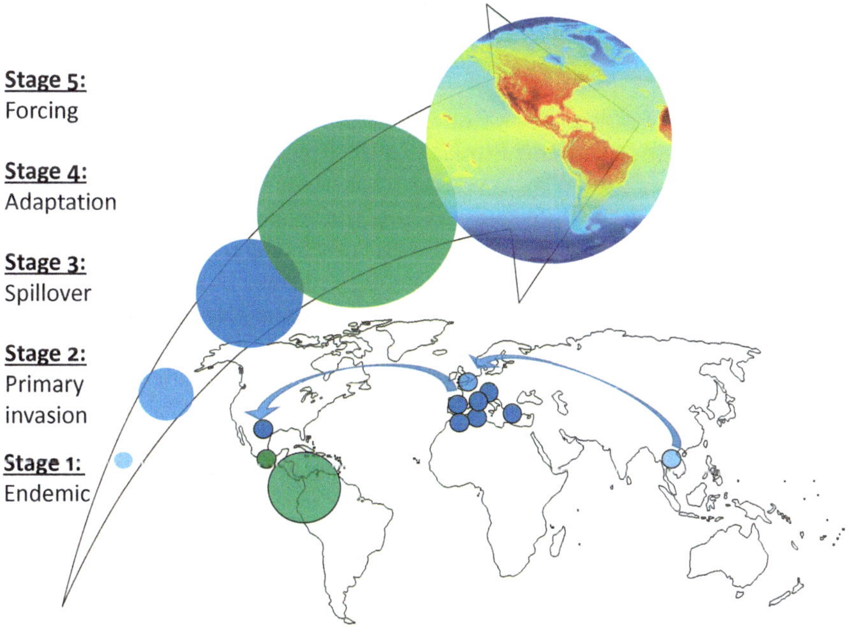

Fig. 2.1 Stages describing the emergence of a pandemic or panzootic fungal infection. In this scenario, an endemic mycosis (stage 1) invades a novel host(s) through a primary invasion event (stage 2). Subsequent spread through anthropogenic trade and/or other mechanisms leads to subsequent spillover events and establishment in new hosts and/or ecosystems (stage 3). Adaptation occurs leading to refinement of the host/pathogen interaction and survival/transmission in new hosts/ecosystems (stage 4). In stage 5, large-scale events such as environmental change may lead to further amplification of disease

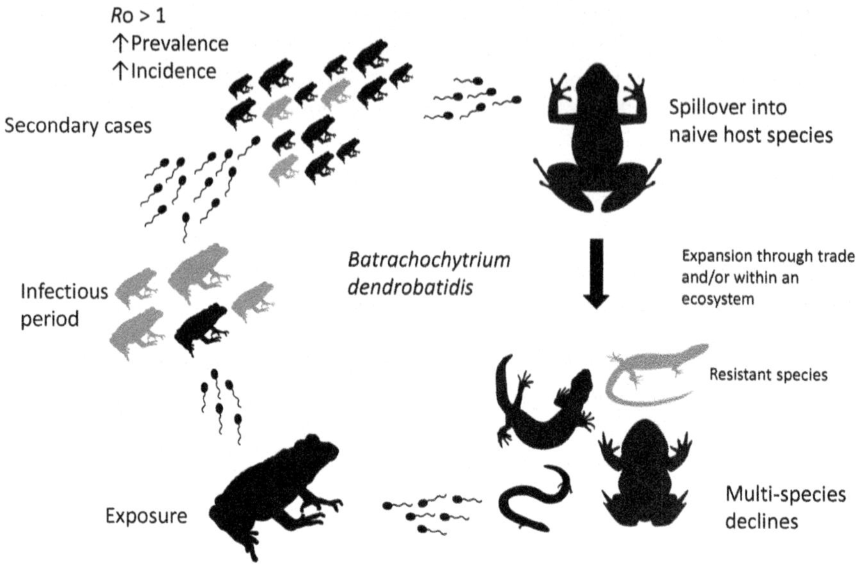

Fig. 2.2 General epidemiological features associated with an outbreak and panzootic of amphibian chytridiomycosis

2.3 Non-transmissible Fungal Infections

Non-transmissible fungal infections are those for which the host is a dead end, and no further onward transmission occurs; this category includes the largest grouping of pathogenic fungi as it includes all those that are **opportunistic infections** and infect immunocompromised hosts. Non-transmissible infections are also referred to as **accidental parasites** or **accidental pathogens** to describe free-living organisms that can multiply in a suitable host, but do not derive obvious evolutionary benefits from the association.

Reservoir hosts are biological hosts that a pathogenic fungus can complete its life cycle in, but are not the primary focus of concern. For example, small mammals are known to be the reservoirs for human mycoses, such as bamboo rats as reservoir hosts for human *Talaromyces marneffei* infection (Cao et al. 2011; Henk et al. 2012) and kangaroo rats as reservoirs of infection for *Coccidioides* sp. and *P. dipodomyis* (Pappagianis 1988; Henk and Fisher 2011). **Environmental reservoirs** are substrates that support the life cycle of a pathogenic fungus outside of susceptible hosts and can act as the source of inocula for susceptible animal hosts. For example, composting sites are an environmental reservoir for very high burdens of *Aspergillus fumigatus* conidia which can lead to human disease if exposure occurs.

Sapronoses is a recently introduced term for the class of pathogens that grow and complete their life cycle as free-living stages in the environment; however, it can use within-host growth as a replication strategy. Sapronoses, also known as

environmentally growing opportunists, comprise a large number of serious fungal infections that affect humans and other animals, for instance, species of *Aspergillus*, *Cryptococcus*, *Coccidioides* and *Histoplasma* are all examples of fungal sapronoses. Sapronoses are generally considered as non-transmissible fungal infections; however, this may not be strictly accurate as there is some evidence to suggest that parasitism of small mammals by *Coccidioides* and *T. marneffei* can result in ongoing transmission following death of the host (Fisher et al. 2001; Henk et al. 2012). The lack of host-to-host transmission causes the epidemiology of sapronoses to manifest dynamics that differ substantially from those that characterise transmissible fungal pathogens. This is because factors that are extrinsic to the host, such as ecological interactions with other free-living organisms leading to resource competition, predation and parasitism, alongside environmental variation in temperature and water availability, will all impact on transmission to susceptible hosts.

'Fungi are the only group of organisms that have been convincingly shown to cause **extinction**'. This remark by Arturo Casadevall (Olsen et al. 2011) is sadly true and relates to an unfortunately rather common combination of traits that characterise fungal infections. While traditional epidemiological theory assumes trade-offs between virulence and transmission owing to the optimisation of the numbers of secondary infections that are caused by a primary infection ($R_0 > 1$), free-living fungi are largely freed from this constraint owing to their ability to replicate outside of the host. When sapronotic life-styles are combined with high virulence, **long-lived environmental stages** and **generalist host ranges**, then high lethality can occur leading to local extirpations or complete host extinctions (Fisher et al. 2012). Such population processes have been observed to have caused the widespread extinction of at least 200 species of amphibian by *B. dendrobatidis* (Fisher et al. 2009; Alroy 2015) (Chap. 14).

References

Alroy J (2015) Current extinction rates of reptiles and amphibians. Proc Natl Acad Sci U S A 112 (42):13003–13008

Ameni G, Terefe W (2004) A cross-sectional study of epizootic lymphangitis in cart-mules in western Ethiopia. Prev Vet Med 66(1–4):93–99

Blehert DS, Hicks AC, Behr M, Meteyer CU, Berlowski-Zier BM, Buckles EL, Coleman JTH, Darling SR, Gargas A, Niver R, Okoniewski JC, Rudd RJ, Stone WB (2009) Bat white-nose syndrome: an emerging fungal pathogen? Science 323(5911):227–227

Briggs CJ, Knapp RA, Vredenburg VT (2010) Enzootic and epizootic dynamics of the chytrid fungal pathogen of amphibians. Proc Natl Acad Sci U S A 107(21):9695–9700

Brown AJ, Brown GD, Netea MG, Gow NA (2014) Metabolism impacts upon *Candida* immunogenicity and pathogenicity at multiple levels. Trends Microbiol 22(11):614–622

Casadevall A, Pirofski LA (2003) The damage-response framework of microbial pathogenesis. Nat Rev Microbiol 1(1):17–24

Cao CW, Liang L, Wang WJ, Luo H, Huang SB, Liu DH, Xu JP, Henk DA, Fisher MC (2011) Common reservoirs for *Penicillium marneffei* infection in humans and rodents, China. Emerg Infect Dis 17(2):209–214

Clare F, Daniel OZ, Garner TWJ, Fisher MC (2016) Assessing the ability of swab data to determine the true burden of infection for the amphibian pathogen *Batrachochytrium dendrobatidis*. Ecohealth 13:360–367

Daszak P, Cunningham AA, Hyatt AD (2000) Emerging infectious diseases of wildlife - threats to biodiversity and human health. Science 287:443–449

Doddington BJ, Bosch J, Oliver JA, Grassly NC, Garcia G, Schmidt BR, Garner TWJ, Fisher MC (2013) Context-dependent amphibian host population response to an invading pathogen. Ecology 94(8):1795–1804

Farrer RA, Desjardins CA, Sakthikumar S, Gujja S, Saif S, Zeng Q, Chen Y, Voelz K, Heitman J, May RC, Fisher MC, Cuomo CA (2015) Genome evolution and innovation across the four major lineages of *Cryptococcus gattii*. MBio 6(5):e00868–e00815

Fehr M (2015) Zoonotic potential of Dermatophytosis in small mammals. J Exot Pet Med 24 (3):308–316

Fisher MC, Garner TWJ (2007) The relationship between the introduction of *Batrachochytrium dendrobatidis*, the international trade in amphibians and introduced amphibian species. Fungal Biol Rev 21:2–9

Fisher MC, Garner TWJ, Walker SF (2009) Global emergence of *Batrachochytrium dendrobatidis* and amphibian chytridiomycosis in space, time, and host. Annu Rev Microbiol 63:291–310

Fisher MC, Henk DA, Briggs CJ, Brownstein JS, Madoff LC, McCraw SL, Gurr SJ (2012) Emerging fungal threats to animal, plant and ecosystem health. Nature 484(7393):186–194

Fisher MC, Koenig GL, White TJ, San-Blas G, Negroni R, Alvarez IG, Wanke B, Taylor JW (2001) Biogeographic range expansion into South America by *Coccidioides immitis* mirrors new world patterns of human migration. Proc Natl Acad Sci U S A 98(8):4558–4562

Fisher MC, Koenig GL, White TJ, Taylor JW (2002) Molecular and phenotypic description of *Coccidioides posadasii* sp nov., previously recognized as the non-California population of *Coccidioides immitis*. Mycologia 94(1):73–84

Gabal MA, Hennager S (1983) Study on the survival of *Histoplasma farciminosum* in the environment. Mykosen 26(9):481–487

Garner TW, Walker S, Bosch J, Leech S, Rowcliffe M, Cunningham AA, Fisher MC (2009) Life history trade-offs influence mortality associated with the amphibian pathogen *Batrachochytrium dendrobatidis*. Oikos 118:783–791

Gugnani H, Fisher MC, Paliwal-Johsi A, Vanittanakom N, Singh I, Yadav PS (2004) Role of *Cannomys badius* as a natural animal host of *Penicillium marneffei* in India. J Clin Microbiol 42 (11):5070–5075

Gurr SJ, Samalova M, Fisher MC (2011) The rise and rise of emerging infectious fungi challenges food security and ecosystem health. Fungal Biol Rev 25(4):181–188

Henk DA, Fisher MC (2011) Genetic diversity, recombination, and divergence in animal associated *Penicillium dipodomyis*. PLoS One 6(8):e22883

Henk DA, Shahar-Golan R, Devi KR, Boyce KJ, Zhan NY, Fedorova ND, Nierman WC, Hsueh PR, Yuen KY, Sieu TPM, Kinh NV, Wertheim H, Baker SG, Day JN, Vanittanakom N, Bignell EM, Andrianopoulos A, Fisher MC (2012) Clonality despite sex: the evolution of host-associated sexual neighborhoods in the pathogenic fungus *Penicillium marneffei*. PLoS Pathog 8(10):e1002851

Heymann DL (2014) Control of communicable diseases manual, 20th edn. Alpha Press, Washington, DC

Jones KE, Patel NG, Levy MA, Storeygard A, Balk D, Gittleman JL, Daszak P (2008) Global trends in emerging infectious diseases. Nature 451(7181):990–993

Joneson S, Stajich JE, Shiu SH, Rosenblum EB (2011) Genomic transition to pathogenicity in Chytrid fungi. PLoS Pathog 7(11):e1002338

Ma HS, Hagen F, Stekel DJ, Johnston SA, Sionov E, Falk R, Polacheck I, Boekhout T, May RC (2009) The fatal fungal outbreak on Vancouver Island is characterized by enhanced intracellular parasitism driven by mitochondrial regulation. Proc Natl Acad Sci U S A 106(31):12980–12985

Martel A, Spitzen-van der Sluijs A, Blooi M, Bert W, Ducatelle R, Fisher MC, Woeltjes A, Bosman W, Chiers K, Bossuyt F, Pasmans F (2013) *Batrachochytrium salamandrivorans* sp nov causes lethal chytridiomycosis in amphibians. Proc Natl Acad Sci U S A 110(38):15325–15329

Mathis A, Weber R, Deplazes P (2005) Zoonotic potential of the microsporidia. Clin Microbiol Rev 18(3):423–445

Moyes DL, Wilson D, Richardson JP, Mogavero S, Tang SX, Wernecke J, Hofs S, Gratacap RL, Robbins J, Runglall M, Murciano C, Blagojevic M, Thavaraj S, Forster TM, Hebecker B, Kasper L, Vizcay G, Iancu SI, Kichik N, Hader A, Kurzai O, Luo T, Kruger T, Kniemeyer O, Cota E, Bader O, Wheeler RT, Gutsmann T, Hube B, Naglik JR (2016) Candidalysin is a fungal peptide toxin critical for mucosal infection. Nature 532(7597):64–68

Olsen LA, Choffnes ER, Relman DA, Pray L (2011) Fungal diseases: an emerging threat to human, animal, and plant health. The National Academies Press, Washington, DC

Pappagianis D (1988) Epidemiology of coccidioidomycosis. Curr Top Med Mycol 2:199–238

Retallick RW, McCallum H, Speare R (2004) Endemic infection of the amphibian chytrid fungus in a frog community post-decline. PLoS Biol 2(11):e351

Rodrigues AM, De Hoog GS, Zhang Y, de Camargo ZP (2014) Emerging sporotrichosis is driven by clonal and recombinant *Sporothrix* species. Emerg Microbes Infect 3:e32

Rosenberg R, Beard CB (2011) Vector-borne infections. Emerg Infect Dis 17(5):769–770

Shelburne SA 3rd, Darcourt J, White AC Jr, Greenberg SB, Hamill RJ, Atmar RL, Visnegarwala F (2005) The role of immune reconstitution inflammatory syndrome in AIDS-related *Cryptococcus neoformans* disease in the era of highly active antiretroviral therapy. Clin Infect Dis 40 (7):1049–1052

Sil A, Andrianopoulos A (2015) Thermally dimorphic human fungal pathogens – polyphyletic pathogens with a convergent pathogenicity trait. Cold Spring Harb Perspect Med 5(8):a019794

Taylor LH, Latham SM, Woolhouse MEJ (2001) Risk factors for human disease emergence. Philos Trans R Soc Lond B 356:983–989. https://doi.org/10.1098/rstb.2001.0888. Published 29 July 2001

Vanittanakom N, Cooper CR Jr, Fisher MC, Sirisanthana T (2006) Penicillium marneffei infection and recent advances in the epidemiology and molecular biology aspects. Clin Microbiol Rev 19 (1):95–110

Vredenburg VT, Knapp RA, Tunstall TS, Briggs CJ (2010) Dynamics of an emerging disease drive large-scale amphibian population extinctions. Proc Natl Acad Sci U S A 107(21):9689–9694

Walker SF, Bosch J, James TY, Litvintseva AP, Valls JAO, Pina S, Garcia G, Rosa GA, Cunningham AA, Hole S, Griffiths R, Fisher MC (2008) Invasive pathogens threaten species recovery programs. Curr Biol 18(18):R853–R854

Part II

Epizootic Mycoses with a Reservoir in Animals, with Occasional Outbreaks

Common and Emerging Dermatophytoses in Animals: Well-Known and New Threats

3

Vit Hubka, Andrea Peano, Adela Cmokova, and Jacques Guillot

Abstract

Zoophilic dermatophytes are frequently responsible for superficial mycoses in mammals worldwide. They comprise approximately ten specialized parasitic fungi belonging to genera *Trichophyton* and *Microsporum*. Due to contagious nature of the disease, the majority of species possess potential to cause outbreaks at least in their principal host(s) and at the same time have the capability to infect a wide spectrum of mammals, including humans. The purpose of this chapter is to trace the current changes in the epidemiology of animal-infecting dermatophytes that show large geographic differences and dynamically alter over time. Emphasis

V. Hubka (✉)
Department of Botany, Faculty of Science, Charles University, Prague, Czech Republic

Laboratory of Fungal Genetics and Metabolism, Institute of Microbiology of the Czech Academy of Sciences, v.v.i., Prague, Czech Republic

First Faculty of Medicine, Charles University, Prague, Czech Republic
e-mail: hubka@biomed.cas.cz

A. Peano
Department of Veterinary Sciences, University of Turin, Turin, Italy
e-mail: andrea.peano@unito.it

A. Cmokova
Department of Botany, Faculty of Science, Charles University, Prague, Czech Republic

Laboratory of Fungal Genetics and Metabolism, Institute of Microbiology of the Czech Academy of Sciences, v.v.i., Prague, Czech Republic
e-mail: cmokova@gmail.com

J. Guillot
Department of Parasitology, Mycology and Dermatology, EnvA, Ecole nationale vétérinaire d'Alfort, UPEC, Maisons-Alfort, France

Dynamyc Research Group, EnvA, Ecole nationale vétérinaire d'Alfort, UPEC, Maisons-Alfort, France
e-mail: jacques.guillot@vet-alfort.fr

S. Seyedmousavi et al. (eds.), *Emerging and Epizootic Fungal Infections in Animals*,
https://doi.org/10.1007/978-3-319-72093-7_3

is given not only to the most important and widespread dermatophyte species representing global issue for both animal and human medicine (*Microsporum canis*, *Trichophyton mentagrophytes*, and *T. verrucosum*) but also to newly emerging pathogens such as *T. benhamiae*, an agent of epidemic dermatophytosis in Europe frequently affecting guinea pigs and their breeders or owners. The methods for identification and molecular typing of dermatophytes are summarized due to their importance for outbreak detection and epidemiological surveillance. Strategies for management and prevention of outbreaks are also presented.

3.1 Introduction

Dermatophytes are the most successful pathogenic fungi causing superficial mycoses (dermatophytosis, also called ringworm) in humans and animals. They encompass ecologically and phylogenetically related fungi belonging to the family Arthrodermataceae (order Onygenales) which are able to use keratin as a sole nutrient source (Gräser et al. 1999a; Sugiyama et al. 2002). In the last several decades, the dermatophytes were usually categorized into three genera, *Trichophyton*, *Epidermophyton*, and *Microsporum*, and associated sexual state used to be classified in *Arthroderma* (Weitzman et al. 1986; Weitzman and Summerbell 1995). With complete abolition of dual nomenclature (McNeill et al. 2012) and availability of multiple gene phylogenies, the number of genera was expanded (de Hoog et al. 2017), but the most important primary pathogenic species remains in the three mentioned genera. Chronology of selected important historical events related to dermatophytosis and dermatophytes is shown in Fig. 3.1.

Dermatophytoses occur frequently in livestock, in companion animals, and also in wildlife. Infections caused by zoophilic dermatophytes are usually benign and self-limiting and respond well to treatment; they are rarely serious or manifest as systemic infection in immunocompromised host (Chermette et al. 2008; Rouzaud et al. 2016). The prevalence in animals shows large geographic differences. It seems to be influenced by trade, exchanges of animals for reproductive purposes, exhibitions (for cats and dogs), and sportive activities (e.g., horse races). Risk of zoonotic transmission depends on the local spectrum of kept animals and epidemiological situation (prevalence of pathogens in animals) but also on local relations between man and animals, hygienic standards, and socioeconomic factors. Dermatophytoses are highly contagious, and animals kept in herds or groups are threatened by the epidemic spread of infection even when one of few infected individuals are introduced into the community. The environment contaminated by arthroconidia and diseased animals (commonly with mild clinical signs or symptomless) represents a potential source of infection to humans.

High financial costs are associated worldwide with treatment, diagnosis, and prevention of dermatophytosis (Drake et al. 1996). Direct economic costs in farming and industry result from unaesthetic aspect of lesions (hide and skin industry), which

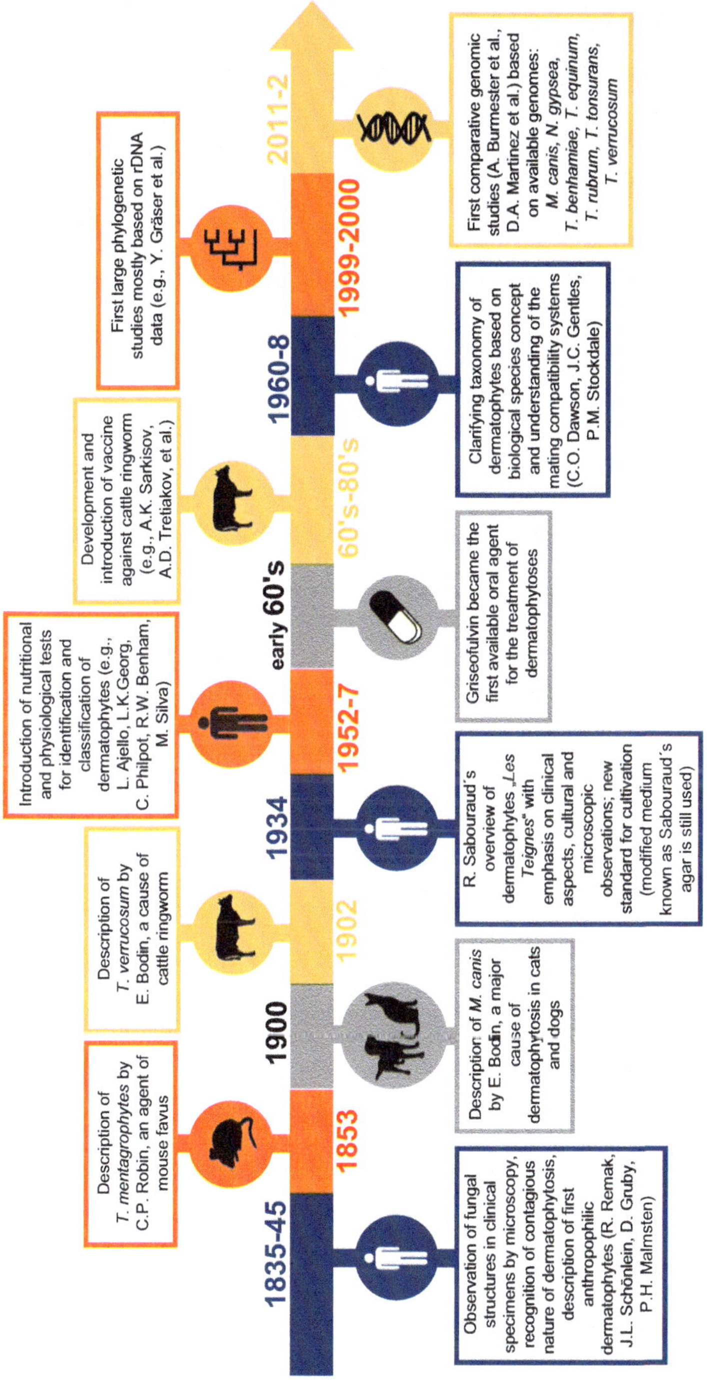

Fig. 3.1 Chronology of important events related to dermatophytosis and taxonomy of dermatophytes

hinder animal trade and are also an obstacle to attend exhibitions and sportive activities (Chermette et al. 2008).

For effective treatment and prevention of the spread of these diseases, it is important to correctly determine the causal agents at the species level, which allows prescription of suitable therapy and at the same time identification of probable source of infection. Correct species identification is an indispensable prerequisite for monitoring changes in the frequency of individual species, helps to evaluate the results of preventive measures and interventions, and is a basic requirement for the preparation of epidemiological studies. Instable taxonomy of dermatophytes, problematic species concept, and phenotypic identification limit comparability of recent epidemiological data with studies from the past. In the last two decades, the advent of molecular methods greatly contributed toward the understanding of the biodiversity of dermatophytes (Gräser et al. 1999b, 2000a, 2000b, 2008; Hubka et al. 2014a), thus challenging the previous classification based on morphology and nutritional tests. However, dermatophytosis in animals is classically based on phenotypic methods, and we have currently no systematic data supported by molecular methods worldwide on the epidemiology of dermatophytes.

3.2 Virulence Factors and Pathogenesis

The dermatophytes are transmitted by direct contact with an infected host or by contact with contaminated objects and environment. The successful initiation of infection is closely related to the capability of the pathogenic fungus to overcome the host resistance mechanisms. The adherence to the keratinized tissue, germination of the arthroconidium, and stratum corneum penetration are the first steps in this process (Hube et al. 2015; Chinnapun 2015). Mechanical disruption of the stratum corneum consequent to microtrauma of different origin (e.g., due to ectoparasites in cats and dogs) appears to be important aspect in facilitating penetration and invasion of hair follicles (Miller et al. 2013). Three crucial steps to overcome cutaneous barriers and cause infection, i.e., adhesion, germination, and invasion of the stratum corneum, are separated by variously long time periods that range from several hours to several days depending on particular species (Aljabre et al. 1993; Rashid et al. 1995; Duek et al. 2004). It has been suggested that pathogenic dermatophytes express carbohydrate-specific adhesins on conidia surface that recognize mannose and galactose on the skin surface during adhesion (Esquenazi et al. 2004). Additionally, secreted proteases are required for the adherence process and participate in invasion of epidermis (Baldo et al. 2008; Baldo et al. 2010; Shi et al. 2016). Keratin is a hard and compact protein, which is unavailable as a nutrient source to the majority of organism in the nature. Its degradation and utilization during penetration and infection are considered a major virulence attribute. The dermatophytes dispose expanded sets of endopeptidases, exopeptidases, and secreted proteases reserved for that purposes (Burmester et al. 2011; Martinez-Rossi et al. 2017). In addition, the process of keratinolysis is facilitated by sulfite efflux, a reducing agent which can cleave keratin-stabilizing cystine bonds and whose production from cystine depends on the cysteine dioxygenase (Grumbt et al. 2013).

Other tentative virulence factors comprise production of siderophores to overcome iron deficiency (Mor et al. 1992; Kröber et al. 2016), melanin with immunomodulatory properties (Youngchim et al. 2011), expression of homologues of multidrug resistance class of ABC transporters modulating susceptibility to antifungals (Fachin et al. 2006; Ghannoum 2016), or production of toxic exometabolites. Genome sequences of dermatophytes revealed a high number of secondary metabolite gene clusters, some of them are upregulated during keratinocyte infection (Burmester et al. 2011; Martinez et al. 2012; Yin et al. 2013). Toxin xanthomegnin was predicted as a virulence factor and can be extracted from infected keratinized tissues but is not detected in uninfected tissues (Gupta et al. 2000; Kandemir et al. 2015).

In general, the stepwise process of host infection is similar across the species; however, the ability to elicit a host reaction upon infection is species-specific. Different dermatophytes induce different immune responses in the host in terms of quality and intensity. They may induce a high level of tissue damage and inflammatory reaction, but in particular host species or individuals, they can manipulate the host's immune response, ensuring survival and prolonging chronic infection (Hube et al. 2015). The interaction of host cells (keratinocytes and immune cells) and dermatophyte and recognition of fungal antigens and secreted enzymes by host are basis for immune response. Cell-mediated immunity and production of various cytokines are widely considered to be involved in modulating the immune response (Achterman and White 2011; Hube et al. 2015; Chinnapun 2015; Martinez-Rossi et al. 2017).

3.3 Clinical Manifestations and Diagnosis

Dermatophytosis can be located on any part of the animal body and usually manifests as a regular and circular focus generally accompanied with alopecia and desquamation. The lesion spread centrifugally from the site of inoculation; multiple lesions can merge into large foci with irregular margins. The lesion can be erythematous over the entire surface but especially on the periphery. It can be covered by pustules and later by exudate and crusts. Other clinical signs may be present, such as pruritus accompanied by scratching and behavior change. Animal dermatophytosis has usually self-limiting nature, and spontaneous regrowth of hairs is generally observed. Infection is usually restricted to the dead cornified layers of the skin because of the inability of the dermatophytes to attack the deeper layers of the skin or organs of immunocompetent hosts. Naive, usually young, animals can be heavily infected with significant impact on health condition and growth. The increased susceptibility of young animals may also reflect differences in the biochemical properties of the skin and skin secretions (especially sebum), growth and replacement of hair, and the physiologic status of the host as related to age. Animals can exhibit subtle signs or act as "asymptomatic" carriers of dermatophytes without history of dermatophytosis as seen relatively commonly in cats infected by *M. canis* or guinea pigs infected by *T. benhamiae*. Regardless of the possibility of more virulent strain emergence, many factors related to the host play a critical role in determining the type of clinical lesions produced and terminating the infection. More comprehensive reviews and monographs are available on this topic (Donnelly et al. 2000; Chermette et al. 2008; Miller et al. 2013; Moretti et al. 2013).

In companion animals (cats, dogs, or small mammals like rabbits, guinea pigs, and chinchillas) as well as in large animals (horses and cattle), dermatophytes are frequently responsible for skin diseases including alopecia and crusts. Dermatophytosis is a contagious disease, and cases are regularly reported in breeding facilities and farms. Human contamination is also indicative. However, experimental diagnosis is systematically required. The use of the Wood's lamp is recommended when *M. canis* infection is suspected in companion animals. The microscopic examination of hairs and scales can be performed from skin scrapings (Fig. 3.2). According to the dermatophytes species, arthroconidia have different dimensions and disposition on the hairs: *M. canis* produces clusters of very small

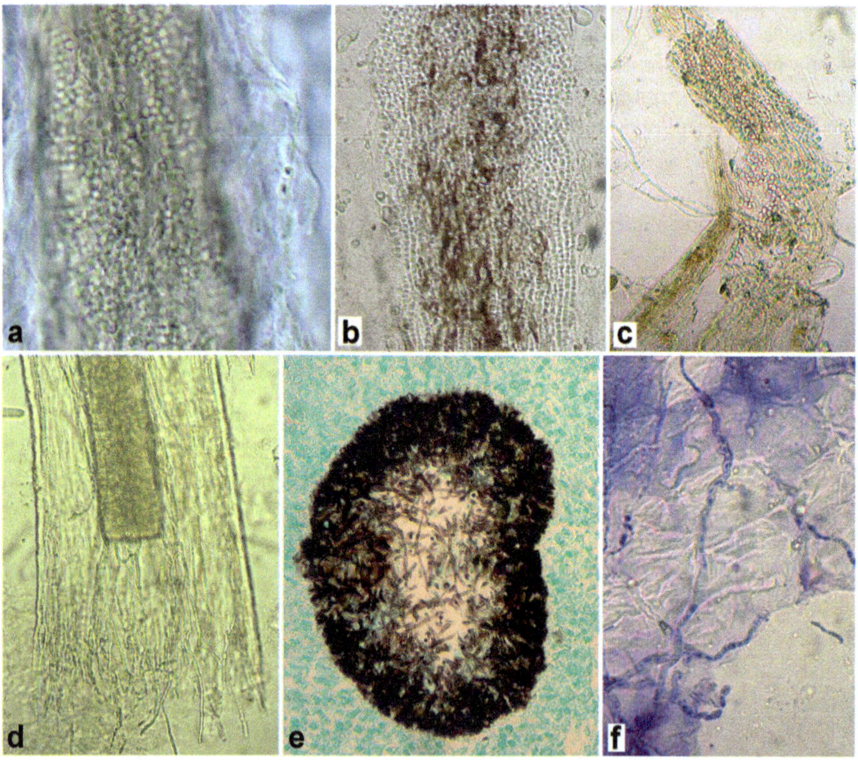

Fig. 3.2 Laboratory diagnosis of dermatophytosis by using direct microscopic examination or histology. (**a**) Arthroconidia of *Microsporum canis* at direct hair examination (digestion in 20% NaOH). (**b**) Arthroconidia of *Trichophyton benhamiae* at direct hair examination (digestion in 20% NaOH). (**c–d**) Arthroconidia of *Trichophyton verrucosum* at direct hair examination (digestion in 20% NaOH). (**e**) Histopathological examination (Gomori-Grocott staining) of pseudomycetoma in a cat due to *M. canis*. (**f**) Cytological examination in case of human dermatophytosis due to zoophilic *Trichophyton mentagrophytes* (scotch tape technique)

(2–4 μm) arthroconidia, whereas members of the genus *Trichophyton* form chains of arthroconidia (Chermette et al. 2008). Fungal culture is still considered as the golden standard for diagnosis. Samples of hairs, crusts, scales, or even cutaneous tissue (in the specific case of pseudomycetoma) should be collected for culture. When a large number of colonies of dermatophyte develop on the culture medium, active infection is demonstrated. When only a few colonies develop on the culture medium, subclinical infection or mechanical carriage (especially in cats or guinea pigs) should be suspected. When unusual clinical presentations are observed, histological examination is recommended (Fig. 3.2).

Sensitivity of both direct microscopic examination and cultivation depends on the combination of host and pathogen. Microscopic examination fails to provide the identification of the agent, whereas culture has usually several weeks turnaround time and the requirement of mycological expertise. The reliance on culture, which may be often unsuccessful for some dermatophyte species, can be substituted by PCR-based tests that are now available for the direct diagnosis of dermatophytosis from clinical samples and significantly reduces the time of diagnosis (hours to days). These methods can suitably supplement or even replace classical diagnostic schemes thanks to their high sensitivity and specificity. They are represented by conventional PCR, RT-PCR, or more complex methodologies (PCR-ELISA, PCR-RLB, microarrays) and usually offer opportunity to detect DNA of any dermatophyte without species identification, but some more recently developed assays are able to identify relatively broad spectrum of species including major zoophilic species (Jensen and Arendrup 2012; Cafarchia et al. 2013b; Dąbrowska et al. 2014; Mehlig et al. 2014; Kupsch et al. 2016).

3.4 Ecology of Dermatophytes and Origin of Zoophilic Species

The Mesozoic era (252–66 million years ago) was associated with a significant ecological morphological diversification of early mammals, a prerequisite for evolutionary success after extinction of the nonavian dinosaurs 66 million years ago. It was assumed based on molecular dating that dermatophytes could radiate approximately 50 million years ago (Harmsen et al. 1995) closely linked to early Cenozoic adaptive explosion of mammals. However, more recently fossil evidence of dermatophytosis in mammals was estimated to be approximately 125 million-year-old (Martin et al. 2015). Despite numerous uncertainties, it seems quite clear that dermatophytes are evolutionary young group of fungi that diverged later than other groups of pathogenic fungi (Wu et al. 2009).

Significant part of the diversity of dermatophytes is represented by geophilic species that act as uncommon causal agents of infections without significant contagious potential. They occur in soil around burrows and nests of terrestrial vertebrates and birds, can be carried in the fur, and therefore cause diagnostic doubts when they are isolated from healthy animals or animals with ambiguous symptoms. Geophilic species are characterized by relatively high intraspecies diversity, and they are all sexual and heterothallic (isolates of two opposite mating types are present in population) with only few exceptions. The sexual process takes place in the soil on the keratinized substrates. Geophilic species are considered an ancestral ("primitive") group of dermatophytes, while "advanced" ecological groups of zoophilic and anthropophilic are derived from geophilic species. The phylogenetic grouping of anthropophilic and zoophilic species of *Trichophyton* and *Microsporum* into monophyletic clades and geophilic species from genera *Arthroderma*, *Nannizzia*, and *Paraphyton* into another monophyletic clades supports the hypothesis that ecology has been crucial driver of the evolution in dermatophytes (Gräser et al. 2008; de Hoog et al. 2017).

Specialized pathogens of animals and human, zoophilic and anthropophilic species, respectively, are primarily associated with one or few related host species but have potential to cause infection in a broad spectrum of animals (Table 3.1). They usually cause mild (chronic) or asymptomatic infections in primary host, often

Table 3.1 Ecology of zoophilic or possibly zoophilic dermatophytes

Species	Principal host(s)/ source, other hosts	Distribution	Epidemic potential in principal hosts	Zoonotic risk for human
Microsporum canis (syn. *M. equinum*)	**Cat, dog, horse**, all mammals	Worldwide	High	High
L. gallinae (syn. *M. vanbreuseghemii*)[a]	**Chicken, soil**, birds, mammals	Worldwide	Low	Low
T. benhamiae	**Guinea pig**, other rodents, dogs	Worldwide	High	High
T. bullosum	**Horse, donkey**, mole?	Syria, Sudan, Tunisia, France, Czech Republic	Insufficient data	Insufficient data
T. equinum	**Horse**	Worldwide	High	Low
T. erinacei	**Hedgehogs** (*Erinaceus europaeus, Atelerix albiventris*)	Europe, New Zealand, Africa (kept as pet worldwide)	High	Probably high
T. eriotrephon	Unknown	Netherlands, Iran	Insufficient data	Insufficient data
T. mentagrophytes	**Rabbits, rodents**, cats, and dogs (especially free roaming and hunting)	Worldwide	High	High
T. quinckeanum	**Rodents** (mice)	Worldwide	High	High
T. simii[a]	Soil, monkeys, chicken, dog	Worldwide	Low	Low
T. verrucosum	**Cattle, other ruminants**, all mammals, birds	Worldwide	High	High

[a]These species are probably geophilic—see Sect. 3.9

being widespread and epidemic/epizootic. In contrast, the infections in less common hosts tend to be acute and highly inflammatory. The population structure of many primary pathogenic species is nearly clonal, and they have unknown sexual state. It was suggested that these "species" spread clonally by asexual propagation in the population of host without having a terrestrial reservoir and thus reducing the probability of encountering a partner of the opposite sex (Summerbell 2002). Consequently, these "species" (clonal offshoots) derived from their sexual ancestors possess predominantly only strains with identical mating type across global population. Kano et al. (2014) demonstrated that strains of *T. verrucosum* have consistently only MAT1-2-1 gene corresponding to the mating type (−). Similarly, single mating is also present in anthropophilic species *T. tonsurans* (mating type −), *T. rubrum* (mating type −), and *T. violaceum* (mating type −) but also in zoophilic *T. equinum* with mating type + (Gräser et al. 2008; Metin and Heitman 2017). The intraspecies variability in these taxa is mostly generated by fixation of neutral mutations. Historically, species and populations tended to remain within limited geographic areas such as continents, a condition leading to structuration of global population. With human migration and animal trade, however, geographically restricted genotypes can be rapidly distributed over large geographic areas, leading to reduction of polymorphism within widely dispersed entities (Gräser et al. 2006).

In contrast to abovementioned species, both mating types are present in population of *T. benhamiae* (Kano et al. 2011; Symoens et al. 2013; Cmokova 2015), *T. mentagrophytes* (Symoens et al. 2011), and *M. canis* (Hironaga et al. 1980) suggesting that sexual reproduction naturally occurs in these species and probably takes place in burrows of wild animals or their close neighborhood rather than in association with dwellings of domestic animals. The distribution of mating-type genes is commonly unequal in mentioned species or at least in some subpopulations. This may be caused by clonal horizontal propagation of single or several clones in the population of the host. Assessment of species boundaries by mating experiments is in general well applicable in geophilic species (Stockdale 1964; Hubka et al. 2015). In contrast, biological compatibilities can considerably disagree with the concept of classical species of anthropophilic and zoophilic dermatophytes. These species are phylogenetically young, and prezygotic reproductive barriers are incomplete resulting in positive mating experiments even between phylogenetically distant species (Anzawa et al. 2010; Kawasaki et al. 2010; Kawasaki 2011). It is however highly unlikely that this kind of hybridization occurs naturally due to different ecological niches of species.

3.5 Phylogenetic Position of Zoophilic Dermatophytes and Their Identification

The dermatophytes have been divided into three genera by using conventional phenotypic taxonomy: *Trichophyton*, *Microsporum*, and *Epidermophyton*. This classification scheme has been widely accepted in the second half of the twentieth century and used till today. Phylogenetic relationships between dermatophytes have been exploited by using numerous genetic loci, including mitochondrial DNA (mtDNA) (Nishio et al. 1992), small subunit of ribosomal DNA (SSU rDNA) (Harmsen et al. 1995), large subunits (LSU) of rDNA (Leclerc et al. 1994), internal transcribed spacer (ITS) region of rDNA (Gräser et al. 1999a, b, 2000b; Makimura et al. 1999; Summerbell et al. 1999; Kawasaki et al. 2011; Pchelin et al. 2016), chitin synthase (CHS) (Kano et al. 1999, 2002; Hirai et al. 2003) and DNA topoisomerase II genes (TOP II), actin, glyceraldehyde-3-phosphate dehydrogenase (GPD) (Kawasaki et al. 2011), translation elongation factor 1-α (TEF 1-α) (Mirhendi et al. 2015), β-tubulin (Rezaei-Matehkolaei et al. 2014; Pchelin et al. 2016), and calmodulin (Ahmadi et al. 2016). The topology of phylogenetic trees based on ITS region (Fig. 3.3) is congruent with those based on protein-coding loci, i.e., *TEF 1-α*, β-tubulin, and calmodulin. The studies with sufficient coverage of species across diversity of Arthrodermataceae consistently showed that both *Trichophyton* and *Microsporum* are polyphyletic prompting reevaluation of the generic concept of dermatophytes (Gräser et al. 1999a; Rezaei-Matehkolaei et al. 2014; Mirhendi et al. 2015; Ahmadi et al. 2016). The new generic concept with monophyly as a main criterion was recently introduced by de Hoog et al. (2017) who increased the number of recognized genera from three to seven. This new concept has only limited consequences for the taxonomy of human and animal pathogenic dermatophytes, while significant changes were made in the taxonomy of geophilic species.

ITS rDNA sequencing is considered a gold standard for species identification. It is still the only genetic locus available for all currently accepted species, is recommended as a barcode, and is widely used for identification in praxis (Irinyi et al. 2015). ITS region sequence has capability to discriminate even closely related species, e.g., *M. canis*, *M. audouinii*, and *M. ferrugineum*; *T. benhamiae* and *T. concentricum*; and *T. quinckeanum* and *T. schoenleinii*. Identification of all mentioned sibling species cannot be realized by amplification of any currently available protein-coding loci. When DNA sequence is available, species identification is usually obtained by comparison with the reference sequence in the database (e.g., GenBank; www.ncbi.nlm.nih.gov/genbank). Because GenBank acts primarily as an archive, many sequences have been annotated incorrectly, the reason why curated databases were established, e.g., ISHAM ITS Database (http://its.mycologylab.org) or Dermatophytes Species Database at Westerdijk Fungal Biodiversity Institute (http://www.westerdijkinstitute.nl/Dermatophytes). A plenty of PCR-based techniques were developed for identification of dermatophytes with variable discriminatory power including nested PCR, PCR-RFLP, RAPD, ISSR-PCR, AFLP, AP-PCR, etc. (Chen et al. 2011; Sharma and Gräser 2011; Cafarchia et al. 2013c). In particular PCR-RFLP targeting ITS or other regions with digestion using various restriction enzymes reached relatively broad popularity due to low financial cost and usability for identification of major pathogens (De Baere et al. 2010; Heidemann et al. 2010; Rezaei-Matehkolaei et al. 2013). Many other methods showed similar or higher potential for species identification, but they have only limited usability due to poor reproducibility or laboriousness and are now obsolete.

While DNA-based molecular methods are usually highly accurate and rapid, they are relatively costly and can be complex to implement. Matrix-assisted laser

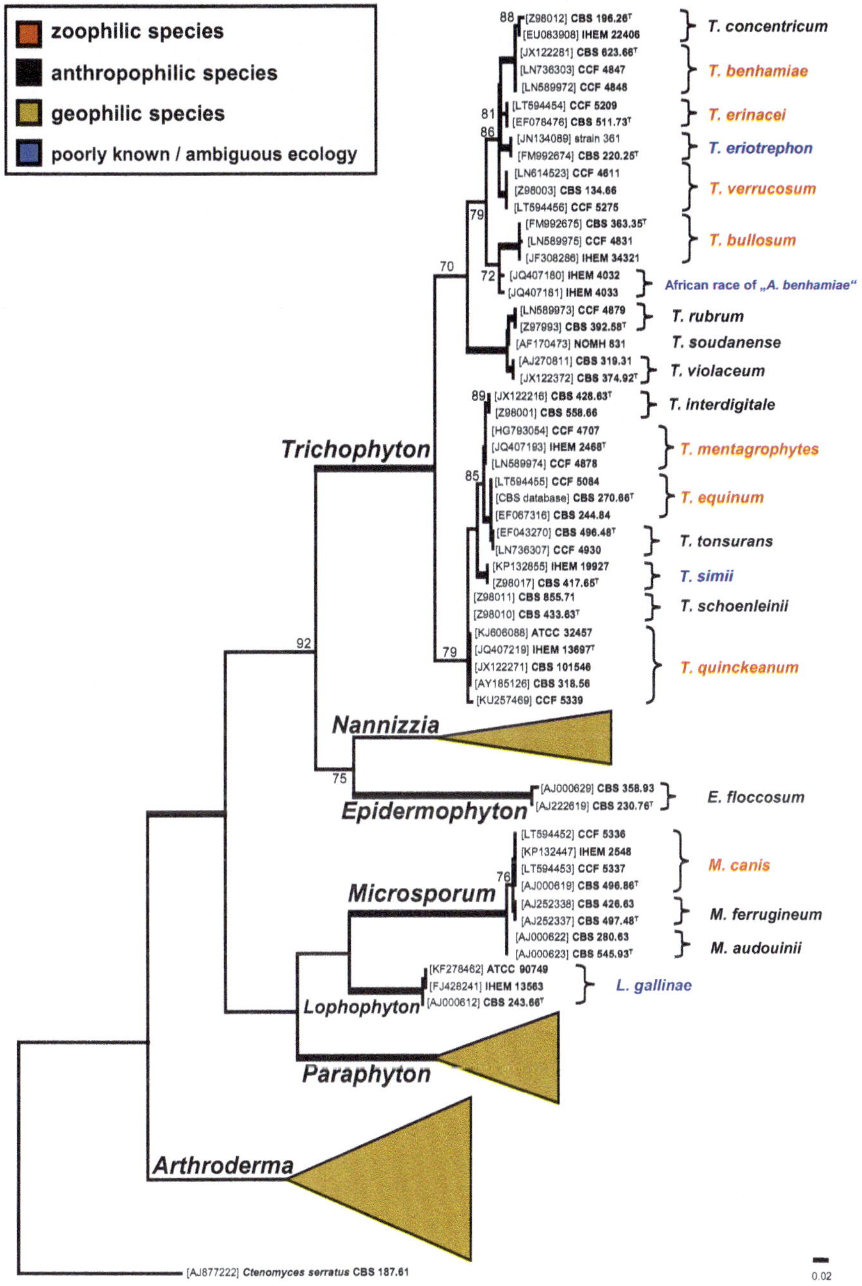

Fig. 3.3 Phylogenetic placement of zoophilic species within dermatophytes. Best scoring maximum likelihood tree-based ITS rDNA constructed with the IQ-TREE version 1.4.0 (Nguyen et al. 2015) by using GTR + I + G4 substitution model. Dataset contained 95 taxa and a total of 731 characters of which 351 were variable and 306 parsimony informative. Support values at

desorption ionization time-of-flight mass spectrometry (MALDI-TOF MS) may represent an alternative to conventional dermatophyte identification. Application of MALDI-TOF MS as an identification procedure for pathogenic dermatophytes has been increasingly used in the laboratories due to its time- and cost-effectiveness (despite the initial cost of investment for obtaining the equipment) (L'Ollivier and Ranque 2017). The method is able to distinguish all major pathogenic dermatophytes; however, the differentiation of phylogenetically closely related taxa is associated with higher level of errors, e.g., *T. quinckeanum* and *T. schoenleinii*, different races of *T. benhamiae*, or *T. interdigitale* and *T. mentagrophytes*. Thus, supplementation of the reference spectra libraries is still required for optimal dermatophyte identification (Gräser 2014).

Despite the advent of molecular methods, conventional species identification of dermatophytes is still the prevailing method of identification worldwide and consists of micro- and macromorphological examination of cultures, sometimes supplemented with various physiological and biochemical tests (e.g., nutritional tests, vitamin requirements tested on T1-T7 *Trichophyton* agars, urease activity, hair perforation test, etc.), mating experiments, etc. The Sabouraud's agar, frequently supplemented by antibiotics chloramphenicol and cycloheximide, is the most commonly used isolation medium. Strains with typical morphology can be identified directly from primary cultures, but subculturing on specific media inducing sporulation or production pigments may be necessary (Robert and Pihet 2008).

3.6 Genotyping Schemes and Population Structure of Major Zoophilic Dermatophytes

Genotyping is often employed to confirm or rule out outbreaks, gain insight into the dynamics of disease transmission, recognize virulent strains and regional and global changes in genotype pattern, determine the source and routes of infections, trace cross-transmission of healthcare-associated pathogens, and evaluate the effectiveness of control measures (Ranjbar et al. 2014). Other common issues in dermatophytes concern differentiation of relapse or reinfection, determining if the infection is caused by one or more strains, if the host can harbor different genotypes varying in their degree of virulence or potential for transmission to other hosts including human, if particular genotypes differ in clinical manifestation, etc. The success in typing of dermatophytes according to phenotype criteria has been limited. Indeed, the same strain may present with different phenotypic characteristics depending on different factors, firstly the conditions of culture (Dhieb et al. 2014).

Fig. 3.3 (continued) branches were obtained from 1000 bootstrap replicates. Only branches with bootstrap support ≥70% are shown; branches with support ≥95% are double thick; ex-type strains are designated by a superscript T. The tree is rooted with *Ctenomyces serratus*

On the other hand, strains with similar colonies may belong to different genetic types.

Because many pathogenic dermatophyte species show nearly clonal population structure and phenotype commonly do not correlate with genotype, DNA-based approaches are supposed to be tools of choice for genotyping. Although **DNA sequencing** allows species identification of dermatophytes, it lacks sufficient discriminatory power to study population structure of most clinically relevant species. Consequently, no **multilocus sequence typing schemes** (MLST) have been evaluated and developed for genotyping of dermatophytes, although MLST has found wide application in many other fungal pathogens (Meyer et al. 2009; Debourgogne et al. 2012; Bernhardt et al. 2013; Maitte et al. 2013). It is worth mentioning that certain level of intraspecies polymorphism was detected by using DNA sequencing of some genetic loci in sexual species, especially *T. mentagrophytes* and *T. benhamiae*. Four **ITS rDNA** sequence genotypes were revealed among 86 isolates of *T. mentagrophytes* sensu lato (Heidemann et al. 2010); genotyping was useful for discrimination between strains of zoophilic origin (*T. mentagrophytes* s. str.) and closely related anthropophilic *T. interdigitale* and correlated with clinical manifestation and phenotype of strains. Additional ITS rDNA genotypes in *T. mentagrophytes s. str.* were revealed by Pchelin et al. (2016) who did not confirm correlation between genotype, origin of strains, and phenotype observed by Heidemann et al. (2010). Combination of ITS rDNA and *GPD* gene revealed five genotypes in global population of *T. benhamiae* (Cmokova 2015) which corresponded with phenotype and in part with geographic origin. Apart from ITS and *GPD* genes, also *TOP II* (Kawasaki et al. 2011) and *TEF-1α* (Mirhendi et al. 2015) have potential for genotyping of *T. mentagrophytes* and *T. benhamiae* but are not available or lack discriminatory power in other pathogenic species. In conclusion, the set of four last mentioned genes has significant potential for typing of *T. mentagrophytes* and *T. benhamiae*, while population of *M. canis* and *T. verrucosum* in domestic animals and humans shows high level of clonality and resists to genotyping by currently available genetic loci. Limited variability was found in ITS region of *M. canis* (Kaneko et al. 2011); unfortunately significant part of the sequence variability in GenBank is caused by sequencing errors as demonstrated by at least four different ITS sequences deposited for the ex-type strain of *M. canis*.

Microsatellite markers are currently the most powerful and effective tool available for subtyping of dermatophytes. The typing schemes have been developed for only limited number of species, i.e., *T. rubrum*, *Nannizzia* (=*Microsporum*) *persicolor*, *M. canis*, and *T. benhamiae*. These methods will be given the most attention in the following paragraphs.

Genotyping attempts were commonly unsuccessful in *M. canis* by using various approaches including RAPD, analysis of NTS region, or PCR-RFLP targeting ITS region (Yu et al. 2004; Leibner-Ciszak et al. 2010; Dobrowolska et al. 2011; Dhieb et al. 2014). In contrast, Spesso et al. (2013) was able to reveal intraspecies variability by using RAPD method among Argentine strains without significant correlation with clinical manifestation or geographic origin. Another method, inter-single-sequence-repeat-PCR (ISSR-PCR), revealed 21 genotypes among a total of 24 strains analyzed (Cano et al. 2005), which may indicate a good discriminatory power of the method employed. However, the stability of the markers was not assessed, and the reproducibility of the method was low. Two microsatellite loci were originally developed by Sharma et al. (2007), while more recently an extended panel of eight loci has been standardized (Pasquetti et al. 2013). By using two microsatellite loci, Sharma et al. (2007) studied genetic variation and dispersal among 101 global *M. canis* strains, distinguished three subpopulations, and found no correlation between genotype, clinical manifestation, and geographic origin. It was suggested that imbalance in the prevalence of particular genotypes among humans and animals was due to the emergence of a virulent genotype with a high potential to infect human host. Extended panel of eight loci proved to have a high discriminating power and revealed the extensive genetic diversity in global population of *M. canis* (Peano et al. 2015). Some multilocus genotypes (ML-GTs) were found with higher frequency, which leads to hypothesize the existence of clonal lines of "major success" due to a stronger parasitic aptitude. Correlation was not found between severity of clinical forms and a particular genotype. Likewise, there was no particular association between specific ML-GTs and zoonotic potential. Some ML-GTs were related to specific geographical contexts. Although it is unlikely that the loci employed for strain typing are connected with phenotypical features of interest (such as virulence, drug resistance etc.), microsatellite analysis has the potential to track these features indirectly, principally due to the clonal mode of reproduction typical of most dermatophytes (genomes are transmitted to the next generation in unaltered condition and thus associated genes – such as virulence genes and microsatellite markers are linked, which may allow tracing the feature of interest within populations of the fungus). Hence, for example, microsatellites were used to demonstrate that isolates of *M. canis* causing pseudomycetoma in cats are genetically related, if not identical, to isolates responsible for superficial ringworm lesions. These results strongly suggest that pseudomycetoma in cats is due to host factors and cannot be attributed to the specific ability of particular genotypes (Pasquetti et al. 2012). The data from two previously published studies (Sharma et al. 2007; da Costa et al. 2013) were reanalyzed here. The results (Fig. 3.4) confirmed that there is no clear association between particular haplotypes, geographical origin, and host.

Polymorphisms in population of *T. benhamiae* (American-European race) were investigated by using RFLP analysis of NTS region which produced 11 different patterns in 46 isolates; this method successfully confirmed laboratory-acquired infection as well as familial outbreaks transmitted from pets (Mochizuki et al. 2002; Takeda et al. 2012). The data indicated that *T. benhamiae* had been brought into Japan with imported animals on several occasions and spreads in Japan by transportation of animals by breeders or pet shops (Takeda et al. 2012; Hiruma et al. 2015). The global population structure of *T. benhamiae* ($n = 326$ isolates) was recently investigated by using sequences of two genetic loci (ITS rDNA and GPD) along with ten microsatellites markers (Cmokova 2015). The combined sequence analysis revealed the presence of five genotypes, while 32 unique genotypes

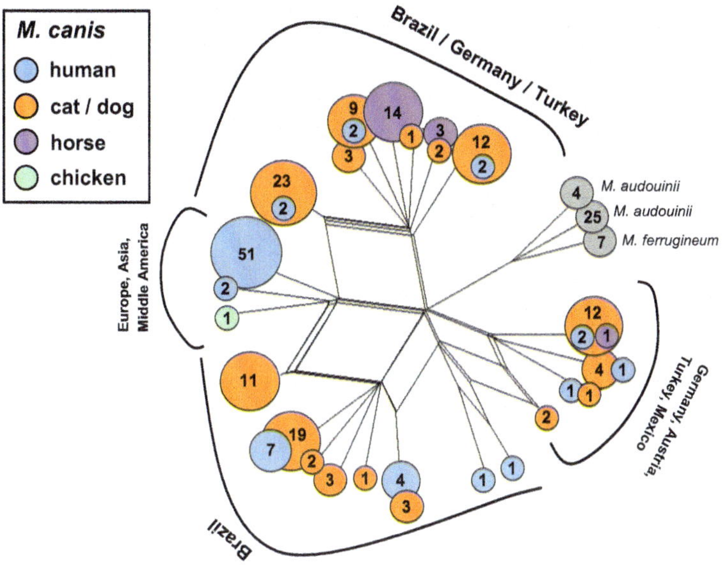

Fig. 3.4 Population structure of *Microsporum canis* complex revealed by analysis of two microsatellite loci—reanalysis of data previously published by Sharma et al. (2007) and da Costa et al. (2013). The dataset included 203 strains of *M. canis*, 29 *M. audouinii*, and 7 *M. ferrugineum*. NeighborNet phylogenetic network built using Jaccard index-based distance matrix with SplitsTree 4.13 (Huson and Bryant 2006). Colored circles correspond to different hosts; numbers in circles indicate number of isolates representing particular haplotype. No clear association between particular haplotypes, geographical origin, and host is evident. However, several haplotypes of "major success" were found in some hosts (haplotype represented by 51 human isolates; haplotype represented by 14 isolates from horses)

arranged into four subpopulations were discovered by microsatellite analysis (Fig. 3.5). The first subpopulation (S1) was most abundant in Europe and associated with guinea pigs; it was characterized by low variability in microsatellite data, yellow colonies with yellow reverse, and exclusively *MAT1-1-1* idiomorph. This clonal subpopulation is currently responsible for the outbreak of infections in the Central Europe. The second subpopulation (S2) comprised strains from North America mostly associated with dogs and typical by highly variable microsatellite data, both mating-type genes were present among strains; the colonies were mostly white, granular, and frequently with red reverse. It is probable that virulent European subpopulation has its origin in closely related subpopulation S2 in North America where the center of genetic variability of *T. benhamiae* is located and where the pathogen probably occurs on wild animals. The third subpopulation (S3) comprised strains from Europe mostly associated with guinea pigs and characterized by the presence of both mating types (predominantly *MAT1-2*). The fourth subpopulation (S4) comprised majority of strains from Japan, and some European strains that were mostly associated with rabbits; all strains had only *MAT1-1* idiomorph. Some of these subpopulations could represent separate taxonomic entities, but more research

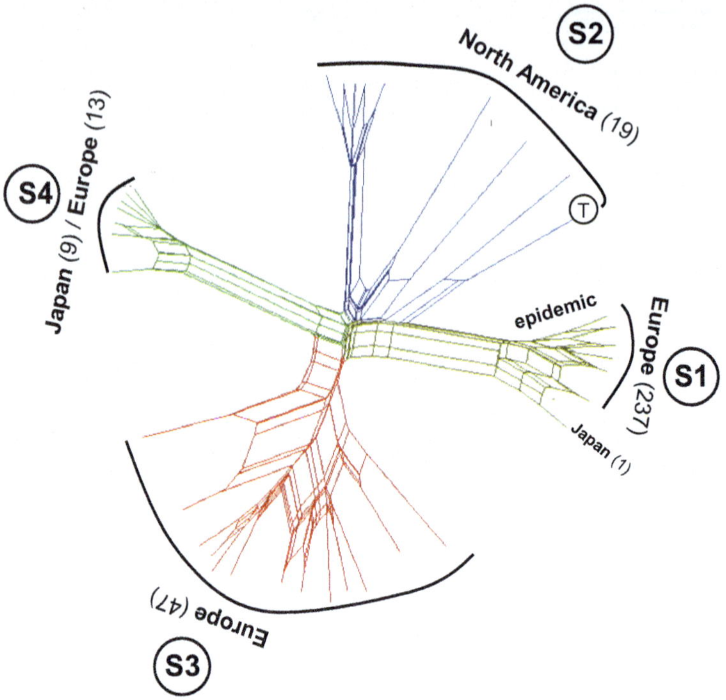

Fig. 3.5 Population structure of *Trichophyton benhamiae* ($n = 326$) revealed by analysis of ten microsatellite loci (Cmokova et al. 2017). NeighborNet phylogenetic network built using Jaccard index-based distance matrix with SplitsTree 4.13 (Huson and Bryant 2006). The data showed that the population of *T. benhamiae* is divided into four distinct subpopulations (designated S1–S4) and 32 genotypes. Numbers in parentheses indicate number of examined isolates from particular subpopulation. The subpopulation S1 was represented by highest number of isolates but at the same time had the lowest number of genotypes (clonal spreading) and is responsible for current outbreak of human and animal infections (mostly transmitted from guinea pigs) in Europe. Based on microsatellite and sequence data (not shown in figure) evidence, this clonal subpopulation is closely related to the subpopulation S2 represented exclusively by strains of North American origin (mostly infections in dogs) including the ex-type strain of *T. benhamiae* (marked with letter T)

is needed to confirm this assumption. Confusions are associated with phenotypes and genotypes recognized in the past among isolates identified as *T. benhamiae*. The isolates are usually designated as "white" or "yellow" phenotype based on macromorphology, and it was anticipated that they are connected with different genotypes (Symoens et al. 2013; Hiruma et al. 2015; Brasch et al. 2016). Apparently, the "white" phenotype which is predominant in Japan and in the USA and minority in Europe is in fact complex of several distinct taxonomic entities.

Random amplification of polymorphic DNA (RAPD) analysis (Kim et al. 2001) and PCR-RFLP targeting ITS rDNA (Heidemann et al. 2010) differentiated zoophilic *T. mentagrophytes* and anthropophilic *T. interdigitale*. Some RAPD analyses were even useful for subtle subtyping of *T. interdigitale* (Kac et al. 1999; Leibner-

Ciszak et al. 2010), but no correlation was detected between obtained profiles and geographic origin of strains. Poor reproducibility of the obtained profiles has reduced the interest in RAPD in favor of microsatellites and more reproducible methods in dermatophytes; but microsatellite markers have not been developed for *T. mentagrophytes*. PCR melting profile (PCR-MP) technique revealed seven genotypes within *T. mentagrophytes* and was able to distinguish zoophilic strains originating from Poland and Denmark (Leibner-Ciszak et al. 2010). Methods focusing on the variability in non-transcribed spacer (NTS) regions of rDNA have currently the highest discriminatory power among the methods applied on *T. mentagrophytes*. Southern blot hybridization-based RFLP analysis of NTS region was useful for subtyping of *T. mentagrophytes* and related *T. interdigitale* (23 subtypes among 60 isolates) (Mochizuki et al. 2003). Analysis of three individual subrepeat elements of the NTS identified 19 molecular types among 42 anthropophilic *T. interdigitale* strains (Jackson et al. 2006), and the method was later applied to 65 clinical strains isolated at one regional hospital in Japan with discrimination of 15 molecular types (Wakasa et al. 2010). Even higher resolution can be expected if the method was used to characterize closely related and sexually reproducing *T. mentagrophytes*.

Limited options are available for genotyping of *T. verrucosum*, *T. erinacei*, and *T. equinum*. It can be expected that at least some microsatellite markers designed for *T. benhamiae* will be useful for subtyping of *T. verrucosum* and *T. erinacei* due to close phylogenetic relatedness of these species (Fig. 3.3). Similarly, molecular subtyping based on single nucleotide polymorphisms (SNPs) developed for *T. tonsurans* (Abdel-Rahman et al. 2010) might work well in closely related *T. equinum*.

The establishment of global databases based on largely comparable data such as microsatellites, SNPs, and DNA sequences is desirable. Such databases would enable to understand global epidemiology of dermatophytes and monitor changes in genotype spectra on global and local scale. Although high throughput sequencing facilities are now widely available and increasingly used even in the epidemiology of fungal infections, this option has not yet been exploited in dermatophytes. SNP detection by using whole-genome sequence typing can be used, instead of MLST, to infer the genetic relatedness of fungal isolates. This ultimate approach will be certainly the method of choice in the future along with decreasing costs (Hadrich and Ranque 2015).

3.7 Major Zoophilic Dermatophytes: Epidemiology and Actual Concerns

3.7.1 *Microsporum canis*

Microsporum canis is the most common dermatophyte in cats and dogs (Fig. 3.6). Cats are considered to be the most important reservoir, but the species is also found regularly in rabbits and horses (Sharma et al. 2007; Cafarchia et al. 2013a; Pasquetti

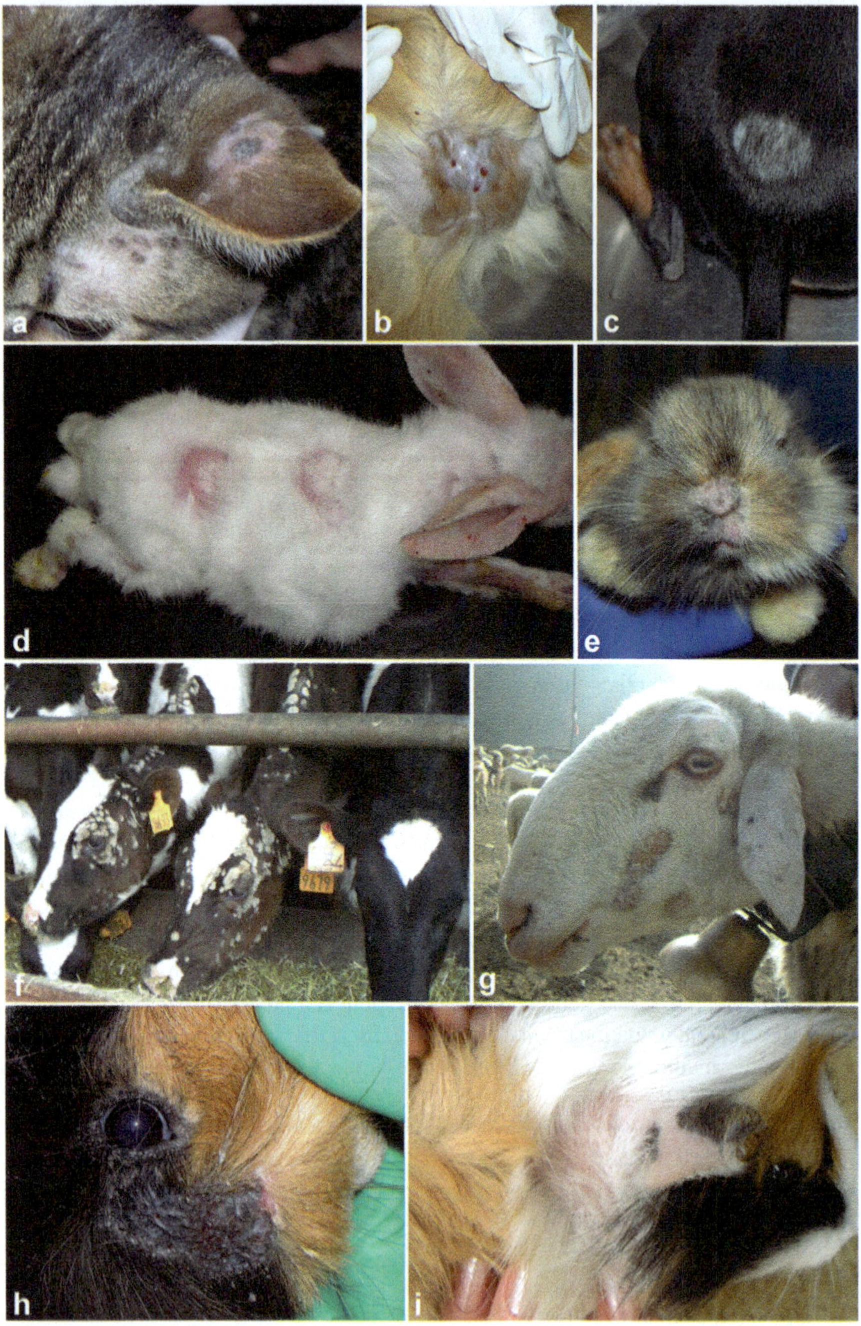

Fig. 3.6 Clinical presentation of infections caused by zoophilic dermatophytes. (**a–c**) *Microsporum canis*: irregular areas of alopecia covered by scales and scabs in a cat (**a**), pseudomycetoma in a cat (**b**), annular area of alopecia in a dog (**c**). (**d–e**) *Trichophyton mentagrophytes*: annular erythematous areas of alopecia on the back, extremities, and ears of a

et al. 2013). It has been occasionally reported in a number of other domestic and wild animals, e.g., cattle, sheep, goats, ferrets, camelids, marmots, eastern cottontails, foxes, etc. (Gallo et al. 2005a, b; Chermette et al. 2008; Pignon and Mayer 2011). Isolates coming from horses were previously classified as *M. equinum*, a species later synonymized with *M. canis* based on phylogenetic and population genetic analyses (Gräser et al. 2000a; Kaszubiak et al. 2004; Sharma et al. 2007). *Microsporum canis* is known to reproduce mainly asexually, although it is also capable to produce sexual state under laboratory conditions (Hironaga et al. 1980).

Microsporum canis is diffused worldwide and plays an important zoonotic role. In some countries it tends to overpass antropophilic dermatophytes as cause of human ringworm episodes (Chermette et al. 2008; Ameen 2010). It is the most frequent agent of tinea capitis in many European countries, the Eastern Mediterranean, South America, and China; it also causes highly inflamed lesions on glabrous skin and infrequently onychomycosis (Ginter-Hanselmayer et al. 2007; Seebacher et al. 2008; Skerlev and Miklić 2010; Uhrlaß et al. 2015; Zhan et al. 2015).

Dermatophytosis by *M. canis* is a pleomorphic and usually not a localized disease in cats despite appearances to the contrary (DeBoer and Moriello 2006). In addition, many infected cats have no or only few lesions. In particular, long-haired breeds can be subclinical carriers or have only minimal clinical symptoms; sometimes lesions become evident after shaving of hair. Isolation of *M. canis* from the haircoat in the absence of obvious lesions indicates either infection or fomite carriage from exposure to a contaminated environment. Distinguishing is often impossible, and Wood's lamp can help to detect minimal lesions invisible at naked eye. A mechanical carriage will only be revealed through fungal cultures (a plate with one or few CFU is usually indicative of fomite carriage). Dogs more often exhibit the classic annular areas of peripherally expanding alopecia, scale, crust, and follicular papules and pustules, with sometimes a central area of hyperpigmentation. It is quite clear that the response to infection in cats, more often than in dogs, tends to resemble that described in human patients with chronic infection by antropophilic dermatophytes. This is illustrated by the high number of cats which develops minimal and persisting lesions, just due to a "tolerant" immune response. This is an evidence of the strong adaptation of the fungus to the feline host. It is known that Persian cats are predisposed to *M. canis* infection (DeBoer and Moriello 2006; Miller et al. 2013) and to the development of more aggressive forms such as pseudomycetoma (Zimmerman et al. 2003; Bianchi et al. 2017). Apart from a genetic predisposition, this may also reflect a less efficient grooming of the haircoat because coat length has been reported as an important factor in the carriage of *M. canis* spores (Sparkes et al. 1993). A genetic predisposition to develop a generalized form of *M. canis* infection seems to exist also in Yorkshire Terrier dogs (Sparkes et al. 1993).

Fig. 3.6 (continued) rabbit (**d**), scaling lesion on the snout and upper lip of a rabbit (**e**). (**f–g**) *Trichophyton verrucosum*: discrete, scaling patches of hair loss located on the head and neck of cattle (**f**) and sheep (**g**). (**h–i**) *Trichophyton benhamiae*: weeping lesion under the eye of a guinea pig (**h**), itchy area of alopecia behind the ear of a guinea pig (**i**)

Available data show that *M. canis* cause >90% of dermatophytoses in cats worldwide. It is generally also the most prevalent dermatophyte isolated from dogs but with greater variation. For instance, in the USA, dermatophytosis was diagnosed in 14.9% of cats and 3.8% of dogs with cutaneous lesions, and *M. canis* is accounted for 92 and 43% of feline and canine cases, respectively (Lewis et al. 1991). In Brazil, dermatophytosis was confirmed in 27.8% of cats presented with dermatological problems, with *M. canis* responsible for 100% of cases; the prevalence in dogs was 9.8%, i.e., 68.5% of all cases of dermatophytosis (Copetti et al. 2006). Similar long-term studies from Europe showed comparable or even higher prevalence of *M. canis* in animals with skin disorders. Dermatophytes were isolated from 40.7% of cats in Croatia, and *M. canis* represented 98.7% of the isolates (Pinter et al. 1999). In Italy, dermatophytosis was diagnosed in 24.7% cats and 18.7% of dogs with *M. canis* representing 97 and 83% of the isolates, respectively (Mancianti et al. 2002). Climatic conditions appear to play a significant role in the diffusion of the pathogen. The prevalence is basically higher in hot and humid climates as shown in studies from climatically different regions of the USA and Italy (Moriello et al. 1994; Romano et al. 1997; Cafarchia et al. 2004; Proverbio et al. 2014). Seasonal differences in the incidence of infection have been found in many countries in animals as well as in humans (Simpanya and Baxter 1996; Cafarchia et al. 2006; Lee et al. 2012; Uhrlaß et al. 2015). With similar climates, prevalence varies in relation to other factors, firstly the lifestyle of animals taken in consideration. In the UK, show cats were reported to have a carrier rate of 12.5%, while in household cats, the isolation rate was 2.2% (Quaife and Womar 1982; Sparkes et al. 1994). In a report from Belgium, 2.1% of pet cats were found to be asymptomatic carriers, while the prevalence in cats in shelters was 16% (Mignon and Losson 1997).

Small outbreaks (usually <20 affected individuals) in household contexts are very common and caused by infected pets adopted from the road or purchased from a breeder or a shop (Alteras and Feuerman 1979; Preiser 1991; de Mendoza et al. 2010; Pasquetti et al. 2013). Regularly, there is an important role of asymptomatic/paucisymptomatic animals and environmental contamination in the diffusion of the disease. Among episodes reported, one of the oldest stands out in terms of number of cases and length. In Montreal (Canada), an exceptionally high number of human infections by *M. canis* (>1000) was reported over an 8-year period (Strachan and Blank 1963). Circumstances that led to this outbreak were not completely clarified, but it was concluded that cats and dogs played a major role in the spread of infection. Other interesting episodes reported in literature are nosocomial epidemics, infections in schools, outbreak in a nursing home for elderly people, and veterinary clinic (Shah et al. 1988; Snider et al. 1993; Drusin et al. 2000; Yu et al. 2004; Gürtler et al. 2005; Grills et al. 2007; Kopel et al. 2012; Pasquetti et al. 2012; Hillary and Suys 2014; Šubelj et al. 2014). In all these contexts, the infection spreads without an animal intervention, which shows that human-to-human transfer of *M. canis*, although considered rare and self-limiting, can be occasionally very efficient. Rarely, *M. canis* causes outbreaks in rabbit farms, and their origin usually came from animals imported from abroad and integrated into the local farms (Gonzalez et al.

1988; Cabañes et al. 1997); outbreaks in laboratory mice (Difonzo et al. 1986) and a porcine farm have also been described (Cabo et al. 1995).

In general, studies of the fungal flora of asymptomatic cats and dogs highlight a very important point. Although many animals have been found to act as healthy carrier of *M. canis*, this fungus should not be considered part of the normal fungal flora. If it was, it would have been isolated routinely from healthy animals regardless of geographical region, lifestyle (indoor or outdoor), or status (pet or stray) (DeBoer and Moriello 2006).

3.7.2 *Trichophyton mentagrophytes*, *T. quinckeanum*, and *T. interdigitale*

In the past, *T. mentagrophytes* complex was divided into numerous anthropophilic or zoophilic varieties associated with multiple sexual states (*A. vanbreuseghemii*, *A. benhamiae*, and *A. simii*) (Takashio 1977; Hironaga and Watanabe 1980; Hejtmánek and Hejtmánková 1989). Phylogenetic analyses resolved the status of these varieties, some of them were elevated to a species level, others became superfluous (Gräser et al. 1999b; de Hoog et al. 2017). The data also confirmed that three sexual states attributed in the past to *T. mentagrophytes* represent separate species complexes. The concept of *T. interdigitale* (former *T. mentagrophytes* var. *interdigitale*) and *T. mentagrophytes s. str.* has changed significantly during the last two decades especially in connection with their neotypification by Gräser et al. (1999b). The selection of neotype (closely phylogenetically related to *T. schoenleinii*) changed significantly the practical use of the well-known name *T. mentagrophytes* that became rare species in clinical practice. The concept of "anthropophilic and zoophilic strains of *T. interdigitale*" was used instead for a transitional period (Nenoff et al. 2007; Heidemann et al. 2010), and consequently, the majority of isolates previously identified as *T. mentagrophytes* were relabeled as zoophilic strains of *T. interdigitale*. The selection of neotype of *T. mentagrophytes* was subsequently disputed by several authors (Sun et al. 2010; Beguin et al. 2012; Chollet et al. 2015a), and in the light of new arguments, an alternative neotype was designated by de Hoog et al. (2017). Although the validity of this neotype may be subject of future nomenclature debate, we follow here the recently designated neotype that returns the name *T. mentagrophytes* back to common use together with the name *T. quinckeanum*, an agent of mouse favus. In contrast to Heidemann et al. (2010), no clear relationship between origin, morphology of the strains, genotype, and clinical manifestation was found in isolates of *T. interdigitale/T. mentagrophytes* from Tunisia (Dhib et al. 2017). These conclusions are in agreement with our observation in strains from Czech patients and may suggest that *T. interdigitale* and *T. mentagrophytes* sensu de Hoog et al. (2017) are conspecific.

The reservoir of *T. mentagrophytes* are rodents, hunting cats (rather than indoor cats), dogs, and less commonly other animals such as ruminants and horses. When transmitted to human, infection usually manifest as inflammatory tinea of glabrous skin (tinea corporis, faciei, barbae), less frequently as tinea capitis (Frealle et al.

2007; Cafarchia et al. 2013c). The isolates are usually typical by colonies in shades of beige and granular/powdery colony texture; numerous microconidia, macroconidia, and spiral hyphae can be usually observed in microscope slides. Mating of isolates with opposite mating types leads to production of a sexual state corresponding to former *A. vanbreuseghemii*. In contrast, anthropophilic *T. interdigitale* represent a clonal lineage (only single mating type) derived from sexual zoophilic lineage that lost the ability to mate with them. The colonies of anthropophilic strains are usually white and cottony; micromorphology is typical by the presence of microconidia and the absence of macroconidia and spiral hyphae. These strains are almost exclusively associated with onychomycosis and tinea pedis in human and absent in animals (Nenoff et al. 2007; Heidemann et al. 2010).

Trichophyton mentagrophytes is distributed worldwide. It is currently a major cause of dermatophytosis in rabbits (Fig. 3.6) and other rodents (except guinea pigs) followed by much less frequent *M. canis*, *T. benhamiae*, and others (Hoppmann and Barron 2007; Cafarchia et al. 2010, 2012; Kraemer et al. 2012). The transmission from pet rabbits to humans (mostly children) is well known, and the individual cases manifest usually as tinea capitis or corporis (Van Rooij et al. 2016; Zhang et al. 2009). Rabbitries constitute an important reservoir of the disease, and their environment can be heavily contaminated. Widespread dermatophytosis in young rabbits impacts profoundly on animal health along with bacterial superinfections (caused mostly by staphylococci). The infected animals have lower indices in prolificacy and growth, and the severely infected individuals usually have to be discarded, thus reducing productivity and causing financial loss (Moretti et al. 2013). Recurring disease is reported in rabbit farm workers (Torres-Rodriguez et al. 1992; Van Rooij et al. 2016). Significant outbreaks due to *T. mentagrophytes* in animals have been infrequently reported (Alteras and Cojocaru 1969; Mesquita et al. 2016) or are underreported. The prevalence in animals is frequently very high and may be neglected due to high percentage of asymptomatic animals. For instance, a study was conducted on 220 Spanish rabbit farms, and 79.5% of the examined animals were positive for *T. mentagrophytes* that corresponds to 98% of all isolated dermatophytes (Torres-Rodriguez et al. 1992). In another study from Spain, dermatophytes were cultured from 83% of rabbits with suspected dermatophytosis, and *T. mentagrophytes* (69.2%) was the most abundant pathogen (Cabañes et al. 1997). Similar situation has been found in Italy, where dermatophytosis was found in ~60–87% rabbit farms and *T. mentagrophytes* represented ~92–93% of isolated dermatophytes. Higher temperature and relative humidity or inadequate and infrequent disinfection practices were identified as the most significant risk factors for dermatophytosis in rabbit farms (Cafarchia et al. 2010). Young animals or animals in fattening and finishing stages are the most frequently infected (Cafarchia et al. 2010; Moretti et al. 2013). It has been shown that the ITS sequences of *T. mentagrophytes* isolates from rabbits from southern Europe countries were identical, but different from those isolated from dogs and cats or rabbits from Asia. These results may suggest that a particular genotype could be prevalent in rabbits in southern Europe (Mesquita et al. 2016). Several outbreaks involving up to dozens of human patients have also been described and transmitted from rabbits (Alteras and Cojocaru 1969;

Veraldi et al. 2012; Mesquita et al. 2016) and horse (Chollet et al. 2015b). Chinchillas are also vulnerable to infection, and *T. mentagrophytes* is isolated from ~5 to 10% of asymptomatic fur-ranched or pet chinchillas and 30% of animals with fur damage (Donnelly et al. 2000; Moretti et al. 2013). The infections in dogs and cats occur regularly through the world, but the infection counts are usually relatively low compared to *M. canis* (Lewis et al. 1991; Mancianti et al. 2002; Khosravi and Mahmoudi 2003; Cafarchia et al. 2004; Seker and Dogan 2011). Significantly higher infection rates are detected in free roaming, feral, and hunting dogs and cats in which *T. mentagrophytes* can even prevail over *M. canis* (Drouot et al. 2009; Duarte et al. 2010).

Trichophyton quinckeanum is a species historically associated with favus in rodents and rarely isolated from human infections (mostly tinea of glabrous skin). It is reasonable that this fungus is even rarer today because of improved living conditions. Its closest relatives are anthropophilic *T. schoenleinii* (an agent of tinea capitis typical by scutula formation) and probably geophilic *T. simii*. The morphology of *T. quinckeanum* is nearly indistinguishable from *T. mentagrophytes*, although the species are phylogenetically distinct (Fig. 3.3). *Trichophyton quinckeanum* has not been detected at all in majority of published epidemiological studies on human and animal dermatophytoses supported by molecular data. The exception is a recent report of 62 infections in human patients in Germany (tinea of glabrous skin and tinea capitis) (Uhrlaß et al. 2018). Interestingly, all cases were diagnosed during a 3-year period (2013–2017) in a single laboratory, and no cases were detected during previous years despite molecular verification of identification. Cats were identified as a source of infection in several patients, and all isolates had a unique ITS genotype that was different from all previously examined European and Asian strains. This suggests a clonal spread of a new genotype.

3.7.3 *Trichophyton verrucosum*

Trichophyton verrucosum is a clonal (Kano et al. 2014), slow-growing species with global distribution. It is typically found in cattle and other ruminants (Fig. 3.6), but it can easily spread to humans and animals, including horses, donkeys, camels, rabbits, dogs, cats, pigs, and even birds (Baudet 1932; Blank 1955; Georg 1960; Dvorak et al. 1965; Ali-Shtayeh et al. 1988; Khosravi and Mahmoudi 2003; Chermette et al. 2008). Modern intensive battery farms are the main reservoir of *T. verrucosum* in developed countries as conditions favor its proliferation. It is mainly transmitted through direct contact with infected animals or contaminated environment, and therefore high prevalence levels often occur in overcrowded stables where the fungus can spread easily among subjects confined in small areas. In cattle, ringworm is usually more widespread in young animals because of their lack of specific immunity against the fungus. The infection is often clearly evident, with alopecic areas covered with thin farinaceous desquamations or with thick crusty lamellar scales difficult to pull off the skin. Lesions are mainly distributed on the head and neck (Fig. 3.6), but in most severe cases, the whole body can be affected. Although frequently considered as a benign self-healing infection, ringworm in cattle may be responsible for economic losses due to the negative impact on milk and meat production. Ringworm also leads to impairments in the hide and skin industries, as lesion scars are evident on leather following tawing and tanning (Chermette et al. 2008; Bond 2010; Hameed et al. 2017).

Trichophyton verrucosum infection can be considered as a cosmopolitan disease as, over time, it has been reported in livestock and people in different countries from all continents. Infection rates in cattle appear variable depending primarily on the geographical and social context. For example, low infection rates (around 2%) were detected in rural areas of Pakistan (Hameed et al. 2017) and Iran (around 5%) (Aghamirian and Ghiasian 2011). In such contexts the fungus probably does not find the conditions which are known to promote its spread, such as overcrowding of animals and high humidity, which are more typically encountered in intensive breeding. Indeed, some surveys, performed instead in intensive and semi-intensive farms in Central Italy, detected much higher infection rates, i.e., 19, 60, and 88% (Moretti et al. 1998; Papini et al. 2009; Agnetti et al. 2014). In particular, the survey which reported the highest prevalence (Papini et al. 2009) was based on the analysis of only young animals living in crowded environments; the authors found high prevalence even in asymptomatic animals (80.4%). In addition, the investigation was carried out during winter months when skin lesions are in general more common because stabled animals are in close contact. A high infection rate (31%) was also found in a study in Jordan which took into account ten large dairy farms (Al-Ani et al. 2002). Other studies reporting high infection rates describe actually outbreak episodes on a limited scale, e.g., study of Dalis et al. (2014) from Nigeria. Likewise, the infection rate reported in a Chinese study (20%) reflects the prevalence within an outbreak in a single farm, with 200 animals infected out of a total of 1000 (Ming et al. 2006). The infections are less frequent in small ruminants (Chermette et al. 2008; Bond 2010), perhaps due to a stronger inherited immune response against the fungus compared with that of cattle or to other factors linked to the breeding systems. An increasing prevalence of the disease and the existence of extensive outbreaks were documented in sheep (Fig. 3.6) in the UK, USA, and Morocco (Pandey et al. 1979; Power and Malone 1987; Sargison et al. 2002). It cannot be excluded that dermatophytosis in sheep and goats is actually an under-diagnosed disease.

The incidence of infections in cattle was decreased in many regions by specific fighting measures, especially by vaccination programs or changes in the agricultural systems, such as reduction of the number of cattle in breeding units; the infections in humans decreases proportionally (Seebacher et al. 2008; Lund et al. 2014). Lack of prophylaxis with *T. verrucosum* vaccination accounts for the high infection rates in Italy (Moretti et al. 2013); in contrast cattle ringworm due to *T. verrucosum* was eradicated in Norway (Lund et al. 2014).

Trichophyton verrucosum is characterized by a high zoonotic potential. People at higher risk of infection are farmers and their families, and veterinaries and technicians involved in animal management. Several week-long sick leaves of employees further increase financial costs (Moretti et al. 2013). Human patients usually develop aggressive inflammatory skin lesions (usually on extremities and head), which may be accompanied by constitutional symptoms, such as fever and lymphadenopathy (Silver et al. 2008; Courtellemont et al. 2017). Tinea barbae and capitis are relatively common clinical forms which can result in irreversible scarring and alopecia. The incidence of human infection among other dermatophyte species is very high in some regions of Africa and Middle East (up to dozen percent) and relatively low in European countries and the USA (usually 0–2%, but up to 4%) (Havlickova et al. 2008; Seebacher et al. 2008; Moretti et al. 2013; Courtellemont et al. 2017).

Zoophilic variety *T. verrucosum* var. *autotrophicum* (described from Karakul sheep, goat, and cattle) and *T. immergens* (from ruminants) have morphology resembling that of *T. verrucosum*, but based on molecular genetic data, they are closely related or identical to *T. mentagrophytes* and *T. tonsurans*, respectively (Gräser et al. 1999b; de Hoog et al. 2017). Infections in horses and donkeys due to *T. verrucosum*-like strains are relatively rare but occur worldwide (Lyskova et al. 2015). It is probable that at least part of these infections is caused by phenotypically similar *T. bullosum* that has been confirmed by molecular data from North Africa, from Middle East, and recently from Europe (Sitterle et al. 2012; Lyskova et al. 2015; Sabou et al. 2018). Large-scale and comprehensive studies have not been performed worldwide on the epidemiology of dermatophytosis in cattle, other ruminants, horses, and donkeys using DNA-based methods for identification of dermatophytes. *Trichophyton verrucosum* was confirmed by DNA sequencing as an agent of infection in recent studies examining *T. verrucosum*-like isolates from cattle and patients infected by cattle from Japan, Czech Republic, and Tunisia (Hubka et al. 2014b; Kano et al. 2014; Neji et al. 2016). In contrast, *T. bullosum* has been confirmed only from horses, donkey and a patient who was likely infected from a donkey. Host specificity of *T. bullosum*, but also the real etiology of infections in uncommon hosts, should be verified in future studies.

3.7.4 *Trichophyton equinum*

Trichophyton equinum is a species strongly associated with horses. It is generally reported as the most common cause of dermatophytosis in horses worldwide and occasionally causes outbreaks in horse farms (English 1961; Connole and Pascoe 1984; Pereira et al. 2006). The infections prevail in young animals, and additional risk factors for dermatophytosis include poor grooming practice, moist conditions, and a high number of animals in the herd (Ahdy et al. 2016). Available data from Italy, Jordan, and Egypt found prevalence of dermatophytosis in horses 9, 18, and 16.8%, respectively, and *T. equinum* represented 66.7, 24 and 58.4% of all isolated dermatophytes (Moretti et al. 1998; Al-Ani et al. 2002; Ahdy et al. 2016). Similarly, the studies on horses with skin lesion from Egypt and Nigeria reported *T. equinum* as a main causal agent of dermatophytosis that was confirmed in 49.2 and 44% culture-positive cases, respectively (Mahmoud 1995; Nweze 2011). Exceptionally, other species were found with higher frequencies in horses such as *M. canis* (Al-Ani et al. 2002) and *T. verrucosum* (Maurice et al. 2016; Balogun et al. 2017). The latter species is usually found when horses are pastured with cattle (Weiss et al. 1984).

Occupational infections in breeders, riders and veterinarians are relatively rare (less than 30 cases reported in the literature) and usually manifest as tinea corporis and tinea capitis, exceptionally as onychomycosis (Veraldi et al. 2018).

The differentiation of *T. equinum* from closely related anthropophilic *T. tonsurans* is possible based on ecological preferences, nutritional requirements, and mating behavior (Woodgyer 2004; Summerbell et al. 2007), while distinguishing based on DNA sequence data may be problematic (de Hoog et al. 2017).

3.8 Emerging Dermatophytes

3.8.1 *Trichophyton benhamiae*: An Emerging Pathogen in Europe

Abandoning the previously accepted nomenclature of an asexual state, i.e., *T. mentagrophytes*, and a sexual state, i.e., *Arthroderma benhamiae* (see above), resulted in situation that no name combined in *Trichophyton* was available for *A. benhamiae* despite its clear phylogenetic position within *Trichophyton* clade (Fig. 3.3). This combination was introduced recently (de Hoog et al. 2017), but during the last decade, the species was usually designated *A. benhamiae* or "*Trichophyton* sp. anamorph of *A. benhamiae*" in the literature. Originally, the sexual state of this species was induced by crossing of two isolates designated by authors as "*T. mentagrophytes* var. *granulosum*" which originated from a dog and a man with dermatophytosis in the USA (Ajello and Cheng 1967). Consequently, the identification of isolates designated as *A. benhamiae* before the molecular era is commonly doubtful and can comprise quite broad spectrum of currently accepted species, i.e., *T. benhamiae s. str.*, *T. erinacei*, *T. mentagrophytes*, and some less common related species (Fig. 3.3). Isolates designated as "African race of *A. benhamiae*" represent an independent monophyletic entity related to *T. bullosum* rather than to *T. benhamiae* (Fig. 3.3). However, mating compatibility between "African race of *A. benhamiae*" and *T. benhamiae* (also called American-European race) was observed in vitro (Takashio 1974). The mating groups were indicated as "races" because no morphological differences could be observed between strains from different continents.

Guinea pigs represent the main host reservoir (Fig. 3.6) in Europe (Drouot et al. 2009; Kraemer et al. 2013; Hubka et al. 2014b); infections occasionally occur also in dogs (mostly USA), pigs (USA), and cats (Belgium) or various rodents (rabbits, chinchilla, mouse, rat, porcupine, degus) (Ajello and Cheng 1967; Takahashi et al. 2008; Takeda et al. 2012; Symoens et al. 2013; Sieklucki et al. 2014; Hiruma et al. 2015; Overgaauw et al. 2017). Our knowledge on spectrum of hosts worldwide supported by molecular data is still insufficient. Based on sequence data, it seems that the majority of infections in guinea pigs is caused by *T. benhamiae*, to a less extent by *T. mentagrophytes* and rarely by *M. canis*; in contrast *T. mentagrophytes* is a frequent pathogen in rabbits followed by *M. canis* and *T. benhamiae* (Arabatzis et al. 2006; Frealle et al. 2007; Drouot et al. 2009; Heidemann et al. 2010; Sun et al. 2010; Cafarchia et al. 2012; Symoens et al. 2013). The prevalence of *T. benhamiae* in various animals probably differs significantly in different geographic areas, and these infections seem to be associated with different genotypes or subpopulations, respectively (Cmokova 2015). A small outbreak of dermatophytosis due to *T. benhamiae* was reported in Canadian porcupines (*Erethizon dorsatum*), a close relative of the guinea pig, housed in a Japanese zoo (Takahashi et al. 2008). A case of infection in Cape porcupine (*Hystrix africaeaustralis*) and a man who handled animals (Marais and Olivier 1965) were reported from Southern Africa.

Trichophyton benhamiae has been reported and confirmed by molecular methods in many European countries, the USA, and Japan. The prevalence and distribution in rodents is largely unknown due to fact that the pathogens were usually identified as *T. mentagrophytes* across published studies. It was determined that the prevalence of dermatophytosis in guinea pigs in Germany was 38.1% of which 91.6% infections were caused by species from *T. mentagrophytes* complex (Kraemer et al. 2012; Kraemer et al. 2013) in agreement with an older study which found prevalence 37% (Weiß et al. 1979). A recent study revealed more than 90% prevalence of *T. benhamiae* in guinea pigs from pet shops in Germany, 9% of which showed visible symptoms (Kupsch et al. 2017). The prevalence in Switzerland was 38.1%, and *T. benhamiae* was confirmed in all mycologically positive samples (Drouot et al. 2009); the prevalence of dermatophytosis in guinea pigs across pet shops in the Netherlands was 16.8% (88% of cases were caused by *T. benhamiae*) that corresponds to 27.3% of pet shops which sell infected but mostly asymptomatic animals (Overgaauw et al. 2017). The retrospective analysis of the activity of a veterinary laboratory from 2010 to 2012 in France demonstrated that dermatophytes were isolated from 41.2% of 148 pet rodents (guinea pigs, rats, mice, hamsters, and chinchillas) and 38.2% of 76 pet rabbits (Guillot et al. 2016). In guinea pigs, *T. benhamiae* was predominant, whereas *T. mentagrophytes* was most frequently isolated from other rodents and from rabbits. In contrast, low prevalence (3.5%) of *T. mentagrophytes* was detected among guinea pigs in Belgium (Vangeel et al. 2000), and no *Trichophyton* spp. were isolated among 200 pet guinea pigs in Italy (d'Ovidio et al. 2014). The lack of reports of *T. benhamiae* from guinea pigs in the USA (a country of origin) is obvious, and guinea pig was determined as a source of infection only in one human case of dermatophytosis (Ajello and Cheng 1967). Although epizootic episodes in guinea pigs colonies and laboratory guinea pigs attributed to *T. mentagrophytes* have been described in the USA (Menges and Georg 1956; Pombier and Kim 1975) and other countries (Rush-Munro et al. 1977; McAller 1980), the identity of the pathogen is unclear in the context of recent taxonomic changes. More recently *T. benhamiae* was confirmed by molecular data in Iran at relatively low frequencies (Abastabar et al. 2013; Ansari et al. 2016; Rezaei-Matehkolaei et al. 2016) contrasting to its absence in the past studies.

Trichophyton benhamiae causes 2.9% of all human dermatophytoses in Germany (Uhrlaß et al. 2015) and 7.2% in the Czech Republic (Hubka et al. 2014b).

Consequently, it is the most important agent of dermatophytosis transmitted from animals in the Czech Republic and causes 22.9% of all infections on glabrous skin and 29.2% of tinea capitis infections; median age of all patients was 12, and women comprised 71% of patients (Hubka et al. 2014b; Hamal et al. 2016). Current status of *T. benhamiae* infection in Japan, the USA, and other countries is almost unknown due to insufficient surveillance and lack of epidemiological studies supported by molecular-based identification. In contrast to Europe, rabbits seem to be a major reservoir of *T. benhamiae* in Japan; 78% of published Japanese cases were reported in women; median age of patients was 26 (Kimura et al. 2015). The infection in guinea pigs does not show any gender predisposition, but the young individuals are more frequently affected and symptomatic, whereas adult animals are mostly symptomless. In symptomatic guinea pigs, the infection manifests as alopecia with scaling and crusting located predominantly on the head, less frequently on the other body parts (Drouot et al. 2009; Kraemer et al. 2012, 2013; Overgaauw et al. 2017).

With respect to the current epidemiological situation, it is clear that *T. benhamiae* is a new emerging pathogen in human clinical material in Europe and Japan. This fact is surprising because guinea pig breeding has a long tradition in these regions, and *T. benhamiae* has been documented among guinea pigs in the past, but the clinical cases in human were absent or very rare. Before widespread dispersal of *T. benhamiae* in Europe, sporadic cases of human and guinea pig infections caused usually by "yellow-pigmented isolates" which macroscopically resembled *M. canis* were reported from different countries. First cases due to *T. benhamiae* in Switzerland were dated in 2002 (Fumeaux et al. 2004), and similarly, first isolates from French cases were collected between 2002 and 2008 (Frealle et al. 2007; Contet-Audonneau and Leyer 2010; Charlent 2011; Khettar and Contet-Audonneau 2012). The first cases were observed in Germany and Czech Republic before 2010, and the pathogen became rapidly epidemic during following years (Hubka et al. 2014b; Nenoff et al. 2014; Skořepová et al. 2014; Uhrlaß et al. 2015). In Japan, the species was first isolated in 1996 from an infected rabbit (Kano et al. 1998); the human cases were reported in following years (Nakamura et al. 2002) and summarized by Kimura et al. (2015). Because *T. benhamiae* became common in companion animals of Japan, increasing number of infections can be expected there.

3.8.2 *Trichophyton erinacei*: An Emerging Pathogen Introduced into New Regions Along with Hedgehogs

Smith and Marples (1964) gave a status of variety *T. mentagrophytes* var. *erinacei* to a "hedgehog fungus" based on morphological and physiological differences from other varieties of *T. mentagrophytes*. It was elevated to a species level by Quaife (1966), and the species rank was supported by phylogeny (Gräser et al. 1999b) (Fig. 3.3).

European hedgehog (*Erinaceus europaeus*) was first reported as a host along with infection of humans who were in contact with animals (Smith and Marples 1964). The four-toed hedgehog (*Atelerix albiventris*) is another host of *T. erinacei* as verified by sequence data and mating experiments (Takahashi et al. 2002). *Trichophyton erinacei* has been recorded also from dogs (Piérard-Franchimont et al. 2008; Kurtdede et al. 2014), wood mice (*Apodemus sylvaticus*) (English 1969), house mice, and rats in New Zealand (Marples 1967); it was suggested that the animals were directly or indirectly infected from hedgehogs living in the same habitat.

The activities associated with direct contact between individuals such as fighting and mating probably represent the main source of cross infection between hedgehogs (Morris and English 1973). The hedgehog mites may act as a vector in the transmission due to common coinfection and presence in the nests and soil. *Trichophyton erinacei* has never been isolated from soil, but it remains viable for up to 1 year in nests. The nests are therefore potential source of infection for other animals and humans (English and Morris 1969).

The presentation in hedgehog ranges from asymptomatic infection to extensive involvement of the body surface. The infection is predominantly located on the head and usually spreads slowly (Morris and English 1973; Takahashi et al. 2002; Schauder et al. 2007). In human, cases of dermatophytosis are localized on contact sites with hedgehog, i.e., extremities (palm, fingers, wrist) are affected in ca 70–80% of reported cases, although tinea corporis, barbae, faciei, capitis, and onychomycosis have been also reported (English et al. 1962; Piérard-Franchimont et al. 2008; Concha et al. 2012).

The natural habitats of the main animal hosts are the UK and Northern and Western Europe (*Erinaceus europaeus*), and the species was introduced into New Zealand with the human immigration. It has been confirmed that Western European hedgehogs became wild in Japan after 1987 or even earlier (Takahashi et al. 2003). The four-toed hedgehog (*Atelerix albiventris*) is widely encountered in savanna and steppe zones of equatorial Africa from Senegal across to Ethiopia and south to the Zambezi River (sporadic in other regions of Africa). It is smaller than the Western European hedgehog, has a white abdomen, and is characterized by lacking the first toe of the hind leg. *Atelerix albiventris* has spread all over the world as a pet animal, most of them are domesticated with wild populations reported to be few (Takahashi et al. 2002; Santana et al. 2010).

Since the description of *T. erinacei*, hundreds of zoonotic infections due to *T. erinacei* have been reported from hedgehogs and humans across many European countries, Middle East, New Zealand, Australia (humans who handled animals in New Zealand), Africa, Japan, Korea, Taiwan, the USA, and Chile (English et al. 1962; Connole and Johnston 1967; Rosen 2000; Schauder et al. 2007; Piérard-Franchimont et al. 2008; Hsieh et al. 2010; Sun et al. 2010; Concha et al. 2012; Rezaei-Matehkolaei et al. 2013; Sieklucki et al. 2014; Drira et al. 2015; Jang et al. 2016). In an epidemiological survey, *T. erinacei* was reported in 44.7% of wild hedgehogs in New Zealand (Smith and Marples 1964), 20–25% of wild hedgehogs in Britain (Morris and English 1969) and 29.5% of hedgehogs in France (Le-Barzic et al. 2017). Skin lesions suggestive of dermatophytosis were observed only in ~6% of infected animals (Le-Barzic et al. 2017). The prevalence in household hedgehogs in Japan was 39% (Takahashi et al. 2003) and 50% in Spain (Abarca et al. 2017). The incidence of human dermatophytosis due to contact with hedgehogs is difficult to estimate. When molecular methods are used for species identification, *T. erinacei* is regularly detected at low frequencies (Sun et al. 2010; Rezaei-Matehkolaei et al. 2013; Hubka et al. 2014b).

Because the hedgehogs are increasingly popular as pets, global prevalence and spread of *T. erinacei* requires close monitoring. The high infection rate in household hedgehogs predicts that the number of patients will have increasing tendency along with growing popularity of hedgehogs as pets.

3.9 Species with Doubtful Ecology and Often Referred to As Zoophilic

Insufficient data are available on distribution of *T. eriotrephon* which is known from four cases of dermatophytosis in man (tinea corporis, Netherlands; tinea manuum and tinea faciei, Iran; tinea barbae, France) (Papegaay 1925; Rezaei-Matehkolaei et al. 2013; Sabou et al. 2018) and a dog (isolate IHEM 24340 from Belgium). Two isolates of unnamed species usually called "African race of *Arthroderma benhamiae*" are known from dermatomycosis in man (Takashio 1974). It is anticipated that both species are zoophilic based on clinical manifestation of infection and their close phylogenetic relationships to pathogenic *Trichophyton* species (Fig. 3.3).

Occasionally, cases of infections in various animals and human are attributed to species from geophilic *Nannizzia* (formerly *Microsporum*) species (*N. gypsea*, *N. fulva*, *N. persicolor*, *N. praecox*, *N. incurvata*, *N. nana*, etc.). Some of these species (mainly *N. persicolor* and *N. nana*) are commonly designated as zoophilic in the medical literature solely based on the evidence of previous clinical cases. Indeed, the boundaries between zoophilic and geophilic species are not always sharp, and this is usually due to insufficient ecological data. Epidemiological features typical for geophilic species involve low host specificity (infections are reported from broad range of animals), clinical cases occur separately without clear connection between themselves (except of contact with soil), geographic origin of clinical isolates do not correlates with distribution of any host, and the dermatophyte species have very limited potential to cause significant outbreaks even if they occur in animals kept in large numbers in a limited space (low contagiousness). Geophilic species *N. persicolor*, relatively commonly associated with infections in dogs (Carlotti and Bensignor 1999; Muller et al. 2011) and humans (Chen et al. 2012; Hubka et al. 2014b) but frequently misidentified with *T. mentagrophytes*, can serve as a good example. The species was considered to be zoophilic in the past with probable reservoir in rodents (Kane et al. 1987). Its geophilic nature was confirmed by subsequent investigations of soil samples that revealed widespread dispersal of *N. persicolor* in soil (Sharma et al. 2008). In addition, poor growth at 37 °C and significant intraspecies variability revealed by using molecular data are typical for geophilic species rather than primarily pathogenic ones (Sharma et al. 2008). Phylogenetic position of *N. persicolor* within geophilic species from *N. gypsea* complex is in agreement with such conclusion because related dermatophytes have usually similar ecology (Gräser et al. 2008).

Similar doubts on ecology exist in *Trichophyton simii* and *Lophophyton* (=*Microsporum*) *gallinae* which were traditionally considered zoophilic species with reservoir in monkeys or chickens, respectively. Both species fulfill many criteria typical for geophiles listed above. *Trichophyton simii* is known from sporadic cases of animal and human mycoses without specific predilection sites. The majority of case reports were summarized by Beguin et al. (2013), and they were described worldwide and involved monkeys, poultry, man, and dog (Okoshi et al. 1966; Clayton 1978; Constantino and Torre Mendoza 1979; Beguin et al. 2013; Yamaguchi et al. 2014). The species is known from soil or sand in India (Padhye and Thirumalachar 1967; Gugnani et al. 1968; Padhye and Carmichael 1968; Jain and Sharma 2011), France (Visset 1973), and Ivory Coast (Beguin et al. 2013) and was also recovered abundantly from asymptomatic small mammals in India, Africa, and Czech Republic (Gugnani et al. 1968, 1975; Ditrich and Otcenasek 1982; Hubálek 2000), from the fur of baboon in Guinea (Mariat and Tapia 1966) and poultry feather (Hubálek 2000) bringing another evidence of probable geophilic origin. No outbreaks were reported except of local epizootic in poultry (Gugnani and Randhawa 1973).

Lophophyton gallinae is well known as a causative agent of avian dermatophytosis (favus) that predominantly manifests in chicks and roosters as white to yellow scales or thick crusts on the comb and wattle; hens are less commonly affected. In severe cases, the infection spread on the other parts of the face, head, and neck with feather loss. The majority of infections was summarized previously by Murata et al. (2013) and older reports by Londero et al. (1969). These cases are again distributed worldwide and involved chickens but also ducks, dogs, monkeys, cats, squirrels, mouse, canary, pigeon, turkey, and man (no apparent predilection site in humans); outbreaks were rarely reported (Londero et al. 1969; Bradley et al. 1993). It is worth noting that some animals were healthy without any signs of infection, and no contact with birds was revealed in history of some patients. The fungus is known also from birds' nests of blue tit (*Cyanistes caeruleus*) and great tit (*Parus major*) (Goodenough and Stallwood 2010). Kawasaki et al. (1995) revealed close relationships between *M. gallinae*, geophilic *M. vanbreuseghemii*, and its sexual state *Arthroderma grubyi*. The conspecificity of these was subsequently supported by analysis of ITS rDNA region (Gräser et al. 1999a, 2008; de Hoog et al. 2017). The synonymization of these taxa was surprising due to different morphology and mating behavior. The ecology of former *M. vanbreuseghemii* is ambiguous (soil, asymptomatic as well as symptomatic animals and humans), similar to *L. gallinae*.

3.10 Outbreaks and Epidemics

Notable epidemic and outbreak episodes due to dermatophytes are considered to be rare, especially compared with other pathogens, such as bacteria and viruses. For instance, GIDEON (Global Infectious Disease and Epidemiology Network) database (http://www.gideononline.com/) reports 1267 outbreaks (with approximately 275,000 patients involved) of diarrhea by *Escherichia coli* in an 82-year period (1934–2016) and 140 outbreaks (with approximately 9600 patients involved) of dermatophytosis in a 133-year period (1882–2015). Of these latter, 25 episodes (18%) were due to *M. canis* and distributed worldwide. This discrepancy between the number of reported outbreaks and real significance of dermatophytes is due to many factors. The episodes occur in different contexts and usually with a limited number (less than 20) of people/animals involved, but in some occasions numbers are much higher. Outbreaks in households and animal communities are frequent worldwide, although they are hardly reported in the literature because dermatophytosis, both in humans and animals, is not a notifiable disease in most countries. Among factors that contribute to the occurrence of such episodes, one of the most important is the poor awareness of people about the role of animals, especially pets, as carriers of dermatophytes. Indeed, in most cases, infected animals are introduced and manipulated without any precaution (e.g., a quarantine period, a veterinary visit, etc.) and left free to stay with animals already present in the house, breeding, pet shop, etc. Significant changes in epidemiological patterns on the large geographic areas take place on relatively long time scale (years and decades) and are detectable only when long-term epidemiological data are available what is an uncommon condition in veterinary dermatology. Additionally, changes in the prevalence of dermatophytosis can be easily neglected in principal hosts of particular pathogens because infections are commonly asymptomatic in high percentage of infected animals. The screening should therefore include ostensibly healthy animals as well.

Exceptionally, extensive changes in the epidemiology happen very quickly. For instance, the emergence and rapid spread of *T. benhamiae* in Europe have been one of major public health events in the field of zoonotic superficial mycoses in recent years that underscored the need for closer collaboration between the veterinary profession, dermatologists, epidemiologists, and public health personnel. Zoonotic infections associated with pet shops are likely to result in individual cases or small familial outbreaks. On the other hand, an infected animal kept in a group in a pet shop can potentially transmit the illness to other animals, and subsequently to a large number of pet owners, who may be geographically dispersed (Halsby et al. 2014). Because of this, pet shops can be the focus of large outbreaks, such as in the case of recent epidemics of *T. benhamiae* infections in pediatric patients. Epidemiological surveys in different European countries showed that ~17–90% of guinea pigs (commonly asymptomatic) in pet shops are infected. This resulted in high incidence of tinea corporis and capitis in children and young adults in affected countries. The first human infections started to occur in different European countries between 2000 and 2010, and it seems that the incidence has not yet reached its peak. For instance, in Germany and Czech Republic, *T. benhamiae* became within several years the most important agent of zoonotic dermatomycoses at all (Hubka et al. 2014b; Uhrlaß et al. 2015).

3.11 Changing Etiology of Zoonotic Dermatophytosis: Perspective of Human Medicine

Dermatophytes are still an important public health problem in both "developed" and "developing" countries, and their prevalence remains high. Zoophilic dermatophytes remain frequent causative agents and should be considered especially in children and adolescents with tinea capitis and tinea of glabrous skin. Changes in epidemiological patterns and prevalence of dermatophytosis in domestic animals are commonly detected secondarily from epidemiological studies on human population because of better surveillance and more complex epidemiological data.

Tinea capitis is a typical clinical entity caused predominantly by zoophilic species in majority of developed countries. *Microsporum canis* is a prevalent causal agent, and this could be related to close association between humans and companion animals and mass tourism to endemic regions (such as the Mediterranean area). Exceptional situation is currently in Germany, where *T. benhamiae* prevails over *M. canis*. Studies in Europe, Asia, and Africa indicate that anthropophilic agents of scalp infections are being almost eradicated and are now more typical of countries with low socioeconomic status. The exception is tinea capitis due to *T. tonsurans* in North America. A shift toward tinea capitis due to anthropophilic dermatophytes (*T. tonsurans*, *T. soudanense*, and *M. audouinii*) is obvious in some urban areas in Europe, e.g., from the UK, France, Sweden, and Switzerland. This pattern seems to be linked to ethnic groups originating from Africa or from the Caribbean. Various anthropophilic species remain major causal agents of tinea capitis in many Asian and African countries, although significant shift toward zoophilic etiology (similar to situation in Europe) was observed in numerous more developed regions during last decades as described in detail, e.g., from China (Havlickova et al. 2008; Seebacher et al. 2008; Nweze 2010; Hayette and Sacheli 2015; Kieliger et al. 2015; Uhrlaß et al. 2015; Zhan et al. 2015; Nweze and Eke 2016). Exceptionally, other zoophilic species such as *T. mentagrophytes* or *T. verrucosum* may represent the major agents of tinea capitis as reported, e.g., from some region of China, Middle East, or Africa (Al-Duboon et al. 1999; Metin et al. 2002; Nweze 2010; Oudaina et al. 2011; Zhan et al. 2015).

Epidemiologic changes in the prevalence and etiology of inflammatory dermatophytoses on bare skin have been less extensively studied. Their prevalence and pathogens responsible for causing them usually reflect local trends in tinea capitis and tinea pedis. In general, broad spectrum of anthropophilic, zoophilic, and geophilic species can be responsible for similar clinical manifestation. Various anthropophilic species cause majority of infections in developed as well as in developing countries, and zoophilic species such as *M. canis*, *T. mentagrophytes*, and *T. verrucosum* supplement the spectrum of the most important causative agents worldwide. *Microsporum canis* still prevails over anthropophilic species in South European countries, although increase of anthropophilic species was observed in the most recent studies. *Trichophyton verrucosum* and *T. mentagrophytes* remain predominant dermatophytes in rural regions of the Middle East (Naseri et al. 2013; Hayette and Sacheli 2015; Chadeganipour et al. 2016).

Zoophilic dermatophytes also importantly contribute to a number of occupational infections in farmers, workers in livestock production, laboratory workers, pet shop workers, and other professions that require contact with animals (McAller 1980; Cafarchia et al. 2013c; Halsby et al. 2014). Incidence of dermatophytosis in farm workers can be strikingly high (Torres-Rodriguez et al. 1992; Agnetti et al. 2014).

3.12 Management of Outbreaks and Their Prevention

In most cases, animal dermatophytoses are self-limiting diseases because innate and/or acquired immunity is strong enough to control the spread of infection. Nevertheless, specific treatment is required to obtain a more rapid clinical cure, to minimize contamination to other hosts (including humans) and to reduce the dissemination of infective arthroconidia in the environment. Recommendations for the treatment are based on both in vivo and in vitro investigations (Moriello et al. 2017). These recommendations systematically include the disinfection of the environment and the limitation of contact between infected animals and healthy ones.

The transmission of zoophilic dermatophyte is through direct contact with infected animals or through contaminated environments, and limiting this kind of contact is a simple way for the prevention of transmission (Chermette et al. 2008). Sometimes prevention is difficult to perform. In case of subclinical infection or mechanical carriage, animals have no cutaneous lesions but may be responsible for contamination. Such a situation is frequently reported in some animal populations (e.g., cats infected by *M. canis* or guinea pigs infected by *T. benhamiae*).

The use of antifungals has been proposed for the prevention of animal dermatophytosis. However, investigations showed that the oral administration of griseofulvin did not allow the prevention of infection in humans. Topical administration of antifungal drugs seems to be more appropriate. When an animal has been in contact with an infected animal or area, it could be useful to use an antifungal shampoo.

Efforts in developing fungal vaccines to prevent dermatophytosis in different animal species are going on (Lund and DeBoer 2008; Mignon et al. 2008; Moriello et al. 2017). Immunoprophylaxis is particularly valuable in large breeding units or when pastures are shared by herds of different origins. Vaccination of animals has become the rule in some countries where bovine dermatophytosis is a notifiable disease (Gudding and Lund 1995). Anti-dermatophyte vaccines have been developed against dermatophytosis in cattle, horses, and less frequently sheep, but they are not available worldwide (Chermette et al. 2008). Both inactivated and live-attenuated vaccines, monovalent or multivalent, have been developed against animal ringworm. Ribosomal fractions of *T. verrucosum* have demonstrated promising immunogenic properties in cattle (Elad and Segal 1995). Recombinant protein and DNA vaccines derived from the heat-shock protein hsp60 of *T. mentagrophytes* allowed to control the clinical course of ringworm in guinea pigs and cattle (Milan et al. 2004). Other studies in horses using an inactivated preparation containing conidia and hyphae of *T. equinum* with adjuvant demonstrated a 75–87% relative protection of vaccinated horses against an infective contact (Pier and Zancanella 1993). Live vaccine against dermatophytes has been developed in the former USSR. A particular strain of *T. verrucosum* (the LTF 130 strain, with abundant production of microconidia in culture) was selected because of a high immunogenicity and attenuated pathogenicity. LTF live vaccines are currently used or have been used for several years in different countries, especially in Europe and on other continents (Canada, Cuba, Kenya, or Mongolia). To obtain optimal results, the vaccination program must concern all the animals of the cattle herd; then only the young calves between 2 weeks and 4 months of age and the newly introduced cattle will be vaccinated. In countries that have achieved successful control and eradication of cattle ringworm, a mass and systematic vaccination of cattle were undertaken. Failure of immunoprophylaxis can be explained by the non-respect of vaccination procedures, the reintroduction of infected cattle, the development of dermatophytes different from *T. verrucosum* and lack of crossed immunity, and the absence of hygienic and disinfection measures in farms. A dramatic reduction of dermatophytosis incidence in cattle, an improvement of the quality of leather and skins, and a decrease in case number of human contaminations have been observed when vaccination is correctly applied (Chermette et al. 2008; Seebacher et al. 2008).

To date, there is no vaccine (with high level of efficacy and safety) for companion animals exposed to *M. canis* infection (Moriello et al. 2017). The vaccine licensed for use in cats in the USA in 1994 provided disappointing results and is not commercialized anymore. Broad-spectrum dermatophyte vaccines (for instance, against *T. mentagrophytes* and *M. canis*) are currently available in some European countries. They have been used in different species including pet carnivores and fur animals.

3.13 Conclusion

Zoophilic dermatophytes remain an important public health concern in both developed and developing countries. *Microsporum canis* is a major cause of dermatophytosis in plenty of domestic animals and tinea capitis in developed countries and urban region of developing ones. Together with other zoophilic species *M. canis* contributes significantly to a number of glabrous skin dermatophytosis worldwide. *Trichophyton verrucosum* and *T. mentagrophytes* cause considerable morbidity and economic losses in rabbit and cattle farms, respectively. Strict compliance with the recommended preventive measures can effectively eliminate these problems. In contrast, *T. benhamiae* and *T. erinacei* are typical emerging zoonotic pathogens associated with pets and small wild mammals. The prevalence and spread of these species require close monitoring, particularly because the infection rates in the principal hosts, guinea pigs, and hedgehogs, respectively, are high, and the hedgehogs are increasingly popular as pets worldwide. There are still numerous questions to resolve in basic research on dermatophytes concerning pathogenesis, species concept, and ecology of some species. Additionally, reproducible and effective genotyping methods are not available for all major zoophilic dermatophytes what limits our understanding of global epidemiological trends, monitoring of changes on the level of genotypes and detecting outbreaks.

Acknowledgments VH was supported by the Charles University Grant Agency (GAUK 8615), Charles University Research Centre program No. 204069, Czech Ministry of Health (AZV 17-31269A), and the project BIOCEV (CZ.1.05/1.1.00/02.0109) provided by the Ministry of Education, Youth and Sports of CR, and ERDF. The contribution of AC was supported by the project of Charles University Grant Agency (GAUK 600217). Vit Hubka is grateful for the support from the Czechoslovak Microscopy Society (CSMS scholarship 2016).

References

Abarca M, Castellá G, Martorell J et al (2017) *Trichophyton erinacei* in pet hedgehogs in Spain: occurrence and revision of its taxonomic status. Med Mycol 55:164–172

Abastabar M, Rezaei-Matehkolaei A, Shidfar MR et al (2013) A molecular epidemiological survey of clinically important dermatophytes in Iran based on specific RFLP profiles of beta-tubulin gene. Iran J Public Health 42:1049–1057

Abdel-Rahman SM, Sugita T, González GM et al (2010) Divergence among an international population of *Trichophyton tonsurans* isolates. Mycopathologia 169:1–13

Achterman RR, White TC (2011) Dermatophyte virulence factors: identifying and analyzing genes that may contribute to chronic or acute skin infections. Int J Microbiol 2012:358305

Aghamirian MR, Ghiasian SA (2011) Dermatophytes as a cause of epizoonoses in dairy cattle and humans in Iran: epidemiological and clinical aspects. Mycoses 54:e52–e56

Agnetti F, Righi C, Scoccia E et al (2014) *Trichophyton verrucosum* infection in cattle farms of Umbria (Central Italy) and transmission to humans. Mycoses 57:400–405

Ahdy AM, Sayed-Ahmed MZ, Younis EE et al (2016) Prevalence and potential risk factors of dermatophytosis in Arabian horses in Egypt. J Equine Vet Sci 37:71–76

Ahmadi B, Mirhendi H, Makimura K et al (2016) Phylogenetic analysis of dermatophyte species using DNA sequence polymorphism in calmodulin gene. Med Mycol 54:500–514

Ajello L, Cheng S-L (1967) The perfect state of *Trichophyton mentagrophytes*. Sabouraudia 5:230–234

Al-Ani F, Younes F, Al-Rawashdeh O (2002) Ringworm infection of cattle and horses in Jordan. Acta Vet Brno 71:55–60

Al-Duboon A, Muhsin T, Al-Rubaiy K (1999) Tinea capitis in Basrah, Iraq. Mycoses 42:331–333

Ali-Shtayeh M, Arda H, Hassouna M et al (1988) Keratinophilic fungi on the hair of cows, donkeys, rabbits, cats, and dogs from the West Bank of Jordan. Mycopathologia 104:109–121

Aljabre S, Richardson M, Scott E et al (1993) Adherence of arthroconidia and germlings of anthropophilic and zoophilic varieties of *Trichophyton mentagrophytes* to human corneocytes as an early event in the pathogenesis of dermatophytosis. Clin Exp Dermatol 18:231–235

Alteras I, Cojocaru I (1969) Human infection by *Trichophyton mentagrophytes* from rabbits. Mycoses 12:543–544

Alteras I, Feuerman EJ (1979) Two outbreaks of *Microsporum canis* ringworm in Israel. Mycopathologia 67:169–172

Ameen M (2010) Epidemiology of superficial fungal infections. Clin Dermatol 28:197–201

Ansari S, Hedayati MT, Zomorodian K et al (2016) Molecular characterization and in vitro antifungal susceptibility of 316 clinical isolates of dermatophytes in Iran. Mycopathologia 181:89–95

Anzawa K, Kawasaki M, Mochizuki T et al (2010) Successful mating of *Trichophyton rubrum* with *Arthroderma simii*. Med Mycol 48:629–634

Arabatzis M, Xylouri E, Frangiadaki I et al (2006) Rapid detection of *Arthroderma vanbreuseghemii* in rabbit skin specimens by PCR–RFLP. Vet Dermatol 17:322–326
Baldo A, Tabart J, Vermout S et al (2008) Secreted subtilisins of *Microsporum canis* are involved in adherence of arthroconidia to feline corneocytes. J Med Microbiol 57:1152–1156
Baldo A, Mathy A, Tabart J et al (2010) Secreted subtilisin Sub3 from *Microsporum canis* is required for adherence to but not for invasion of the epidermis. Br J Dermatol 162:990–997
Balogun R, Jegede H, Jibril A, Kwanashie C, Kazeem H (2017) Prevalence and distribution of dermatophytes among domestic horses in Kwara state, Nigeria. Sokoto J Vet Sci 15:1–6
Baudet EARF (1932) Recherches experimentales sur les *Trichophyton* animaux a cultures favifornes. Ann Parasitol Hum Comp 10:520–541
Beguin H, Pyck N, Hendrickx M et al (2012) The taxonomic status of *Trichophyton quinckeanum* and *T. interdigitale* revisited: a multigene phylogenetic approach. Med Mycol 50:871–882
Beguin H, Goens K, Hendrickx M et al (2013) Is *Trichophyton simii* endemic to the Indian subcontinent? Med Mycol 51:444–448
Bernhardt A, Sedlacek L, Wagner S et al (2013) Multilocus sequence typing of *Scedosporium apiospermum* and *Pseudallescheria boydii* isolates from cystic fibrosis patients. J Cyst Fibros 12:592–598
Bianchi MV, Laisse CJ, Vargas TP et al (2017) Intra-abdominal fungal pseudomycetoma in two cats. Rev Iberoam Micol 34:112–115
Blank F (1955) Dermatophytes of animal origin transmissible to man. Am J Med Sci 229:302–316
Bond R (2010) Superficial veterinary mycoses. Clin Dermatol 28:226–236
Bradley FA, Bickford A, Walker RL (1993) Diagnosis of favus (avian dermatophytosis) in Oriental breed chickens. Avian Dis 37:1147–1150
Brasch J, Beck-Jendroschek V, Voss K et al (2016) *Arthroderma benhamiae* strains in Germany. Morphological and physiological characteristics of the anamorphs. Hautarzt 67:700–705
Burmester A, Shelest E, Glöckner G et al (2011) Comparative and functional genomics provide insights into the pathogenicity of dermatophytic fungi. Genome Biol 12:R7
Cabañes F, Abarca ML, Bragulat MR (1997) Dermatophytes isolated from domestic animals in Barcelona, Spain. Mycopathologia 137:107–113
Cabo JG, Asensio MB, Rodriguez FG et al (1995) An outbreak of dermatophytosis in pigs caused by *Microsporum canis*. Mycopathologia 129:79–80
Cafarchia C, Romito D, Sasanelli M et al (2004) The epidemiology of canine and feline dermatophytoses in southern Italy. Mycoses 47:508–513
Cafarchia C, Romito D, Capelli G et al (2006) Isolation of *Microsporum canis* from the hair coat of pet dogs and cats belonging to owners diagnosed with *M. canis* tinea corporis. Vet Dermatol 17:327–331
Cafarchia C, Camarda A, Coccioli C et al (2010) Epidemiology and risk factors for dermatophytoses in rabbit farms. Med Mycol 48:975–980
Cafarchia C, Weigl S, Figueredo LA et al (2012) Molecular identification and phylogenesis of dermatophytes isolated from rabbit farms and rabbit farm workers. Vet Microbiol 154:395–402
Cafarchia C, Figueredo LA, Otranto D (2013a) Fungal diseases of horses. Vet Microbiol 167:215–234
Cafarchia C, Gasser RB, Figueredo LA et al (2013b) An improved molecular diagnostic assay for canine and feline dermatophytosis. Med Mycol 51:136–143
Cafarchia C, Iatta R, Latrofa MS et al (2013c) Molecular epidemiology, phylogeny and evolution of dermatophytes. Infect Genet Evol 20:336–351
Cano J, Rezusta A, Solé M et al (2005) Inter-single-sequence-repeat-PCR typing as a new tool for identification of *Microsporum canis* strains. J Dermatol Sci 39:17–21
Carlotti D, Bensignor E (1999) Dermatophytosis due to *Microsporum persicolor* (13 cases) or *Microsporum gypseum* (20 cases) in dogs. Vet Dermatol 10:17–27
Chadeganipour M, Mohammadi R, Shadzi S (2016) A 10-year study of dermatophytoses in Isfahan, Iran. J Clin Lab Anal 30:103–107

Charlent A-L (2011) Le complexe *Trichophyton mentagrophytes*, caractérisation mycologique et moléculaire d'un nouveau variant: *Trichophyton mentagrophytes* var. *porcellae*. Master Thesis. Université Henri Poincaré, Nancy

Chen SCA, Ellis D, Sorrell TC et al (2011) *Trichophyton*. In: Liu D (ed) Molecular detection of human fungal pathogens. CRC Press, Boca Raton, pp 357–375

Chen W, Seidl HP, Ring J et al (2012) Two pediatric cases of *Microsporum persicolor* infection. Int J Dermatol 51:204–206

Chermette R, Ferreiro L, Guillot J (2008) Dermatophytoses in animals. Mycopathologia 166:385–405

Chinnapun D (2015) Virulence factors involved in pathogenicity of dermatophytes. Walailak J Sci Tech 12:573–580

Chollet A, Cattin V, Fratti M et al (2015a) Which fungus originally was *Trichophyton mentagrophytes*? Historical review and illustration by a clinical case. Mycopathologia 180:1–5

Chollet A, Wespi B, Roosje P et al (2015b) An outbreak of *Arthroderma vanbreuseghemii* dermatophytosis at a veterinary school associated with an infected horse. Mycoses 58:233–238

Clayton Y (1978) The changing pattern of tinea capitis in London schoolchildren. Mykosen Suppl 1:104–107

Cmokova A (2015) Molecular typization of isolates from *Arthroderma benhamiae* complex, a zoonotic agent of epidemic dermatophytosis in Europe. Master Thesis. Charles University, Prague

Cmokova A, Kolarik M, Dobiasova S et al (2017) Outbreak of children's dermatophytosis due to highly virulent population of *Trichophyton benhamiae*. Mycoses 60(Suppl. 2):134

Concha M, Nicklas C, Balcells E et al (2012) The first case of tinea faciei caused by *Trichophyton mentagrophytes* var. *erinacei* isolated in Chile. Int J Dermatol 51:283–285

Connole M, Johnston L (1967) A review of animal mycoses in Australia. Vet Bull Weybridge 37:145–153

Connole M, Pascoe R (1984) Recognition of *Trichophyton equinum* var. *equinum* infection of horses. Aust Vet J 61:94–94

Constantino MFH, Torre Mendoza C (1979) Isolation of dermatophytes from clinical cases of dermatomycosis in the game fowl. Philipp J Vet Med 18:79–91

Contet-Audonneau N, Leyer C (2010) Émergence d'un dermatophyte transmis par le cochon d'Inde et proche de *Trichophyton mentagrophytes* var. *erinacei: T. mentagrophytes* var. *porcellae*. J Mycol Med 20:321–325

Copetti MV, Santurio JM, Cavalheiro AS et al (2006) Dermatophytes isolated from dogs and cats suspected of dermatophytosis in southern Brazil. Acta Sci Vet 34:119–124

Courtellemont L, Chevrier S, Degeilh B et al (2017) Epidemiology of *Trichophyton verrucosum* infection in Rennes University Hospital, France: a 12-year retrospective study. Med Mycol 55:720–724

d'Ovidio D, Grable S, Ferrara M et al (2014) Prevalence of dermatophytes and other superficial fungal organisms in asymptomatic guinea pigs in Southern Italy. J Small Anim Pract 55:355–358

da Costa FV, Farias MR, Bier D et al (2013) Genetic variability in *Microsporum canis* isolated from cats, dogs and humans in Brazil. Mycoses 56:582–588

Dąbrowska I, Dworecka-Kaszak B, Brillowska-Dąbrowska A (2014) The use of a one-step PCR method for the identification of *Microsporum canis* and *Trichophyton mentagrophytes* infection of pets. Acta Biochim Pol 61:375–378

Dalis J, Kazeem H, Kwaga J et al (2014) An outbreak of ringworm caused by *Trichophyton verrucosum* in a group of calves in Vom, Nigeria. Afr J Microbiol Res 8:783–787

De Baere T, Summerbell R, Theelen B et al (2010) Evaluation of internal transcribed spacer 2-RFLP analysis for the identification of dermatophytes. J Med Microbiol 59:48–54

de Hoog GS, Dukik K, Monod M et al (2017) Toward a novel multilocus phylogenetic taxonomy for the dermatophytes. Mycopathologia 182:5–31

de Mendoza MH, de Mendoza JH, Alonso JM et al (2010) A zoonotic ringworm outbreak caused by a dysgonic strain of *Microsporum canis* from stray cats. Rev Iberoam Micol 27:62–65
DeBoer D, Moriello K (2006) Dermatophytosis. In: Greene C (ed) Infectious diseases of the dog and cat, 3rd edn. Elsevier, St. Louis, pp 550–565
Debourgogne A, Gueidan C, de Hoog S et al (2012) Comparison of two DNA sequence-based typing schemes for the *Fusarium solani* species complex and proposal of a new consensus method. J Microbiol Methods 91:65–72
Dhib I, Khammari I, Yaacoub A et al (2017) Relationship between phenotypic and genotypic characteristics of *Trichophyton mentagrophytes* strains isolated from patients with dermatophytosis. Mycopathologia 182:487–493
Dhieb C, Essghaier B, El Euch D et al (2014) Phenotypical and molecular characterization of *Microsporum canis* strains in North-Tunisia. Pol J Microbiol 63:307–315
Difonzo E, Palleschi G, Vannini P et al (1986) *Microsporum canis* epidemic in laboratory mice. Mycoses 29:591–595
Ditrich O, Otcenasek M (1982) *Microsporum vanbreuseghemii* and *Trichophyton simii* in Czechoslovakia. Ces Mykol 36:236–242
Dobrowolska A, Dębska J, Kozłowska M et al (2011) Strains differentiation of *Microsporum canis* by RAPD analysis using $(GACA)_4$ and $(ACA)_5$ primers. Pol J Microbiol 60:145–148
Donnelly TM, Rush EM, Lackner PA (2000) Ringworm in small exotic pets. Semin Avian Exotic Pet Med 9:82–93
Drake LA, Dinehart SM, Farmer ER et al (1996) Guidelines of care for superficial mycotic infections of the skin: tinea corporis, tinea cruris, tinea faciei, tinea manuum, and tinea pedis. J Am Acad Dermatol 34:282–286
Drira I, Neji S, Hadrich I et al (2015) Tinea manuum due to *Trichophyton erinacei* from Tunisia. J Mycol Med 25:200–203
Drouot S, Mignon B, Fratti M et al (2009) Pets as the main source of two zoonotic species of the *Trichophyton mentagrophytes* complex in Switzerland, *Arthroderma vanbreuseghemii* and *Arthroderma benhamiae*. Vet Dermatol 20:13–18
Drusin LM, Ross BG, Rhodes KH et al (2000) Nosocomial ringworm in a neonatal intensive care unit: a nurse and her cat. Infect Control Hosp Epidemiol 21:605–607
Duarte A, Castro I, da Fonseca IMP et al (2010) Survey of infectious and parasitic diseases in stray cats at the Lisbon Metropolitan Area, Portugal. J Feline Med Surg 12:441–446
Duek L, Kaufman G, Ulman Y et al (2004) The pathogenesis of dermatophyte infections in human skin sections. J Infect 48:175–180
Dvorak J, Otcenasek M, Komarek J (1965) Das Spektrum der aus Tierläsionen in Ostböhmen in den Jahren 1962–1964 isolierten Dermatophyten. Mycoses 8:126–127
Elad D, Segal E (1995) Immunogenicity in calves of a crude ribosomal fraction of *Trichophyton verrucosum*: a field trial. Vaccine 13:83–87
English M (1961) An outbreak of equine ringworm due to *Trichophyton equinum*. Vet Rec 73:578–580
English MP (1969) Ringworm in wild mammals: further investigations. J Zool 159:515–522
English MP, Morris P (1969) *Trichophyton mentagrophytes* var. *erinacei* in hedgehog nests. Sabouraudia 7:118–121
English MP, Evans CD, Hewitt M et al (1962) Hedgehog ringworm. Br Med J 1:149–151
Esquenazi D, Alviano CS, de Souza W et al (2004) The influence of surface carbohydrates during in vitro infection of mammalian cells by the dermatophyte *Trichophyton rubrum*. Res Microbiol 155:144–153
Fachin AL, Ferreira-Nozawa MS, Maccheroni W Jr et al (2006) Role of the ABC transporter TruMDR2 in terbinafine, 4-nitroquinoline N-oxide and ethidium bromide susceptibility in *Trichophyton rubrum*. J Med Microbiol 55:1093–1099
Frealle E, Rodrigue M, Gantois N et al (2007) Phylogenetic analysis of *Trichophyton mentagrophytes* human and animal isolates based on MnSOD and ITS sequence comparison. Microbiology 153:3466–3477

Fumeaux J, Mock M, Ninet B et al (2004) First report of *Arthroderma benhamiae* in Switzerland. Dermatology 208:244–250
Gallo M, Lanfranchi P, Poglayen G et al (2005a) Seasonal 4-year investigation into the role of the alpine marmot (*Marmota marmota*) as a carrier of zoophilic dermatophytes. Med Mycol 43:373–379
Gallo M, Tizzani P, Peano A et al (2005b) Eastern cottontail (*Sylvilagus floridanus*) as carrier of dermatophyte fungi. Mycopathologia 160:163–166
Georg LK (1960) Animal ringworm in public health: diagnosis and nature. US Government Printing Office, Washington, DC
Ghannoum M (2016) Azole resistance in dermatophytes: prevalence and mechanism of action. J Am Podiatr Med Assoc 106:79–86
Ginter-Hanselmayer G, Weger W, Ilkit M et al (2007) Epidemiology of tinea capitis in Europe: current state and changing patterns. Mycoses 50(Suppl. 2):6–13
Gonzalez J, Solans C, Latre M (1988) *Microsporum canis* productor de dermatofitosis en conejos. Rev Iberoam Micol 5:84–89
Goodenough AE, Stallwood B (2010) Intraspecific variation and interspecific differences in the bacterial and fungal assemblages of blue tit (*Cyanistes caeruleus*) and great tit (*Parus major*) nests. Microb Ecol 59:221–232
Gräser Y (2014) Species identification of dermatophytes by MALDI-TOF MS. Curr Fungal Infect Rep 8:193–197
Gräser Y, El Fari M, Vilgalys R et al (1999a) Phylogeny and taxonomy of the family Arthrodermataceae (dermatophytes) using sequence analysis of the ribosomal ITS region. Med Mycol 37:105–114
Gräser Y, Kuijpers AFA, Presber W et al (1999b) Molecular taxonomy of *Trichophyton mentagrophytes* and *T. tonsurans*. Med Mycol 37:315–330
Gräser Y, Kuijpers AFA, El Fari M et al (2000a) Molecular and conventional taxonomy of the *Microsporum canis* complex. Med Mycol 38:143–153
Gräser Y, Kuijpers AFA, Presber W et al (2000b) Molecular taxonomy of the *Trichophyton rubrum* complex. J Clin Microbiol 38:3329–3336
Gräser Y, De Hoog S, Summerbell R (2006) Dermatophytes: recognizing species of clonal fungi. Med Mycol 44:199–209
Gräser Y, Scott J, Summerbell R (2008) The new species concept in dermatophytes—a polyphasic approach. Mycopathologia 166:239–256
Grills CE, Bryan PL, O'Moore E et al (2007) *Microsporum canis*: Report of a primary school outbreak. Australas J Dermatol 48:88–90
Grumbt M, Monod M, Yamada T et al (2013) Keratin degradation by dermatophytes relies on cysteine dioxygenase and a sulfite efflux pump. J Investig Dermatol 133:1550–1555
Gudding R, Lund A (1995) Immunoprophylaxis of bovine dermatophytosis. Can Vet J 36:302–306
Gugnani H, Randhawa H (1973) An epizootic of dermatophytosis caused by *Trichophyton simii* in poultry. Sabouraudia 11:1–3
Gugnani H, Shrivastav J, Gupta N (1968) Occurrence of *Arthroderma simii* in soil and on hair of small mammals. Sabouraudia 6:77–80
Gugnani H, Wattal B, Sandhu R (1975) Dermatophytes and other keratinophilic fungi recovered from small mammals in India. Mycoses 18:529–538
Guillot J, Johannsen C, Guechi R et al (2016) Dermatophytes species isolated from companion animals in France. In: Proceedings of the 8th world congress of veterinary dermatology, Bordeaux, France
Gupta AK, Ahmad I, Borst I et al (2000) Detection of xanthomegnin in epidermal materials infected with *Trichophyton rubrum*. J Investig Dermatol 115:901–905
Gürtler TGR, Diniz LM, Nicchio L (2005) Tinea capitis micro-epidemic by *Microsporum canis* in a day care center of Vitória-Espírito. An Bras Dermatol 80:267–272
Hadrich I, Ranque S (2015) Typing of fungi in an outbreak setting: lessons learned. Curr Fungal Infect Rep 9:314–323

Halsby KD, Walsh AL, Campbell C et al (2014) Healthy animals, healthy people: zoonosis risk from animal contact in pet shops, a systematic review of the literature. PLoS One 9:e89309

Hamal P, Hubka V, Cmokova A et al (2016) Molecular epidemiology of dermatophytosis: results of a three year multicentric survey in the Czech Republic. Mycoses 59(Suppl. 1):14

Hameed K, Riaz FC, Nawaz MA et al (2017) *Trichophyton verrucosum* infection in livestock in the Chitral District (Pakistan). J Infect Dev Ctries 11:326–333

Harmsen D, Schwinn A, Weig M et al (1995) Phylogeny and dating of some pathogenic keratinophilic fungi using small subunit ribosomal RNA. J Med Vet Mycol 33:299–303

Havlickova B, Czaika V, Friedrich M (2008) Epidemiological trends in skin mycoses worldwide. Mycoses 51(Suppl. 4):2–15

Hayette M-P, Sacheli R (2015) Dermatophytosis, trends in epidemiology and diagnostic approach. Curr Fungal Infect Rep 9:164–179

Heidemann S, Monod M, Gräser Y (2010) Signature polymorphisms in the internal transcribed spacer region relevant for the differentiation of zoophilic and anthropophilic strains of *Trichophyton interdigitale* and other species of *T. mentagrophytes* sensu lato. Br J Dermatol 162:282–295

Hejtmánek M, Hejtmánková N (1989) Hybridization and sexual stimulation in *Trichophyton mentagrophytes*. Folia Microbiol 34:77–79

Hillary T, Suys E (2014) An outbreak of tinea capitis in elderly patients. Int J Dermatol 53:e101–e103

Hirai A, Kano R, Nakamura Y et al (2003) Molecular taxonomy of dermatophytes and related fungi by chitin synthase 1 (CHS1) gene sequences. Antonie Leeuwenhoek 83:11–20

Hironaga M, Watanabe S (1980) Mating behavior of 334 Japanese isolates of *Trichophyton mentagrophytes* in relation to their ecological status. Mycologia 72:1159–1170

Hironaga M, Nozaki K, Watanabe S (1980) Ascocarp production by *Nannizzia otae* on keratinous and non-keratinous agar media and mating behavior of *N. otae* and 123 Japanese isolates of *Microsporum canis*. Mycopathologia 72:135–141

Hiruma J, Kano R, Harada K et al (2015) Occurrence of *Arthroderma benhamiae* genotype in Japan. Mycopathologia 179:219–223

Hoppmann E, Barron HW (2007) Rodent dermatology. J Exot Pet Med 16:238–255

Hsieh C-W, Sun P-L, Wu Y-H (2010) *Trichophyton erinacei* infection from a hedgehog: a case report from Taiwan. Mycopathologia 170:417–421

Hubálek Z (2000) Keratinophilic fungi associated with free-living mammals and birds. In: Kushwaha RKS, Guarro J (eds) Biology of dermatophytes. Revista Iberoamericana de Micología, Bilbao, pp 93–103

Hube B, Hay R, Brasch J et al (2015) Dermatomycoses and inflammation: the adaptive balance between growth, damage, and survival. J Mycol Med 25:e44–e58

Hubka V, Dobiasova S, Dobias R et al (2014a) *Microsporum aenigmaticum* sp. nov. from *M. gypseum* complex, isolated as a cause of tinea corporis. Med Mycol 52:387–396

Hubka V, Větrovský T, Dobiášová S et al (2014b) Molecular epidemiology of dermatophytoses in the Czech Republic – two year study results. Čes-Slov Dermatol 89:167–174

Hubka V, Nissen C, Jensen R et al (2015) Discovery of a sexual stage in *Trichophyton onychocola*, a presumed geophilic dermatophyte isolated from toenails of patients with a history of *T. rubrum* onychomycosis. Med Mycol 53:798–809

Huson DH, Bryant D (2006) Application of phylogenetic networks in evolutionary studies. Mol Biol Evol 23:254–267

Irinyi L, Serena C, Garcia-Hermoso D et al (2015) International Society of Human and Animal Mycology (ISHAM)-ITS reference DNA barcoding database—the quality controlled standard tool for routine identification of human and animal pathogenic fungi. Med Mycol 53:313–337

Jackson CJ, Mochizuki T, Barton RC (2006) PCR fingerprinting of *Trichophyton mentagrophytes* var. *interdigitale* using polymorphic subrepeat loci in the rDNA nontranscribed spacer. J Med Microbiol 55:1349–1355

Jain N, Sharma M (2011) Distribution of dermatophytes and other related fungi in Jaipur city, with particular reference to soil pH. Mycoses 54:52–58

Jang MS, Park JB, Jang JY et al (2016) Kerion celsi caused by *Trichophyton erinacei* from a hedgehog treated with terbinafine. J Dermatol 44(9):1070–1071.

Jensen RH, Arendrup MC (2012) Molecular diagnosis of dermatophyte infections. Curr Opin Infect Dis 25:126–134

Kac G, Bougnoux M, Feuilhade De Chauvin M et al (1999) Genetic diversity among *Trichophyton mentagrophytes* isolates using random amplified polymorphic DNA method. Br J Dermatol 140:839–844

Kandemir H, Ilkit M, Çürük A (2015) Xanthomegnin detection does not discriminate between *Trichophyton rubrum* and *T. mentagrophytes* complexes. J Microbiol Methods 111:122–126

Kane J, Sigler L, Summerbell R (1987) Improved procedures for differentiating *Microsporum persicolor* from *Trichophyton mentagrophytes*. J Clin Microbiol 25:2449–2452

Kaneko T, Kaneko M, Makimura K (2011) Cluster analysis of *Microsporum canis* isolated from a patient with tinea corporis and an infected cat based on the DNA sequences of nuclear ribosomal internal transcribed spacer 1. Mycoses 54:e867–e869

Kano R, Nakamura Y, Yasuda K et al (1998) The first isolation of *Arthroderma benhamiae* in Japan. Microbiol Immunol 42:575–578

Kano R, Nakamura Y, Watanabe S et al (1999) Phylogenetic relation of *Epidermophyton floccosum* to the species of *Microsporum* and *Trichophyton* in chitin synthase 1 (CHS1) gene sequences. Mycopathologia 146:111–113

Kano R, Nakamura Y, Watanabe S et al (2002) Chitin synthase 1 and 2 genes of dermatophytes. Stud Mycol 47:49–55

Kano R, Yamada T, Makimura K et al (2011) *Arthroderma benhamiae* (the teleomorph of *Trichophyton mentagrophytes*) mating type-specific genes. Mycopathologia 171:333–337

Kano R, Yoshida E, Yaguchi T et al (2014) Mating type gene (MAT1-2) of *Trichophyton verrucosum*. Mycopathologia 177:103–112

Kaszubiak A, Klein S, de Hoog GS et al (2004) Population structure and evolutionary origins of *Microsporum canis*, *M. ferrugineum* and *M. audouinii*. Infect Genet Evol 4:179–186

Kawasaki M (2011) Verification of a taxonomy of dermatophytes based on mating results and phylogenetic analyses. Med Mycol J 52:291–295

Kawasaki M, Aoki M, Ishizaki H (1995) Phylogenetic relationships of some *Microsporum* and *Arthroderma* species inferred from mitochondrial DNA analysis. Mycopathologia 130:11–21

Kawasaki M, Anzawa K, Wakasa A et al (2010) Matings among three teleomorphs of *Trichophyton mentagrophytes*. Jpn J Med Mycol 51:143–152

Kawasaki M, Anzawa K, Ushigami T et al (2011) Multiple gene analyses are necessary to understand accurate phylogenetic relationships among *Trichophyton* species. Med Mycol J 52:245–254

Khettar L, Contet-Audonneau N (2012) Cochon d'Inde et dermatophytose. Ann Dermatol Venereol 139:631–635

Khosravi A, Mahmoudi M (2003) Dermatophytes isolated from domestic animals in Iran. Mycoses 46:222–225

Kieliger S, Glatz M, Cozzio A et al (2015) Tinea capitis and tinea faciei in the Zurich area–an 8-year survey of trends in the epidemiology and treatment patterns. J Eur Acad Dermatol Venereol 29:1524–1529

Kim J, Takahashi Y, Tanaka R et al (2001) Identification and subtyping of *Trichophyton mentagrophytes* by random amplified polymorphic DNA. Mycoses 44:157–165

Kimura U, Yokoyama K, Hiruma M et al (2015) Tinea faciei caused by *Trichophyton mentagrophytes* (molecular type *Arthroderma benhamiae*) mimics impetigo: a case report and literature review of cases in Japan. Med Mycol J 56:E1–E5

Kopel E, Amitai Z, Sprecher H et al (2012) Tinea capitis outbreak in a paediatric refugee population, Tel Aviv, Israel. Mycoses 55:e36–e39

Kraemer A, Mueller R, Werckenthin C et al (2012) Dermatophytes in pet guinea pigs and rabbits. Vet Microbiol 157:208–213
Kraemer A, Hein J, Heusinger A et al (2013) Clinical signs, therapy and zoonotic risk of pet guinea pigs with dermatophytosis. Mycoses 56:168–172
Kröber A, Scherlach K, Hortschansky P et al (2016) HapX mediates iron momeostasis in the pathogenic dermatophyte *Arthroderma benhamiae* but is dispensable for virulence. PLoS One 11:e0150701
Kupsch C, Ohst T, Pankewitz F et al (2016) The agony of choice in dermatophyte diagnostics–performance of different molecular tests and culture in the detection of *T. rubrum* and *T. interdigitale*. Clin Microbiol Infect 22:735.e711–735.e717
Kupsch C, Berlin M, Gräser Y (2017) Dermophytes and guinea pigs: an underestimated danger? Hautarzt 68:827–830
Kurtdede A, Haydardedeoglu A, Alihosseini H et al (2014) Dermatophytosis caused by *Trichophyton mentagrophytes* var. *erinacei* in a dog: a case report. Vet Med (Praha) 59:349–351
L'Ollivier C, Ranque S (2017) MALDI-TOF-based dermatophyte identification. Mycopathologia 182:183–192
Le-Barzic C, Denaes C, Arné P et al (2017) *Trichophyton erinacei*, un incontournable compagnon des hérissons… Résultats d'une enquête épidémiologique dans un centre de soins en Île-de-France. J Mycol Med 27:e39
Leclerc M, Philippe H, Guého E (1994) Phylogeny of dermatophytes and dimorphic fungi based on large subunit ribosomal RNA sequence comparisons. J Med Vet Mycol 32:331–341
Lee WJ, Song CH, Lee S-J et al (2012) Decreasing prevalence of *Microsporum canis* infection in Korea: through analysis of 944 cases (1993–2009) and review of our previous data (1975–1992). Mycopathologia 173:235–239
Leibner-Ciszak J, Dobrowolska A, Krawczyk B et al (2010) Evaluation of a PCR melting profile method for intraspecies differentiation of *Trichophyton rubrum* and *Trichophyton interdigitale*. J Med Microbiol 59:185–192
Lewis DT, Foil CS, Hosgood G (1991) Epidemiology and clinical features of dermatophytosis in dogs and cats at Louisiana State University: 1981–1990. Vet Dermatol 2:53–58
Londero A, Ramos CD, Fischman O (1969) Four epizooties of *Trichophyton gallinae* infection on chickens in Brasil. Mycoses 12:31–38
Lund A, DeBoer DJ (2008) Immunoprophylaxis of dermatophytosis in animals. Mycopathologia 166:407–424
Lund A, Bratberg AM, Næss B et al (2014) Control of bovine ringworm by vaccination in Norway. Vet Immunol Immunopathol 158:37–45
Lyskova P, Hubka V, Petricakova A et al (2015) Equine dermatophytosis due to *Trichophyton bullosum*, a poorly known zoophilic dermatophyte masquerading as *T. verrucosum*. Mycopathologia 180:407–419
Mahmoud A-L (1995) Dermatophytes and other keratinophilic fungi causing ringworm of horses. Folia Microbiol 40:293–296
Maitte C, Leterrier M, Le Pape P et al (2013) Multilocus sequence typing of *Pneumocystis jirovecii* from clinical samples: how many and which loci should be used? J Clin Microbiol 51:2843–2849
Makimura K, Tamura Y, Mochizuki T et al (1999) Phylogenetic classification and species identification of dermatophyte strains based on DNA sequences of nuclear ribosomal internal transcribed spacer 1 regions. J Clin Microbiol 37:920–924
Mancianti F, Nardoni S, Cecchi S et al (2002) Dermatophytes isolated from symptomatic dogs and cats in Tuscany, Italy during a 15-year-period. Mycopathologia 156:13–18
Marais V, Olivier DL (1965) Isolation of *Trichophyton mentagrophytes* from a porcupine. Sabouraudia 4:49–52
Mariat F, Tapia G (1966) Denombrement des champignons keratinophiles d'une population de cynocephales (*Papio papio*). Ann Parasitol Hum Comp 41:627–634

Marples MJ (1967) Non-domestic animals in New Zealand and in Rarotonga as reservoir of the agents of ringworm. N Z Med J 66:299–302
Martin T, Marugán-Lobón J, Vullo R et al (2015) A Cretaceous eutriconodont and integument evolution in early mammals. Nature 526:380–384
Martinez DA, Oliver BG, Gräser Y et al (2012) Comparative genome analysis of *Trichophyton rubrum* and related dermatophytes reveals candidate genes involved in infection. MBio 3: e00259–e00212
Martinez-Rossi NM, Peres NT, Rossi A (2017) Pathogenesis of dermatophytosis: sensing the host tissue. Mycopathologia 182:215–227
Maurice MN, Kazeem HM, Kwanashie CN et al (2016) Equine dermatophytosis: a survey of its occurrence and species distribution among horses in Kaduna State, Nigeria. Scientifica 2016:8309253
McAller R (1980) An epizootic in laboratory guinea pigs due to *Trichophyton mentagrophytes*. Aust Vet J 56:234–236
McNeill J, Barrie FR, Buck WR et al (2012) International code of nomenclature for algae, fungi, and plants (Melbourne code) adopted by the eighteenth international botanical congress, Melbourne, Australia, July 2011. Regnum vegetabile 154. Koeltz Scientific Books, Königstein
Mehlig L, Garve C, Ritschel A et al (2014) Clinical evaluation of a novel commercial multiplex-based PCR diagnostic test for differential diagnosis of dermatomycoses. Mycoses 57:27–34
Menges RW, Georg LK (1956) An epizootic of ringworm among guinea pigs caused by *Trichophyton mentagrophytes*. J Am Vet Med Assoc 128:395–398
Mesquita JR, Vasconcelos-Nóbrega C, Oliveira J et al (2016) Epizootic and epidemic dermatophytose outbreaks caused by *Trichophyton mentagrophytes* from rabbits in Portugal, 2015. Mycoses 59:668–673
Metin B, Heitman J (2017) Sexual reproduction in dermatophytes. Mycopathologia 182:45–55
Metin A, Subaşi Ş, Bozkurt H et al (2002) Tinea capitis in Van, Turkey. Mycoses 45:492–495
Meyer W, Aanensen DM, Boekhout T et al (2009) Consensus multi-locus sequence typing scheme for *Cryptococcus neoformans* and *Cryptococcus gattii*. Med Mycol 47:561–570
Mignon B, Losson B (1997) Prevalence and characterization of *Microsporum canis* carriage in cats. J Med Vet Mycol 35:249–256
Mignon B, Tabart J, Baldo A et al (2008) Immunization and dermatophytes. Curr Opin Infect Dis 21:134–140
Milan R, Alois R, Josef C et al (2004) Recombinant protein and DNA vaccines derived from hsp60 *Trichophyton mentagrophytes* control the clinical course of trichophytosis in bovine species and guinea-pigs. Mycoses 47:407–417
Miller WH, Griffin CE, Campbell KL (2013) Fungal and algal skin diseases: dermatophytosis. In: Miller WH et al (eds) Muller and Kirk's small animal dermatology, 7th edn. Elsevier, St. Louis, pp 231–243
Ming PX, Ti YLX, Bulmer GS (2006) Outbreak of *Trichophyton verrucosum* in China transmitted from cows to humans. Mycopathologia 161:225–228
Mirhendi H, Makimura K, de Hoog GS et al (2015) Translation elongation factor 1-α gene as a potential taxonomic and identification marker in dermatophytes. Med Mycol 53:215–224
Mochizuki T, Watanabe S, Kawasaki M et al (2002) A Japanese case of tinea corporis caused by *Arthroderma benhamiae*. J Dermatol 29:221–225
Mochizuki T, Ishizaki H, Barton RC et al (2003) Restriction fragment length polymorphism analysis of ribosomal DNA intergenic regions is useful for differentiating strains of *Trichophyton mentagrophytes*. J Clin Microbiol 41:4583–4588
Mor H, Kashman Y, Winkelmann G et al (1992) Characterization of siderophores produced by different species of the dermatophytic fungi *Microsporum* and *Trichophyton*. Biometals 5:213–216
Moretti A, Boncio L, Pasquali P et al (1998) Epidemiological aspects of dermatophyte infections in horses and cattle. J Vet Med Ser B 45:205–208

Moretti A, Agnetti F, Mancianti F et al (2013) Dermatophytosis in animals: epidemiological, clinical and zoonotic aspects. G Ital Dermatol Venereol 148:563–572

Moriello KA, Coyner K, Paterson S, Mignon B (2017) Diagnosis and treatment of dermatophytosis in dogs and cats. Vet Dermatol 28:266–268

Moriello KA, Kunkle G, Deboer DJ (1994) Isolation of dermatophytes from the haircoats of stray cats from selected animal shelters in two different geographic regions in the United States. Vet Dermatol 5:57–62

Morris P, English MP (1969) *Trichophyton mentagrophytes* var. *erinacei* in British hedgehogs. Sabouraudia 7:122–128

Morris P, English MP (1973) Transmission and course of *Trichophyton erinacei* infections in British hedgehogs. Sabouraudia 11:42–47

Muller A, Guaguère E, Degorce-Rubiales F et al (2011) Dermatophytosis due to *Microsporum persicolor*: a retrospective study of 16 cases. Can Vet J 52:385–388

Murata M, Takahashi H, Takahashi S et al (2013) Isolation of *Microsporum gallinae* from a fighting cock (*Gallus gallus domesticus*) in Japan. Med Mycol 51:144–149

Nakamura Y, Kano R, Nakamura E et al (2002) Case report. First report on human ringworm caused by *Arthroderma benhamiae* in Japan transmitted from a rabbit. Mycoses 45:129–131

Naseri A, Fata A, Najafzadeh MJ et al (2013) Surveillance of dermatophytosis in northeast of Iran (Mashhad) and review of published studies. Mycopathologia 176:247–253

Neji S, Trabelsi H, Hadrich I et al (2016) Molecular characterization of strains of the *Trichophyton verrucosum* complex from Tunisia. Med Mycol 54:787–793

Nenoff P, Herrmann J, Gräser Y (2007) *Trichophyton mentagrophytes sive interdigitale?* A dermatophyte in the course of time. J Dtsch Dermatol Ges 5:198–202

Nenoff P, Uhrlaß S, Krüger C et al (2014) *Trichophyton* species von *Arthroderma benhamiae* – a new infectious agent in dermatology. J Dtsch Dermatol Ges 12:571–582

Nguyen L-T, Schmidt HA, von Haeseler A et al (2015) IQ-TREE: a fast and effective stochastic algorithm for estimating maximum-likelihood phylogenies. Mol Biol Evol 32:268–274

Nishio K, Kawasaki M, Ishizaki H (1992) Phylogeny of the genera *Trichophyton* using mitochondrial DNA analysis. Mycopathologia 117:127–132

Nweze E (2010) Dermatophytosis in Western Africa: a review. Pak J Biol Sci 13:649–656

Nweze EI (2011) Dermatophytoses in domesticated animals. Rev Inst Med Trop Sao Paulo 53:94–99

Nweze I, Eke I (2016) Dermatophytosis in northern Africa. Mycoses 59:137–144

Okoshi S, Takashio M, Hasegawa A (1966) Ringworm caused by *Trichophyton simii* in a captive Chimpanzee. Jpn J Med Mycol 7:204–207

Oudaina W, Biougnach H, Riane S et al (2011) Epidemiology of tinea capitis in outpatients at the Children's Hospital in Rabat (Morocco). J Mycol Med 21:1–5

Overgaauw P, van Avermaete K, Mertens C, Meijer M, Schoemaker N (2017) Prevalence and zoonotic risks of *Trichophyton mentagrophytes* and *Cheyletiella* spp. in guinea pigs and rabbits in Dutch pet shops. Vet Microbiol 205:106–109

Padhye A, Carmichael J (1968) Mating reactions of *Trichophyton simii* and *T. mentagrophytes* strains from poultry farm soil in India. Sabouraudia 6:238–240

Padhye A, Thirumalachar M (1967) Isolation of *Trichophyton simii* and *Cryptococcus neoformans* from soil in India. Hindustan Antibiot Bull 9:155–157

Pandey V, Makin L, Vanbreusegham R (1979) Prevalence and distribution of ringworm by *Trichophyton verrucosum* in sheep in the high atlas of Moracco. Ann Soc Belg Med Trop 54:385–389

Papegaay J (1925) Over pathogene huidschimmels in Amsterdam voorkomend bij den mensch. Ned Tijdschr Geneeskd 69:879–890

Papini R, Nardoni S, Fanelli A et al (2009) High infection rate of *Trichophyton verrucosum* in calves from Central Italy. Zoonoses Public Health 56:59–64

Pasquetti M, Gräser Y, Mialot M et al (2012) Use of microsatellite markers for typing of *Microsporum canis* isolates causing pseudomycetoma in cats. Mycoses 55(Suppl. 4):152–153

Pasquetti M, Peano A, Soglia D et al (2013) Development and validation of a microsatellite marker-based method for tracing infections by *Microsporum canis*. J Dermatol Sci 70:123–129
Pchelin IM, Zlatogursky VV, Rudneva MV et al (2016) Reconstruction of phylogenetic relationships in dermatomycete genus *Trichophyton* Malmsten 1848 based on ribosomal internal transcribed spacer region, partial 28S rRNA and beta-tubulin genes sequences. Mycoses 59:566–575
Peano A, Pasquetti M, Guillot J et al (2015) Use of eight microsatellite markers to reveal the genetic variability of *Microsporum canis* from different hosts and different geographic origins. In: Proceedings of the 19th congress of the international society for human and animal mycology, Sydney, Australia
Pereira DIB, Oliveira LSS, Bueno A et al (2006) Outbreak of *Trichophyton equinum* var. *equinum* in horses in the south of Brazil. Cienc Rural 36:1849–1853
Pier A, Zancanella P (1993) Immunization of horses against dermatophytosis caused by *Trichophyton equinum*. Equine Pract 15:23–27
Piérard-Franchimont C, Hermanns J-F, Collette C et al (2008) Hedgehog ringworm in humans and a dog. Acta Clin Belg 63:322–324
Pignon C, Mayer J (2011) Zoonoses of ferrets, hedgehogs, and sugar gliders. Vet Clin North Am Exot Anim Pract 14:533–549
Pinter L, Jurak Z, Ukalovic M et al (1999) Epidemiological and clinical features of dermatophytoses in dogs and cats in Croatia between 1990 and 1998. Vet Arh 69:261–270
Pombier E, Kim J (1975) An epizootic outbreak of ringworm in a guinea-pig colony caused by *Trichophyton mentagrophytes*. Lab Anim 9:215–221
Power S, Malone A (1987) An outbreak of ringworm in sheep in Ireland caused by *Trichophyton verrucosum*. Vet Rec 121:218–220
Preiser G (1991) An outbreak of tinea corporis among schoolgirls. Pediatrics 88:327–328
Proverbio D, Perego R, Spada E et al (2014) Survey of dermatophytes in stray cats with and without skin lesions in Northern Italy. Vet Med Int 2014:565470
Quaife R (1966) Human infection due to the hedgehog fungus, *Trichophyton mentagrophytes* var. *erinacei*. J Clin Pathol 19:177–178
Quaife R, Womar S (1982) *Microsporum canis* isolations from show cats. Vet Rec 110:333–334
Ranjbar R, Karami A, Farshad S et al (2014) Typing methods used in the molecular epidemiology of microbial pathogens: a how-to guide. New Microbiol 37:1–15
Rashid A, Scott E, Richardson M (1995) Early events in the invasion of the human nail plate by *Trichophyton mentagrophytes*. Br J Dermatol 133:932–940
Rezaei-Matehkolaei A, Makimura K, de Hoog S et al (2013) Molecular epidemiology of dermatophytosis in Tehran, Iran, a clinical and microbial survey. Med Mycol 51:203–207
Rezaei-Matehkolaei A, Mirhendi H, Makimura K et al (2014) Nucleotide sequence analysis of beta tubulin gene in a wide range of dermatophytes. Med Mycol 52:674–688
Rezaei-Matehkolaei A, Rafiei A, Makimura K et al (2016) Epidemiological aspects of dermatophytosis in Khuzestan, southwestern Iran, an update. Mycopathologia 181:547–553
Robert R, Pihet M (2008) Conventional methods for the diagnosis of dermatophytosis. Mycopathologia 166:295–306
Romano C, Valenti L, Barbara R (1997) Dermatophytes isolated from asymptomatic stray cats. Mycoses 40:471–472
Rosen T (2000) Hazardous hedgehogs. South Med J 93:936–938
Rouzaud C, Hay R, Chosidow O et al (2016) Severe dermatophytosis and acquired or innate immunodeficiency: a review. J Fungi 2:4
Rush-Munro F, Woodgyer A, Hayter M (1977) Ringworm in guinea-pigs. Mycoses 20:292–296
Sabou M, Denis J, Boulanger N et al (2018) Molecular identification of *Trichophyton benhamiae* in Strasbourg, France: a 9-year retrospective study. Med Mycol 56, https://doi.org/10.1093/mmy/myx100
Santana EM, Jantz HE, Best TL (2010) *Atelerix albiventris* (Erinaceomorpha: Erinaceidae). Mamm Species 42:99–110

Sargison N, Thomson J, Scott P et al (2002) Ringworm caused by *Trichophyton verrucosum*-an emerging problem in sheep flocks. Vet Rec 150:755–756
Schauder S, Kirsch-Nietzki M, Wegener S et al (2007) Von Igeln auf Menschen: Zoophile Dermatomykose durch *Trichophyton erinacei* bei 8 Patienten. Hautarzt 58:62–67
Seebacher C, Bouchara J-P, Mignon B (2008) Updates on the epidemiology of dermatophyte infections. Mycopathologia 166:335–352
Seker E, Dogan N (2011) Isolation of dermatophytes from dogs and cats with suspected dermatophytosis in Western Turkey. Prev Vet Med 98:46–51
Shah P, Krajden S, Kane J et al (1988) Tinea corporis caused by *Microsporum canis*: report of a nosocomial outbreak. Eur J Epidemiol 4:33–38
Sharma R, Gräser Y (2011) Microsporum. In: Liu D (ed) Molecular detection of human fungal pathogens. CRC Press, Boca Raton, pp 285–298
Sharma R, De Hoog S, Presber W et al (2007) A virulent genotype of *Microsporum canis* is responsible for the majority of human infections. J Med Microbiol 56:1377–1385
Sharma R, Presber W, Rajak RC et al (2008) Molecular detection of *Microsporum persicolor* in soil suggesting widespread dispersal in central India. Med Mycol 46:67–73
Shi Y, Niu Q, Yu X et al (2016) Assessment of the function of SUB6 in the pathogenic dermatophyte *Trichophyton mentagrophytes*. Med Mycol 54:59–71
Sieklucki U, Oh SH, Hoyer LL (2014) Frequent isolation of *Arthroderma benhamiae* from dogs with dermatophytosis. Vet Dermatol 25:39–41
Silver S, Vinh DC, Embil JM (2008) The man who got too close to his cows. Diagn Microbiol Infect Dis 60:419–420
Simpanya M, Baxter M (1996) Isolation of fungi from the pelage of cats and dogs using the hairbrush technique. Mycopathologia 134:129–133
Sitterle E, Frealle E, Foulet F et al (2012) *Trichophyton bullosum*: a new zoonotic dermatophyte species. Med Mycol 50:305–309
Skerlev M, Miklić P (2010) The changing face of *Microsporum* spp infections. Clin Dermatol 28:146–150
Skořepová M, Hubka V, Polášková S et al (2014) Our first experiences with Infections caused by *Arthroderma benhamiae* (*Trichophyton* sp.) Čes-Slov Dermatol 89:192–198
Smith J, Marples MJ (1964) *Trichophyton mentagrophytes* var. *erinacei*. Sabouraudia 3:1–10
Snider R, Landers S, Levy ML (1993) The ringworm riddle: an outbreak of *Microsporum canis* in the nursery. Pediatr Infect Dis J 12:145–148
Sparkes A, Gruffydd-Jones T, Shaw S et al (1993) Epidemiological and diagnostic features of canine and feline dermatophytosis in the United Kingdom from 1956 to 1991. Vet Rec 133:57–61
Sparkes A, Werrett G, Stokes C et al (1994) *Microsporum canis*: inapparent carriage by cats and the viability of arthrospores. J Small Anim Pract 35:397–401
Spesso M, Nuncira C, Burstein V et al (2013) Microsatellite-primed PCR and random primer amplification polymorphic DNA for the identification and epidemiology of dermatophytes. Eur J Clin Microbiol Infect Dis 32:1009–1015
Stockdale PM (1964) The *Microsporum gypseum* complex (*Nannizzia incurvata* Stockd., *N. gypsea* (Nann.) comb. nov., *N. fulva* sp. nov.) Sabouraudia 3:114–126
Strachan A, Blank F (1963) On 1117 *Microsporum canis* infections in Montreal (1954–1961). Dermatology 126:271–290
Šubelj M, Marinko JS, Učakar V (2014) An outbreak of *Microsporum canis* in two elementary schools in a rural area around the capital city of Slovenia, 2012. Epidemiol Infect 142:2662–2666
Sugiyama M, Summerbell R, Mikawa T (2002) Molecular phylogeny of onygenalean fungi based on small subunit (SSU) and large subunit (LSU) ribosomal DNA sequences. Stud Mycol 47:5–23
Summerbell R (2002) What is the evolutionary and taxonomic status of asexual lineages in the dermatophytes? Stud Mycol 47:97–101

Summerbell R, Haugland R, Li A et al (1999) rRNA gene internal transcribed spacer 1 and 2 sequences of asexual, anthropophilic dermatophytes related to *Trichophyton rubrum*. J Clin Microbiol 37:4005–4011
Summerbell RC, Moore MK, Starink-Willemse M et al (2007) ITS barcodes for *Trichophyton tonsurans* and *T. equinum*. Med Mycol 45:193–200
Sun PL, Hsieh HM, Ju YM et al (2010) Molecular characterization of dermatophytes of the *Trichophyton mentagrophytes* complex found in Taiwan with emphasis on their correlation with clinical observations. Br J Dermatol 163:1312–1318
Symoens F, Jousson O, Planard C et al (2011) Molecular analysis and mating behaviour of the *Trichophyton mentagrophytes* species complex. Int J Med Microbiol 301:260–266
Symoens F, Jousson O, Packeu A et al (2013) The dermatophyte species *Arthroderma benhamiae*: intraspecies variability and mating behaviour. J Med Microbiol 62:377–385
Takahashi Y, Haritani K, Sano A et al (2002) An isolate of *Arthroderma benhamiae* with *Trichophyton mentagrophytes* var. *erinacei* anamorph isolated from a four-toed hedgehog (*Atelerix albiventris*) in Japan. Jpn J Med Mycol 43:249–255
Takahashi Y, Sano A, Takizawa K et al (2003) The epidemiology and mating behavior of *Arthroderma benhamiae* var. *erinacei* in household four-toed hedgehogs (*Atelerix albiventris*) in Japan. Jpn J Med Mycol 44:31–38
Takahashi H, Takahashi-Kyuhachi H, Takahashi Y et al (2008) An intrafamilial transmission of *Arthroderma benhamiae* in Canadian porcupines (*Erethizon dorsatum*) in a Japanese zoo. Med Mycol 46:465–473
Takashio M (1974) Observations on African and European strains of *Arthroderma benhamiae*. Int J Dermatol 13:94–101
Takashio M (1977) The *Trichophyton mentagrophytes* complex. In: Iwata K (ed) Recent advances in medical and veterinary mycology. University of Tokyo Press, Tokyo, pp 271–276
Takeda K, Nishibu A, Anzawa K et al (2012) Molecular epidemiology of a major subgroup of *Arthroderma benhamiae* isolated in Japan by restriction fragment length polymorphism analysis of the non-transcribed spacer region of ribosomal RNA gene. Jpn J Infect Dis 65:233–239
Torres-Rodriguez J, Dronda M, Rossell J et al (1992) Incidence of dermatophytoses in rabbit farms in Catalonia, Spain, and its repercussion on human health. Eur J Epidemiol 8:326–329
Uhrlaß S, Krüger C, Nenoff P (2015) *Microsporum canis*: Aktuelle Daten zur Prävalenz des zoophilen Dermatophyten im mitteldeutschen Raum. Hautarzt 66:855–862
Uhrlaß S, Schroedl W, Mehlhorn C et al (2018) Molecular epidemiology of *Trichophyton quinckeanum* – a zoophilic dermatophyte on the rise. J Dtsch Dermatol Ges 16:21–32
Vangeel I, Pasmans F, Vanrobaeys M et al (2000) Prevalence of dermatophytes in asymptomatic guinea pigs and rabbits. Vet Rec 146:440–441
Van Rooij P, Detandt M, Nolard N (2016) *Trichophyton mentagrophytes* of rabbit origin causing family incidence of kerion: an environmental study. Mycoses 49:426–430
Veraldi S, Genovese G, Peano A (2018) Tinea corporis caused by *Trichophyton equinum* in a rider and review of the literature Infection 46:135–137
Veraldi S, Guanziroli E, Schianchi R (2012) Epidemic of tinea corporis due to *Trichophyton mentagrophytes* of rabbit origin. Pediatr Dermatol 29:392–393
Visset M (1973) *Arthroderma simii* dans les sables de La Baule, premier isolement en Europe. Bull Soc Mycol Fr 11:151–152
Wakasa A, Anzawa K, Kawasaki M et al (2010) Molecular typing of *Trichophyton mentagrophytes* var. *interdigitale* isolated in a university hospital in Japan based on the non-transcribed spacer region of the ribosomal RNA gene. J Dermatol 37:431–440
Weiß R, Böhm KH, Mumme J et al (1979) 13 Jahre Veterinärmedizinische Mykologische Routinediagnostik. Dermatophytennachweise in den Jahren 1965 Bis 1977. Sabouraudia 17:345–353
Weiss R, Földy I, Christoph H (1984) Trichophyton-verrucosum-Infektion bei Pferden. Tierarztl Prax 12:49–53
Weitzman I, Summerbell RC (1995) The dermatophytes. Clin Microbiol Rev 8:240–259

Weitzman I, McGinnis M, Padhye A et al (1986) The genus *Arthroderma* and its later synonym *Nannizzia*. Mycotaxon 25:505–518

Woodgyer A (2004) The curious adventures of *Trichophyton equinum* in the realm of molecular biology: a modern fairy tale. Med Mycol 42:397–403

Wu Y, Yang J, Yang F et al (2009) Recent dermatophyte divergence revealed by comparative and phylogenetic analysis of mitochondrial genomes. BMC Genomics 10:238

Yamaguchi S, Sano A, Hiruma M et al (2014) Isolation of dermatophytes and related species from domestic dowl (*Gallus gallus domesticus*). Mycopathologia 178:135–143

Yin W-B, Chooi YH, Smith AR et al (2013) Discovery of cryptic polyketide metabolites from dermatophytes using heterologous expression in *Aspergillus nidulans*. ACS Synth Biol 2:629–634

Youngchim S, Pornsuwan S, Nosanchuk JD et al (2011) Melanogenesis in dermatophyte species in vitro and during infection. Microbiology 157:2348–2356

Yu J, Wan Z, Chen W et al (2004) Molecular typing study of the *Microsporum canis* strains isolated from an outbreak of tinea capitis in school. Mycopathologia 157:37–41

Zhan P, Li D, Wang C et al (2015) Epidemiological changes in tinea capitis over the sixty years of economic growth in China. Med Mycol 53:691–698

Zhang H, Ran Y, Liu Y et al (2009) *Arthroderma vanbreuseghemii* infection in three family members with kerion and tinea corporis. Med Mycol 47:539–544

Zimmerman K, Feldman B, Robertson J et al (2003) Dermal mass aspirate from a Persian cat. Vet Clin Pathol 32:213–217

Coccidioidomycosis in Animals

4

Bridget Marie Barker

Abstract

A broad diversity of animals is susceptible to infection by *Coccidioides* species. However severe or disseminated disease in animals other than pet dogs is not commonly reported in the literature. It is unclear if these cases are indeed rare or if they are not diagnosed and reported. The awareness of coccidioidomycosis is increasing in the Central Valley of California and southern Arizona, but outside of these areas the disease is not often diagnosed. Cases outside the endemic region frequently have delayed diagnosis, and as summarized here for many animals, the diagnosis was not made until after euthanasia. Frequently, a fungal infection is not considered as a primary cause of death or disease, in spite of the fact that hundreds of thousands of these infections occur every year. In the USA, coccidioidomycosis cases reported rival the number of cases of tuberculosis and Lyme disease. Considering that it is likely that only 10% of infections nationwide are reported, this disease has significant burden in the USA. Disease burden in the rest of the Americas remains unknown. Clearly, better diagnostics, effective treatments, and development of vaccines would greatly improve public health and reduce economic costs associated with coccidioidomycosis. Additionally, a great deal of work remains to fully understand the ecology and basic biology of the causative agent.

B. M. Barker (✉)
Northern Arizona University, Flagstaff, AZ, USA
e-mail: bridget.barker@nau.edu

S. Seyedmousavi et al. (eds.), *Emerging and Epizootic Fungal Infections in Animals*,
https://doi.org/10.1007/978-3-319-72093-7_4

4.1 Phylogenetics and Population Structure

There are two distinct cryptic species within the genus *Coccidioides* (*Ascomycota*, *Pezizomycotina*, *Eurotiomycetes*, *Onygenales*, *Onygenaceae*): *Coccidioides immitis* and *C. posadasii* (Fisher et al. 2002). Previous phylogenetic analyses and morphological characterization showed that *Uncinocarpus reesii*, a keratinophilic saprotroph, is one of the closest related fungi to *Coccidioides* (Sigler et al. 1998; Pan et al. 1994). However, recent work reveals that *Amauroascus mutatus*, *A. niger*, *Byssoonygena ceratinophila*, and *Chrysosporium queenslandicum* are phylogenetically closer to *Coccidioides* than *U. reesii* (Whiston and Taylor 2015). Within the *Onygenaceae*, no other dimorphic human pathogens have been identified; however other animal pathogens exist (Sigler et al. 1998; Herr et al. 2001; Sigler et al. 2013). The *Ajellomycetaceae*, which are distinct from the *Onygenaceae*, include *Blastomyces dermatitidis*, *Histoplasma capsulatum*, *Paracoccidioides brasiliensis* (Untereiner et al. 2004), and *Emergomyces* species (Dukik et al. 2017).

The *Onygenales* contain at least these two families as well as the *Gymnoascaceae* and *Arthrodermataceae*, but these are still preliminary assignments, and more work remains to be done to understand the complete picture of phylogenetic relationships. The *Onygenales* are a sister order to the *Eurotiales*, and the class containing these orders, the *Eurotiomycetes*, includes a number of species that cause disease in both animals and plants (Wang et al. 2009). Several genomes of *Eurotiomycetes* have been sequenced, which allows for comparative genomic studies.

The current understanding of genetic population structure within *C. immitis* suggests the existence of two populations: San Joaquin/Central Valley California (SJV) and Southern California/Mexico (SDMX) (Fisher et al. 2001, 2002). The genetic population structure of *C. posadasii* suggests three main populations: Texas/South America (TXSA), Mexico, and Arizona. Limited gene flow occurs among populations. More effort to understand genetic diversity in Mexico and Central and South America is needed. Recent evidence for even smaller-scale population structure within Arizona was reported (Teixeira and Barker 2016). Yuma and Phoenix isolates are distinct from Tucson patient and soil isolates, which suggests there may be specific ecological adaptations between these two areas. The Sonoran Desert is a highly variable landscape, which ranges from desert upland/grassland (Tucson) to the lower elevation desert biome (Phoenix/Yuma). *Coccidioides* growing in the soil in these areas would experience different abiotic and/or biotic stressors (Fisher et al. 2007; Lacy and Swatek 1974).

Additionally, genetic diversity in environmental and veterinary isolates differs from the genetic diversity among isolates infecting humans (Teixeira and Barker 2016). Human clinical isolates therefore provide valuable insights into population structure; however, isolates obtained directly from the environment are necessary to truly understand fine-scale population structure and determine if certain regions (i.e., Tucson vs. Phoenix) support the growth and survival of specific genotypes. Perhaps more intriguing is that the pool of diversity in the environment is higher than what is observed to cause disease to the human population, even with a small number of isolates available for genetic comparisons. Environmental isolates of *Coccidioides* exist, but they are difficult to obtain (Barker et al. 2012; Johnson et al. 2014; Litvintseva et al. 2015; Lauer et al. 2012; Brillhante et al. 2012; Baptista-Rosas et al. 2007; Fisher et al. 2007). Greater effort is needed to assess the environmental reservoir of *Coccidioides* in the environment and true population structure.

The first reported case of the disease was described by Alejandro Posadas over 120 years ago in Argentina (Posadas 1892). Granulomas in skin lesions resembling a protist were observed. In 1896, the organism was named *Coccidioides immitis*: "*Coccidioides*" for the suspected coccidium protozoan and "*immitis*" which is Latin for "not mild" (Rixford and Gilchrist 1896). In 1900, it was clearly shown by researchers working in California that the causative organism was a fungus. Ophuls (1905) named the protozoan-like structure a spherule, a parasitic stage of the life cycle of the fungus. The disease was considered rare and fatal, as these were the first category of cases to be recognized (Morris 1924; Ryfkogel 1908). However, this view was changed by five cases of acute infections from which patients fully recovered and proved that *Coccidioides* exposure could result in nonlethal illness (Dickson 1937).

4.2 Life Cycle of *Coccidioides*

Coccidioides immitis and *C. posadasii* are dimorphic fungi that switch between a mycelial phase and a spherule phase (Fig. 4.1). A switch from polar to isotropic growth occurs when a susceptible host inhales clonal arthroconidia, and the development of the unique infectious structure is initiated. The spherule matures and releases endospores, which may develop into new spherules or arrest growth. Mild or asymptomatic infections generally stay localized to the lungs. More severe infections can disseminate to other body sites (spleen, synovial joints, liver, kidneys, etc.), and endospores can cross the blood-brain barrier (Nguyen et al. 2013). The fungus can initiate mycelial growth from excised tissue or other biosamples (biopsy, sputum, synovial fluid, etc.) even at 37 °C, although how and when this occurs in nature are not known. High temperature, elevated CO_2 concentration/low oxygen, and specific nutrients all play a role in the formation, growth, and maintenance of the spherule/endospore cycle (Converse and Besemer 1959). Conditions consistent with development of spherules include shaking cultures at 37 °C, under 6–20% CO_2, with a liquid medium containing glucose, ammonium acetate, potassium phosphate, magnesium sulfate, and zinc sulfate at a pH of 6.3 (Breslau and Kubota 1964; Brooks and Northey 1963; Northey and Brooks 1962; Converse 1955). There is variation among strains, but generally mature spherules develop in 3–6 days (Pappagianis et al. 1956; Huppert et al. 1982).

Whereas *U. reesii* has a defined sexual life cycle, the sexual cycle of *Coccidioides* is unknown (Sigler et al. 1998). However, typical ascomycete mating-type (*MAT*) loci were identified in *Coccidioides* using comparative genomics methods (Mandel et al. 2007; Fraser et al. 2007). *MAT1-1* and *MAT1-2* are present in a 1:1 Mendelian ratio in over 400 strains of *Coccidioides immitis* and *C. posadasii*, which suggests the sexual recombination is frequent in these species (Mandel et al. 2007). Moreover,

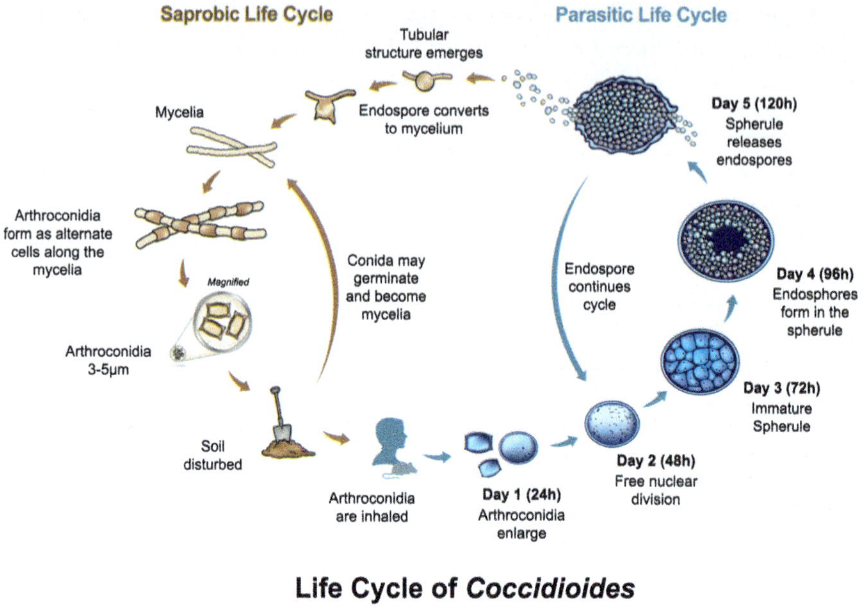

Fig. 4.1 Dimorphic asexual life cycle of *Coccidioides*. From Lewis et al. 2015, used with attribution

mRNA is transcribed for genes in the *MAT* locus, which supports the hypothesis of a functional sexual cycle. All data obtained to date are consistent with the prediction that both species of *Coccidioides* are highly recombining sexual organisms (Burt et al. 1996; Engelthaler et al. 2015).

4.3 Coccidioidomycosis

All work with *Coccidioides* organism requires a biosafety-level 3 containment, primarily because infectious particles are easily produced and aerosolized (Stevens et al. 2009). The organism was previously designated as a select agent but was removed in 2012 (Oct 5 2012 Federal Register; Dixon 2001). Healthy dogs, humans, or other mammals living in or visiting endemic areas can easily become infected (Cairns et al. 2000; Nguyen et al. 2013). It is thought that approximately 60% of infections are asymptomatic based on conversion data from skin testing new military recruits from non-endemic areas that were stationed in Arizona and California (Drips and Smith 1964; Smith et al. 1956). This has been supported by new data collected from skin testing prisoners in California with the delayed hypersensitivity-based skin test now commercially available (Wack et al. 2015; de Perio et al. 2015).

In Arizona and California, an increase in coccidioidomycosis (CM) has been reported since 1995 (Thompson et al. 2015; Twarog and Thompson 2015). Because CM severity is so variable, not all infections are diagnosed and reported, and the overall infection rate is closer to 200,000 per year in the USA (Nguyen et al. 2013; Galgiani 2007). Mild CM in humans presents with general symptoms, such as coughing, fever, and malaise. Indeed, many community-acquired pneumonias caused by *Coccidioides* in endemic regions are misdiagnosed as viral or bacterial pneumonia (Valdivia et al. 2006). Normally, acute disease is self-limiting and antifungal therapy is not necessary. However, some of these patients can experience symptoms for many months, and medical intervention may be recommended (Chen et al. 2011; Sunenshine et al. 2007). Although severe disease manifests in less than 5% of cases, it can result in life-threatening disease, which may require surgery, antifungal drug therapy, and hospitalization (Sondermeyer et al. 2013; Flaherman et al. 2007; Galgiani 2007).

Variable exposures could also play a role in differential severity of CM. Infectious dosages of 50 arthroconidia of a highly virulent strain have an LD_{50} of 17 days in a murine model of CM (Sorensen et al. 1999; Kirkland and Fierer 1983). The infectious dose of arthroconidia administered to cattle determined the level of infection (Reed 1960). Dogs infected with between 10^6 and 10^4 conidia died or had severe disease; whereas infection with 10^3 conidia resulted in mild CM (Hugenholtz et al. 1958). Similarly, monkeys exposed to 10^4 conidia had 80% fatality, and infection with 50 conidia produced nonfatal infection (Converse et al. 1962b).

In addition to infectious dose, phenotypic variation among isolates of *Coccidioides* may play a role in CM disease progression. Strain morphology varies from floccus non-pigmented to flat glabrous pigmented colonies, which suggests that the strains may produce different secondary metabolites (Baker et al. 1943). In one classic study, arthroconidia production was assessed in 47 strains (Friedman et al. 1953). The majority had typical 3–5 μm barrel-shaped arthroconidia; however, the number of conidia produced was highly variable. Growth media type affected this phenotype: some strains grown on glucose yeast extract produced conidia, but not on Sabouraud's agar, and vice versa. Three strains produced no conidia on either media. This variation in conidial production could result in variable host exposure.

Variation in virulence has been described (Friedman and Smith 1957; Berman et al. 1956; Friedman et al. 1955; Huppert et al. 1967). Three human clinical strains' LD_{50} values ranged from 17 to 90 days after infection with 100 arthroconidia (Friedman et al. 1955). The same group further assessed 27 clinical isolates with average LD_{50} from 17 to 41 days; however, 10 strains did not reach an LD_{50} after 90 days (Berman et al. 1956). Upon necropsy, mice showed infection for four of the nonfatal strains, but the mice did not exhibit outward signs of illness. Interestingly, one of the four strains was obtained from a fatal human infection. This strain was reassessed and showed similar virulence results. Additionally, the strain did not produce typical 3 × 5 micron arthroconidia and produced fewer conidia than other "normal" strains (Friedman and Smith 1957).

4.4 Epidemiology of Human and Animal Coccidioidomycosis

In the 1940s, one of the first investigations into the source of infection for a localized outbreak of CM was conducted (Davis et al. 1942). In April 1940 on a field trip to San Benito County, California, several Stanford students and faculty spent 2 days collection specimens and camping. Ten days after return to campus, a student was seen at the health center with fever, chest congestion, and malaise. Five more students from the same field trip reported similar illness within the week. Although the other students recovered, the first student that had become ill was diagnosed with coccidioidomycosis. Eventually, the group exposure was determined to the result of students digging a rattlesnake out of ground squirrel burrow. *Coccidioides immitis* was cultured out of the soils that were collected at the presumed exposure site.

Phylogenomic analyses reveal that *C. immitis* and *C. posadasii* speciated between five and ten million years ago (Engelthaler et al. 2015; Sharpton et al. 2009). The genus *Coccidioides* is possibly as old as 40–50 million years (Bowman et al. 1992; Fisher et al. 2002). Many mammalian orders rapidly diversified during the early Cenozoic, and the appearance of rodent fossils in North America corresponds to the proposed emergence of the *Coccidioides* genus (Fisher et al. 2002; Tapaltsyan et al. 2015; Saarinen et al. 2014). The emergence of both the Sierra Madres and Rocky Mountains in North America corresponds with divergence among early North American mammals (Saarinen et al. 2014; Bally and Palmer 1989). As South America was a separate continent until approximately three million years ago, recent introduction of *Coccidioides posadasii* to that region, and the origin of *Coccidioides* in the Sonoran Desert, is supported by both genomics and geological records (Engelthaler et al. 2015).

4.4.1 Climate

Climate clearly influences the incidence of CM and presence of the organism in the environment. Early investigations into the role of climate at Williams Air Force Base in Maricopa County during WW2 showed two seasonal increases in infection rates after winter and summer precipitation (Hugenholtz 1957). A similar trend among Arizona residents has been observed more recently (Tamerius and Comrie 2011; Kolivras and Comrie 2003). In Kern County California, only a single increase in infection rate occurs after the winter rainy season (Talamantes et al. 2007). Climatic factors that might affect rates in other endemic area, specifically in Mexico and South America, are unknown at this time (Vargas-Gastelum et al. 2015; Baptista-Rosas et al. 2007, 2012). The recent autochthonous infections in Washington state raise concerns, as there is not a consensus on the effect climate change will have on the spread of the organism into new areas (Litvintseva et al. 2015). The fungus has been found sporadically at Dinosaur National Monument in northern Utah, which indicates suitable habitat can be found in the Great Basin Desert (Johnson et al. 2014; Fisher et al. 2007; Petersen et al. 2004). Climate change predicts less frequent but more intense precipitation, and higher mean temperatures in the western USA, which may expand the endemic region.

4.4.2 Environmental Niche

Current understanding of the distribution of the fungus in the environment is based on extensive skin testing using a delayed-type hypersensitive reaction to various *Coccidioides* antigens (Fiese 1958). Using either spherulin (antigen derived from parasitic growth phase) or coccidioidin (antigens derived from saprobic growth phase), researchers mapped highest disease prevalence in the southwestern USA (Pappagianis 1988; Ajello 1971). Long-term Arizona residents in Maricopa County (Phoenix), Pima County (Tucson), and Pinal County (Casa Grande, Florence) had over 70% positive skin test rates, when compared to surrounding counties with only 10–40% reactivity rates (Maddy 1957, 1958b; Palmer et al. 1957; Edwards and Palmer 1957). Similar testing completed in California shows that Kern County (Bakersfield), Tulare County (Visalia), and Kings County (Hanford) residents had 50–70% skin test positive rates, while in surrounding counties positive skin tests dropped to 10% (Edwards and Palmer 1957). Additionally, in Mexico, Central America, and South America, similar distributions of positive skin tests have been found (Campins 1970; Mayorga and Espinoza 1970). However, in sparsely populated regions, this approach may not reflect the distribution of *Coccidioides* in the environment. Moreover, human migration over large and small spatial scales confounds fine spatial-scale analyses.

Several groups have worked to understand the ecology of *Coccidioides* (Whiston and Taylor 2014; Barker et al. 2012; Baptista-Rosas et al. 2007; Lacy and Swatek 1974; Swatek and Omieczynski 1970; Teel et al. 1970; Elconin et al. 1964). Defining factors that determine the presence of *Coccidioides*, as well as distribution of the fungus in the soil at local sites, has been investigated (Litvintseva et al. 2015; Johnson et al. 2014; Barker et al. 2012; Baptista-Rosas et al. 2007, 2012; Kolivras and Comrie 2003; Greene et al. 2000; Lacy and Swatek 1974; Swatek 1970; Swatek and Omieczynski 1970; Egeberg and Ely 1956). Environmental isolates of *Coccidioides* are usually obtained via inoculation of soil extracts in a susceptible rodent model (Davis et al. 1942; Levine and Winn 1964). The distribution of the fungus in the environment has been determined to be sporadic and highly localized, on the order of a square meter or less in area (Maddy 1958b, 1965). Additionally, trapping of rodents at the positive sites was conducted, with low overall levels of infectivity (Emmons 1942, 1943; Emmons and Ashburn 1942).

Defining key factors that explain presence of *Coccidioides* in the environment remains an elusive problem. Associations with saline and alkaline soils are the pattern in California (Plunkett et al. 1963; Egeberg et al. 1964; Elconin et al. 1964). However, in Arizona sandy and porous soil along with rodent burrows appears to be the strongest association (Barker et al. 2012; Emmons 1942; Maddy 1959, 1965; Smith 1971; Swatek 1970). However, caution is warranted when attempting to generalize the results, due to few studies and variable design of collection. Overall, most soil samples tested are negative (ranges from 99 to 80%), and completely randomized sampling approaches have resulted in predominately negative results (Lacy and Swatek 1974; Greene et al. 2000; Lauer et al. 2012;

Barker et al. 2012; Baptista-Rosas et al. 2012). Complex relationships among microbial organisms that share the same habitats have been investigated (Egeberg et al. 1964; Orr 1968). Direct plating from soil often results in overgrowth by more rapidly growing fungi under laboratory growth conditions (Swatek and Omieczynski 1970; Greene et al. 2000; Barker et al. 2012). However, *Coccidioides* is competitive under certain circumstances and may persist for decades the same location (Barker et al. 2012; Greene et al. 2000).

Distinct population and species boundaries, both within and among each species, are still unclear due to evidence of hybridization and introgression and that both species have been recovered among patients in southern California and northern Mexico (Neafsey et al. 2010; Fisher et al. 2001, 2002; Canteros et al. 2015; Johnson et al. 2014; Litvintseva et al. 2015; Lauer et al. 2012). MLST analysis of over 600 clinical and environmental isolates reveals population structure within Arizona and that clinical isolates are distinct from environmental isolates (Teixeira and Barker 2016).

Additionally, techniques to detect the fungus in the environment are being developed, which will help to understand and define the environmental niche of *Coccidioides*. New work on air sampling has provided a needed tool to monitor seasonal fluctuations (Chow et al. 2016). Extracting DNA from soil and dust has become much more common and molecular methods to detect the organism more robust (Johnson et al. 2014; Lauer et al. 2012; Baptista-Rosas et al. 2012; Litvintseva et al. 2015). Although detecting *Coccidioides* DNA in soil does not prove the presence of infectious arthroconidia, it is a method for screening a large number of soil samples, which would be necessary to model the environmental niche of the fungus.

4.4.3 Substrate Preferences

Comparative genomic studies have revealed functional differences associated with pathogenic, saprobic, or commensal lifestyles. *Coccidioides* species can digest keratin and other animal products and appear to have lost many genes associated with plant-derived carbon sources (Whiston and Taylor 2014, 2015; Sharpton et al. 2008). This suggests that this genus has specialized on animal-derived nutrients. This could be either acquired from tissue digestion during the parasitic phase in vivo or could be dead, decaying, or other sources of keratin such as skin, feathers, and hair (Lange et al. 2016; Lopes et al. 2008). It is likely that the primary nutritional mode is closely associated with animal-derived sources and may have led to the evolution of the parasitic lifestyle (Whiston and Taylor 2014; Sharpton et al. 2009).

Specifically, comparisons between *Onygenales* and *Eurotiales* (sister orders) show a reduction of cellulose-binding domain containing proteins, tannases, cellulases, cutinases, melibiases, pectin lyase, and pectin esterases among the *Onygenales* (Sharpton et al. 2009). These are all classes of genes associated with plant-derived nutritional sources. Two families appear to be expanded only in *Coccidioides*, and in particular the M35 class of deuterolysin metalloproteases, which contains at least one known virulence factor Mep1 (Hung et al. 2005; Whiston and Taylor 2015). This family of proteases has a preferred substrate of histones and protamines, which are arginine-rich molecules (Doi et al. 2003). This class is also under positive selection, supporting the idea that these genes are associated with the evolution of *Coccidioides* (Li et al. 2012). It is suggested that keratin degradation is associated with M35 class of metalloproteases, and thus *Coccidioides* may be associated preferentially with animal-derived nutritional sources, rather than plant-derived nutrients. This remains to be experimentally proven, and it is possible that the role of M35 deuterolysin metalloproteases in *Coccidioides* has diverged into new, and as yet unexplored, functions.

Other animal-associated nutrition sources include dung and frass. Fecal samples taken from lizards (*Uta stansburiana*, *Gerrhonotus* spp., *Sceloporus occidentalis*, *Crotaphytus wislizeni*, *Cnemidophorus tigris*), skunks (*Spilogale gracilis*), black-tailed deer (*Dama hemionus*), goats (*Capra* spp.), sheep (*Ovis* spp.), and burros (*Equus asinus*) near positive soil locations were subjected to culture to attempt to recover *Coccidioides*; however, none grew the organism (Swatek et al. 1967). *Coccidioides* has been cultured from bat guano in a single report (Krutzsch and Watson 1978).

4.4.4 Range Expansion

Coccidioides spp. are found in arid or semi-arid regions throughout the Americas but are thought to be at highest prevalence in southern Arizona (Fig. 4.2) (Fisher et al. 2007). *C. immitis* is found in central and southern California (Fisher et al. 2001). The range extends into northern Mexico, and recent work has found *C. immitis* in Yakima and Benton counties of Washington, at Dinosaur National Monument in Utah, and from a patient in Colombia with no travel history (Marsden-Haug et al. 2014; Litvintseva et al. 2015; Johnson et al. 2014; Canteros et al. 2015). *Coccidioides posadasii* is found in Arizona, Nevada, Utah, New Mexico, Texas, and throughout Mexico, with dispersed populations in Central and South America (Whiston and Taylor 2014; Duarte-Escalante et al. 2013; Brilhante et al. 2013; Campins 1970). Hybrid strains indicate that the two species coexist in nature (Neafsey et al. 2010). To determine if this is a recent or ancient phenomenon, environmental sampling is needed to accurately assess the prevalence of both species, and hybrid offspring, at a given location.

Direct isolations from soil throughout the range of both species in North and South America will clarify population structure and species boundaries. Clinical isolates are still needed to track emergence of any virulent strains or new point source outbreaks (Litvintseva et al. 2015). With greater surveillance and awareness, it is predicted that potential habitat for and cases of *Coccidioides* may be found throughout the western USA (Baddley et al. 2011). In fact, recent reports of cases in

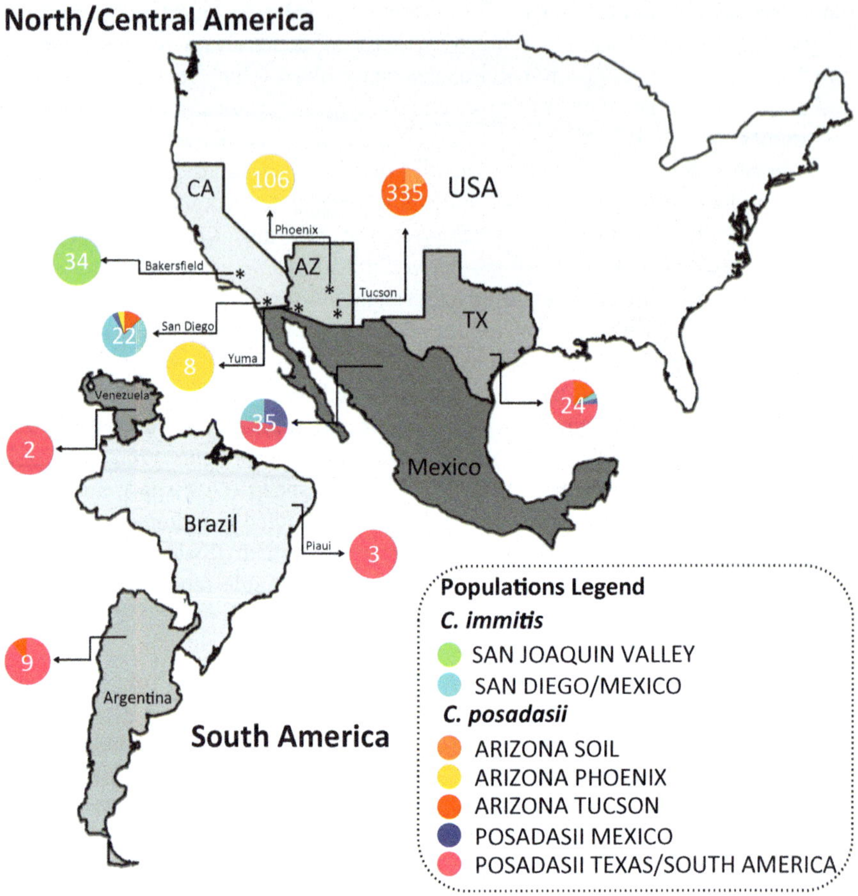

Fig. 4.2 Distribution and prevalence of *Coccidioides* spp. in arid or semi-arid regions throughout the America

Missouri northeast of the endemic area are concerning (Turabelidze et al. 2015). Although many of these cases could be the result of travel to endemic areas, the possibility must be considered that these are locally acquired infections as a result of drier soil and dust storms, fomites, or even unrecognized small foci of growth of the fungus (Hage et al. 2012; Stagliano et al. 2007; Desai et al. 2001).

4.5 Pathophysiology and Clinical Signs of Coccidioidomycosis

Both *Coccidioides* species cause the disease coccidioidomycosis (CM) also referred to as San Joaquin Valley fever, valley fever, desert rheumatism, or "cocci/coccy."

Nonfatal disease after exposure opened the possibility of vaccine development (Levine et al. 1965; Converse et al. 1962a; Swatek 1970). Starting in 1960s, the first vaccine was made with killed spherules (Levine and Kong 1965, 1966; Levine et al. 1965; Converse 1965; Castleberry et al. 1965). Unfortunately, severe side effects and lack of significant protection at the dosage given complicated further development of the killed vaccine (Pappagianis 1993). Work continues to identify candidates for vaccine development (Yoon and Clemons 2013; Hung et al. 2012; Cole et al. 2012; Xue et al. 2009; Awasthi 2007; Johnson et al. 2007; Awasthi et al. 2005). The development of an effective vaccine would provide needed protection to anyone in endemic areas (Nguyen et al. 2013; Cole et al. 2012).

Prior to the 1950s, there was no effective treatment for coccidioidomycosis (Einstein 1975). The first drug to be found effective was Amphotericin B; however, long-term treatment is complicated by nephrotoxic side effects (Longley and Mendenhall 1960; Fiese 1957; Halde et al. 1957; Lawrence and Hoeprich 1976). The current recommended treatment of CM is fluconazole (Catanzaro et al. 1990; Fierer et al. 1990; Galgiani et al. 1988; Finquelievich et al. 1988; Stevens 1977). Although these drugs are generally well-tolerated, toxicity and drug interactions are still concerns (Stevens and Clemons 2007). Even with treatment, infections may not be cleared for patients who have disseminated disease, although recent work with nikkomycin Z shows promise (Shubitz et al. 2013; Galgiani 2007; Hector et al. 1990). Frequently, lifelong therapeutics and monitoring of disease are required, particularly with coccidioidal meningitis (Antony et al. 2006).

Otherwise healthy people and animals living in or visiting endemic areas contract CM via the inhalation of conidia. Rarely, infection has occurred by dermal invasion, in laboratory accidents, and from bandaged subcutaneous lesions (Gaidici and Saubolle 2009; Smith and Harrell 1948; Fischer and Kane 1973). Although most human infections are asymptomatic, symptoms can range from mild to severe (Nguyen et al. 2013). It is argued that the primary reason for this variation is host genotype (Galgiani 2014). This is an inherently unsatisfying argument, given variation in disease presentation as a result of both inoculum levels, as well as isolate/strain virulence in various laboratory models of infection (Cox and Magee 1998, 2004; Cox and Vivas 1977; Hugenholtz et al. 1958; Friedman and Smith 1957; Berman et al. 1956; Friedman et al. 1953, 1955). Infectious dose, variation among strains, and variation among hosts all play a role in disease outcome. Genome wide association studies (GWAS) could define genetic-based differences in fungal virulence and host response (Muller et al. 2011). Additionally, defining conditions that influence *Coccidioides*' growth and reproduction will assist with preventing exposure.

4.5.1 Coccidioidomycosis in Primates

Progression and severity of disease in humans are influenced by host response to infection. An increase in disease burden has been detected as an increasing number of naïve hosts travel to endemic areas (Fig. 4.3). North American census data shows

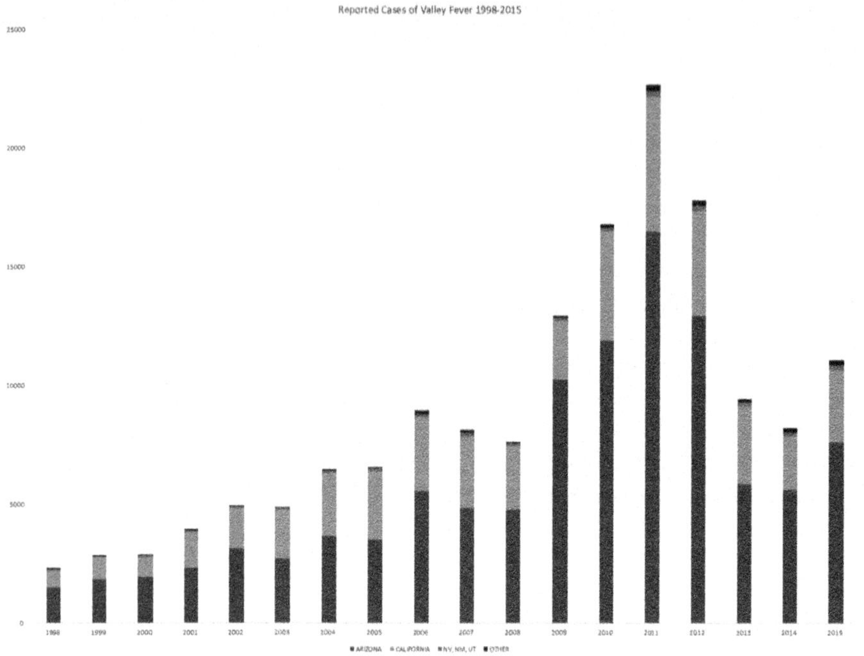

Fig. 4.3 Reported cases of coccidioidomycosis in the USA between 1998 and 2015. Data retrieved from https://www.cdc.gov/fungal/diseases/coccidioidomycosis/statistics.html

Arizona population has increased from 3,665,228 in 1990 to 6,731,484 in 2014, with a median age of 36.5 and around one million people aged 65 and older. In California population has increased over the same time period, from 29,760,021 to 38,802,500, with a median age of 35.6 and 4,617,907 over 65. Severe disease and negative outcomes are more common in elderly vs. younger patients (Blair et al. 2008). Persons with underlying disease such as HIV infection or diabetes can have greater risk of disseminated infections (Ampel 2007; Wheeler et al. 2015) The third trimester of pregnancy is a risk factor for disseminated CM (Crum and Ballon-Landa 2006). Transplant patients are at risk of severe disease, and antifungal prophylactic therapy and screening for patients in the endemic area is recommended (Mendoza et al. 2015; Kahn et al. 2015; Kauffman et al. 2014). Additionally, there is an indication that certain ethnicities have more severe disease and higher rates of dissemination, specifically African Americans (Wheeler et al. 2015; Ruddy et al. 2011; Pappagianis et al. 1979; Sievers 1974). Finally, certain professions or living conditions may expose people to higher inoculum, such as construction, landscaping and living in rural dusty areas (Yau 2016; Das et al. 2012; Cowper and Emmett 1953; Sievers and Fisher 1982).

Among genetic determinants associated with higher possibility of dissemination are mutations in STAT1, STAT3, IL-12Rβ-1, and IFN-γR1. The STAT1 mutation conferred a constitutively active gain of function, which likely results in a dysregulation of IFN-γ mediated inflammation, and was associated with both severe CM and histoplasmosis infections (Sampaio et al. 2013). The STAT3 mutations associated with Job's syndrome result in hyper-IgE levels in serum and dysregulation of IL-17 and Th17 response to infection, and patients are at risk of severe fungal infections (Odio et al. 2015). IL-12Rβ-1 novel missense mutations in two patients with disseminated infection reveals the importance of the IL12:IL23: IFNγ intersection as a risk factor in the human host for severe disease (Vinh et al. 2011). Finally an IFNγ receptor 1 autosomal dominant mutation resulted in severe and long-term disease in a pediatric case of CM, culminating to a coinfection with nontuberculosis *Mycobacterium* (Vinh et al. 2009). Upon determination of the root cause of the susceptibility and aggressive antimicrobial treatment, the patient improved.

These limited but important studies highlight the axis of IL12:IL23:IFNγ deficiencies and dysregulation as one cause of severe disseminated CM in humans. However, it does not explain a majority of cases or the cause of severe acute disease. These studies remain to be completed, and highlight needed future work. In addition, nonhuman primates are susceptible to infection, and several documented cases of captive macaques, chimpanzees, and baboons have been reported (Johnson et al. 1998; Bellini et al. 1991; Rosenberg et al. 1984; Rapley and Long 1974; Hoffman et al. 2007; Herrin et al. 2005; Ginocchio et al. 2013; Beaman et al. 1980; Breznock et al. 1975; Blundell et al. 1961). These reported infections are frequently severe disseminated valley fever; however it is not known if this is a common manifestation of the disease in nonhuman primates and how this informs us about human disease. Certainly, primates have been used in vaccine research, and immune profiles have been observed when compared to humans.

4.5.2 Coccidioidomycosis in Dogs

Domestic dogs (*Canis familiaris*) have been proposed as sentinels of disease and a way to map the organism in the environment in endemic regions where the disease is not reported to the CDC or occurrence outside the endemic region (Gautam et al. 2013). The approach has had success in California and Arizona and has been useful in guiding environmental collections of *Coccidioides* (Barker et al. 2012; Shubitz 2007; Butkiewicz et al. 2005; Shubitz et al. 2005; Grayzel et al. 2016). Canine prevalence data would be a useful tool for monitoring emergence of new regions of endemicity.

Certain breeds may have higher risk of disseminated disease. Data currently are more suggestive than definitive. Early work that studied 100 dogs postmortem found that male dogs, boxers, and Doberman pinschers had higher prevalence of severe disease that resulted in death (Maddy 1958a). A more recent study showed that boxers, beagles, pointers, Australian shepherds, and Scottish terriers were over-represented in comparison to general population in a retrospective study of 218 dogs based in Davis, California (Davidson and Pappagianis 1994). A prospective study on 124 dogs and a cross-sectional study of 381 dogs in Arizona grouped dogs according to American Kennel Club guidelines and found no association with breed group (Butkiewicz et al. 2005). Major impediments to understanding breed specificity are changing breed popularity, unknown pedigrees, and lack of reliable demographic data. Certain breeds of dogs tend to be more popular and are thus more likely to have high incidence within a single clinic. Backyard breeders and unreliable pedigrees can confound the correct breed assignment. Finally, demographic data currently relies on owners licensing dogs, and for certain breeds that are deemed dangerous (pit bulls, German shepherds, Rottweilers, etc.), the breed may not be disclosed (i.e., listed as "mix" or "unknown").

Conversion to a positive serology in dogs occurs at a similar rate compared to humans: around 70% of infections are asymptomatic (Shubitz et al. 2005). Approximately 24% of 124 dogs enrolled in a prospective study that were followed for 2 years converted positive for *Coccidioides* exposure, with 6% having clinical disease (Butkiewicz et al. 2005). Another benefit to using naturally infected dogs as models to understand disease in humans is that severity and disease manifestations are similar. The disease generally stays localized to the lungs and hilar lymph nodes (Graupmann-Kuzma et al. 2008). Dogs can develop extrapulmonary complications, such as dissemination to bone, meninges, and other internal organs in about 20% of symptomatic cases, which is higher than in humans (Davidson and Pappagianis 1994).

Treatment options are similar to what is available for people. Amphotericin B and azoles are the drugs of choice with similar toxicities and side effects. Generally, it is recommended that any animal that is clinically ill should be treated with antifungals (Graupmann-Kuzma et al. 2008). Diagnosis is often based on clinical manifestations consistent with disease combined with serological testing, travel history, and radiographic findings (Shubitz and Dial 2005; Ajithdoss et al. 2011). The decision on ending antifungal treatment is also complicated by the fact that in some animals IgG titers do not drop below the threshold that would allow a determination of a cure or remission and lifelong antifungal therapy is common for dogs with disseminated disease (Graupmann-Kuzma et al. 2008). New antifungals, such as nikkomycin Z, are being developed and show efficacy in dogs (Shubitz et al. 2013, 2015).

Other canids that live in endemic regions that may be naturally infected include Mexican gray wolves (*Canis lupus baileyi*), coyotes (*Canis latrans*), kit foxes (*Vulpes macrotis*), and the endangered San Joaquin kit fox (*Vulpes macrotis mutica*). A standard method of testing wild animals for any infectious disease is blood collection, which is followed by serological analysis. For CM, an antibody immunodiffusion assay or ELISA to detect *Coccidioides* + IgG or +IgM is standard. However, a negative result may be uninformative, and certainly in wild animals where testing has not been optimized, results should be viewed with caution. No reports of valley fever in wolves were found in the literature; however, wolves were eliminated from most of the endemic region prior to awareness of the disease and were not likely residents of lower deserts (Carroll et al. 2014). One report of an investigation of valley fever among coyotes reveals that three of five coyotes captured in the Tucson AZ area were subclinically infected and two had culturable fungus from tracheobronchial lymph node tissue (Straub et al. 1961). The fungi were then tested in a mouse model, and only one of the two cultures produced an infection consistent with valley fever. Finally, a survey of San Joaquin kit foxes showed a very low level of positive *Coccidioides* serology; however, the exact testing method was not reported (McCue and O'Farrell 1988).

A greater understanding of disease prevalence and severity among all canids found in endemic regions could provide valuable information on the genetic basis of host response to disease and distribution in the environment.

4.5.3 Coccidioidomycosis in Cats

The prevalence of valley fever in domestic cats (*Felis catus*) and other felines is thought to be lower than canines and with a very different presentation (Greene and Troy 1995). An interesting case presentation was reported in a wild mountain lion (*Felis concolor*) that had been trapped in Texas and transported to Florida for research purposes related to Florida panther reintroduction (Clyde et al. 1990). The animal was otherwise healthy; however, lung function and blood work were identified as abnormal during an examination and minor surgery. A communicable disease needed to be rapidly diagnosed as other mountain lions were to be released into the wild, and this animal had been in proximity to them. The animal was euthanized, and upon postmortem examination, disseminated CM with peritoneal involvement was discovered and likely represented a natural infection in a wild animal. Another peritoneal CM in a mountain lion was reported in a captured animal taking up residence in a tree in a private yard in Kern County, California (Adaska 1999). The animal was extremely lethargic and emaciated. Recent reports of plague in the area were cause for concern, and the animal was shot. Postmortem examination revealed several granulomatous structures throughout the peritoneum consistent with *Coccidioides* spherules. Infection was confirmed via histopathology and culture.

Other large felids in captivity have been reported infected by *Coccidioides* including a Indochinese tiger in the El Paso Zoo (*Panthera tigris corbetti*) and two Bengal tigers (*Panthera tigris tigris*) (Helmick et al. 2006; Henrickson and Biberstein 1972). In the case of the tiger, other animals tested at the facility did not have positive serology, and few cases of valley fever had been detected at the facility. The tiger had other comorbidities: chronic renal disease and pancreatic adenocarcinoma (Helmick et al. 2006). The other case report detailed that both Bengal tigers were male and contained at the same location in southern California. The tigers also had severe hepatic disease, which may have predisposed them to disseminated CM. In these two cases, infection with *Coccidioides* may have been asymptomatic for several years and only surfaced when other disease processes occurred.

In domestic cats, valley fever has variable presentation with the first report of disease in two cats appearing in 1963 (Reed et al. 1963). Both animals were euthanized and upon necropsy confirmed to have significant disseminated CM and disease manifestations similar to canine. However, cats are thought to have fewer fungal infections, and few reports of feline CM are found in the literature. Cats tend to be diagnosed with valley fever later in life, with an average age of 6.2 years at diagnosis, and the most common clinical manifestation was skin involvement (Greene and Troy 1995). This study of 48 cats also revealed that respiratory distress was not common. The primary treatment was ketoconazole; however fluconazole or itraconazole was also used. In 44 of the 48 cats treated, 25% failed treatment and were euthanized or died soon after diagnosis. Of the remaining, long-term (10 months) therapy was necessary with relapse after removal from treatment in a few cats.

4.5.4 Coccidioidomycosis in Armadillos

In South America, CM outbreaks have been associated with armadillo hunters (Eulalio et al. 2001; Wanke et al. 1999). Armadillos have lower body temperature than many other mammals and harbor other pathogens, primarily *Mycobacterium leprae*, the causative agent of leprosy (Duthie et al. 2011; Loughry et al. 2009). An investigation of 26 captured armadillos (*Dasypus novemcinctus*) revealed that three were subclinically infected. No evidence of gross pathology or histopathology was discovered upon necropsy; however, macerated spleen and lung tissue grew colonies consistent with *Coccidioides*. These colonies were subsequently used to infect mice, and these mice developed CM.

The role of armadillos in the ecology, distribution, and prevalence of *Coccidioides posadasii* in the environment is unknown. These animals have been implicated in association with another fungal disease, paracoccidioidomycosis (Bagagli et al. 2006; Arantes et al. 2016). Additionally, whether or not this association is restricted to South America, or also occurs in Texas, remains to be assessed. An interesting observation however, is the fact that Texas and South America fungal isolates are more genotypically related than Texas isolates are to Arizona isolates, despite being geographically more distant. Armadillos are not found commonly in Arizona and Central California.

4.5.5 Coccidioidomycosis in Rodents

The role of desert rodents, specifically the Heteromyidae, in the life cycle of *Coccidioides* is an area of research that has provided confusing data. Early researchers noticed a correlation of higher rates of infection with soil disturbance around rodent burrows (Emmons 1942; Stiles and Davis 1942), which suggests a rodent reservoir for *Coccidioides* (Ashburn and Emmons 1942). Trapping of 1942 animals occurred in Lordsburg, NM, and Wilcox, Tucson, Casa Grande, and Phoenix AZ. A range of species were obtained, including pocket mice, kangaroo rats, grasshopper mice, deer mice, pack rats, ground squirrels, rabbits, and harvest mice. No animals from Lordsburg or Wilcox were infected with *Coccidioides*. Even in Tucson, Phoenix, and Casa Grande, infection rates in trapped wild rodents were low.

Seven out of 207 animals (pocket mice and kangaroo rats only) investigated had confirmed *Coccidioides* infection, with three additional animals having lesions but no fungal growth (Emmons 1943). Wild-caught rodents are susceptible to infection by *Coccidioides*; however, route and dosage of infection did not mimic a natural infection (Swatek and Plunkett 1957). *Onygenales* fungi degrade keratin; thus it is possible that *Coccidioides* is associated with hair and skin in rodent burrows (Untereiner et al. 2004; Sharpton et al. 2009).

It is possible that a rodent is highly susceptible to CM. Because a sick rodent could be susceptible to predation, a severely ill animal may die in the burrow. Two lines of evidence support this hypothesis. Tissue from infected mice fed to predators did not produce fungal colonies from fecal or pellet material (Swatek et al. 1967), and when infected animals were sacrificed and buried in soil negative for *Coccidioides*, the soil was subsequently positive for growth (Maddy and Crecelius 1965). *Coccidioides* does not survive gut passage in foxes, coyotes, or owls (Swatek et al. 1967). Fecal samples from various predators and owl pellets were collected for 5 years near known positive soil sites, but no *Coccidioides* was recovered (Swatek and Omieczynski 1970; Swatek et al. 1967). Non-predators such as lizards, skinks, a burro, deer, goat, and sheep were also analyzed. Interestingly, *Coccidioides* does survive in the gut of laboratory and wild mice (Lubarsky and Plunkett 1954). This supports the possibility of transmission of *Coccidioides* through the gut of rodents, which frequently cannibalize.

Laboratory mice have also shown variable resistance to infection (Kirkland and Fierer 1983). Inbred mice in particular are highly susceptible to severe disease at low dosage of conidia in an intraperitoneal infection model. Comparing DBA/2N (outbred) mice to Balb/c or C57b6 (inbred) mice, the mean lethal dose for outbred was greater than 10^5 vs. less than or equal to 10^3 per mouse. IL-10 deficiencies have been implicated as the source of this susceptibility (Fierer 2007; Fierer et al. 1998).

4.5.6 Coccidioidomycosis in Captive Animals and Other Wildlife

Many wild and domestic mammals are susceptible to CM. The first report of a California sea lion (*Zalophus californianus*) infected with *Coccidioides* was a captive animal that was housed in a zoo in Tucson, AZ (Reed et al. 1976). Several years later, a naturally infected dolphin (*Tursiops truncatus*) was found upon necropsy to be infected by *C. immitis* (Reidarson et al. 1998). In a recent retrospective report, 36 wild marine mammals that had beached along the Central California coast between 1998 and 2012, including sea lions, sea otters, and harbor seals, were found to be infected with *Coccidioides* (Huckabone et al. 2015). How marine mammals are exposed remains an open question; however, it is most likely from dust and airborne particulate moving from endemic areas in California to the ocean.

Captive animals in endemic regions are a source of unusual infections but provide information about the range of susceptible hosts. A black rhinoceros (*Diceros bicornis*) that was moved from Texas to the Milwaukee County Zoo developed severe progressive lameness that was determined upon necropsy to have been caused by disseminated CM (Wallace et al. 2009). Interestingly, between 1984 and 1994, the Phoenix Zoo had several wallabies and kangaroos infected and perish from valley fever (Reed et al. 1996). In another endemic region, a case of a giant red kangaroo being infected, as well as positive soil in its enclosure, was also reported (Hutchinson et al. 1973). One recent case of a koala at the San Diego Zoo succumbing to infection was also reported (Burgdorf-Moisuk et al. 2012). Whether marsupials are particularly susceptible to severe disease remains unknown.

Other reported wildlife native to the Sonoran Desert that has been found to be naturally infected and suffer disease includes bats, desert bighorn sheep, and javelina (peccary). Bats appear to be incidentally infected by *Coccidioides* (Cordeiro Rde et al. 2012; Krutzsch and Watson 1978). In a recent survey looking for another fungus *Histoplasma capsulatum*, *Coccidioides posadasii* was discovered. An experimental infection of *Macrotus californicus* (California leaf-nosed bat) showed this animal to be susceptible to a range of dosages: 50, 100, 200, and 400 conidia (Krutzsch and Watson 1978). However, the pallid bat (*Antrozous pallidus*) was infected only at the highest dosage of 400 conidia.

In the one reported case of desert bighorn sheep (*Ovis canadensis nelsoni*), the ram, along with 12 other sheep, was captured in November 1984 from the Marble Mountains in San Bernardino County as part of a relocation effort (Jessup et al. 1989). Animals were penned over the winter in a 17-acre enclosure in the Whipple Mountains near the Colorado River. The herd was released in the February 1985. In mid-September 1985, the ram was observed to have a respiratory infection and died shortly after capture. Postmortem revealed severe disseminated CM, with lungs being heavily infected and damaged. Continued follow-up with the rest of the herd revealed no additional evidence of CM, and no other reports have appeared. Thus it is unknown how frequently this animal is infected and what the burden of disease is for this native desert dweller.

The javelina, or the collared peccary (*Pecari (Tayassu) tajacu*), is a common resident of the desert southwest and behaviorally likely to be exposed to large inocula. However, only a single report was found in this animal (Lochmiller et al. 1985). The 25 animals in the report were captured throughout Texas and housed in an outdoor enclosure at Texas A&M University. In January 1984, a female exhibited neurological symptoms that resulted in euthanasia, and a blood analysis indicated an infection. The two other animals in the same pen were also euthanized as prevention. Necropsy revealed several granulomas in the lungs, kidney, and spleen verified as *Coccidioides* via histopathology. It is unclear how the infection occurred, as the animals that were ill were harvested outside the endemic range, and Brazos County (where animals were held captive) is considered to be outside the endemic range. Of interest is that both reports occurred in 1984–1985. It is unknown if this was concurrent with particularly high level of disease in humans as well, as cases were not nationally reported before 1994.

4.5.7 Coccidioidomycosis in Livestock

Greater assessment of CM in livestock has been reported, likely because of potential economic impact. Reports in pigs, cattle, sheep, horses, and llamas have revealed high rates of infection with rare cases of disseminated disease, although llamas and horses appear to have more frequent severe disease than other livestock.

A first report of CM in cattle (*Bos taurus*) was in 1918, after a slaughterhouse observed infected thoracic lymph tissue and submitted the tissue to a USDA pathologist (Giltner 1918). The pathologist grew *Coccidioides* from the tissue and tested the organism in several animals: guinea pigs, rabbits, dogs, cattle, sheep, and swine. In calves, subcutaneous infection resulted in lesions at the site of infection but no apparent disease. However, intravenous inoculation resulted in rapid death and involvement of lung tissue.

Later observations by a veterinary meat inspector in Los Angeles slaughterhouses of 3173 cattle from the southwestern USA revealed a similar trend of natural infections resulting in thoracic lymph node involvement (Maddy 1954b). Between 1947 and 1951, *Coccidioides* infected 1.8% of all cattle (calves excluded), 2.9% of steers and heifers, and 7.3% of steers and heifers from San Joaquin Valley feedlots. Of interest was an assessment of animals shipped direct for slaughter from other areas. Upon arrival, 23 animals from west Texas (Amarillo, Lubbock), 36 animals from eastern New Mexico (Clovis, Tatum), 17 from southeastern Colorado (Springfield), and 4 from southern Oregon (Medford) were found to be infected. No animal had evidence of acute CM.

Skin test surveys of 11,643 range cattle in Arizona revealed a high rate of infection that overlapped with the disease prevalence and distribution seen in humans (Maddy et al. 1960a). Between 1954 and 1959, cattle in Arizona between 1 and 6 years of age were skin tested using the coccidioidin skin test developed for humans (Palmer et al. 1957; Edwards and Palmer 1957). Pinal County had the highest rate of infection with 42% of cattle being skin test positive. However, assessments made during slaughter at a southwestern feedlot reveal that although coccidioidal granulomas are visible in thoracic viscera, no other signs of disease were noted (Reed 1960). In addition, laboratory infection experiments with dosages ranging from 5×10^5 to 1.5×10^6 arthroconidia and mycelia intratracheally showed that cattle did not develop disease, with few granulomas in lungs or thoracic lymph tissue (Maddy et al. 1960b). Serological testing was not confirmatory; however infected animals did eventually convert to skin test positive using coccidioidin. Thus cattle are susceptible to infection but do not develop severe disease.

Reports of new world camelids infected and suffering disseminated valley fever are few; however, it appears that this animal does develop severe disease (Fowler et al. 1992). The first case report was a single 8-year-old female llama (*Lama glama*) with severe disease which was euthanized and upon necropsy was discovered to have disseminated valley fever (Muir and Pappagianis 1982). The rest of the herd was tested for infection using a complement fixation antibody test, and 3 of the 11 other llamas showed evidence of infection but no disease. This initial disease report was expanded in a report on 19 retrospective cases between 1981 and 1989 from California and Arizona (Fowler et al. 1992). All animals but one had disseminated valley fever with multiple affected organs. Clinical signs varied widely, with and without cough or dyspnea, and dissemination to kidney, liver, intestine, adrenal gland, and meninges. Dermal infection was more common in llamas from California, but no gender or age correlations were observed. Diagnosis in llamas is confounded by highly variable clinical presentation and lack of reliable serological testing. Only one additional recent report was found, which described ocular disease in a 7-year-old male llama, which disseminated and resulted in euthanasia (Coster et al. 2010). Of particular interest in this case is a lack of travel or residence in the endemic region. Llamas appear to have higher rate of complicated disease than other agricultural animals, but overall disease burden is not high.

CM in naturally infected sheep (*Ovis aries*) has been reported (Maddy 1954a; Beck 1929). Upon slaughter, in both cases, the animals had lesions in the mediastinal and bronchiolar lymph nodes. In the earlier case, *Coccidioides* was grown from the tissue and verified in a guinea pig model of infection (Beck 1929). In the second case, caseous lymphadenitis was suspected, which could be caused by a contagious bacterial infection common in sheep and goats, so tissue was sent for pathology (Maddy 1954a). *Coccidioides* was determined to be the causative agent, although no evidence of illness was present, and no validation was performed in a rodent model of infection. One experimental infection of sheep has been conducted (Giltner 1918). Two animals were infected, one intravenously and the other subcutaneously. Both animals appeared well nourished, and no evidence of outward disease was reported. However, the intravenous infection resulted in multiple organ involvement, and fungus grew from the liver, lymph, and lung tissue. The animal infected subcutaneously had no fungal structures in any tissue.

One experimental case of infection in a swine (*Sus scrofa*) is reported (Giltner 1918). Two animals were infected via the right marginal ear vein. Upon necropsy, miliary lung nodules and lesions in spleen and liver were observed. A single report of a young pig succumbing to infection was found (Prchal and Crecelius 1966). A survey of both young (6-month-old butcher hogs) and older (3-year-old breeding sows) animals in Tucson, Arizona, revealed no disease but many granulomatous lesions in the bronchiolar lymph nodes (Prchal and Crecelius 1966). *Coccidioides* was confirmed first by microscopy, followed by infection in mice. It appears that although animals are susceptible to infection, complicated disease rarely develops.

CM in horses (*Equus caballus*) is reported with higher frequency that other agricultural animals, possibly because the animals are often considered pets, rather than livestock. One review states that pulmonary involvement and weight loss is a common manifestation of disease, along with osteomyelitis and lesions in thoracic lymph nodes and liver (Ziemer et al. 1992). All 15 animals died or were euthanized, and treatment was not effective in the 4 animals where it was attempted. These cases were also interesting in that several of the animals resided in areas not considered to be highly endemic. Subsequent treatment of animals was more successful due to earlier diagnosis and treatment (Higgins et al. 2006). A case of a 14-day-old foal with acute CM that was euthanized due to severe disease suggests that some horses are more susceptible to infection and suffer disease as a result (Maleski et al. 2002).

Interestingly, the wild Przewalski's horse (*Equus przewalskii*) appears to be even more susceptible to severe disease than domesticated horses (*E. caballus*). These animals appear to be resistant to most common infectious disease of horses; however, they are susceptible to disseminated CM (Terio et al. 2003). Thirty Przewalski horses housed in the San Diego Wild Animal Park as part of a breeding and reintroduction program died over a 16-year period, with 10 deaths attributed to CM. No other exotic equids at the same facility had reported valley fever deaths. The cause of susceptibility to valley fever could be from a defect in *Coccidioides*-specific immune response, but results were inconclusive.

Specific non-mammalian hosts infected intraperitoneally, including crayfish, goldfish, and amphibians, developed mycelia in various tissues, and lizards can develop spherules in the lung (Swatek and Plunkett 1957). A naturally infected Sonoran gopher snake with pulmonary lesions and histological and microscopic evidence of *Coccidioides* infection was reported (Timm et al. 1988). Although morphology consistent with *Coccidioides* was observed, the fungus was not validated by mouse infection, and genotyping was not yet available. A second case of an infected Sonoran lyre snake at the Phoenix Zoo suggests that natural infection of reptiles deserves more attention (Reed et al. 1996). Again microscopy consistent with *Coccidioides* was observed, but in this case the fungal agent was confirmed by infection in a mouse. An incidental finding of a lung granuloma in a Gila monster was reported by the Arizona Veterinary Diagnostic Laboratory, but the animal perished from another cause (Reed et al. 1996).

No avian species have been reported to have infection. No detailed investigation of invertebrate associations, such as nematodes in soil or soil-burrowing insects, has been conducted. An interesting account of mice injected with chromium-51 and buried in soil was relayed at an international CM meeting (Egeberg 1985). Following the movement of the radiation, insects were implicated in the digestion of mouse tissue in both cases.

4.6 Conclusion and Future Directions

Over the last 30 years, reported cases of valley fever have increased dramatically (Sondermeyer et al. 2013; Huang et al. 2012; Hector et al. 2011; Lewis et al. 2015; Twarog and Thompson 2015). Data show that strains recovered in a recent study of Arizona isolates were highly variable with no clonal structure; therefore, a pathogenic clone was not responsible for the rise in cases (Jewell et al. 2008). A central question regarding the increase of disease remains unanswered. Possible non-mutually exclusive causes include climate change, increased host susceptibility, and changes in reporting and awareness. Another complication is the high levels of recombination, admixture, and genetic diversity and an as yet undiscovered sexual life cycle that could produce an alternative infectious morphology: the ascospore. Genetic variation in the fungus and ability to adapt to novel hosts and colonize new environments are additional unanswered questions. The potential emergence of antifungal resistance is cause for concern, and the lack of early accurate diagnosis and treatment recommendations for humans is troublesome. Furthermore, treatment and diagnostic development for our four-legged friends is based on clinical trials in humans. Thus, many clinicians are forced to use "wait and see" approaches to treatment. Greater research effort to understand the organisms as well as the disease is needed, particularly in response to the potential for the disease to expand into new areas.

Acknowledgments This work was supported in part by the Arizona Biomedical Research Commission grant (ABRC/ADHS 14-082975).

References

Adaska JM (1999) Peritoneal coccidioidomycosis in a mountain lion in California. J Wildl Dis 35 (1):75–77

Ajello L (1971) Coccidioidomycosis and histoplasmosis. A review of their epidemiology and geographical distribution. Mycopathol Mycol Appl 45(3):221–230

Ajithdoss DK, Trainor KE, Snyder KD, Bridges CH, Langohr IM, Kiupel M, Porter BF (2011) Coccidioidomycosis presenting as a heart base mass in two dogs. J Comp Pathol 145 (2-3):132–137. https://doi.org/10.1016/j.jcpa.2010.12.014

Ampel NM (2007) Coccidioidomycosis in persons infected with HIV-1. Ann N Y Acad Sci 1111:336–342. https://doi.org/10.1196/annals.1406.033

Antony SJ, Jurczyk P, Brumble L (2006) Successful use of combination antifungal therapy in the treatment of *Coccidioides* meningitis. J Natl Med Assoc 98(6):940–942

Arantes TD, Theodoro RC, Teixeira Mde M, Bosco Sde M, Bagagli E (2016) Environmental mapping of *Paracoccidioides* spp. in Brazil reveals new clues into genetic diversity, biogeography and wild host association. PLoS Negl Trop Dis 10(4):e0004606. https://doi.org/10.1371/journal.pntd.0004606

Ashburn LL, Emmons CW (1942) Spontaneous coccidioidal granuloma in the lungs of wild rodents. Arch Pathol 34(5):791–800

Awasthi S (2007) Dendritic cell-based vaccine against *Coccidioides* infection. Ann N Y Acad Sci 1111:269–274. https://doi.org/10.1196/annals.1406.013

Awasthi S, Awasthi V, Magee DM, Coalson JJ (2005) Efficacy of antigen 2/proline-rich antigen cDNA-transfected dendritic cells in immunization of mice against *Coccidioides posadasii*. J Immunol 175(6):3900–3906

Baddley JW, Winthrop KL, Patkar NM, Delzell E, Beukelman T, Xie F, Chen L, Curtis JR (2011) Geographic distribution of endemic fungal infections among older persons, United States. Emerg Infect Dis 17(9):1664–1669. https://doi.org/10.3201/eid1709.101987

Bagagli E, Bosco SM, Theodoro RC, Franco M (2006) Phylogenetic and evolutionary aspects of *Paracoccidioides brasiliensis* reveal a long coexistence with animal hosts that explain several biological features of the pathogen. Infect Genet Evol 6(5):344–351. https://doi.org/10.1016/j.meegid.2005.12.002

Baker EE, Mrak EM, Smith CE (1943) The morphology, taxonomy, and distribution of *Coccidioides immitis* Rixford and Gilchrist 1896. Farlowia 1(2):199–244

Bally A, Palmer A (1989) The geology of North America—an overview. Geological Society of America, Boulder

Baptista-Rosas RC, Hinojosa A, Riquelme M (2007) Ecological niche modeling of *Coccidioides* spp. in western north American deserts. Ann N Y Acad Sci 1111:35–46. https://doi.org/10.1196/annals.1406.003

Baptista-Rosas RC, Catalán-dibene J, Romero-olivares AL, Hinojosa A, Cavazos T, Riquelme M (2012) Molecular detection of *Coccidioides* spp. from environmental samples in Baja

California: linking Valley Fever to soil and climate conditions. Fungal Ecol 5(2):177–190. https://doi.org/10.1016/j.funeco.2011.08.004
Barker BM, Tabor JA, Shubitz LF, Perrill R, Orbach MJ (2012) Detection and phylogenetic analysis of *Coccidioides posadasii* in Arizona soil samples. Fungal Ecol 5(2):163–176. https://doi.org/10.1016/j.funeco.2011.07.010
Beaman L, Holmberg C, Henrickson R, Osburn B (1980) The incidence of coccidioidomycosis among nonhuman primates housed outdoors at the California primate research center. J Med Primatol 9(4):254–261
Beck MD (1929) Occurrence of *Coccidioides immitis* in lesions of slaughtered animals. Exp Biol Med 26(6):534–536. https://doi.org/10.3181/00379727-26-4385
Bellini S, Hubbard GB, Kaufman L (1991) Spontaneous fatal coccidioidomycosis in a native-born hybrid baboon (*Papio cynocephalus anubis/Papio cynocephalus cynocephalus*). Lab Anim Sci 41(5):509–511
Berman RJ, Friedman L, Roessler WG, Smith CE (1956) The virulence and infectivity of twenty-seven strains of *Coccidioides immitis*. Am J Hyg 64(2):198–210
Blair JE, Mayer AP, Currier J, Files JA, Wu Q (2008) Coccidioidomycosis in elderly persons. Clin Infect Dis 47(12):1513–1518. https://doi.org/10.1086/593192
Blundell GP, Castleberry MW, Lowe EP, Converse JL (1961) The pathology of *Coccidioides immitis* in the *Macaca mulatta*. Am J Pathol 39:613–630
Bowman BH, Taylor JW, White TJ (1992) Molecular evolution of the fungi: human pathogens. Mol Biol Evol 9(5):893–904
Breslau AM, Kubota MY (1964) Continuous in vitro cultivation of spherules of *Coccidioides immitis*. J Bacteriol 87:468–472
Breznock AW, Henrickson RV, Silverman S, Schwartz LW (1975) Coccidioidomycosis in a rhesus monkey. J Am Vet Med Assoc 167(7):657–661
Brilhante RS, de Lima RA, Ribeiro JF, de Camargo ZP, Castelo-Branco Dde S, Grangeiro TB, Cordeiro Rde A, Gadelha Rocha MF, Sidrim JJ (2013) Genetic diversity of *Coccidioides posadasii* from Brazil. Med Mycol 51(4):432–437. https://doi.org/10.3109/13693786.2012.731711
Brillhante RS, Moreira Filho RE, Rocha MF, Castelo-Branco Dde S, Fechine MA, Lima RA, Picanco YV, Cordeiro Rde A, Camargo ZP, Queiroz JA, Araujo RW, Mesquita JR, Sidrim JJ (2012) Coccidioidomycosis in armadillo hunters from the state of Ceara, Brazil. Mem Inst Oswaldo Cruz 107(6):813–815
Brooks LD, Northey WT (1963) Studies on Coccidioides immitis. II. Physiological studies on in vitro spherulation. J Bacteriol 85:12–15
Burgdorf-Moisuk A, Stalis IH, Pye GW (2012) Disseminated Coccidioidomycosis in a Koala (*Phascolarctos cinereus*). J Zoo Wildl Med 43(1):197–199. https://doi.org/10.1638/2011-0180.1
Burt A, Carter DA, Koenig GL, White TJ, Taylor JW (1996) Molecular markers reveal cryptic sex in the human pathogen *Coccidioides immitis*. Proc Natl Acad Sci U S A 93(2):770–773
Butkiewicz CD, Shubitz LE, Dial SM (2005) Risk factors associated with *Coccidioides* infection in dogs. J Am Vet Med Assoc 226(11):1851–1854
Cairns L, Blythe D, Kao A, Pappagianis D, Kaufman L, Kobayashi J, Hajjeh R (2000) Outbreak of coccidioidomycosis in Washington state residents returning from Mexico. Clin Infect Dis 30 (1):61–64. https://doi.org/10.1086/313602
Campins H (1970) Coccidioidomycosis in South America. A review of its epidemiology and geographic distribution. Mycopathol Mycol Appl 41(1):25–34
Canteros CE, Velez HA, Toranzo AI, Suarez-Alvarez R, Tobon OA, Del Pilar Jimenez AM, Restrepo MA (2015) Molecular identification of *Coccidioides immitis* in formalin-fixed, paraffin-embedded (FFPE) tissues from a Colombian patient. Med Mycol 53(5):520–527. https://doi.org/10.1093/mmy/myv019

Carroll C, Fredrickson RJ, Lacy RC (2014) Developing metapopulation connectivity criteria from genetic and habitat data to recover the endangered Mexican wolf. Conserv Biol 28(1):76–86. https://doi.org/10.1111/cobi.12156

Castleberry MW, Converse JL, Sinski JT, Lowe EP, Pakes SP, Delfavero JE (1965) Coccidioidomycosis: studies of canine vaccination and therapy. J Infect Dis 115:41–48

Catanzaro A, Fierer J, Friedman PJ (1990) Fluconazole in the treatment of persistent coccidioidomycosis. Chest 97(3):666–669

Chen S, Erhart LM, Anderson S, Komatsu K, Park B, Chiller T, Sunenshine R (2011) Coccidioidomycosis: knowledge, attitudes, and practices among healthcare providers—Arizona, 2007. Med Mycol 49(6):649–656. https://doi.org/10.3109/13693786.2010.547995

Chow NA, Griffin DW, Barker BM, Loparev VN, Litvintseva AP (2016) Molecular detection of airborne *Coccidioides* in Tucson, Arizona. Med Mycol 54(6):584–592. https://doi.org/10.1093/mmy/myw022

Clyde VL, Kollias GV, Roelke ME, Wells MR (1990) Disseminated Coccidioidomycosis in a western cougar (*Felis concolor*). J Zoo Wildl Med 21(2):200–205

Cole GT, Hurtgen BJ, Hung CY (2012) Progress toward a human vaccine against Coccidioidomycosis. Curr Fungal Infect Rep 6(4):235–244. https://doi.org/10.1007/s12281-012-0105-y

Converse JL (1955) Growth of spherules of *Coccidioides immitis* in a chemically defined liquid medium. Proc Soc Exp Biol Med 90(3):709–711

Converse J (1965) The effect of nonviable and viable vaccines in experimental coccidioidomycosis. Am Rev Respir Dis 92(6):159–174

Converse JL, Besemer AR (1959) Nutrition of the parasitic phase of *Coccidioides immitis* in a chemically defined liquid medium. J Bacteriol 78(2):231–239

Converse JL, Castleberry MW, Besemer AR, Snyder EM (1962a) Immunization of mice against coccidioidomycosis. J Bacteriol 84:46–52

Converse JL, Lowe EP, Castleberry MW, Blundell GP, Besemer AR (1962b) Pathogenesis of *Coccidioides immitis* in monkeys. J Bacteriol 83:871–878

Coster ME, Ramos-Vara JA, Vemulapalli R, Stiles J, Krohne SG (2010) *Coccidioides posadasii* keratouveitis in a llama (*Lama glama*). Vet Ophthalmol 13(1):53–57. https://doi.org/10.1111/j.1463-5224.2009.00747.x

Cowper HH, Emmett J (1953) Coccidioidomycosis in San Fernando Valley; report of a study carried out in 1951–1952. Calif Med 79(2):97–98

Cox RA, Magee DM (1998) Protective immunity in coccidioidomycosis. Res Immunol 149 (4-5):417–428. discussion 506-417

Cox RA, Magee DM (2004) Coccidioidomycosis: host response and vaccine development. Clin Microbiol Rev 17(4):804–839. https://doi.org/10.1128/CMR.17.4.804-839.2004

Cox RA, Vivas JR (1977) Spectrum of *in vivo* and *in vitro* cell-mediated immune responses in coccidioidomycosis. Cell Immunol 31(1):130–141

Crum NF, Ballon-Landa G (2006) Coccidioidomycosis in pregnancy: case report and review of the literature. Am J Med 119(11):993 e911–993 e997. https://doi.org/10.1016/j.amjmed.2006.04.022

Das R, McNary J, Fitzsimmons K, Dobraca D, Cummings K, Mohle-Boetani J, Wheeler C, McDowell A, Iossifova Y, Bailey R, Kreiss K, Materna B (2012) Occupational coccidioidomycosis in California: outbreak investigation, respirator recommendations, and surveillance findings. J Occup Environ Med 54(5):564–571. https://doi.org/10.1097/JOM.0b013e3182480556

Davidson AP, Pappagianis D (1994) Canine coccidioidomycosis: 1970–1993. In: 5th international conference on Coccidioidomycosis. Stanford University, Stanford, pp 155–162

Davis BL Jr, Smith R, Smith C (1942) An epidemic of coccidioidal infection (coccidioidomycosis). J Am Med Assoc 118(14):1182–1186. https://doi.org/10.1001/jama.1942.02830140012004

de Aguiar Cordeiro R, e Silva KR, Brilhante RS, Moura FB, Duarte NF, Marques FJ, Filho RE, de Araujo RW, Bandeira Tde J, Rocha MF, Sidrim JJ (2012) *Coccidioides posadasii* infection in bats, Brazil. Emerg Infect Dis 18(4):668–670. https://doi.org/10.3201/eid1804.111641

de Perio MA, Niemeier RT, Burr GA (2015) *Coccidioides* exposure and Coccidioidomycosis among prison employees, California, United States. Emerg Infect Dis 21(6):1031–1033. https://doi.org/10.3201/eid2106.141201

Desai SA, Minai OA, Gordon SM, O'Neil B, Wiedemann HP, Arroliga AC (2001) Coccidioidomycosis in non-endemic areas: a case series. Respir Med 95(4):305–309. https://doi.org/10.1053/rmed.2000.1039

Dickson EC (1937) "Valley Fever" of the San Joaquin Valley and fungus *Coccidioides*. Cal West Med 47(3):151–155

Dixon DM (2001) *Coccidioides immitis* as a select agent of bioterrorism. J Appl Microbiol 91 (4):602–605

Doi Y, Lee BR, Ikeguchi M, Ohoba Y, Ikoma T, Tero-Kubota S, Yamauchi S, Takahashi K, Ichishima E (2003) Substrate specificities of deuterolysin from *Aspergillus oryzae* and electron paramagnetic resonance measurement of cobalt-substituted deuterolysin. Biosci Biotechnol Biochem 67(2):264–270. https://doi.org/10.1271/bbb.67.264

Drips W Jr, Smith CE (1964) Epidemiology of Coccidioidomycosis. A contemporary military experience. JAMA 190:1010–1012

Duarte-Escalante E, Zuniga G, Frias-De-Leon MG, Canteros C, Castanon-Olivares LR, Reyes-Montes Mdel R (2013) AFLP analysis reveals high genetic diversity but low population structure in *Coccidioides posadasii* isolates from Mexico and Argentina. BMC Infect Dis 13:411. https://doi.org/10.1186/1471-2334-13-411

Dukik K, Munoz JF, Jiang Y, Feng P, Sigler L, Stielow JB et al (2017) Novel taxa of thermally dimorphic systemic pathogens in the *Ajellomycetaceae (Onygenales)*. Mycoses 60(5):296–309

Duthie MS, Truman RW, Goto W, O'Donnell J, Hay MN, Spencer JS, Carter D, Reed SG (2011) Insight toward early diagnosis of leprosy through analysis of the developing antibody responses of *Mycobacterium leprae*-infected armadillos. Clin Vaccine Immunol 18(2):254–259. https://doi.org/10.1128/CVI.00420-10

Edwards PQ, Palmer CE (1957) Prevalence of sensitivity to coccidioidin, with special reference to specific and nonspecific reactions to coccidioidin and to histoplasmin. Dis Chest 31(1):35–60

Egeberg RO (1985) Socioeconomic impact of Coccidioidomycosis. In: Einstein HE (ed) The 4th international conference on Coccidioidomycosis, San Diego, 1984. National Foundation for Infectious Diseases, Washington

Egeberg RO, Ely AF (1956) *Coccidioides immitis* in the soil of the southern San Joaquin Valley. Am J Med Sci 231(2):151–154

Egeberg RO, Elconin AE, Egeberg MC (1964) Effect of salinity and temperature on *Coccidioides immitis* and three antagonistic soil saprophytes. J Bacteriol 88:473–476

Einstein HE (1975) Amphotericin for coccidioidomycosis. West J Med 123(1):45

Elconin AF, Egeberg RO, Egeberg MC (1964) Significance of soil salinity on the ecology of *Coccidioides immitis*. J Bacteriol 87:500–503

Emmons CW (1942) Isolation of *Coccidioides* from soil and rodents. Public Health Rep 57 (4):109–111. https://doi.org/10.2307/4583988

Emmons CW (1943) Coccidioidomycosis in wild rodents. A methods of determining the extent of endemic areas. Public Health Rep 58(1):1–5. https://doi.org/10.2307/4584326

Emmons CW, Ashburn LL (1942) The isolation of *Haplosporangium parvum n.* sp. and *Coccidioides immitis* from wild rodents. Their relationship to coccidioidomycosis. Public Health Rep 57(46):1715–1727. https://doi.org/10.2307/4584276

Engelthaler D, Roe CC, Litvintseva AP, Driebe EM, Schupp JM, Gade L, Waddell V, Komatsu K, Arathoon E, Logemann H, Thompson GR, Chiller T, Barker B, Keim P (2015) Local population structure and patterns of western hemisphere dispersal for *Coccidioides* spp., the fungal cause of valley fever. MBio 7(2):e00550–e00516

Eulalio KD, de Macedo RL, Cavalcanti MA, Martins LM, Lazera MS, Wanke B (2001) *Coccidioides immitis* isolated from armadillos (*Dasypus novemcinctus*) in the state of Piaui, northeast Brazil. Mycopathologia 149(2):57–61

Fierer J (2007) The role of IL-10 in genetic susceptibility to coccidioidomycosis on mice. Ann N Y Acad Sci 1111:236–244. https://doi.org/10.1196/annals.1406.048
Fierer J, Kirkland T, Finley F (1990) Comparison of fluconazole and SDZ89-485 for therapy of experimental murine coccidioidomycosis. Antimicrob Agents Chemother 34(1):13–16
Fierer J, Walls L, Eckmann L, Yamamoto T, Kirkland TN (1998) Importance of interleukin-10 in genetic susceptibility of mice to *Coccidioides immitis*. Infect Immun 66(9):4397–4402
Fiese MJ (1957) Treatment of disseminated coccidioidomycosis with amphotericin B; report of a case. Calif Med 86(2):119–120
Fiese MJ (1958) Coccidioidomycosis. Thomas, Springfield
Finquelievich JL, Negroni R, Arechavala A (1988) Treatment with itraconazole of experimental coccidioidomycosis in the Wistar rat. Mycoses 31(2):80–86
Fischer JB, Kane J (1973) *Coccidioides immitis:* a hospital hazard. Can J Public Health 64 (3):276–278
Fisher MC, Koenig GL, White TJ, San-Blas G, Negroni R, Alvarez IG, Wanke B, Taylor JW (2001) Biogeographic range expansion into South America by *Coccidioides immitis* mirrors new world patterns of human migration. Proc Natl Acad Sci U S A 98(8):4558–4562. https://doi.org/10.1073/pnas.071406098
Fisher MC, Koenig GL, White TJ, Taylor JW (2002) Molecular and phenotypic description of *Coccidioides posadasii* sp. nov., previously recognized as the non-California population of *Coccidioides immitis*. Mycologia 94(1):73–84
Fisher FS, Bultman MW, Johnson SM, Pappagianis D, Zaborsky E (2007) *Coccidioides* niches and habitat parameters in the southwestern United States: a matter of scale. Ann N Y Acad Sci 1111:47–72. https://doi.org/10.1196/annals.1406.031
Flaherman VJ, Hector R, Rutherford GW (2007) Estimating severe coccidioidomycosis in California. Emerg Infect Dis 13(7):1087–1090. https://doi.org/10.3201/eid1307.061480
Fowler ME, Pappagianis D, Ingram I (1992) Coccidioidomycosis in llamas in the United States: 19 cases (1981–1989). J Am Vet Med Assoc 201(10):1609–1614
Fraser JA, Stajich JE, Tarcha EJ, Cole GT, Inglis DO, Sil A, Heitman J (2007) Evolution of the mating type locus: insights gained from the dimorphic primary fungal pathogens *Histoplasma capsulatum*, *Coccidioides immitis*, and *Coccidioides posadasii*. Eukaryot Cell 6(4):622–629. https://doi.org/10.1128/EC.00018-07
Friedman L, Smith CE (1957) The comparison of four strains of *Coccidioides immitis* with diverse histories. Mycopathol Mycol Appl 8(1):47–53
Friedman L, Papagianis D, Berman RJ, Smith CE (1953) Studies on *Coccidioides immitis*: morphology and sporulation capacity of forty-seven strains. J Lab Clin Med 42(3):438–444
Friedman L, Smith CE, Gordon LE (1955) The assay of virulence of *Coccidioides* in white mice. J Infect Dis 97(3):311–316
Gaidici A, Saubolle MA (2009) Transmission of coccidioidomycosis to a human via a cat bite. J Clin Microbiol 47(2):505–506. https://doi.org/10.1128/JCM.01860-08
Galgiani JN (2007) Coccidioidomycosis: changing perceptions and creating opportunities for its control. Ann N Y Acad Sci 1111:1–18. https://doi.org/10.1196/annals.1406.041
Galgiani JN (2014) How does genetics influence valley fever? Research underway now to answer this question. AZ Medicine Fall. pp 30–33
Galgiani JN, Stevens DA, Graybill JR, Dismukes WE, Cloud GA (1988) Ketoconazole therapy of progressive coccidioidomycosis. Comparison of 400- and 800-mg doses and observations at higher doses. Am J Med 84(3 Pt 2):603–610
Gautam R, Srinath I, Clavijo A, Szonyi B, Bani-Yaghoub M, Park S, Ivanek R (2013) Identifying areas of high risk of human exposure to coccidioidomycosis in Texas using serology data from dogs. Zoonoses Public Health 60(2):174–181. https://doi.org/10.1111/j.1863-2378.2012.01526.x
Giltner LT (1918) Occurrence of coccidioidal granuloma (Oidiomycosis) in cattle. J Agric Res 14:533–542

Ginocchio CC, Lotlikar M, Li X, Elsayed HH, Teng Y, Dougherty P, Kuhles DJ, Chaturvedi S, St George K (2013) Identification of endogenous *Coccidioides posadasii* contamination of commercial primary rhesus monkey kidney cells. J Clin Microbiol 51(4):1288–1290. https://doi.org/10.1128/JCM.00132-13

Graupmann-Kuzma A, Valentine BA, Shubitz LF, Dial SM, Watrous B, Tornquist SJ (2008) Coccidioidomycosis in dogs and cats: a review. J Am Anim Hosp Assoc 44(5):226–235

Grayzel SE, Martinez-Lopez B, Sykes JE (2016) Risk factors and spatial distribution of canine Coccidioidomycosis in California, 2005–2013. Transbound Emerg Dis 64(4):1110–1119. https://doi.org/10.1111/tbed.12475

Greene RT, Troy GC (1995) Coccidioidomycosis in 48 cats: a retrospective study (1984–1993). J Vet Intern Med 9(2):86–91

Greene DR, Koenig G, Fisher MC, Taylor JW (2000) Soil isolation and molecular identification of *Coccidioides immitis*. Mycologia 92(3):406–410. https://doi.org/10.2307/3761498

Hage CA, Knox KS, Wheat LJ (2012) Endemic mycoses: overlooked causes of community acquired pneumonia. Respir Med 106(6):769–776. https://doi.org/10.1016/j.rmed.2012.02.004

Halde C, Mc NE, Newcomer VD, Sternberg TH (1957) Properdin levels in mice and man with coccidioidomycosis during soluble amphotericin B administration. Antibiot Annu 5:598–601

Hector RF, Zimmer BL, Pappagianis D (1990) Evaluation of nikkomycins X and Z in murine models of coccidioidomycosis, histoplasmosis, and blastomycosis. Antimicrob Agents Chemother 34(4):587–593

Hector RF, Rutherford GW, Tsang CA, Erhart LM, McCotter O, Anderson SM, Komatsu K, Tabnak F, Vugia DJ, Yang Y, Galgiani JN (2011) The public health impact of coccidioidomycosis in Arizona and California. Int J Environ Res Public Health 8(4):1150–1173. https://doi.org/10.3390/ijerph8041150

Helmick KE, Koplos P, Raymond J (2006) Disseminated coccidioidomycosis in a captive Indochinese tiger (*Panthera tigris corbetti*) with chronic renal disease. J Zoo Wildl Med 37 (4):542–544

Henrickson RV, Biberstein EL (1972) Coccidioidomycosis accompanying hepatic disease in two Bengal tigers. J Am Vet Med Assoc 161(6):674–677

Herr RA, Tarcha EJ, Taborda PR, Taylor JW, Ajello L, Mendoza L (2001) Phylogenetic analysis of *Lacazia loboi* places this previously uncharacterized pathogen within the dimorphic Onygenales. J Clin Microbiol 39(1):309–314. https://doi.org/10.1128/JCM.39.1.309-314.2001

Herrin KV, Miranda A, Loebenberg D (2005) Posaconazole therapy for systemic coccidioidomycosis in a chimpanzee (*Pan troglodytes*): a case report. Mycoses 48(6):447–452. https://doi.org/10.1111/j.1439-0507.2005.01155.x

Higgins JC, Leith GS, Pappagianis D, Pusterla N (2006) Treatment of *Coccidioides immitis* pneumonia in two horses with fluconazole. Vet Rec 159(11):349–351

Hoffman K, Videan EN, Fritz J, Murphy J (2007) Diagnosis and treatment of ocular coccidioidomycosis in a female captive chimpanzee (*Pan troglodytes*): a case study. Ann N Y Acad Sci 1111:404–410. https://doi.org/10.1196/annals.1406.018

Huang JY, Bristow B, Shafir S, Sorvillo F (2012) Coccidioidomycosis-associated deaths, United States, 1990–2008. Emerg Infect Dis 18(11):1723–1728. https://doi.org/10.3201/eid1811.120752

Huckabone SE, Gulland FM, Johnson SM, Colegrove KM, Dodd EM, Pappagianis D, Dunkin RC, Casper D, Carlson EL, Sykes JE, Meyer W, Miller MA (2015) Coccidioidomycosis and other systemic mycoses of marine mammals stranding along the central California, USA coast: 1998–2012. J Wildl Dis 51(2):295–308. https://doi.org/10.7589/2014-06-143

Hugenholtz P (1957) Climate and Coccidioidomycosis. In: Ajello L (ed) Symposium on Coccidioidomycosis. U.S. Public Health Service, Phoenix, pp 136–143

Hugenholtz PG, Reed RE, Maddy KT, Trautman RJ, Barger JD (1958) Experimental coccidioidomycosis in dogs. Am J Vet Res 19(71):433–437

Hung CY, Seshan KR, Yu JJ, Schaller R, Xue J, Basrur V, Gardner MJ, Cole GT (2005) A metalloproteinase of *Coccidioides posadasii* contributes to evasion of host detection. Infect Immun 73(10):6689–6703. https://doi.org/10.1128/IAI.73.10.6689-6703.2005

Hung CY, Hurtgen BJ, Bellecourt M, Sanderson SD, Morgan EL, Cole GT (2012) An agonist of human complement fragment C5a enhances vaccine immunity against *Coccidioides* infection. Vaccine 30(31):4681–4690. https://doi.org/10.1016/j.vaccine.2012.04.084

Huppert M, Levine HB, Sun SH, Peterson ET (1967) Resistance of vaccinated mice to typical and atypical strains of *Coccidioides immitis*. J Bacteriol 94(4):924–927

Huppert M, Sun SH, Harrison JL (1982) Morphogenesis throughout saprobic and parasitic cycles of *Coccidioides immitis*. Mycopathologia 78(2):107–122

Hutchinson LR, Duran F, Lane CD, Robertstad GW, Portillo M (1973) Coccidioidomycosis in a Giant red kangaroo (*Macropus rufus*). J Zoo Anim Med 4(1):22–24. https://doi.org/10.2307/20094175

Jessup DA, Kock N, Berbach M (1989) Coccidioidomycosis in a desert bighorn sheep (*Ovis canadensis nelsoni*) from California. J Zoo Wildl Med 20(4):471–473

Jewell K, Cheshier R, Cage GD (2008) Genetic diversity among clinical *Coccidioides spp.* isolates in Arizona. Med Mycol 46(5):449–455. https://doi.org/10.1080/13693780801961337

Johnson JH, Wolf AM, Edwards JF, Walker MA, Homco L, Jensen JM, Simpson BR, Taliaferro L (1998) Disseminated coccidioidomycosis in a mandrill baboon (*Mandrillus sphinx*): a case report. J Zoo Wildl Med 29(2):208–213

Johnson SM, Lerche NW, Pappagianis D, Yee JL, Galgiani JN, Hector RF (2007) Safety, antigenicity, and efficacy of a recombinant coccidioidomycosis vaccine in cynomolgus macaques (Macaca fascicularis). Ann N Y Acad Sci 1111:290–300. https://doi.org/10.1196/annals.1406.042

Johnson SM, Carlson EL, Fisher FS, Pappagianis D (2014) Demonstration of *Coccidioides immitis* and *Coccidioides posadasii* DNA in soil samples collected from dinosaur National Monument, Utah. Med Mycol 52(6):610–617. https://doi.org/10.1093/mmy/myu004

Kahn A, Carey EJ, Blair JE (2015) Universal fungal prophylaxis and risk of coccidioidomycosis in liver transplant recipients living in an endemic area. Liver Transpl 21(3):353–361. https://doi.org/10.1002/lt.24055

Kauffman CA, Freifeld AG, Andes DR, Baddley JW, Herwaldt L, Walker RC, Alexander BD, Anaissie EJ, Benedict K, Ito JI, Knapp KM, Lyon GM, Marr KA, Morrison VA, Park BJ, Patterson TF, Schuster MG, Chiller TM, Pappas PG (2014) Endemic fungal infections in solid organ and hematopoietic cell transplant recipients enrolled in the Transplant-Associated Infection Surveillance Network (TRANSNET). Transpl Infect Dis 16(2):213–224. https://doi.org/10.1111/tid.12186

Kirkland TN, Fierer J (1983) Inbred mouse strains differ in resistance to lethal *Coccidioides immitis* infection. Infect Immun 40(3):912–916

Kolivras KN, Comrie AC (2003) Modeling valley fever (coccidioidomycosis) incidence on the basis of climate conditions. Int J Biometeorol 47(2):87–101. https://doi.org/10.1007/s00484-002-0155-x

Krutzsch PH, Watson RH (1978) Isolation of *Coccidioides immitis* from bat guano and preliminary findings on laboratory infectivity of bats with *Coccidioides immitis*. Life Sci 22(8):679–684

Lacy GH, Swatek FE (1974) Soil ecology of *Coccidioides immitis* at Amerindian middens in California. Appl Microbiol 27(2):379–388

Lange L, Huang Y, Busk PK (2016) Microbial decomposition of keratin in nature-a new hypothesis of industrial relevance. Appl Microbiol Biotechnol 100(5):2083–2096. https://doi.org/10.1007/s00253-015-7262-1

Lauer A, Baal JD, Baal JC, Verma M, Chen JM (2012) Detection of *Coccidioides immitis* in Kern County, California, by multiplex PCR. Mycologia 104(1):62–69. https://doi.org/10.3852/11-127

Lawrence RM, Hoeprich PD (1976) Comparison of amphotericin B and amphotericin B methyl ester: efficacy in murine coccidioidomycosis and toxicity. J Infect Dis 133(2):168–174

Levine HB, Kong YM (1965) Immunity development in mice receiving killed *Coccidioides immitis* spherules: effect of removing residual vaccine. Sabouraudia 4(3):164–170

Levine HB, Kong YC (1966) Immunologic impairment in mice treated intravenously with killed *Coccidioides immitis* spherules: suppressed response to intramuscular doses. J Immunol 97 (3):297–305

Levine HB, Winn WA (1964) Isolation of *Coccidioides immitis* from soil. Health Lab Sci 1:29–32

Levine HB, Kong YC, Smith C (1965) Immunization of mice to *Coccidioides immitis*: dose, regimen and spherulation stage of killed spherule vaccines. J Immunol 94:132–142

Lewis ER, Bowers JR, Barker BM (2015) Dust devil: the life and times of the fungus that causes valley fever. PLoS Pathog 11(5):e1004762. https://doi.org/10.1371/journal.ppat.1004762

Li J, Yu L, Tian Y, Zhang KQ (2012) Molecular evolution of the deuterolysin (M35) family genes in *Coccidioides*. PLoS One 7(2):e31536. https://doi.org/10.1371/journal.pone.0031536

Litvintseva AP, Marsden-Haug N, Hurst S, Hill H, Gade L, Driebe EM, Ralston C, Roe C, Barker BM, Goldoft M, Keim P, Wohrle R, Thompson GR, Engelthaler DM, Brandt ME, Chiller T (2015) Valley fever: finding new places for an old disease: *Coccidioides immitis* found in Washington state soil associated with recent human infection. Clin Infect Dis 60(1):E1–E3. https://doi.org/10.1093/cid/ciu681

Lochmiller RL, Hellgren EC, Hannon PG, Grant WE, Robinson RM (1985) Coccidioidomycosis (*Coccidioides immitis*) in the collared peccary (*Tayassu tajacu*: Tayassuidae) in Texas. J Wildl Dis 21(3):305–309. https://doi.org/10.7589/0090-3558-21.3.305

Longley BJ, Mendenhall JT (1960) The successful treatment of pulmonary coccidioidomycosis with intravenous amphotericin B. Am Rev Respir Dis 81:574–578

Lopes BG, Santos AL, de Bezerra CC, Wanke B, Dos Santos Lazera M, Nishikawa MM, Mazotto AM, Kussumi VM, Haido RM, Vermelho AB (2008) A 25-kDa serine peptidase with keratinolytic activity secreted by *Coccidioides immitis*. Mycopathologia 166(1):35–40. https://doi.org/10.1007/s11046-008-9116-1

Loughry WJ, Truman RW, McDonough CM, Tilak MK, Garnier S, Delsuc F (2009) Is leprosy spreading among nine-banded armadillos in the southeastern United States? J Wildl Dis 45 (1):144–152. https://doi.org/10.7589/0090-3558-45.1.144

Lubarsky R, Plunkett OA (1954) Survival of *C. immitis* in passage through the digestive tract of mice. Public Health Rep 69(5):494–497

Maddy KT (1954a) Coccidioidomycosis in a sheep. J Am Vet Med Assoc 124(927):465

Maddy KT (1954b) Coccidioidomycosis of cattle in the southwestern United States. J Am Vet Med Assoc 124(927):456–464

Maddy KT (1957) Ecological factors of the geographic distribution of *Coccidioides immitis*. J Am Vet Med Assoc 130(11):475–476

Maddy KT (1958a) Disseminated coccidioidomycosis of the dog. J Am Vet Med Assoc 132 (11):483–489

Maddy KT (1958b) The geographic distribution of *Coccidioides immitis* and possible ecologic implications. Ariz Med 15(3):178–188

Maddy KT (1959) A study of a site in Arizona where a dog apparently acquired a *Coccidioides immitis* infection. Am J Vet Res 20(77):642–646

Maddy KT (1965) Observations on *Coccidioides immitis* found growing naturally in soil. Ariz Med 22:281–288

Maddy KT, Crecelius HG (1965) Establishment of *Coccidioides immitis* in negative soil following burial of infected animal tissues. In: Ajello L (ed) The second symposium on Coccidioidomycosis. The University of Arizona Press, Phoenix, pp 309–312

Maddy KT, Crecelius HG, Cornell RG (1960a) Distribution of *Coccidioides immitis* determined by testing in cattle. Public Health Rep 75(10):955–962. https://doi.org/10.2307/4590968

Maddy KT, Reed RE, Trautman RJ, Snell VN (1960b) Experimental bovine coccidioidomycosis. Am J Vet Res 21:748–752

Maleski K, Magdesian KG, LaFranco-Scheuch L, Pappagianis D, Carlson GP (2002) Pulmonary coccidioidomycosis in a neonatal foal. Vet Rec 151(17):505–508

Mandel MA, Barker BM, Kroken S, Rounsley SD, Orbach MJ (2007) Genomic and population analyses of the mating type loci in *Coccidioides* species reveal evidence for sexual reproduction and gene acquisition. Eukaryot Cell 6(7):1189–1199. https://doi.org/10.1128/EC.00117-07

Marsden-Haug N, Hill H, Litvintseva AP, Engelthaler DM, Driebe EM, Roe CC, Ralston C, Hurst S, Goldoft M, Gade L, Wohrle R, Thompson GR, Brandt ME, Chiller T (2014) *Coccidioides immitis* identified in soil outside of its known range—Washington, 2013. MMWR Morb Mortal Wkly Rep 63(20):450

Mayorga RP, Espinoza H (1970) Coccidioidomycosis in Mexico and central America. Mycopathol Mycol Appl 41(1):13–23

McCue PM, O'Farrell TP (1988) Serological survey for selected diseases in the endangered San Joaquin kit fox (*Vulpes macrotis mutica*). J Wildl Dis 24(2):274–281. https://doi.org/10.7589/0090-3558-24.2.274

Mendoza N, Noel P, Blair JE (2015) Diagnosis, treatment, and outcomes of coccidioidomycosis in allogeneic stem cell transplantation. Transpl Infect Dis 17(3):380–388. https://doi.org/10.1111/tid.12372

Morris M (1924) Coccidioides of the central nervous system. Cal West Med 22(10):483–485

Muir S, Pappagianis D (1982) Coccidioidomycosis in the llama: case report and epidemiologic survey. J Am Vet Med Assoc 181(11):1334–1337

Muller LA, Lucas JE, Georgianna DR, McCusker JH (2011) Genome-wide association analysis of clinical vs. nonclinical origin provides insights into *Saccharomyces cerevisiae* pathogenesis. Mol Ecol 20(19):4085–4097. https://doi.org/10.1111/j.1365-294X.2011.05225.x

Neafsey DE, Barker BM, Sharpton TJ, Stajich JE, Park DJ, Whiston E, Hung CY, McMahan C, White J, Sykes S, Heiman D, Young S, Zeng Q, Abouelleil A, Aftuck L, Bessette D, Brown A, FitzGerald M, Lui A, Macdonald JP, Priest M, Orbach MJ, Galgiani JN, Kirkland TN, Cole GT, Birren BW, Henn MR, Taylor JW, Rounsley SD (2010) Population genomic sequencing of *Coccidioides* fungi reveals recent hybridization and transposon control. Genome Res 20 (7):938–946. https://doi.org/10.1101/gr.103911.109

Nguyen C, Barker BM, Hoover S, Nix DE, Ampel NM, Frelinger JA, Orbach MJ, Galgiani JN (2013) Recent advances in our understanding of the environmental, epidemiological, immunological, and clinical dimensions of coccidioidomycosis. Clin Microbiol Rev 26(3):505–525. https://doi.org/10.1128/CMR.00005-13

Northey WT, Brooks LD (1962) Studies on *Coccidioides immitis*. I. A simple medium for in vitro spherulation. J Bacteriol 84:742–746

Odio CD, Milligan KL, McGowan K, Rudman Spergel AK, Bishop R, Boris L, Urban A, Welch P, Heller T, Kleiner D, Jackson MA, Holland SM, Freeman AF (2015) Endemic mycoses in patients with STAT3-mutated hyper-IgE (job) syndrome. J Allergy Clin Immunol 136 (5):1411–1413. https://doi.org/10.1016/j.jaci.2015.07.003

Ophuls W (1905) Further observations on a pathogenic mould formerly described as a protozoon (*Coccidioides Immitis*, *Coccidioides Pyogenes*). J Exp Med 6(4–6):443–485

Orr GF (1968) The use of bait in isolating *Coccidioides immitis* from soil: a preliminary study. Mycopathol Mycol Appl 36(1):28–32

Palmer CE, Edwards PQ, Allfather WE (1957) Characteristics of skin reactions to coccidioidin and histoplasmin, with evidence of an unidentified source of sensitization. Am J Hyg 66(2):196–213

Pan S, Sigler L, Cole GT (1994) Evidence for a phylogenetic connection between *Coccidioides immitis* and *Uncinocarpus reesii* (Onygenaceae). Microbiology 140(Pt 6):1481–1494

Pappagianis D (1988) Epidemiology of coccidioidomycosis. Curr Top Med Mycol 2:199–238

Pappagianis D (1993) Evaluation of the protective efficacy of the killed *Coccidioides immitis* spherule vaccine in humans. The valley fever vaccine study group. Am Rev Respir Dis 148 (3):656–660. https://doi.org/10.1164/ajrccm/148.3.656

Pappagianis D, Smith CE, Kobayashi GS (1956) Relationship of the *in vivo* form of *Coccidioides immitis* to virulence. J Infect Dis 98(3):312–319

Pappagianis D, Lindsay S, Beall S, Williams P (1979) Ethnic background and the clinical course of coccidioidomycosis. Am Rev Respir Dis 120(4):959–961

Petersen LR, Marshall SL, Barton-Dickson C, Hajjeh RA, Lindsley MD, Warnock DW, Panackal AA, Shaffer JB, Haddad MB, Fisher FS, Dennis DT, Morgan J (2004) Coccidioidomycosis among workers at an archeological site, northeastern Utah. Emerg Infect Dis 10(4):637–642. https://doi.org/10.3201/eid1004.030446

Plunkett OA, Walker L, Huppert M (1963) An unusual isolate of *Coccidioides immitis* from the Los Banos area of California. Sabouraudia 3(1):16–20

Posadas A (1892) Un nuevo caso de micosis fungoidea con posrospemias. An Cir Med Argentina 15:585–597

Prchal CJ, Crecelius HG (1966) Coccidioidomycosis in swine. J Am Vet Med Assoc 148 (10):1168–1169

Rapley WA, Long JR (1974) Coccidioidomycosis in a baboon recently imported from California. Can Vet J 15(2):39–41

Reed RE (1960) Coccidioidomycosis in animals. Ariz Med 17:26–27

Reed RE, Hoge RS, Trautman RJ (1963) Coccidioidomycosis in two cats. J Am Vet Med Assoc 143:953–956

Reed RE, Migaki G, Cummings JA (1976) Coccidioidomycosis in a California sea lion (*Zalophus californianus*). J Wildl Dis 12(3):372–375. https://doi.org/10.7589/0090-3558-12.3.372

Reed RE, Ingram KA, Reggiardo C, Shupe MR (1996) Coccidioidomycosis in domestic and wild animals. In: 5th international conference on Coccidioidomycosis. National Foundation for Infectious Disease, Bethesda, pp 146–154

Reidarson TH, Griner LA, Pappagianis D, McBain J (1998) Coccidioidomycosis in a bottlenose dolphin. J Wildl Dis 34(3):629–631

Rixford E, Gilchrist TC (1896) Two cases of protozoan (coccidioidal) infection of the skin and other organs. Johns Hopkins Hosp Rep 10:209–268

Rosenberg DP, Gleiser CA, Carey KD (1984) Spinal coccidioidomycosis in a baboon. J Am Vet Med Assoc 185(11):1379–1381

Ruddy BE, Mayer AP, Ko MG, Labonte HR, Borovansky JA, Boroff ES, Blair JE (2011) Coccidioidomycosis in African Americans. Mayo Clin Proc 86(1):63–69. https://doi.org/10.4065/mcp.2010.0423

Ryfkogel HA (1908) Fungus *Coccidioides*. Cal State J Med 6(6):200–202

Saarinen JJ, Boyer AG, Brown JH, Costa DP, Ernest SKM, Evans AR, Fortelius M, Gittleman JL, Hamilton MJ, Harding LE, Lintulaakso K, Lyons SK, Okie JG, Sibly RM, Stephens PR, Theodor J, Uhen MD, Smith FA (2014) Patterns of maximum body size evolution in Cenozoic land mammals: eco-evolutionary processes and abiotic forcing. Proc R Soc Lond B 281 (1784):20132049

Sampaio EP, Hsu AP, Pechacek J, Bax HI, Dias DL, Paulson ML, Chandrasekaran P, Rosen LB, Carvalho DS, Ding L, Vinh DC, Browne SK, Datta S, Milner JD, Kuhns DB, Long Priel DA, Sadat MA, Shiloh M, De Marco B, Alvares M, Gillman JW, Ramarathnam V, de la Morena M, Bezrodnik L, Moreira I, Uzel G, Johnson D, Spalding C, Zerbe CS, Wiley H, Greenberg DE, Hoover SE, Rosenzweig SD, Galgiani JN, Holland SM (2013) Signal transducer and activator of transcription 1 (STAT1) gain-of-function mutations and disseminated coccidioidomycosis and histoplasmosis. J Allergy Clin Immunol 131(6):1624–1634. https://doi.org/10.1016/j.jaci.2013.01.052

Sharpton TJ, Neafsey DE, Galagan JE, Taylor JW (2008) Mechanisms of intron gain and loss in *Cryptococcus*. Genome Biol 9(1):R24. https://doi.org/10.1186/gb-2008-9-1-r24

Sharpton TJ, Stajich JE, Rounsley SD, Gardner MJ, Wortman JR, Jordar VS, Maiti R, Kodira CD, Neafsey DE, Zeng Q, Hung CY, McMahan C, Muszewska A, Grynberg M, Mandel MA, Kellner EM, Barker BM, Galgiani JN, Orbach MJ, Kirkland TN, Cole GT, Henn MR, Birren BW, Taylor JW (2009) Comparative genomic analyses of the human fungal pathogens *Coccidioides* and their relatives. Genome Res 19(10):1722–1731. https://doi.org/10.1101/gr.087551.108

Shubitz LF (2007) Comparative aspects of coccidioidomycosis in animals and humans. Ann N Y Acad Sci 1111:395–403. https://doi.org/10.1196/annals.1406.007

Shubitz LF, Dial SM (2005) Coccidioidomycosis: a diagnostic challenge. Clin Tech Small Anim Pract 20(4):220–226. https://doi.org/10.1053/j.ctsap.2005.07.002
Shubitz LE, Butkiewicz CD, Dial SM, Lindan CP (2005) Incidence of *Coccidioides* infection among dogs residing in a region in which the organism is endemic. J Am Vet Med Assoc 226 (11):1846–1850
Shubitz LF, Roy ME, Nix DE, Galgiani JN (2013) Efficacy of Nikkomycin Z for respiratory coccidioidomycosis in naturally infected dogs. Med Mycol 51(7):747–754. https://doi.org/10.3109/13693786.2013.770610
Shubitz LF, Trinh HT, Galgiani JN, Lewis ML, Fothergill AW, Wiederhold NP, Barker BM, Lewis ER, Doyle AL, Hoekstra WJ, Schotzinger RJ, Garvey EP (2015) Evaluation of VT-1161 for treatment of Coccidioidomycosis in murine infection models. Antimicrob Agents Chemother 59 (12):7249–7254. https://doi.org/10.1128/AAC.00593-15
Sievers ML (1974) Disseminated coccidioidomycosis among southwestern American Indians. Am Rev Respir Dis 109(6):602–612
Sievers ML, Fisher JR (1982) Decreasing incidence of disseminated coccidioidomycosis among Piman and San Carlos apache Indians. A probable environmental basis. Chest 82(4):455–460
Sigler L, Flis AL, Carmichael JW (1998) The genus *Uncinocarpus* (Onygenaceae) and its synonym Brunneospora: new concepts, combinations and connections to anamorphs in *Chrysosporium*, and further evidence of relationship with *Coccidioides immitis*. Can J Bot 76(9):1624–1636. https://doi.org/10.1139/b98-110
Sigler L, Hambleton S, Pare JA (2013) Molecular characterization of reptile pathogens currently known as members of the *Chrysosporium* anamorph of *Nannizziopsis vriesii* complex and relationship with some human-associated isolates. J Clin Microbiol 51(10):3338–3357. https://doi.org/10.1128/JCM.01465-13
Smith JW (1971) Coccidioidomycosis. Tex Med 67(11):117–123
Smith DT, Harrell ER Jr (1948) Fatal coccidioidomycosis; a case of a laboratory infection. Am Rev Tuberc 57(4):368–374
Smith CE, Saito MT, Simons SA (1956) Pattern of 39,500 serologic tests in coccidioidomycosis. J Am Med Assoc 160(7):546–552
Sondermeyer G, Lee L, Gilliss D, Tabnak F, Vugia D (2013) Coccidioidomycosis-associated hospitalizations, California, USA, 2000–2011. Emerg Infect Dis 19(10):1590–1597. https://doi.org/10.3201/eid1910.130427
Sorensen KN, Clemons KV, Stevens DA (1999) Murine models of blastomycosis, coccidioidomycosis, and histoplasmosis. Mycopathologia 146(2):53–65
Stagliano D, Epstein J, Hickey P (2007) Fomite-transmitted coccidioidomycosis in an immunocompromised child. Pediatr Infect Dis J 26(5):454–456. https://doi.org/10.1097/01.inf.0000259231.95285.bc
Stevens DA (1977) Miconazole in coccidioidomycosis. West J Med 126(4):315–316
Stevens DA, Clemons KV (2007) Azole therapy of clinical and experimental coccidioidomycosis. Ann N Y Acad Sci 1111:442–454. https://doi.org/10.1196/annals.1406.039
Stevens DA, Clemons KV, Levine HB, Pappagianis D, Baron EJ, Hamilton JR, Deresinski SC, Johnson N (2009) Expert opinion: what to do when there is *Coccidioides* exposure in a laboratory. Clin Infect Dis 49(6):919–923. https://doi.org/10.1086/605441
Stiles GW, Davis CL (1942) Coccidioidal granuloma (Coccidioidomycosis) its incidence in man and animals and its diagnosis in animals. J Am Med Assoc 119(10):765–769. https://doi.org/10.1001/jama.1942.02830270001001
Straub M, Trautman RJ, Greene JW (1961) Coccidioidomycosis in 3 coyotes. Am J Vet Res 22:811–813
Sunenshine RH, Anderson S, Erhart L, Vossbrink A, Kelly PC, Engelthaler D, Komatsu K (2007) Public health surveillance for coccidioidomycosis in Arizona. Ann N Y Acad Sci 1111:96–102. https://doi.org/10.1196/annals.1406.045
Swatek FE (1970) Ecology of *Coccidioides immitis*. Mycopathol Mycol Appl 41(1):3–12

Swatek FE, Omieczynski DT (1970) Isolation and identification of *Coccidioides immitis* from natural sources. Mycopathol Mycol Appl 41(1):155–166
Swatek FE, Plunkett OA (1957) Experimental infections of wild rodents and animals other than mammals. In: Ajello L (ed) Proceedings of the first symposium on Coccidioidomycosis. University of Arizona Press, Phoenix, pp 161–167
Swatek FE, Omieczynski DT, Plunkett OA (1967) *Coccidioides immitis* in California. In: Ajello L (ed) The second symposium on Coccidioidomycosis. University of Arizona Press, Tucson, pp 255–264
Talamantes J, Behseta S, Zender CS (2007) Fluctuations in climate and incidence of coccidioidomycosis in Kern County, California: a review. Ann N Y Acad Sci 1111:73–82. https://doi.org/10.1196/annals.1406.028
Tamerius JD, Comrie AC (2011) Coccidioidomycosis incidence in Arizona predicted by seasonal precipitation. PLoS One 6(6):e21009. https://doi.org/10.1371/journal.pone.0021009
Tapaltsyan V, Eronen Jussi T, Lawing AM, Sharir A, Janis C, Jernvall J, Klein Ophir D (2015) Continuously growing rodent molars result from a predictable quantitative evolutionary change over 50 million years. Cell Rep 11(5):673–680. https://doi.org/10.1016/j.celrep.2015.03.064
Teel KW, Yow MD, Williams TW Jr (1970) A localized outbreak of coccidioidomycosis in southern Texas. J Pediatr 77(1):65–73
Teixeira MM, Barker BM (2016) Use of population genetics to assess the ecology, evolution, and population structure of *Coccidioides*. Emerg Infect Dis 22(6):1022–1030. https://doi.org/10.3201/eid2206.151565
Terio KA, Stalis IH, Allen JL, Stott JL, Worley MB (2003) Coccidioidomycosis in Przewalski's horses (*Equus przewalskii*). J Zoo Wildl Med 34(4):339–345
Thompson GR 3rd, Stevens DA, Clemons KV, Fierer J, Johnson RH, Sykes J, Rutherford G, Peterson M, Taylor JW, Chaturvedi V (2015) Call for a California coccidioidomycosis consortium to face the top ten challenges posed by a recalcitrant regional disease. Mycopathologia 179 (1–2):1–9. https://doi.org/10.1007/s11046-014-9816-7
Timm KI, Sonn RJ, Hultgren BD (1988) Coccidioidomycosis in a Sonoran gopher snake, *Pituophis melanoleucus affinis*. J Med Vet Mycol 26(2):101–104
Turabelidze G, Aggu-Sher RK, Jahanpour E, Hinkle CJ, Centers for Disease Control and Prevention (2015) Coccidioidomycosis in a state where it is not known to be endemic-Missouri, 2004–2013. MMWR Morb Mortal Wkly Rep 64(23):636–639
Twarog M, Thompson GR 3rd (2015) Coccidioidomycosis: recent updates. Semin Respir Crit Care Med 36(5):746–755. https://doi.org/10.1055/s-0035-1562900
Untereiner WA, Scott JA, Naveau FA, Sigler L, Bachewich J, Angus A (2004) The Ajellomycetaceae, a new family of vertebrate-associated Onygenales. Mycologia 96 (4):812–821
Valdivia L, Nix D, Wright M, Lindberg E, Fagan T, Lieberman D, Stoffer T, Ampel NM, Galgiani JN (2006) Coccidioidomycosis as a common cause of community-acquired pneumonia. Emerg Infect Dis 12(6):958–962
Vargas-Gastelum L, Romero-Olivares AL, Escalante AE, Rocha-Olivares A, Brizuela C, Riquelme M (2015) Impact of seasonal changes on fungal diversity of a semi-arid ecosystem revealed by 454 pyrosequencing. FEMS Microbiol Ecol 91(5):fiv044. https://doi.org/10.1093/femsec/fiv044
Vinh DC, Masannat F, Dzioba RB, Galgiani JN, Holland SM (2009) Refractory disseminated coccidioidomycosis and mycobacteriosis in interferon-gamma receptor 1 deficiency. Clin Infect Dis 49(6):e62–e65. https://doi.org/10.1086/605532
Vinh DC, Schwartz B, Hsu AP, Miranda DJ, Valdez PA, Fink D, Lau KP, Long-Priel D, Kuhns DB, Uzel G, Pittaluga S, Hoover S, Galgiani JN, Holland SM (2011) Interleukin-12 receptor beta1 deficiency predisposing to disseminated Coccidioidomycosis. Clin Infect Dis 52(4):e99–e102. https://doi.org/10.1093/cid/ciq215

Wack EE, Ampel NM, Sunenshine RH, Galgiani JN (2015) The return of delayed-type hypersensitivity skin testing for Coccidioidomycosis. Clin Infect Dis 61(5):787–791. https://doi.org/10.1093/cid/civ388

Wallace RS, Clyde VL, Steinberg H (2009) Coccidioidomycosis in a black rhinoceros (*Diceros bicornis*). J Zoo Wildl Med 40(2):365–368

Wang H, Xu Z, Gao L, Hao B (2009) A fungal phylogeny based on 82 complete genomes using the composition vector method. BMC Evol Biol 9:195. https://doi.org/10.1186/1471-2148-9-195

Wanke B, Lazera M, Monteiro PC, Lima FC, Leal MJ, Ferreira Filho PL, Kaufman L, Pinner RW, Ajello L (1999) Investigation of an outbreak of endemic coccidioidomycosis in Brazil's northeastern state of Piaui with a review of the occurrence and distribution of *Coccidioides immitis* in three other Brazilian states. Mycopathologia 148(2):57–67

Wheeler C, Lucas KD, Mohle-Boetani JC (2015) Rates and risk factors for Coccidioidomycosis among prison inmates, California, USA, 2011. Emerg Infect Dis 21(1):70–75. https://doi.org/10.3201/eid2101.140836

Whiston E, Taylor JW (2014) Genomics in *Coccidioides*: insights into evolution, ecology, and pathogenesis. Med Mycol 52(2):149–155. https://doi.org/10.1093/mmy/myt001

Whiston E, Taylor JW (2015) Comparative phylogenomics of pathogenic and nonpathogenic species. G3 (Bethesda) 6(2):235–244. https://doi.org/10.1534/g3.115.022806

Xue J, Chen X, Selby D, Hung CY, Yu JJ, Cole GT (2009) A genetically engineered live attenuated vaccine of Coccidioides posadasii protects BALB/c mice against coccidioidomycosis. Infect Immun 77(8):3196–3208. https://doi.org/10.1128/IAI.00459-09

Yau AA (2016) Risk factors and epidemiology of Coccidioidomycosis demonstrated by a case of spontaneous pulmonary rupture of cavitary Coccidioidomycosis. Case Rep Infect Dis 2016:8165414. https://doi.org/10.1155/2016/8165414

Yoon HJ, Clemons KV (2013) Vaccines against *Coccidioides*. Korean J Intern Med 28(4):403–407. https://doi.org/10.3904/kjim.2013.28.4.403

Ziemer EL, Pappagianis D, Madigan JE, Mansmann RA, Hoffman KD (1992) Coccidioidomycosis in horses: 15 cases (1975–1984). J Am Vet Med Assoc 201(6):910–916

Histoplasmosis in Animals

5

Jacques Guillot, Christine Guérin, and René Chermette

Abstract

Histoplasma capsulatum is a dimorphic fungus that is widely distributed in the tropical or subtropical areas of the world and infects numerous mammalian hosts. The outcome of the disease depends on many factors including the immune status of the host, the inoculum size and the virulence of the isolate. The single species *H. capsulatum* is supposed to include three distinct subspecies which do not share exactly the geographical distribution and which are responsible for variable clinical signs in different animal species. *Histoplasma capsulatum* var. *capsulatum* may be found in many regions all over the world; it is responsible for pulmonary and systemic infections with small-sized yeast-form cells in humans and many animal species, including companion animals. *Histoplasma capsulatum* var. *duboisii* is reported in Western and Central Africa and develops as large-sized yeasts with lymphadenopathy and dissemination to the skin and bones in primates. *Histoplasma capsulatum* var. *farciminosum* develops in the skin and the subcutaneous lymphatic system. In horses, the disease is called epizootic lymphangitis. It has been eradicated from large areas of the world but

J. Guillot (✉)
Department of Parasitology, Mycology and Dermatology, EnvA, Ecole nationale vétérinaire d'Alfort, UPEC, Maisons Alfort, France

Dynamyc Research Group, EnvA, Ecole nationale vétérinaire d'Alfort, UPEC, Maisons-Alfort, France
e-mail: jacques.guillot@vet-alfort.fr

C. Guérin
Galapagos company, Romainville, France
e-mail: guerinc.paris@free.fr

R. Chermette
Department of Parasitology, Mycology and Dermatology, Ecole nationale vétérinaire d'Alfort, Dynamyc Research Group, EnvA, UPEC, Maisons-Alfort, France
e-mail: rene.chermette@vet-alfort.fr

S. Seyedmousavi et al. (eds.), *Emerging and Epizootic Fungal Infections in Animals*,
https://doi.org/10.1007/978-3-319-72093-7_5

is still a major cause of morbidity and mortality in various countries particularly in Africa. The condition has a serious effect on the health and welfare of severely affected animals.

5.1 Causative Agents

Like other dimorphic fungi, *Histoplasma* organisms are characterised by their temperature-dependent transition from a saprophytic mould phase to a parasitic yeast form in host tissues. The ability to convert from the mould to the yeast form is a prerequisite for virulence. In the tissues of mammalian hosts, *Histoplasma* yeasts are oval and small in size (2 × 4 μm) with a thin wall and a narrow-based budding process. They are extracellular or may be found inside host macrophages (Fig. 5.1). In culture or in environmental conditions, *Histoplasma* organisms form branched septate filaments (1–2 μm in diameter). From these filaments, two types of conidia can be produced: round- or pear-shaped micro-aleurioconidia (2 × 4 μm) and tuberculate and refringent wall macro-aleurioconidia (6 × 15 μm) (de Hoog et al. 2009).

Histoplasma organisms can be found in soils in temperate and subtropical areas, but some regions are recognised as areas of hyperendemicity. This is the case for midwestern and southern USA and regions along the rivers (Missouri, Ohio or Mississippi) (Kauffman 2009). Traditionally, the single species *H. capsulatum* comprises three distinct subspecies with variable geographical distribution, host preferences and associated clinical signs (de Hoog et al. 2009). *Histoplasma capsulatum* var. *capsulatum* may be found in many different regions all over the world. The subspecies *capsulatum* is responsible for pulmonary and systemic infections with small-sized yeastlike cells in the macrophages of many mammals, including humans. *Histoplasma capsulatum* var. *duboisii* is present in Western and Central Africa. It develops as large (6 × 15 μm) yeasts with a thick wall in tissues of primates. The subspecies *duboisii* is responsible for lymphadenopathy with a possible dissemination to the skin and bones. *Histoplasma capsulatum* var. *farciminosum*

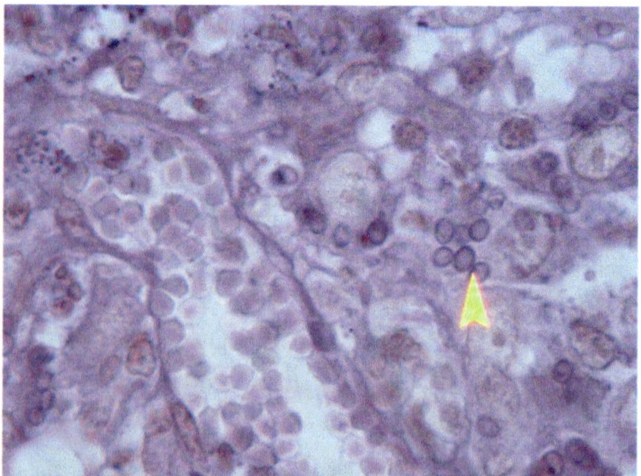

Fig. 5.1 Photomicrograph of subcutaneous tissue in a cat. Large numbers of *Histoplasma capsulatum* organisms filling the cytoplasm of macrophages are visible (arrow). Periodic acid stain (courtesy from Georges Plassiart, Metz, France)

can be found in many countries from Africa, Asia and South America. This subspecies infects the skin and the subcutaneous lymphatic system of horses, donkeys and mules. It has also been recovered from humans, dogs, cats and badgers. Using multilocus sequence typing (MLST), Kasuga et al. (2003) divided the species *H. capsulatum* into the following eight geographically separate groups: North American-1, North American-2, Latin American (=A), Latin American (=B), Australian, Netherlands (of Indonesian origin), Eurasian and African. The subspecies *farciminosum* was composed of three phylogenetic groups, supporting the hypothesis that this taxon is a collection of strains from different clades rather than a true phylogenetic species. Many strains from Europe and Asia were supposed to represent a single clone because they shared the same alleles at all four investigated loci. As no resolution of the branching order of the clades could be obtained, Kasuga et al. (2003) suggested that *H. capsulatum* radiated rapidly over a short period, most probably 3–13 million years ago.

Isolates of *H. capsulatum* were classified in two chemotypes according to the composition of the cell wall (Reiss et al. 1977). Chemotype I includes isolates without polysaccharide α-(1,3)-glucan in their cell wall. All isolates from this chemotype belong to the North American-2 clade defined by Kasuga et al. (2003). Isolates from other *H. capsulatum* clades are corresponding to chemotype II. Their cell wall contains α-(1,3)-glucan, which might be required for virulence and immune evasion. However isolates from chemotype I are recovered from lesions despite the absence of α-(1,3)-glucan (Rappleye et al. 2007).

5.2 Epidemiology of Histoplasmosis

The main historical facts and pioneering events that contributed to our knowledge on epidemiology of human and animal histoplasmosis are summarised in Fig. 5.2. A large diversity of domestic and wild mammals can be infected by *H. capsulatum* (Chermette and Guillot 2010): non-human primates, equids (including horses, donkeys and mules), cattle, dromedaries, rabbits (Brandoa et al. 2014), hedgehogs (Snider et al. 2008), pigs, dogs, cats, other carnivores such as grey (*Urocyon cinereoargenteus*) and red foxes (*Vulpes vulpes*), brown bears (*Ursus arctos*), raccoons (*Procyon lotor*), striped (*Mephitis mephitis*) and spotted (*Spilogale putorius*) skunks, European badgers (*Meles meles*) and sea otters (*Enhydra lutris*) (Morita et al. 2001). Spontaneous or experimentally induced histoplasmosis has also been reported in various species of rodents among which common (*Mus musculus*) and white-footed mice (*Peromyscus leucopus*); black (*Rattus rattus*), grey (*Rattus norvegicus*) and spiny rats (*Proechimys semispinosus*); and opossums of the genera *Didelphis* and *Philander*. *Histoplasma* organisms have also been isolated from many different species of bats whose guano is considered the best substrate for the proliferation and survival of the filamentous form of the subspecies *capsulatum*.

Inhalation is supposed to be the primary mode of entry of *H. capsulatum* var. *capsulatum*. Animals that are immunocompromised or were subjected to a large dose of infectious fungal elements are at a greater risk to develop infection with dissemination. Infection is caused by the inhalation of propagules from the saprobic

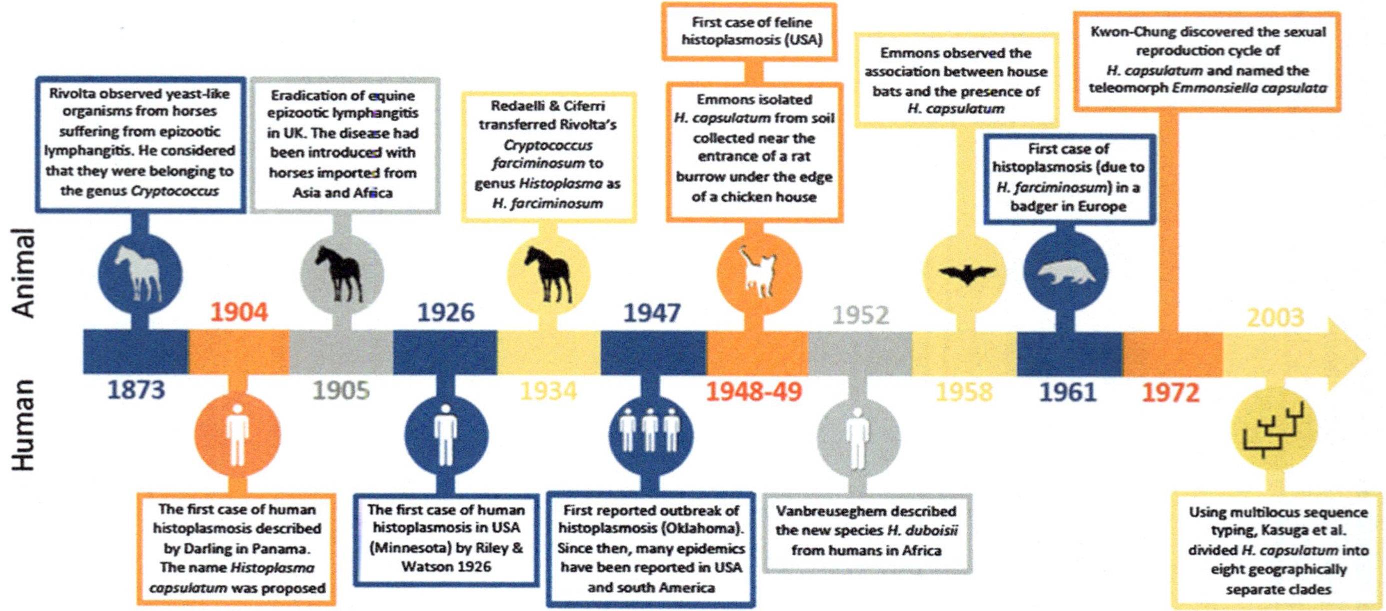

Fig. 5.2 Chronological order of the main facts and pioneering events that contributed to the epidemiological studies on human and animal histoplasmosis

filamentous phase. This phase develops in special habitats, particularly on bat guano that has accumulated in confined spaces such as caves. A second possible mode of entry by ingestion of infective material has been suggested. This could account for cases of primary gastrointestinal histoplasmosis in dogs and cats (Brömel and Sykes 2005; Stark 1982). A moderate climate with constant humidity seems to be the most appropriate combination for the development and survival of *Histoplasma*. In companion animals, infection due to *H. capsulatum* var. *capsulatum* is most frequently observed in endemic/enzootic regions of the Mississippi and Ohio River valleys in North America (Brömel and Sykes 2005; Sykes and Guillot 2015). Cases were also reported from South America (Forjaz and Fischman 1985), Italy (Mantovani et al. 1968; Reginato et al. 2014), Greece (Mavropoulou et al. 2010), Japan (Murata et al. 2007; Ueda et al. 2003) and Australia (Mackie et al. 1997). Recently, cases of feline histoplasmosis were reported in Colorado, California, New Mexico and Texas, locations that were traditionally considered as non-enzootic (Balajee et al. 2013). A massive *H. capsulatum* infection in juvenile raccoons from northern California was reported by Clothier et al. (2014). In 2005, an infected northern sea otter (*Enhydra lutris kenyoni*) was found on Kodiak Island, Alaska (Burek-Huntington et al. 2014). Histological examination revealed the presence of *Histoplasma* yeasts and the subspecies *capsulatum* could be identified by direct sequencing. The authors suggested that the sea otter was contaminated by migratory birds or through aerosol transmission.

Europe is usually considered non-endemic/enzootic for *H. capsulatum*. However, in a review about humans with histoplasmosis in Europe over a 5-year period (from January 1995 to December 1999), Ashbee et al. (2008) identified eight patients (from Germany, Italy and Turkey) who had never travelled abroad and hence may correspond to autochthonous cases. Histoplasmosis has also been reported in wild and domestic animals in Europe. Several investigations demonstrated that *H. capsulatum* may be responsible for cutaneous lesions in Eurasian badgers (*Meles meles*) in Switzerland (Burgisser et al. 1961), Denmark (Jensen et al. 1992), Germany (Grosse et al. 1997; Eisenberg et al. 2013) and Austria (Bauder et al. 2001). Infection was limited to the skin and subcutaneous lymph nodes. The badger's habitat and its rummaging and omnivorous mode of life are potential predisposing factors to infectious diseases, including tuberculosis and histoplasmosis. Recently, *H. capsulatum* var. *capsulatum* DNA was detected from the lungs of a bat (*Nyctalus noctula*) trapped in France (González-González et al. 2013). Disseminated histoplasmosis was described in a cat in Greece (Mavropoulou et al. 2010) and in dogs in Italy (Mantovani et al. 1968; Reginato et al. 2014).

The subspecies *farciminosum* was epidemiologically investigated between January 2003 and June 2004 in 19,082 carthorses in Ethiopia (Ameni 2006). A mean prevalence of 18.8% was reported. The highest prevalence (39%) was observed in the Mojo region. The prevalence of infection was associated with average annual temperatures rather than mean annual rainfall. Statistically significant association was also observed between the altitude and the prevalence of infection: histoplasmosis was more frequently reported in humid and hot regions with an altitude from 1500 to 2300 m (Ameni 2006). Equine histoplasmosis has been reported from China, India, Indonesia, Iraq, Israel, Japan, Pakistan and Syria. In Africa, cases are reported in many countries: Algeria, Angola, Cameroon, Chad,

Egypt, Ethiopia, Ghana, Morocco, Nigeria, Togo, Tunisia, Senegal and Sudan (Guérin 2010). Official disease distribution maps from OIE (Office International des Epizooties) indicate that equine histoplasmosis in Africa is restricted to Ethiopia, Senegal and South Africa (OIE WAHID Maps 2005). The disease is also reported from South and Central America (Al-Ani 1999). Compared to major equine epizootic diseases, infection due the subspecies *farciminosum* is not frequently diagnosed. The confusion with other infectious diseases may probably account for this situation.

Infections limited to the skin were also reported in cats residing in Switzerland (Fischer et al. 2013) and in eastern France (Fantini et al. 2014). In these cases, histopathological examination revealed the presence of microorganisms consistent with *Histoplasma* (Fig. 5.1), and the subspecies *farciminosum* was identified by PCR and MLST approach directly from the tissues (Guillot et al. 2015).

Inoculation is considered the primary mode of entry of *H. capsulatum* var. *farciminosum*, probably mostly by contamination of cutaneous wounds in horses (Guérin 2010). Nevertheless experimental transmission of the disease by subcutaneous or intradermal inoculation of pus containing *H. capsulatum* var. *farciminosum* yeasts gave inconstant results. Some cases of direct transmission between infected and healthy wounded animals or after mating have been suspected (Al-Ani and Al Delaimi 1986). The subspecies *farciminosum* has also been isolated from the digestive tract of hematophagous flies (Gabal and Hennager 1983). Ameni and Terefe (2004) indicated that there was a significant association between histoplasmosis and the presence of ticks in mules. Seasonal changes may have a significant effect upon direct transmission efficiency, as rain increases mud projections on wounds, delays healing processes and increases the risk of infection in horses (Guérin et al. 1992). Telluric contamination of exposed and vulnerable body areas such as limbs, eyes or nostrils could explain the main localisations of primary lesions.

5.3 Infections Due to *Histoplasma capsulatum* var. *capsulatum*

Histoplasmosis due to *H. capsulatum* var. *capsulatum* has a wide range of clinical manifestations, presenting as mild respiratory distress, acute respiratory infection or life-threatening disseminated disease. The outcome is variable according to the inoculum size and the strain virulence. Once inhaled, *Histoplasma* propagules convert to yeasts in the lungs. A granulomatous inflammatory response occurs that consists primarily of macrophages, variable numbers of lymphocytes and sometimes fibrosis. Some animals control the initial infection but remain latently infected with small numbers of yeast cells. Subsequent immune suppression can lead to reactivation of infection years later. As a consequence, the incubation period may range from 2 to 3 weeks to several years. *Histoplasma* organisms may be found in pulmonary lymph nodes, in reticuloendothelial organs (liver, spleen and bone marrow), in the skin, in the central nervous system and in the small and/or large intestinal tract.

Clinical signs of histoplasmosis in cats and dogs are commonly non-specific, such as weight loss, inappetence, weakness, dehydration and fever (Brömel and Sykes 2005). *Histoplasma capsulatum* was first reported as a pathogen in a cat in 1949. In a review of 571 cats with deep mycotic infections in the USA, histoplasmosis was with 16.7% the second most commonly reported fungal disease after cryptococcosis (46.1%) (Davies and Troy 1996). Approximately 40% of infected cats have respiratory signs such as dyspnoea and tachypnoea and to a lesser extent cough and nasal discharge. Respiratory signs may be absent in some animals with dissemination of *Histoplasma* to non-pulmonary sites. Thoracic radiographs usually reveal diffuse, linear, nodular or miliary interstitial patterns, but mixed interstitial-alveolar-bronchial patterns and an absence of abnormal findings have also been reported. Ocular signs (chorioretinitis, retinal detachment, optic neuritis, anterior uveitis or panophthalmitis) occur quite frequently (in more or less one-quarter of cats). In dogs, chronic diarrhoea (often with hematochezia or melena) and wasting (pale mucous membranes, weight loss and weakness) are frequently observed. Other clinical signs include cutaneous nodules, joint pain, lesions on the tongue, myosis, splenomegaly and ocular signs (such as anterior uveitis, chorioretinitis, optic neuritis and retinal detachment) (Brömel and Sykes 2005).

Systemic histoplasmosis with intermittent cough, dyspnoea, lymphadenopathy, hyperthermia, anorexia, weight loss and diarrhoea has been described, mostly in horses (Rezabek et al. 1993), more rarely in cattle (Morgado et al. 1976) and the dromedary (Chandel and Khere 1994). Occasionally, *H. capsulatum* var. *capsulatum* is also responsible for abortion in mares with lesions of placentitis and invasion of the foetus; perinatal death of foals, usually due to severe granulomatous pneumonia (Saunders et al. 1983); or ocular mycosis with keratitis in horses (Richter et al. 2003).

5.4 Infections Due to *Histoplasma capsulatum* var. *duboisii*

Infections due to *H. capsulatum* var. *duboisii* have been reported in baboons originating from West Africa after transfer to other locations (Gugnani and Muotoe-Okafor 1997). Secondary infections of the skin, subcutaneous tissues and the lymph nodes in the form of small papules and ulcerative granulomas have been reported in absence of involvement of the lungs and internal viscera. The subspecies *duboisii* has not been isolated from livestock but is recognised in humans and in some wild mammals including bats (*Nycteris hispida*, *Tadarida pumila*), baboons (*Papio cynocephalus*) and aardvarks (*Orycteropus afer*), which could be markers of endemic foci (Chermette and Guillot 2010). The virulence of *H. capsulatum* var. *duboisii* seems to be lower than that of *H. capsulatum* var. *capsulatum*, which is consistent with the tendency of the variety *duboisii* to form mainly localised cutaneous and subcutaneous infections (Rippon 1988).

5.5 Infections Due to *Histoplasma capsulatum* var. *farciminosum*

Following traumatic inoculation, the development of the subspecies *farciminosum* is usually responsible for nodular cutaneous lesions. In the skin, the fungus induces an inflammatory lesion containing granulocytes, macrophages and multinucleated giant cells. A fibrous and oedematous peripheral reaction further surrounds the nodule. An abscess progressively develops and starts to discharge yellow pus containing yeasts, macrophages and granulocytes. Finally, a granulation tissue appears and an ulcer with inverted borders is formed. Infected leucocytes or yeasts are able to spread the infection via lymphatic vessels to adjacent tissues. Bacterial superinfection may occur.

In equids, the subspecies *farciminosum* is responsible for a debilitating disease called epizootic lymphangitis (Guérin 2010; Scantlebury and Reed 2009). The incubation period ranges from 1 and 7 months. The classical presentation is a superficial lymphangitis. Sometimes genital organs or bones are involved. Hyperthermia is rarely reported. The lesions are present on the skin and more rarely on the mucous membranes (lips, conjunctiva, nasal or respiratory epithelium). Five types of lesions have been described:

1. The initial lesion is a cutaneous ulcer with inverted borders and painful outline; thick, yellowish and sometimes bloody pus is produced, mostly observed on the limbs, thorax, chest, neck and head.
2. The cord is a congested or knotted rope lymph vessel (Fig. 5.3).
3. The spots are hemispheric nodules, 0.5–3 cm diameter, tough and painless, isolated or lined on a cord.
4. The tumours are located on lymph nodes, up to 5–30 cm diameter, and may turn into fistula.
5. The engorgements are due to a diffuse reaction within the conjunctive tissue and are observed around the lower limbs.

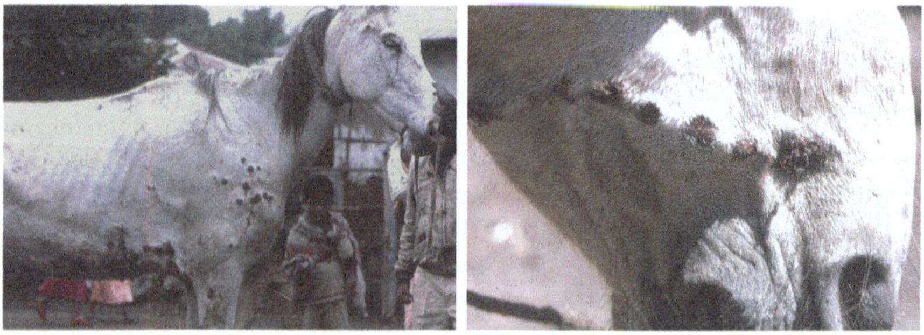

Fig. 5.3 Lesions of equine epizootic lymphangitis in Ethiopia. (Left) Ulcers and nodules; (right) cordlike lesions (courtesy from Christine Guérin)

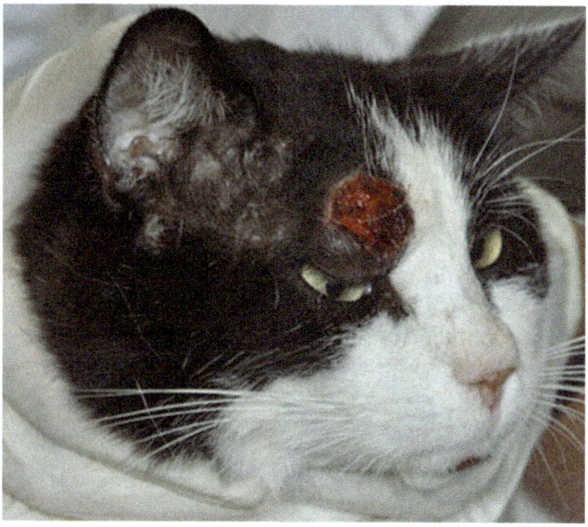

Fig. 5.4 A lesion of feline histoplasmosis due to *Histoplasma capsulatum* var. *farciminosum* (Fischer et al. 2013)

Infected horses become restless because of the disturbance due to numerous flies attracted by cutaneous lesions. There is a progressive loss of appetite and condition as severity of the disease increases. Because of its debilitating nature and its high case fatality, some horse-owners in Ethiopia have started to refer to equine histoplasmosis as "horse AIDS" (Ameni 2006). In a study investigating the economics of the carthorse industry in Ethiopia, Aklilu and Zerfu (2010) reported that losses to the owner due to morbidity of a horse with histoplasmosis resulted in more than a 50% reduction in daily earnings.

Infection due to *H. capsulatum* var. *farciminosum* has been reported in animals other than equids. Histoplasmosis in four dogs diagnosed in Japan lacked pulmonary or gastrointestinal lesions and was characterised by multiple granulomatous or ulcerated lesions on the skin and in the mouth. The subspecies *farciminosum* was reported as the causative agent in these cases (Murata et al. 2007). In Europe, similar observations have been made in badgers in Germany (Eisenberg et al. 2013) but also in cats in Switzerland (Fischer et al. 2013) and eastern France (Fantini et al. 2014, Guillot et al. 2015) (Fig. 5.4).

5.6 Diagnosis of Histoplasmosis in Animals

A definitive diagnosis of histoplasmosis is made by cytologic or histopathologic identification of *Histoplasma* yeasts. The latter are typically observed within macrophages or, less frequently, free in pyogranulomatous exudates. *Histoplasma* yeasts are small (2 × 4 μm), oval or globose elements surrounded by a clear halo. Histopathological sections should be examined very carefully because the yeasts are small and quite difficult to visualise upon regular staining, like HE. Sometimes

Histoplasma yeasts stain poorly with PAS, making the silver staining GMS more valuable. Differential diagnosis includes other fungi (*Sporothrix* spp., *Blastomyces dermatitidis*, *Cryptococcus* spp.), which develop as yeasts in tissues, but also protozoa like *Leishmania* spp.

Culture is possible on Sabouraud dextrose agar without cycloheximide or on brain-heart infusion with blood. With an incubation temperature of 27–30 °C, *Histoplasma* fungi form whitish and slow-growing colonies. Incubation periods of 2–4 weeks may be required before growth is appreciated. Under these conditions, infective conidia (thick-walled, large, tuberculate macroconidia and small oval microconidia) are produced. Because these conidia may cause infection in laboratory personnel, the growth of the mycelial phase of *Histoplasma* poses a health hazard and should be performed in specialised laboratories with adapted levels of confinement.

PCR techniques have been developed for the diagnosis of human histoplasmosis (Muraosa et al. 2016). These techniques could also be applied in animals. Several serological techniques, including immunodiffusion and complement fixation tests, have been used to detect human antibodies to *H. capsulatum*, but serology does not seem to be a reliable diagnostic tool (Kauffman 2009). This also applies to dogs and cats. Serology for *H. capsulatum* antibodies in nine cases of canine disseminated histoplasmosis revealed a titer of 1:8 in one case (Mitchell and Stark 1980). Serology to detect antibodies against *H. capsulatum* was positive in only four out of nine feline cases (Davies and Troy 1996).

Antigen detection tests are now widely used for the diagnosis of human histoplasmosis. The tests are usually performed on urine specimens. In a retrospective study, Cook et al. (2012) compared the results of a urine antigen assay with standard diagnostic methods in cats with clinical signs of histoplasmosis. Antigenuria was detected in 17 out of 18 infected cats. The histofarcin skin test was developed by Soliman et al. (1985) for horses. This test proved to be a valuable tool in diagnosing epizootic lymphangitis in the field (Ameni et al. 2006). A delayed, intradermal, type IV hypersensitivity reaction indicates previous exposure to *Histoplasma*.

5.7 Control of Histoplasmosis in Animals

In dogs and cats infected by the subspecies *capsulatum* or *farciminosum*, the treatment of choice is oral itraconazole. The recommended dose is 10 mg/kg q 12–24 h for a minimum of 4–6 months. Treatment should be continued for at least 2 months after resolution of clinical signs (Brömel and Sykes 2005). In dogs, amphotericin B has been used successfully to treat local and disseminated histoplasmosis, but relapses are common.

In horses infected by the variety *farciminosum*, most of reported treatments date back to the beginning of the twentieth century. Some of them (iodide, mercuric, arsenic or imidazole-derived drugs) were efficient, but were always long, relatively expensive and toxic. In combination with surgical removal of the lesion or cauterisation, these treatments led to recovery within 4–6 weeks (Guérin 2010).

Amphotericin B is the listed drug of choice for the treatment of clinical cases of epizootic lymphangitis by the OIE (Anon 2004).

Elimination of the infection can be achieved by culling infected horses and application of strict hygiene practices to prevent spread of the organism. However, culling is hardly acceptable in highly endemic areas where the use of horse-drawn taxis and carts to generate an income is a means of survival for a significant number of families.

Boquet and Nègre developed a vaccine against epizootic lymphangitis obtained from a yeast culture inactivated by heat (cited by Curasson 1942). Subcutaneous injections every 7 days during 5 weeks proved to be efficient for the treatment of horses. More recently, a live attenuated vaccine was developed and tested in China. This vaccine was reported to protect 75.5% of horses inoculated, with immunity persisting for more than 2 years (Zhang et al. 1986). This vaccine is not commercially available, and there were some issues of adverse reactions that would need addressing (Scantlebury et al. 2015).

5.8 Public Health Considerations

The potential role of bats in spreading *H. capsulatum* var. *capsulatum* remains unclear. However, the high risk of natural bat infection with this fungus in caves has been well-documented. In a recent investigation in Mexico, lung samples from 122 bats were examined (Gonzalez-Gonzalez et al. 2014). A total of 98 samples revealed *Histoplasma* infection. Benedict and Mody (2016) provided an update on the epidemiologic features of 105 documented human outbreaks in the USA. The presence of bats (or bat droppings) was reported in 24 (23%) outbreaks and the presence of birds or bird droppings in 59 (56%). Birds most frequently involved were chickens (41% of bird-related outbreaks) and blackbirds (starlings, grackles), pigeons and gulls.

Transmission of *H. capsulatum* from companion animals to humans has never been reported. However, infected pets may be a sentinel for human exposure, and this is especially relevant if the animal resides with immunocompromised human beings. Concurrent infections of owners and companion animals were reported after exposure to the same environment or source of infective material (Dillon et al. 1982, Davies and Colbert 1990). A common-source environmental exposure was also suggested in a study evaluating the geographical specificity of *H. capsulatum* var. *capsulatum* isolates in Brazil (de Medeiros Muniz et al. 2001). Using the MLST typing technique, Balajee et al. (2013) demonstrated that isolates from feline and human cases were genetically distinct. However, the feline cases were from California, Colorado and Texas, whereas the human cases were from different regions (Missouri and Georgia). This result might indicate that the genetic differences were related to geographic distance rather than to host specificity.

References

Aklilu N, Zerfu A (2010) Socio-economic impact of epizootic lymphangitis on horse-drawn taxi business in central Ethiopia. In: Proceedings of the 6th international colloquium on working equids, New Delhi

Al-Ani FK (1999) Epizootic lymphangitis in horses: a review of literature. Rev Sci Tech 18:691–699

Al-Ani FK, Al Delaimi K (1986) Epizootic lymphangitis in horses: clinical, epidemiological and haematological studies. Pak Vet J 6:96–100

Ameni G (2006) Epidemiology of equine histoplasmosis (epizootic lymphangitis) in carthorses in Ethiopia. Vet J 172:160–165

Ameni G, Terefe W (2004) A cross-sectional study of epizootic lymphangitis in cart-mules in western Ethiopia. Prev Vet Med 66:93–99

Ameni G, Terefe W, Hailu A (2006) Histofarcin test for the diagnosis of epizootic lymphangitis in Ethiopia: development, optimisation and validation in the field. Vet J 171:358–362

Anon D (2004) Epizootic lymphangitis. In: OIE manual of diagnostic tests and vaccines for terrestrial animals, Chapter 2.5.13, updated 23.07.2004. http://www.oie.int/eng/normes/mmanual/A_00091.htm

Ashbee HR, Evans EG, Viviani MA et al (2008) Histoplasmosis in Europe: report on an epidemiological survey from the European confederation of medical mycology working group. Med Mycol 46:57–65

Balajee SA, Hurst SF, Chang LS et al (2013) Multilocus sequence typing of *Histoplasma capsulatum* in formalin-fixed paraffin-embedded tissues from cats living in non-endemic regions reveals a new phylogenetic clade. Med Mycol 51:345–351

Bauder B, Steineck T, Kübber-Heiss A (2001) First evidence of histoplasmosis in a badger (*Meles meles*) in Austria. In: Wissenschaftliche Mitteilungen aus dem Niederösterreichischen Landesmuseum. Vetmeduni, Vienna, pp 198–199

Benedict K, Mody RK (2016) Epidemiology of histoplasmosis outbreaks, United States, 1938–2013. Emerg Infect Dis 22(3):370–378

Brandao J, Woods S, Fowlkes N et al (2014) Disseminated histoplasmosis (*Histoplasma capsulatum*) in a pet rabbit: case report and review of the literature. J Vet Diagn Investig 26:158–162

Brömel C, Sykes JE (2005) Histoplasmosis in dogs and cats. Clin Tech Small Anim Pract 20:227–232

Burek-Huntington KA, Gill V, Bradway DS (2014) Locally acquired disseminated histoplasmosis in a northern sea otter (*Enhydra lutris kenyoni*) in Alaska, USA. J Wildl Dis 50:389–392

Burgisser H, Fankhauser R, Kaplan W et al (1961) Mycoses in a badger in Switzerland: histologically histoplasmosis. Pathol Microbiol (Basel) 24:794–802

Chandel BS, Khere HN (1994) Occurrence of histoplasmosis like-disease in camel (*Camelus dromedarius*). Indian Vet J 71:521–523

Chermette R, Guillot J (2010) Mycoses due to dimorphic fungi. In: Lefevre PC, Blancou J, Chermette R, Uilenberg G (eds) Infectious and parasitic diseases of livestock. Lavoisier, Paris, pp 1385–1394

Clothier KA, Villanueva M, Torain A et al (2014) Disseminated histoplasmosis in two juvenile raccoons (*Procyon lotor*) from a nonendemic region of the United States. J Vet Diagn Investig 26:297–301

Cook AK, Cunningham LY, Cowell AK, Wheat LJ (2012) Clinical evaluation of urine *Histoplasma capsulatum* antigen measurement in cats with suspected disseminated histoplasmosis. J Fel Med Surg 14:512–515

Curasson G (1942) Traité de pathologie exotique vétérinaire et comparée. In: Tome II—Maladies microbiennes. Vigot Frères Editeurs [in French]

Davies SF, Colbert RL (1990) Concurrent human and canine histoplasmosis from cutting decayed wood. Ann Intern Med 113:252–253

Davies C, Troy GC (1996) Deep mycotic infections in cats. J Am Anim Hosp Assoc 32:380–391
de Hoog GS, Guarro J, Gene J et al (2009) Atlas of clinical fungi: the ultimate benchtool for diagnostics. Centraalbureau voor Schimmelcultures, KNAW fungal. Universitat Rovira i Virgili, Utrecht
de Medeiros Muniz M, Pizzini CV, Peralta JM et al (2001) Genetic diversity of *Histoplasma capsulatum* strains isolated from soil, animals, and clinical specimens in Rio de Janeiro state, Brazil, by a PCR-based random amplified polymorphic DNA assay. J Clin Microbiol 39:4487–4494
Dillon AR, Teer PA, Powers RD et al (1982) Canine abdominal histoplasmosis: a report of four cases. J Am Anim Hosp Assoc 18:498–502
Eisenberg T, Seeger H, Kasuga T et al (2013) Detection and characterization of *Histoplasma capsulatum* in a German badger (*Meles meles*) by ITS sequencing and multilocus sequencing analysis. Med Mycol 15:1–8
Fantini O, Guillot J, Gier S et al (2014) A case of histoplasmosis in a cat limited to the skin. In: Proceedings of the annual congress of the European Society of Veterinary Dermatology, Salzburg
Fischer NM, Favrot C, Monod M et al (2013) A case in Europe of feline histoplasmosis apparently limited to the skin. Vet Dermatol 24:635–638
Forjaz MHH, Fischman O (1985) Animal histoplasmosis in Brazil: isolation of *Histoplasma capsulatum* from a dog on the northern coast of São Paulo. Mykosen 28:191–194
Gabal MA, Hennager S (1983) Study on the survival of *Histoplasma farciminosum* in the environment. Mycosen 1983(26):481–487
González-González AE, Ramírez JA, Aliouat-Denis CM et al (2013) Molecular detection of *Histoplasma capsulatum* in the lung of a free-ranging common noctule (*Nyctalus noctula*) from France using the Hcp100 gene. J Zoo Wildl Med 44:15–20
González-González AE, Aliouat-Denis CM, Ramírez-Bárcenas JA et al (2014) *Histoplasma capsulatum* and *Pneumocystis* spp. co-infection in wild bats from Argentina, French Guyana, and Mexico. BMC Microbiol 14:23
Grosse G, Staib F, Rapp J et al (1997) Pathological and epidemiological aspects of skin lesions in histoplasmosis. Observations in an AIDS patient and badgers outside endemic areas of histoplasmosis. Zentralbl Bakteriol 285:531–539
Guérin C (2010) Equine epizootic lymphangitis. In: Lefevre PC, Blancou J, Chermette R, Uilenberg G (eds) Infectious and parasitic diseases of livestock. Lavoisier, Paris, pp 1395–1401
Guérin C, Abebe S, Touati F (1992) Lymphangite épizootique du cheval en Éthiopie. J Mycol Med 2:1–5
Gugnani HC, Muotoe-Okafor F (1997) African histoplasmosis: a review. Rev Iberoam Micol 14:155–159
Guillot J, Pin D, Chabé M et al (2015) Feline histoplasmosis due to *Histoplasma capsulatum* var. *farciminosum* in eastern France. In: Proceedings of the annual congress of the international society of human and animal mycology, Melbourne
Jensen HE, Bloch B, Henriksen P et al (1992) Disseminated histoplasmosis in a badger (*Meles meles*) in Denmark. APMIS 100:586–592
Kasuga T, White TJ, Koenig G et al (2003) Phylogeography of the fungal pathogen *Histoplasma capsulatum*. Mol Ecol 12:3383–3401
Kauffman CA (2009) Histoplasmosis. Clin Chest Med 30:217–225
Mackie JT, Kaufman L, Ellis D (1997) Confirmed histoplasmosis in an Australian dog. Aust Vet J 75:362–363
Mantovani A, Mazzoni A, Ajello L (1968) Histoplasmosis in Italy. Isolation of *Histoplasma capsulatum* from dogs in the province of Bologna. Sabouraudia 6:163–164
Mavropoulou A, Grandi G, Calvi L et al (2010) Disseminated histoplasmosis in a cat in Europe. J Small Anim Pract 51:176–180
Mitchell M, Stark DR (1980) Disseminated canine histoplasmosis: a clinical survey of 24 cases in Texas. Can Vet J 21:95–100

Morgado A, Rusch K, Luengo M et al (1976) Histoplasmosis bovina. Primera comunicacion en Chile. Arch Med Vet 8:106–107

Morita T, Kishimoto M, Shimada A et al (2001) Disseminated histoplasmosis in a sea otter (*Enhydra lutris*). J Comp Pathol 125:219–223

Muraosa Y, Toyotome T, Yahiro M et al (2016) Detection of *Histoplasma capsulatum* from clinical specimens by cycling probe-based real-time PCR and nested real-time PCR. Med Mycol 54:433–438

Murata Y, Sano A, Ueda Y et al (2007) Molecular epidemiology of canine histoplasmosis in Japan. Med Mycol 45:233–247

OIE WAHID Interface (2005) http://www.oie.int/wahis_2

Rappleye CA, Eissenberg LG, Goldman WE (2007) *Histoplasma capsulatum* alpha-(1,3)-glucan blocks innate immune recognition by the beta-glucan receptor. Proc Natl Acad Sci U S A 104:1366–1370

Reginato A, Giannuzzi P, Ricciardi M et al (2014) Extradural spinal cord lesion in a dog: first case study of canine neurological histoplasmosis in Italy. Vet Microbiol 170:451–455

Reiss E, Miller SE, Kaplan W et al (1977) Antigenic, chemical, and structural properties of cell walls of *Histoplasma capsulatum* yeast-form chemotypes 1 and 2 after serial enzymatic hydrolysis. Infect Immun 16:690–700

Rezabek GB, Donahue JM, Giles RC et al (1993) Histoplasmosis in horses. J Comp Pathol 109:47–55

Richter M, Hauser B, Kaps S et al (2003) Keratitis due to *Histoplasma* spp. in a horse. Vet Ophtalmol 6:99–103

Rippon JW (1988) Medical mycology: the pathogenic fungi and the pathogenic actinomycetes, 3rd edn. Saunders, Philadelphia

Saunders JR, Matthiesen RJ, Kaplan W (1983) Abortion due to histoplasmosis in a mare. J Am Vet Med Assoc 183:1097–1099

Scantlebury C, Reed K (2009) Epizootic lymphangitis. In: Mair T (ed) Infectious diseases of the horse. EVJ Ltd, Ely, pp 390–406

Scantlebury CE, Zerfu A, Pinchbeck GP et al (2015) Participatory appraisal of the impact of epizootic lymphangitis in Ethiopia. Prev Vet Med 120:265–276

Snider TA, Joyner PH, Clinkenbeard KD (2008) Disseminated histoplasmosis in an African pygmy hedgehog. J Am Vet Med Assoc 232:74–76

Soliman R, Saad MA, Refai M (1985) Studies on histoplasmosis farciminosii (epizootic lymphangitis) in Egypt. Application of a skin test ('histofarcin') in the diagnosis of epizootic lymphangitis in horses. Mykosen 28:457–461

Stark DR (1982) Primary gastrointestinal histoplasmosis in a cat. J Am Anim Hosp Assoc 18:154–156

Sykes J, Guillot J (2015) Cryptococcosis and other systemic mycoses. In: Parasitoses and vector borne diseases in cats. Merial, Lyon, pp 288–300

Ueda Y, Sano A, Tamura M et al (2003) Diagnosis of histoplasmosis by detection of the internal transcribed spacer region of fungal rRNA gene from a paraffin-embedded skin sample from a dog in Japan. Vet Microbiol 94:219–224

Zhang WT, Wang ZR, Liu YP et al (1986) Attenuated vaccine against epizootic lymphangitis of horses. Chin J Vet Sci Technol 7:3–5

Paracoccidioidomycosis in Animals and Humans

6

Sandra de Moraes Gimenes Bosco and Eduardo Bagagli

Abstract

Paracoccidioidomycosis (PCM) is the most important systemic mycosis in Latin American countries, especially in Brazil, Colombia, Venezuela, and Argentina. The disease is caused by *Paracoccidioides brasiliensis* and *P. lutzii*, which are dimorphic fungi belonging to the *Ajellomycetaceae*. The disease is relatively common in humans, but poorly known in animals. The finding that nine-banded armadillos (*Dasypus novemcinctus*) are naturally infected by *P. brasiliensis*, systematically recovered from their tissues, has opened up new opportunities to better comprehend the fungus' eco-epidemiology. Armadillos are ancient South American mammals belonging to the order Xenarthra, which also comprises anteaters and sloths. The fungus was detected by nested PCR and histopathology in the anteater *Myrmecophaga tridactyla* and in the two-toed sloth *Choloepus didactylus*, respectively. The pathogen was also detected by molecular assays in different species of wild road-killed animals originating from PCM-endemic areas. In domestic animals, asymptomatic PCM infection has been detected by intradermal tests and serological surveys, e.g., in dogs, cats, sheep, horses, cattle, pigs, and chickens. Naturally acquired PCM disease was reported in three dogs with generalized lymphadenomegaly and hepatosplenomegaly. Certainly, PCM has been underdiagnosed in animals. Veterinarians should always keep this fungal infection in mind when dealing with animals from regions endemic/enzootic for PCM.

S. d. M. G. Bosco · E. Bagagli (✉)
Department of Microbiology and Immunology, Institute of Biosciences, UNESP, Botucatu, São Paulo, Brazil
e-mail: smgbosco@ibb.unesp.br; bagagli@ibb.unesp.br

S. Seyedmousavi et al. (eds.), *Emerging and Epizootic Fungal Infections in Animals*,
https://doi.org/10.1007/978-3-319-72093-7_6

6.1 Introduction

Paracoccidioidomycosis (PCM) is an endemic/enzootic mycosis acquired by airborne inhalation of infective conidia of *Paracoccidioides* spp. present in the environment (Brummer et al. 1993; Bocca et al. 2013). Once inhaled, the fungus may spread by the lymphatic-hematogenic route to other tissues. Described in 1908 by Adolfo Lutz in São Paulo state, Brazil, PCM was believed to be caused by a single agent, *Paracoccidioides brasiliensis* (Lutz 1908; Splendore 1912; Almeida 1930). A new species, *Paracoccidioides lutzii*, was recently described as an additional etiological agent of PCM. This has affected epidemiological, clinical, and diagnostic approaches (Teixeira et al. 2014a, b). PCM is the major systemic mycosis in Latin American countries and ranks eighth among causes of human death from infectious and parasitic diseases in Brazil (Coutinho et al. 2002, 2015). While the immunological and clinical aspects of PCM are relatively well studied in humans, little is known about the ecology of the fungus and the role of nonhuman hosts.

6.2 Causative Agents

6.2.1 Morphology

Mycelia of *Paracoccidioides* spp. present slow growth when cultured on Sabouraud, Mycosel®, or potato dextrose agars. Initially the colony displays a white, cotton-like surface. Over time, the colony becomes white to beige, while the surface takes a wrinkled aspect with the presence of fissures, resembling popcorn. The microscopy shows thin, septate hyphae, which might present the infective arthroconidia (also referred to as arthroaleurioconidia or planoconidia). These conidia may vary in abundance and in sizes, according to its genotypes or cryptic species (Theodoro et al. 2012).

The macroscopic morphology of the yeast colony shows a cerebriform aspect and beige color. The yeast form grows within 5–7 days of culturing at 37 °C on Sabouraud, Mycosel®, and GPY (glucose, peptone, yeast extract) agars. Microscopically a large rounded yeastlike cell appears (6–30 μm in diameter), with thick cell walls, birefringent, with buds varying in number and size. The buds generally range from 2 to 10 μm in diameter and are connected to the parent cell by a narrow isthmus, different from that observed in *Blastomyces dermatitidis*, in which the single bud is usually broad-based. The presence of large cells, surrounded by numerous small buddings, is classically known as "pilot wheel." A typical "Mickey mouse cap" can also be observed in culture and in histopathological slides.

6.2.2 Classification and Molecular Aspects

Molecular and ultrastructural studies have positioned *Paracoccidioides* spp. taxonomically in the phylum *Ascomycota* and order *Onygenales*. The order comprises numerous pathogenic fungi. The thermally dimorphic genus *Coccidioides* is classified in the

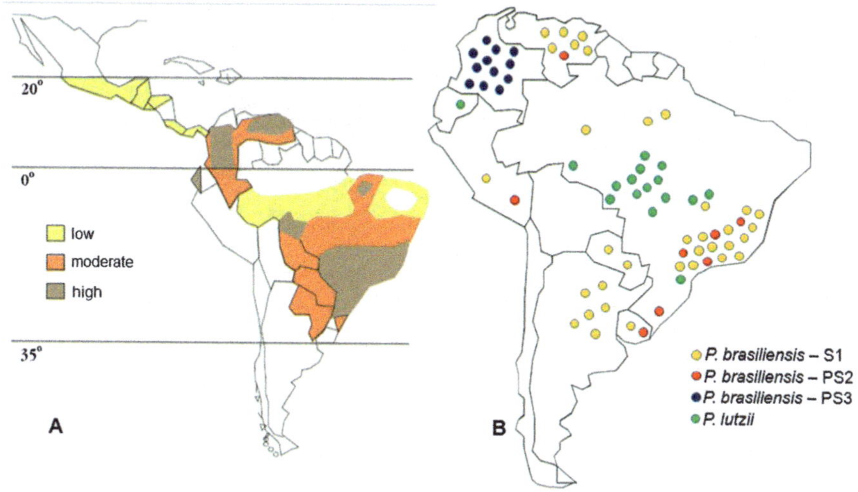

Fig. 6.1 (**a**) Endemic areas of paracoccidioidomycosis, according to Shikanai-Yasuda et al. (2006). (**b**) Geographic distribution of the cryptic species of *P. brasiliensis* and *P. lutzii*, based on Theodoro et al. (2012)

family, *Onygenaceae*, while *Blastomyces* and *Histoplasma* are with *Paracoccidioides* in *Ajellomycetaceae* (Untereiner et al. 2004). Species of these groups have a saprophytic phase that occurs in soil containing animal remains (feces and carcasses) in relatively protected locations, such as animal burrows, caves, buildings liners, and other enclosed spaces. Members of the family *Ajellomycetaceae* also comprise other, less common species which are also associated with animals, such as *Lacazia loboi*, *Emmonsia*-like (Herr et al. 2001; Vilela et al. 2009; Theodoro et al. 2011), now described in *Emergomyces* (Dukik et al. 2017).

Molecular phylogenetic studies have indicated that *P. brasiliensis* is a species complex with distinct genotypes and/or cryptic species, such as S1 (species 1, which occurs in Brazil, Argentina, Paraguay, Peru, and Venezuela), PS2 (phylogenetic species 2, found in Brazil and Venezuela), and PS3 (phylogenetic species 3, found mainly in Colombia) (Matute et al. 2006a, b; Theodoro et al. 2011) (Fig. 6.1). Additional studies have indicated a genetically more distant group of isolates that was formally described as a separate species, *Paracoccidioides lutzii*, as a tribute to Adolfo Lutz who first described the disease (Theodoro et al. 2012; Teixeira et al. 2014a, b). It has been also suggested that this group can be divided into separate taxonomic species: *P. brasiliensis sensu stricto* (S1), *P. americana* sp. nov. (PS2), *P. restrepiensis* sp. nov. (PS3) (Turissini et al. 2017). The phylogenetic species PS4, described by Teixeira et al. (2014a), received the name of *P. venezuelensis* sp. nov. (Turissini et al. 2017).

6.3 Epidemiology of Paracoccidioidomycosis

PCM is restricted to Latin American countries (Figs. 6.1 and 6.2). In Brazil, which accounts for the majority of the cases, three large endemic areas are found: (1) São Paulo, Rio de Janeiro, Espírito Santo and Minas Gerais (southeast region), Goiás and

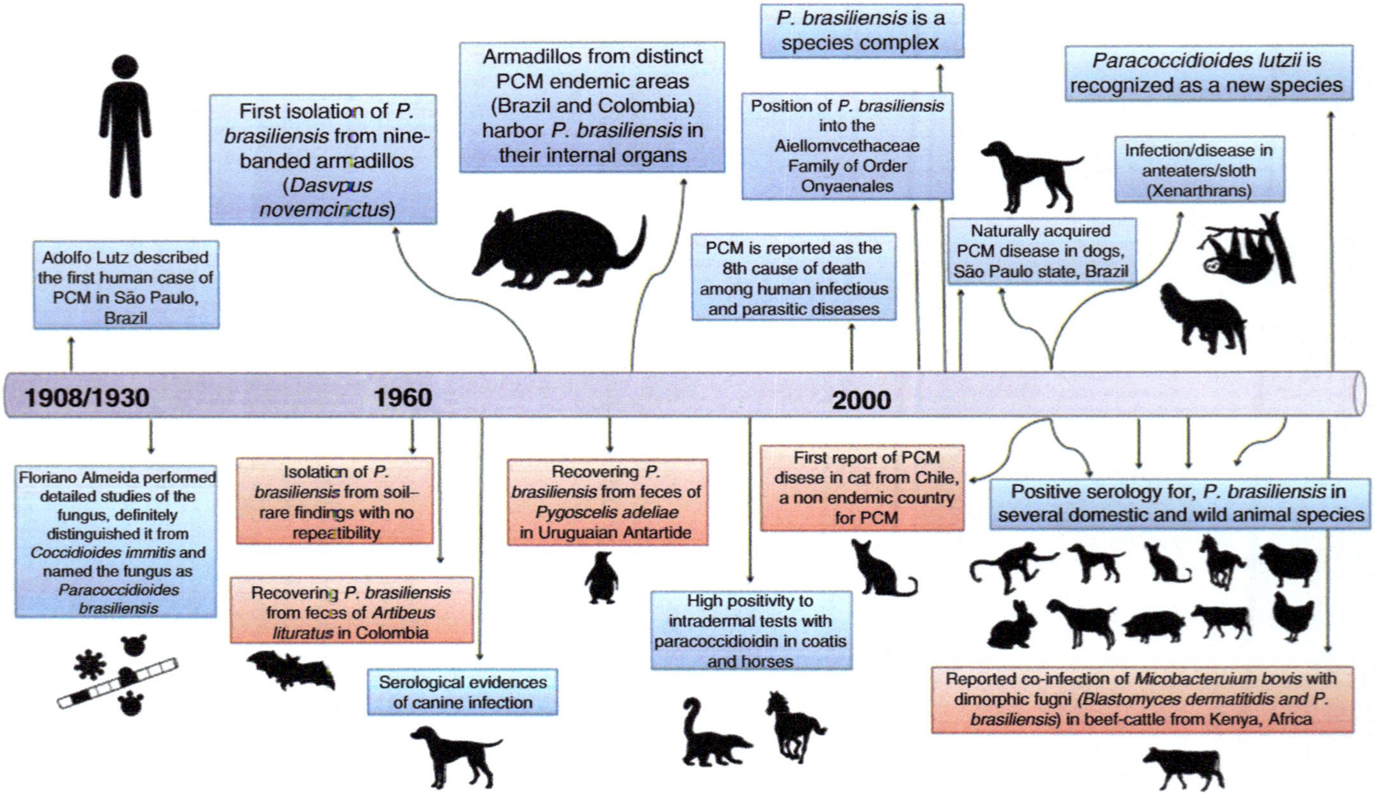

Fig. 6.2 The timeline of *Paracoccidioides* spp. and paracoccidioidomycosis eco-epidemiology. In red box, there are anecdotal findings with no repetition and/or confirmation

Mato Grosso do Sul (midwest region), and Paraná to northern Rio Grande do Sul (southern region); (2) Pará, Maranhão, and Tocantins (eastern of the Amazon region); and (3) Rondônia and the western Amazon region. Low endemicity of PCM is observed in the Brazilian northeast, a semiarid region where no autochthonous cases appear to exist in this semiarid region of Brazil. The endemic area of Mexico is located in its southern region, between the Gulf of Mexico and the Pacific coast. In Venezuela, areas endemic for PCM correspond to the northern region and in the state of Bolivar. Endemicity is also observed in Ecuador (Rio Cuenca Valley), in the region of the Peruvian Amazon forest, in the northeast and northwest of Argentina, in the eastern half of Paraguay, and in most of the Bolivian territory with a tropical climate. The PCM endemicity is low in Uruguay, while no autochthonous cases of the disease have been reported from Chile, Guyana, Surinam, French Guiana, Belize, or Nicaragua (Wanke and Londero 1994; Martinez 2015).

The ecology of *Paracoccidioides* species is still poorly known, and the fungus continues to challenge researchers in relation to its ecological niche (Restrepo 1985; Restrepo-Moreno 1994; Restrepo et al. 2001; Bagagli et al. 2012). While it is known that the fungus exists in nature as a saprobe in soil, it has only rarely been cultured from soil or soil-related products, including feces of both frugivorous bats (*Artibeus lituratus*), penguins (*Pygoscelis adeliae*), and dog food contaminated with soil (Franco et al. 2000; Restrepo et al. 2000). These reports were difficult to reproduce and are not conclusive with respect to transmission of the fungus (Bagagli et al. 2012).

Experimental PCM infection has been well studied and documented in several animal models, such as in mice, rats, hamsters, and guinea pigs (McEwen et al. 1987; Tani et al. 1987; Singer-Vermes et al. 1993; Coelho et al. 1994), as well as in dogs and cattle (Mós et al. 1974; Costa et al. 1978; Ono et al. 2003; Eisele et al. 2004). Natural PCM infection in domestic and wild animals has been detected by intradermal tests, serological surveys, histopathological analysis, molecular biology, and by isolation of the pathogen in culture. The first studies in dogs that employed complement fixation and precipitation tests revealed seropositivity in almost 80% of the animals evaluated, which originated from recognized human endemic areas of PCM (Mós and Netto 1974; Mós et al. 1974). Intradermal tests using paracoccidioidin as antigen among domestic animals, cattle, horses, and sheep, yielded positive results in 40.2, 63.8, and 40.8%, respectively (Costa et al. 1978, 1995a; Costa and Fava Netto 1978). Intradermal tests in terrestrial wild animals (coatimundis and Felidae) present higher positivity than in arboreal animals (weeping capuchins and marmosets), i.e., 82.98 and 22.45%, respectively (Costa et al. 1995b).

Additional epidemiological studies using sensitive serological approaches, such as ELISA, Western blotting, and immunodiffusion, as well as histopathology and nested PCR, have confirmed that natural PCM infection occurs frequently in several domestic and wild animals from endemic areas. As to dogs, high positivity was confirmed among animals from PCM-endemic areas of Brazil (Silveira et al. 2006; Teles et al. 2015) and Argentina (Canteros et al. 2010), especially when in rural areas, where positivity may reach 80.5–89.5% (Ono et al. 2001; Fontana et al. 2010). It is important to emphasize that serological surveys were carried out on puppies and young and adult dogs of both sexes and of different breeds. No significantly different prevalence was observed with sex, breed, or age; the origin of the dog (rural vs. urban) was the most important parameter. PCM infection was also observed in cats (de Oliveira et al. 2013), chickens (Oliveira et al. 2011), dairy cattle (Silveira et al. 2008), pigs (Belitardo et al. 2014a), dairy goats (Ferreira et al. 2013), rabbits (Belitardo et al. 2014b), and sheep (Oliveira et al. 2012), as well as in wild mammals, such as the capuchin monkeys (*Cebus* sp.) and golden howler monkeys (*Alouatta caraya*) (Corte et al. 2007). Infections were also noted in the small rodents (*Akodon* sp., *Thaptomys nigrita*, *Euryoryzomys russatus*, and *Oligoryzomys nigripes*) (Sbeghen et al. 2015), road-killed nine-banded armadillos (*D. novemcinctus*), seven-banded armadillos (*D. septemcinctus*), guinea pigs (*Cavia aperea*), porcupines (*Sphiggurus spinosus*), grisons (*Galictis vittata*), and raccoons (*Procyon cancrivorus*) (Richini-Pereira et al. 2008b).

The nine-banded armadillos and the naked-tailed armadillo (*Cabassous centralis*) carry *P. brasiliensis* in their internal organs (Naiff et al. 1986; Bagagli et al. 1998, 2003; Corredor et al. 1999, 2005; Silva-Vergara et al. 2000; Richini-Pereira et al. 2008a). In some endemic areas, all evaluated armadillos provided positive culture, especially from fragments of mesenteric lymph nodes and to a less extent from spleen and liver tissues (Bagagli et al. 2003) (Fig. 6.3). There is no clear evidence as to whether infected armadillos develop signs of active PCM disease, despite few granulomas containing fungal elements observed in lymph nodes, liver, and lungs of some animals (Bagagli and Bosco 2008). Armadillos originated in South America in the same geographic area as PCM. Besides intense contact with soil where they live as terrestrial animals, armadillos present a low body temperature and a weak cell-mediated immunity, which render them particularly susceptible to fungal infection (Bagagli and Bosco 2008). Other xenarthran species, such as the seven-banded armadillo, anteater, and sloth, are also infected, being positive by nested PCR with primers specific for *P. brasiliensis* or histopathology (Richini-Pereira et al. 2008b, 2009, Trejo-Chávez et al. 2011).

6.4 Pathogenicity

Once inhaled, the fungus may be destroyed in the lung parenchyma by non-specific phagocytic cells, or it may multiply and produce a primary focus that drains into the regional lymph node located in the pulmonary hilum, compounding the primary complex in PCM. The fungus may spread via the lympho-hematogenic route and develop metastatic foci in any organ or host tissue. At this time, the infection is asymptomatic or oligosymptomatic and can develop into one of the following conditions: (1) complete resolution and healing, (2) involution maintenance of viable fungi in quiescent foci, or (3) progression to pulmonary and/or disseminated progressive disease (Franco et al. 1993; Montenegro and Franco 1994).

In host tissue an inflammatory response culminates in granuloma formation. The granuloma is composed of multinucleated giant cells and epithelioid cells, and its center contains one or more yeast cells in contact with polymorphonuclear

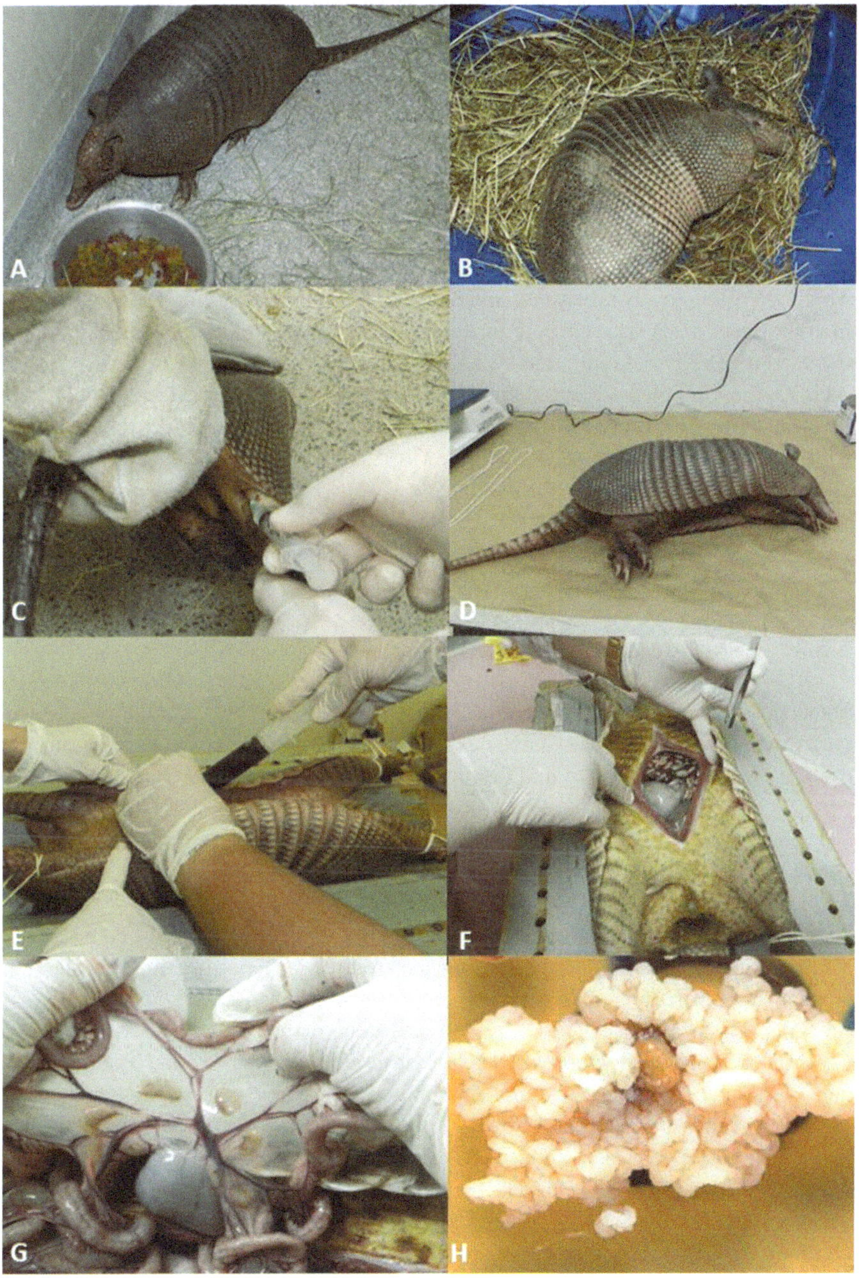

Fig. 6.3 Overview of armadillo's handling procedure for *P. brasiliensis* screening. Courtesy of Bosco and Bagagli. (**a**) Photograph of an armadillo in the captive environment, (**b**) an armadillo sheltered in plastic container containing hay, (**c**) intramuscular administration of anesthetic agents, (**d**) animal under anesthesia, (**e**) collection of blood through cardiac puncture to induce euthanasia, (**f**) opening of the abdominal cavity to harvest mesenteric lymph nodes, liver, spleen, and other internal organs, (**g**) gross pathology of mesenteric lymph node chain, (**h**) plate macroscopy of *P. brasiliensis* at Mycosel agar after 15 days of incubation at 35 °C recovered from mesenteric lymph nodes (the typical cerebriform aspect)

leukocytes. A halo of mononuclear cells surrounds the granuloma. It can be present in both acute and chronic PCM (Franco et al. 1997).

Several studies have shown that depressions of the cellular immunity enhance development of severe forms, which generally tend to resolve with antifungal treatment (Peraçoli et al. 1982; Musatti et al. 1994; Soares et al. 2000). PCM may be included in the bipolar disease model, similar to leprosy: in the anergic pole patients with disseminated and severe dysfunction of cellular immunity are grouped, while the hyperegic pole harbors patients with localized manifestation and preserved cellular immunity. Patients with severe forms have lesions with few granulomas which are often poorly structured and have large amounts of fungal elements inside. In contrast, patients with chronic forms produce typical granulomas with few fungi (Mota et al. 1985). The immune system dysfunction appears to be related to factors such as the presence of circulating immune complexes interacting with T cell-specific antibodies, fungal antigens (even at low concentrations), or cytokine release with inhibiting effects (Mota et al. 1988; Moscardi-Bacchi et al. 1989; Peraçoli et al. 2003; de Castro et al. 2013).

6.5 Clinical Signs

6.5.1 PCM Infection

Paracoccidioidin-positive infected individuals who do not have symptoms or laboratory evidence of the disease have developed the primary complex, but do not show disease progression (Wanke and Londero 1994).

6.5.2 PCM Disease in Humans

The disease occurs as a consequence of the progression of the primary complex, the reactivation of a quiescent focus (endogenous reinfection), or exogenous reinfection. Three main forms are observed: (1) acute-subacute, also called juvenile type or juvenile, which affects mainly children and young adults under 30 years of age; (2) chronic, also called adult type, usually observed in patients over 30 years of age; and (3) sequelae, usually chronic obstructive pulmonary disease, dysphonia, and laryngeal scarring (Franco et al. 1987).

Acute-subacute PCM (juvenile type) affects both sexes and represents about 10% of the general distribution of the disease. Among these patients, about 5% are children between 10 and 14 years old. Although the disease is rare in very young children, its occurrence in that age range is serious and potentially fatal. The juvenile form can be subdivided into mild or severe, according to the degree of dissemination. Clinical manifestations are caused by rapid and progressive involvement of the mononuclear phagocyte system with diffuse lymphadenopathy, hepatosplenomegaly, and bone marrow dysfunction in severe cases. Cutaneous manifestations and bone lesions may also be observed in acute PCM. Fever and weight loss also accompany this clinical picture that quickly leads to general commitment of the patient (Del Negro et al. 1994; Mendes 1994; Shikanai-Yasuda et al. 2006).

Chronic adult PCM is responsible for most cases of PCM in humans (about 90%). It is observed mainly in males between 30 and 60 years old, mainly rural workers with frequent contact with or disturbance of soil, e.g., farmers, construction workers, tractor drivers, and armadillo hunters (Cadavid and Restrepo 1993; Wanke and Londero 1994; Shikanai-Yasuda et al. 2006). Among women the disease is less frequent (F/M ratio 1:10), since the female hormone 17-β estradiol provides protection by preventing the transformation of inhaled conidia in yeast in the lung parenchyma (Restrepo et al. 1984; Salazar et al. 1988). Factors such as smoking, alcohol consumption, and malnutrition are also associated with chronic adult PCM. Probability of patients with PCM clinically becoming ill was 14 times higher among smokers and 3.6 times among individuals who ingested >50 g/day of alcohol (Martinez and Moya 1992; Santos et al. 2003; Bellissimo-Rodrigues et al. 2011). Chronic PCM can be subdivided into unifocal and multifocal with mild, moderate, and severe degrees of disease (Franco et al. 1987; Shikanai-Yasuda et al. 2006).

6.5.3 PCM Associated with Immunosuppression

Association of PCM with human immunodeficiency virus (HIV) infection is reduced when compared, for example, with histoplasmosis. In general patients with PCM/AIDS have prolonged fever, substantial weight loss, lymphadenopathy, hepatosplenomegaly, lung damage, and skin and neurological injuries. A prevalence of PCM/AIDS in 1.4% of patients has been reported in the southeastern region of Brazil; most of them have CD4 counts lower than 200 cells/mL, they are young, and few of them work in agriculture (Goldani and Sugar 1995; Bellissimo-Rodrigues et al. 2011).

6.5.4 PCM Sequelae

PCM induces a granulomatous inflammatory response in the host, which often leads to fibrosis. Advanced stages of the disease present substantial increase in cytokines, such as TNF-α and TGF-β, which induce the accumulation of collagen. Fibrosis has been observed in the lungs of ~50% of patients, of whom a small percentage develop chronic obstructive pulmonary disease (COPD). Studies show that 15–50% of patients show reduced adrenal gland function and about 3% develop Addison's disease and require hormone replacement therapy (Do Valle et al. 1993; Faiçal et al. 1996; Oñate et al. 2002). The CNS is also affected by PCM sequelae in about 6–27% of patients, leading to motoric deficits, seizures (epilepsy), and/or hydrocephalus (Nóbrega and Spina-França 1994; Shikanai-Yasuda et al. 2006).

Another important aspect in PCM sequelae is the presence of fibrotic lesions in mucosa and skin. Accordingly, humans with this disease have presented chronic changes of the voice (dysphonia due to the injury of the vocal cords), laryngeal obstruction that leads to tracheostomy, and microstomia due to perioral lesions (Paniago et al. 2003; Weber et al. 2006). Abdominal lymphatic obstruction, reported in acute PCM, has led to malabsorption syndromes, protein loss, and obstructive jaundice (Cazzo et al. 2015).

6.5.5 PCM Disease in Animals

In armadillos, animals that are frequently infected, occasional granuloma formation was observed in the liver and lungs, which may indicate that the animals develop active PCM disease (Bagagli and Bosco 2008). The generalized PCM reported in one two-toed sloth leads to progressive lethargy, anorexia, dehydration, and fatality. Necropsy revealed granulomatous lesions in the lungs, liver, spleen, and kidneys containing numerous fungal cells as shown in silver-stained histopathological sections (Trejo-Chávez et al. 2011).

As already mentioned, *P. brasiliensis* infection in dogs seems to be common; however, PCM disease with clinical symptoms has been reported only in three dogs, in which the main clinical sign was generalized lymphadenomegaly. In the first report, an adult female, non-neutered, Doberman, presented swelling of submandibular lymph nodes and poor general condition. The diagnosis was confirmed by histopathology, immunohistochemistry, and molecular detection of GP43 gene in lymph node biopsy. The animal was treated with ketoconazole producing total regression of lymphadenopathy. However, clinical recurrence was observed after 18 months, and the dog was euthanized without being subjected to autopsy (Ricci et al. 2004). The second case, also in an adult female non-neutered Doberman, showed lymphnodes enlargement (Fig. 6.4), and the diagnosis was confirmed by culturing *P. brasiliensis* from popliteal lymph node biopsy. The animal was successfully treated with itraconazolc for 2 years (Farias et al. 2011). The third report was described in a 5-year-old female Labrador with lymphnodes enlargement and skin lesion in the left side of the upper lip. The diagnosis was confirmed by cytological, serological, and molecular identification of *P. brasiliensis* isolated from fine needle aspirate from lymphnode. The dog was treated with itraconazole and is under observation for 18 months with remission of the symptoms (Headley et al. 2017).

A single case of feline PCM was reported in a cat from Chile, which developed neurological and renal symptoms. The host was an 8-month-old Persian cat with anorexia, weakness, fever, and neurological signs (depression, nystagmus, and tremors). The diagnosis was confirmed by cytopathological detection of the fungus in cerebrospinal fluid and urine, with no fungal culture (Gonzalez et al. 2010). This report should be considered with caution, since (1) the animal is from Chile, a country that PCM has never been recorded for human, and (2) the images of micromorphology of the fungus in the publication are not completely compatible with *P. brasiliensis*. Likewise, the same precaution should be considered to the reported case of *Mycobacterium bovis* and dimorphic fungi (*P. brasiliensis* and *B. dermatitidis*) co-infection in beef cattle in Kenya, Africa, by the observation of yeasts cells in Ziehl-Neelsen smears (Kuria and Gathogo 2013).

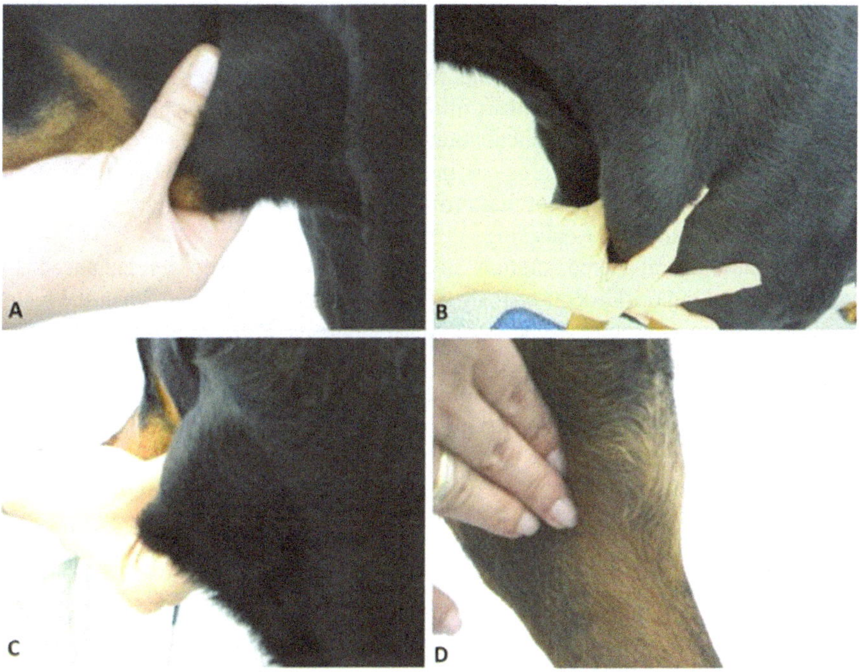

Fig. 6.4 Generalized lymphadenomegaly in a 6-year-old female Doberman. (**a**) Submandibular lymph node; (**b**) prescapular lymph node; (**c**) inguinal lymph node; (**d**) popliteal lymph node. "Copyright Mycopathologia (172(2):147–152, 2011), reprinted with permission"

P. lutzii has never been isolated directly from armadillos or any other animal species. To date, there is one study showing serological evidences about *P. lutzii* infection in domestic and wild animals (Mendes et al. 2017). The authors evaluated 481 animals (horses, dogs, and wild mammals) from South region of Brazil, and 105 reacted positively for *P. lutzii*. Among these seropositive animals, 54 showed cross-reaction with *P. brasiliensis* antigens and only 51 animals (11 horses, 30 dogs, and 10 wild mammals) were considered infected by *P. lutzii*, confirming the existence of this species in South region of Brazil (Mendes et al. 2017).

6.6 Diagnosis

The diagnosis of PCM requires the identification of the fungus in any clinical specimen by direct mycological examination, fungal culture, and histopathology or by molecular detection. Clinical materials consist of scrapings from a lesion, respiratory secretions, bronchoalveolar lavage, pus, cerebrospinal fluid, lymph node aspirates, synovial fluid, ascites and other fluids, or biopsied tissue fragments.

For direct examination, better results are obtained when the sample is clarified with KOH (10%) demonstrating large yeast cells with thick, birefringent walls and multipolar budding. Isolation of the fungus may be obtained by culturing on Sabouraud glucose agar, brain-heart infusion agar, Mycosel®, or glucose-peptone-yeast extract (GPY) agar with incubation at 25 and 35 °C to confirm dimorphism.

Fine needle aspiration cytology is indicated for solid tissues, and the puncture should be guided by equipment that provides a high-resolution image. The aspirate may also be employed for culturing.

Histopathological demonstration of *Paracoccidioides* spp. is enhanced by staining with Gomori-Grocott or PAS. Special attention is required when the histological sections show only some small yeast elements with single buddings, since small forms of *P. brasiliensis* exist. In this situation, culturing, immunohistochemistry, or molecular biology should be employed for differential diagnosis.

Serological tests are important, both for diagnosis and for monitoring treatment. The most employed techniques are double immunodiffusion in agar gel (ID) and immunoenzymatic assays, such as ELISA (enzyme-linked immunosorbent assay), which present high sensitivity and specificity. Cryptic species in *Paracoccidioides* may have an impact on immunodiagnosis of PCM. PCM patients from the southeast of Brazil were 100% positive in serology when tested with B339 antigen from *P. brasiliensis*. When the same sera were tested using an antigen preparation from *P. lutzii* 510B isolate, positivity decreased to 41%. Conversely, patients from the midwest of Brazil tested 92% positive with antigens from *P. lutzii* and 26% with *P. brasiliensis* antigen (Batista et al. 2010; Gegembauer et al. 2014).

Molecular techniques have high sensitivity and specificity and have a great potential for diagnosis and environmental detection. The most frequently used technique is nested PCR with primers derived from rDNA. Also panfungal primers ITS1-ITS4 or ITS4-ITS5 (White et al. 1990) or species-specific primers for *P. brasiliensis* and *P. lutzii* can be used (Theodoro et al. 2005; Arantes et al. 2013).

6.7 Public Health Concerns

Prevention and control of new infections by *Paracoccidioides* spp. are difficult to adopt, since the exact ecological niche of the fungus is still unknown. Inhalation of aerosols from soil in hyperendemic areas particularly armadillo burrows might be a risk factor. Hunting dogs with digging habits are probably more exposed to infection, and it is recommended to monitor the dog for possible clinical changes. Early diagnosis improves prognosis.

PCM is a serious zoonosis in a public health context. The disease can be classified as a saprozoonosis, transmission occurring by abiotic material (soil) impacted by the natural host animal. Direct transmission from an animal or among humans is not known to occur. *Paracoccidioides* spp. are classified as a risk group 2 organism, and handling of fungal cultures should be done under appropriate safety conditions, especially when dealing with the infective mycelial form of the fungus.

PCM is a non-reportable neglected disease, and, since it is the most important systemic mycosis in Latin America, health authorities should encourage programs and campaigns with the aim of educating the public about the risks and potential manners of acquiring the infection.

References

Almeida F (1930) Estudos comparativos do granuloma coccidioidico nos Estados Unidos e no Brasil. Novo gênero para o parasito brasileiro. An Fac Med Sao Paulo 5:125–142

Arantes TD, Theodoro RC, Macoris SADG, Bagagli E (2013) Detection of *Paracoccidioides* spp. in environmental aerosol samples. Med Mycol 51:83–92

Bagagli E, Bosco SMG (2008) Armadillos and dimorphic pathogenic fungi: ecological and evolutionary aspects. In: Lowghry WJ, Viscaino SF (eds) The biology of the Xenarthra. University Press of Florida, Gainesville, pp 103–110

Bagagli E, Sano A, Coelho KI et al (1998) Isolation of *Paracoccidioides brasiliensis* from armadillos (*Dasypus noveminctus*) captured in an endemic area of paracoccidioidomycosis. Am J Trop Med Hyg 58:505–512

Bagagli E, Franco M, Bosco SDM et al (2003) High frequency of *Paracoccidioides brasiliensis* infection in armadillos (*Dasypus novemcinctus*): an ecological study. Med Mycol 41:217–223

Bagagli E, Arantes T, Theodoro R (2012) *Paracoccidioides brasiliensis*: ecology and evolution. Mycoses 55:54–55

Batista J, de Camargo ZP, Fernandes GF et al (2010) Is the geographical origin of a *Paracoccidioides brasiliensis* isolate important for antigen production for regional diagnosis of paracoccidioidomycosis? Mycoses 53:176–180. https://doi.org/10.1111/j.1439-0507.2008.01687.x

Belitardo DR, Calefi AS, Borges IK et al (2014a) Detection of antibodies against *Paracoccidioides brasiliensis* in free-range domestic pigs (*Sus scrofa*). Mycopathologia 177:91–95. https://doi.org/10.1007/s11046-013-9724-2

Belitardo DR, Calefi AS, Sbeghen MR et al (2014b) *Paracoccidioides brasiliensis* infection in domestic rabbits (*Oryctolagus cuniculus*). Mycoses 57:222–227. https://doi.org/10.1111/myc.12146

Bellissimo-Rodrigues F, Machado AA, Martinez R (2011) Paracoccidioidomycosis epidemiological features of a 1,000-cases series from a hyperendemic area on the southeast of Brazil. Am J Trop Med Hyg 85:546–550. https://doi.org/10.4269/ajtmh.2011.11-0084

Bocca AL, Amaral AC, Teixeira MM et al (2013) Paracoccidioidomycosis: eco-epidemiology, taxonomy and clinical and therapeutic issues. Future Microbiol 8:1177–1191. https://doi.org/10.2217/fmb.13.68

Brummer E, Castaneda E, Restrepo A (1993) Paracoccidioidomycosis: an update. Clin Microbiol Rev 6:89–117

Cadavid D, Restrepo A (1993) Factors associated with *Paracoccidiodes brasiliensis* infection among permanent residents of three endemic areas in Colombia. Epidemiol Infect 111:121–133

Canteros CE, Madariaga MJ, Lee W et al (2010) Endemic fungal pathogens in a rural setting of Argentina: seroepidemiological study in dogs. Rev Iberoam Micol 27:14–19. https://doi.org/10.1016/j.riam.2009.11.002

Cazzo E, Ferrer JAP, Chaim EA (2015) Obstructive jaundice secondary to paracoccidioidomycosis. Trop Gastroenterol 36:46–47

Coelho KIR, Defaveri J, Rezkallah-Iwasso MT, Peraçoli MTS (1994) Experimental paracoccidioidomycosis. In: Franco M, Lacaz CDS, Restrepo-Moreno A, Negro GD (eds) Paracoccidioidomycosis. CRC Press, Boca Raton, pp 87–107

Corredor GG, Castaño JH, Peralta LA et al (1999) Isolation of *Paracoccidioides brasiliensis* from the nine-banded armadillo *Dasypus novemcinctus*, in an endemic area for paracoccidioidomycosis in Colombia. Rev Iberoam Micol 16:216–220

Corredor GG, Peralta LA, Castaño JH et al (2005) The naked-tailed armadillo *Cabassous centralis* (Miller 1899): a new host to *Paracoccidioides brasiliensis*. Molecular identification of the isolate. Med Mycol 43:275–280

Corte AC, Svoboda WK, Navarro IT et al (2007) Paracoccidioidomycosis in wild monkeys from Paraná state, Brazil. Mycopathologia 164:225–228. https://doi.org/10.1007/s11046-007-9059-y

Costa EO, Fava Netto C (1978) Contribution to the epidemiology of paracoccidioidomycosis and histoplasmosis in the state of São Paulo, Brazil. Paracoccidioidin and histoplasmin intradermic tests in domestic animals. Sabouraudia 16:93–101

Costa EO, Netto CF, Rodrigues A, Brito T (1978) Bovine experimental paracoccidioidomycosis intradermic test standardization. Sabouraudia 16:103–113

Costa EO, Diniz LS, Netto CF (1995a) The prevalence of positive intradermal reactions to paracoccidioidin in domestic and wild animals in São Paulo, Brazil. Vet Res Commun 19:127–130

Costa EO, Diniz LS, Netto CF et al (1995b) Delayed hypersensitivity test with paracoccidioidin in captive Latin American wild mammals. J Med Vet Mycol 33:39–42

Coutinho ZF, da Silva D, Lazera M et al (2002) Paracoccidioidomycosis mortality in Brazil (1980–1995). Cad Saúde Pública 18:1441–1454

Coutinho ZF, Wanke B, Travassos C et al (2015) Hospital morbidity due to paracoccidioidomycosis in Brazil (1998–2006). Trop Med Int Health 20:673–680. https://doi.org/10.1111/tmi.12472

de Castro LF, Ferreira MC, da Silva RM et al (2013) Characterization of the immune response in human paracoccidioidomycosis. J Infect 67:470–485. https://doi.org/10.1016/j.jinf.2013.07.019

de Oliveira GG, Belitardo DR, Balarin MRS et al (2013) Serological survey of paracoccidioidomycosis in cats. Mycopathologia 176:299–302. https://doi.org/10.1007/s11046-013-9681-9

Del Negro G, Lacaz CS, Zamith VA, Siqueira AM (1994) General clinical aspects: polar forms of Paracoccidioidomycosis, the disease in childhood. In: Franco M, Lacaz CDS, Restrepo-Moreno A, Negro GD (eds) Paracoccidioidomycosis. CRC Press, Boca Raton, pp 225–232

Do Valle AC, Guimaraes MR, Cuba J et al (1993) Recovery of adrenal function after treatment of paracoccidioidomycosis. Am J Trop Med Hyg 48:626–629

Dukik K, Muñoz JF, Jiang Y, Feng P et al (2017) Novel taxa of thermally dimorphic systemic pathogens in the Ajellomycetaceae (Onygenales). Mycoses 60(5):296–309. https://doi.org/10.1111/myc.12601

Eisele RC, Juliani LC, Belitardo DR et al (2004) Immune response in dogs experimentally infected with *Paracoccidioides brasiliensis*. Med Mycol 42:549–553

Faiçal S, Borri ML, Hauache OM, Ajzen S (1996) Addison's disease caused by *Paracoccidioides brasiliensis*: diagnosis by needle aspiration biopsy of the adrenal gland. AJR Am J Roentgenol 166:461–462. https://doi.org/10.2214/ajr.166.2.8553971

Farias MR, Condas LAZ, Ribeiro MG et al (2011) Paracoccidioidomycosis in a dog: case report of generalized lymphadenomegaly. Mycopathologia 172:147–152

Ferreira JB, Navarro IT, Freire RL et al (2013) Evaluation of *Paracoccidioides brasiliensis* infection in dairy goats. Mycopathologia 176:95–99. https://doi.org/10.1007/s11046-013-9644-1

Fontana FF, dos Santos CTB, Esteves FM et al (2010) Seroepidemiological survey of paracoccidioidomycosis infection among urban and rural dogs from Uberaba, Minas Gerais, Brazil. Mycopathologia 169:159–165. https://doi.org/10.1007/s11046-009-9241-5

Franco M, Montenegro MR, Mendes RP et al (1987) Paracoccidioidomycosis: a recently proposed classification of its clinical forms. Rev Soc Bras Med Trop 20:129–132

Franco M, Peracoli MT, Soares A et al (1993) Host-parasite relationship in paracoccidioidomycosis. Curr Top Med Mycol 5:115–149

Franco MV, Goes AM, Koury MC (1997) Model of in vitro granulomatous hypersensitivity in human paracoccidioidomycosis. Mycopathologia 137:129–136

Franco M, Bagagli E, Scapolio S, Lacaz CDS (2000) A critical analysis of isolation of *Paracoccidioides brasiliensis* from soil. Med Mycol 38:185–191

Gegembauer G, Araujo LM, Pereira EF et al (2014) Serology of paracoccidioidomycosis due to *Paracoccidioides lutzii*. PLoS Negl Trop Dis 8:e2986. https://doi.org/10.1371/journal.pntd.0002986
Goldani LZ, Sugar AM (1995) Paracoccidioidomycosis and AIDS: an overview. Clin Infect Dis 21:1275–1281
Gonzalez JF, Montiel NA, Maass RL (2010) First report on the diagnosis and treatment of encephalic and urinary paracoccidioidomycosis in a cat. J Feline Med Surg 12:659–662. https://doi.org/10.1016/j.jfms.2010.03.016
Headley SA, Pretto-Giordano LG, Di Santis GW, Gomes LA, Macagnan R, da Nóbrega DF, Leite KM, de Alcântara BK, Itano EN, Alfieri AA, Ono MA (2017) Paracoccidioides brasiliensis-associated dermatitis and lymphadenitis in a dog. Mycopathologia 182(3–4):425–434
Herr RA, Tarcha EJ, Taborda PR et al (2001) Phylogenetic analysis of *Lacazia loboi* places this previously uncharacterized pathogen within the dimorphic Onygenales. J Clin Microbiol 39:309–314. https://doi.org/10.1128/JCM.39.1.309-314.2001
Kuria JN, Gathogo SM (2013) Concomitant fungal and *Mycobacterium bovis* infections in beef cattle in Kenya. Onderstepoort J Vet Res 80(1):585. https://doi.org/10.4102/ojvr.v80i1.585
Lutz A (1908) Uma mycose pseudococcidica localisada na bocca e observada no Brasil. Contribuição ao conhecimento das hyphoblastomycoses americanas. Brasil Méd 22:121–124
Martinez R (2015) Epidemiology of paracoccidioidomycosis. Rev Inst Med Trop São Paulo 57 (Suppl 19):11–20. https://doi.org/10.1590/S0036-46652015000700004
Martinez R, Moya MJ (1992) The relationship between paracoccidioidomycosis and alcoholism. Rev Saúde Pública 26:12–16
Matute DR, McEwen JG, Puccia R et al (2006a) Cryptic speciation and recombination in the fungus *Paracoccidioides brasiliensis* as revealed by gene genealogies. Mol Biol Evol 23:65–73. https://doi.org/10.1093/molbev/msj008
Matute DR, Sepulveda VE, Quesada LM et al (2006b) Microsatellite analysis of three phylogenetic species of *Paracoccidioides brasiliensis*. J Clin Microbiol 44:2153–2157. https://doi.org/10.1128/JCM.02540-05
McEwen JG, Bedoya V, Patiño MM et al (1987) Experimental murine paracoccidiodomycosis induced by the inhalation of conidia. J Med Vet Mycol 25:165–175
Mendes RP (1994) The gamut of clinical manifestations. In: Franco M, Lacaz CDS, Restrepo-Moreno A, Negro GD (eds) Paracoccidioidomycosis. CRC Press, Boca Raton, pp 233–257
Mendes JF, Klafke GB, Albano APN, Cabana AL, Teles AJ, Camargo ZP, Xavier MO, Meireles MCA (2017) Paracoccidioidomycosis infection in domestic and wild mammals by *Paracoccidioides lutzii*. Mycoses 60:402–406
Montenegro MR, Franco MF (1994) Pathology. In: Franco M, Lacaz CDS, Restrepo-Moreno A, Negro GD (eds) Paracoccidioidomycosis. CRC Press, Boca Raton, pp 131–150
Mós EDN, Netto CF (1974) Contribution to the study of paracoccidioidomycosis. I. Possible epidemiological role of dogs. Serological and anatomo-pathological study. Rev Inst Med Trop São Paulo 16:154–159
Mós EN, Netto CF, Saliba AM, de Brito T (1974) Contribution to the study of paracoccidioidomycosis. II. Experimental infection of dogs. Rev Inst Med Trop São Paulo 16:232–237
Moscardi-Bacchi M, Soares A, Mendes R et al (1989) In situ localization of T lymphocyte subsets in human paracoccidioidomycosis. J Med Vet Mycol 27:149–158
Mota NG, Rezkallah-Iwasso MT, Peraçoli MT et al (1985) Correlation between cell-mediated immunity and clinical forms of paracoccidioidomycosis. Trans R Soc Trop Med Hyg 79:765–772
Mota NG, Peraçoli MT, Mendes RP et al (1988) Mononuclear cell subsets in patients with different clinical forms of paracoccidioidomycosis. J Med Vet Mycol 26:105–111
Musatti CC, Peraçoli MTS, Soares AMVC, Rezkallah-Iwasso MT (1994) Cell-mediated immunity in patients with Paracoccidioidomycosis. In: Franco M, Lacaz CDS, Restrepo-Moreno A, Negro GD (eds) Paracoccidioidomycosis. CRC Press, Boca Raton, pp 175–186
Naiff RD, Ferreira LC, Barrett TV et al (1986) Enzootic paracoccidioidomycosis in armadillos (*Dasypus novemcinctus*) in the state of Pará. Rev Inst Med Trop São Paulo 28:19–27

Nóbrega JPS, Spina-França A (1994) Neuroparacocidioidomycosis. In: Franco M, Lacaz CDS, Restrepo-Moreno A, Negro GD (eds) Paracoccidioidomycosis. CRC Press, Boca Raton, pp 321–330

Oliveira GG, Silveira LH, Itano EN et al (2011) Serological evidence of *Paracoccidioides brasiliensis* infection in chickens from Paraná and Mato Grosso do Sul states, Brazil. Mycopathologia 171:197–202. https://doi.org/10.1007/s11046-010-9366-6

Oliveira GG, Navarro IT, Freire RL et al (2012) Serological survey of Paracoccidioidomycosis in sheep. Mycopathologia 173:63–68. https://doi.org/10.1007/s11046-011-9463-1

Oñate JM, Tobón AM, Restrepo A (2002) Adrenal gland insufficiency secondary to paracoccidioidomycosis. Biomédica 22:280–286

Ono MA, Bracarense AP, Morais HS et al (2001) Canine paracoccidioidomycosis: a seroepidemiologic study. Med Mycol 39:277–282

Ono MA, Kishima MO, Itano EN et al (2003) Experimental paracoccidioidomycosis in dogs. Med Mycol 41:265–268

Paniago AMM, Aguiar JIA, Aguiar ES et al (2003) Paracoccidioidomycosis: a clinical and epidemiological study of 422 cases observed in Mato Grosso do Sul. Rev Soc Bras Med Trop 36:455–459

Peraçoli MT, Mota NG, Montenegro MR (1982) Experimental paracoccidioidomycosis in the Syrian hamster. Morphology and correlation of lesions with humoral and cell-mediated immunity. Mycopathologia 79:7–17

Peraçoli MTS, Kurokawa CS, Calvi SA et al (2003) Production of pro- and anti-inflammatory cytokines by monocytes from patients with paracoccidioidomycosis. Microbes Infect 5:413–418

Restrepo A (1985) The ecology of *Paracoccidioides brasiliensis*: a puzzle still unsolved. Sabouraudia 23:323–334

Restrepo A, Salazar ME, Cano LE et al (1984) Estrogens inhibit mycelium-to-yeast transformation in the fungus *Paracoccidioides brasiliensis*: implications for resistance of females to paracoccidioidomycosis. Infect Immun 46:346–353

Restrepo A, Baumgardner D, Bagagli E et al (2000) Clues to the presence of pathogenic fungi in certain environments. Med Mycol 38:67–77

Restrepo A, McEwen JG, Castañeda E (2001) The habitat of *Paracoccidioides brasiliensis*: how far from solving the riddle? Med Mycol 39:233–241

Restrepo-Moreno A (1994) Ecology of *Paracoccidioides brasiliensis*. In: Franco M, Lacaz CDS, Restrepo-Moreno A, Negro GD (eds) Paracoccidioidomycosis. CRC Press, Boca Raton, pp 121–129

Ricci G, Mota FT, Wakamatsu A et al (2004) Canine paracoccidioidomycosis. Med Mycol 42:379–383

Richini-Pereira VB, Bosco SDMG, Macoris SDG et al (2008a) Detecção de *Paracoccidioides brasiliensis* em tatus (*Dasypus novemcinctus*) provenientes de uma reserva de Cerrado do Instituto Lauro de Souza Lima (Bauru, SP). BEPA Bol Epidemiol Paul 5:4–8

Richini-Pereira VB, Bosco SDMG, Griese J et al (2008b) Molecular detection of *Paracoccidioides brasiliensis* in road-killed wild animals. Med Mycol 46:35–40

Richini-Pereira VB, Bosco SM, Theodoro RC et al (2009) Importance of xenarthrans in the eco-epidemiology of *Paracoccidioides brasiliensis*. BMC Res Notes 2:228

Salazar ME, Restrepo A, Stevens DA (1988) Inhibition by estrogens of conidium-to-yeast conversion in the fungus *Paracoccidioides brasiliensis*. Infect Immun 56:711–713

Santos WA, da Silva BM, Passos ED et al (2003) Association between smoking and paracoccidioidomycosis: a case-control study in the state of Espírito Santo, Brazil. Cad Saúde Pública 19:245–253

Sbeghen MR, Zanata TB, Macagnan R et al (2015) *Paracoccidioides brasiliensis* infection in small wild mammals. Mycopathologia 180:435–440. https://doi.org/10.1007/s11046-015-9928-8

Shikanai-Yasuda MA, Telles Filho FDQ, Mendes RP et al (2006) Guidelines in paracoccidioidomycosis. Rev Soc Bras Med Trop 39:297–310

Silva-Vergara ML, Martinez R, Camargo ZP et al (2000) Isolation of *Paracoccidioides brasiliensis* from armadillos (*Dasypus novemcinctus*) in an area where the fungus was recently isolated from soil. Med Mycol 38:193–199

Silveira LH, Domingos IH, Kouchi K et al (2006) Serological detection of antibodies against *Paracoccidioides brasiliensis* in dogs with leishmaniasis. Mycopathologia 162:325–329. https://doi.org/10.1007/s11046-006-0046-5

Silveira LH, Paes RCS, Medeiros EV et al (2008) Occurrence of antibodies to *Paracoccidioides brasiliensis* in dairy cattle from Mato Grosso do Sul, Brazil. Mycopathologia 165:367–371. https://doi.org/10.1007/s11046-008-9095-2

Singer-Vermes LM, Burger E, Russo M et al (1993) Advances in experimental paracoccidioidomycosis using an isogenic murine model. Arch Med Res 24:239–245

Soares AM, Peraçoli MT, Dos Santos RR (2000) Correlation among immune response, morphogenesis of the granulomatous reaction and spleen lymphoid structure in murine experimental paracoccidioidomycosis. Med Mycol 38:371–377

Splendore A (1912) Zimonematosi con localizzazione nella cavita della bocca osservata nel Brasile. Bull Soc Path 5:313–319

Tani EM, Franco M, Peraçoli MT, Montenegro MR (1987) Experimental pulmonary paracoccidioidomycosis in the Syrian hamster: morphology and correlation of lesions with the immune response. J Med Vet Mycol 25:291–300

Teixeira MDM, Theodoro RC, Nino-Vega G et al (2014a) *Paracoccidioides* species complex: ecology, phylogeny, sexual reproduction, and virulence. PLoS Pathog 10:e1004397. https://doi.org/10.1371/journal.ppat.1004397

Teixeira MDM, Theodoro RC, Oliveira FFM et al (2014b) *Paracoccidioides lutzii* sp. nov.: biological and clinical implications. Med Mycol 52:19–28. https://doi.org/10.3109/13693786.2013.794311

Teles AJ, Klafke GB, Cabana ÂL et al (2015) Serological investigation into *Paracoccidioides brasiliensis* infection in dogs from southern Rio Grande do Sul, Brazil. Mycopathologia 181 (3-4):323–328. https://doi.org/10.1007/s11046-015-9972-4

Theodoro RC, Candeias JMG, Araújo JP et al (2005) Molecular detection of *Paracoccidioides brasiliensis* in soil. Med Mycol 43:725–729

Theodoro RC, Volkmann G, Liu X-Q, Bagagli E (2011) PRP8 intein in Ajellomycetaceae family pathogens: sequence analysis, splicing evaluation and homing endonuclease activity. Fungal Genet Biol 48:80–91. https://doi.org/10.1016/j.fgb.2010.07.010

Theodoro RC, Teixeira MDM, Felipe MSS et al (2012) Genus *Paracoccidioides*: species recognition and biogeographic aspects. PLoS One 7:e37694. https://doi.org/10.1371/journal.pone.0037694

Trejo-Chávez A, Ramírez-Romero R, Ancer-Rodríguez J et al (2011) Disseminated paracoccidioidomycosis in a southern two-toed sloth (*Choloepus didactylus*). J Comp Pathol 144:231–234. https://doi.org/10.1016/j.jcpa.2010.08.012

Turissini DA, Gomez OM, Teixeira MM, Mcewen JG, Matute DR (2017) Species boundaries in the human pathogen *Paracoccidioides*. Fungal Genet Biol 106:9–25

Untereiner WA, Scott JA, Naveau FA et al (2004) The Ajellomycetaceae, a new family of vertebrate-associated Onygenales. Mycologia 96:812–821

Vilela R, Rosa PS, Belone AFF et al (2009) Molecular phylogeny of animal pathogen *Lacazia loboi* inferred from rDNA and DNA coding sequences. Mycol Res 113:851–857. https://doi.org/10.1016/j.mycres.2009.04.007

Wanke B, Londero AT (1994) Epidemiology and Paracoccidioidomycosis infection. In: Franco M, Lacaz CDS, Restrepo-Moreno A, Negro GD (eds) Paracoccidioidomycosis. CRC Press, Boca Raton, pp 109–119

Weber SAT, Brasolotto A, Rodrigues L et al (2006) Dysphonia and laryngeal sequelae in paracoccidioidomycosis patients: a morphological and phoniatric study. Med Mycol 44:219–225. https://doi.org/10.1080/13693780500340320

White TJ, Bruns T, Lee S, Taylor JW (1990) Amplification and direct sequencing of fungal ribosomal RNA genes for phylogenetics. In: Innis MA, Gelfand DH, Sninsky JJ, White TJ (eds) PCR protocols: a guide to methods and applications. Academic Press, San Diego, pp 315–322

Adiaspiromycosis and Diseases Caused by Related Fungi in *Ajellomycetaceae*

7

Andrew M. Borman, Yanping Jiang, Karolina Dukik, Lynne Sigler, Ilan S. Schwartz, and G. Sybren de Hoog

Abstract

Adiaspiromycosis, classically caused by *Emmonsia* species, is primarily a pulmonary disease affecting small mammals, especially members of the orders Rodentia and Carnivora. The disease name derives from the tissue form of the fungus (adiaspores), which develops when the inhaled conidia of the mycelial form of the fungus present in soil enlarge in lungs to produce thick-walled non-replicative structures. *Emmonsia crescens* has apparently worldwide

A. M. Borman
UK National Mycology Reference Laboratory, Public Health England, Bristol, UK
e-mail: Andy.Borman@uhBristol.nhs.uk

Y. Jiang
Department of Dermatology, The Affiliated Hospital, Guizhou Medical University, Guiyang, China

Center of Expertise in Mycology Radboudumc/CWZ, Nijmegen, The Netherlands
e-mail: jiangyanping119@163.com

K. Dukik
Westerdijk Fungal Biodiversity Institute, Utrecht, The Netherlands
e-mail: k.dukik@westerdijkinstitute.nl

L. Sigler
Microfungus Collection and Herbarium [now UAMH Centre for Global Microfungal Diversity] and Biological Sciences, University of Alberta, Edmonton, Canada
e-mail: Lynne.Sigler@ualberta.ca

I. S. Schwartz
Division of Infectious Diseases, Department of Medicine, University of Alberta, Edmonton, AB, Canada
e-mail: ilan.schwartz@ualberta.ca

G. S. de Hoog (✉)
Center of Expertise in Mycology RadboudUMC/CWZ, Nijmegen, The Netherlands

Westerdijk Fungal Biodiversity Institute, Utrecht, The Netherlands
e-mail: s.hoog@westerdijkinstitute.nl

S. Seyedmousavi et al. (eds.), *Emerging and Epizootic Fungal Infections in Animals*,
https://doi.org/10.1007/978-3-319-72093-7_7

distribution and is associated with very large adiaspores (up to 400 μm). The type species of *Emmonsia*, *E. parva*, however, is now recognised to be a *Blastomyces* species and its 'small adiaspores' are comparable to the broad-based budding cells current in that genus. This chapter briefly summarises current knowledge concerning taxonomy, epidemiology, biology and clinical syndromes of the principal etiological agents of adiaspiromycosis in mammals.

7.1 Causative Agents of Adiaspiromycosis and Relatives

Adiaspiromycosis is a pulmonary infection by members of the family Ajellomycetaceae (order Onygenales) where the fungus is present in pulmonary tissue as very large, thick-walled resting cells known as adiaspores (Fig. 7.1). The genus *Emmonsia* was described for fungi producing adiaspores and until recently comprised two species: *Emmonsia parva* and *E. crescens* (Ciferri and Montemartini 1959; Emmons and Jellison 1960). *Emmonsia pasteuriana* was later added to the genus (Drouhet et al. 1998) despite the absence of adiaspores, instead having a pathogenic phase with principally small budding cells. Recent molecular phylogenetic studies have recognised the polyphyletic nature of the genus *Emmonsia*, and consequent revisions have changed the taxonomic landscape considerably (Dukik et al. 2017; Jiang et al. 2018). Classical mating experiments demonstrated that *E. crescens* has an *Ajellomyces* teleomorph (Sigler 1996), and DNA sequence comparisons already had suggested a sister species relationship between *Emmonsia parva* (the type species of *Emmonsia*) and *Blastomyces dermatitidis* (Peterson and Sigler 1998; Sigler 2005). With the validation of the name *Blastomyces* (de Hoog et al. 2017), the generic name *Emmonsia* has been discarded as a synonym. *Emmonsia parva* was thus reclassified as a *Blastomyces* species, *B. parvus* (Jiang et al. 2018). *Emmonsia crescens* was described as the type species of a separate genus, *Emmonsia* (Jiang et al. 2018). *Emmonsia crescens* is responsible for most animal cases of adiaspiromycosis (Sigler 2005), a disease which is occasionally observed in humans (Anstead et al. 2012). *Emmonsia pasteuriana* was also reclassified as the type species of a new genus, *Emergomyces (E. pasteurianus)*, alongside *Emergomyces africanus* (Kenyon et al. 2014; Dukik et al. 2017), *Emergomyces orientalis* (Wang et al. 2017), *Emergomyces canadensis* and *Emergomyces europaeus* (Jiang et al. 2018). *Emergomyces pasteurianus* and *E. africanus*, which both have a yeast rather than an adiaspore tissue form, principally cause disseminated infections in humans with T-cell immune defects (Drouhet et al. 1998; Feng et al. 2015; Schwartz et al. 2015a; Malik et al. 2016; Dukik et al. 2017).

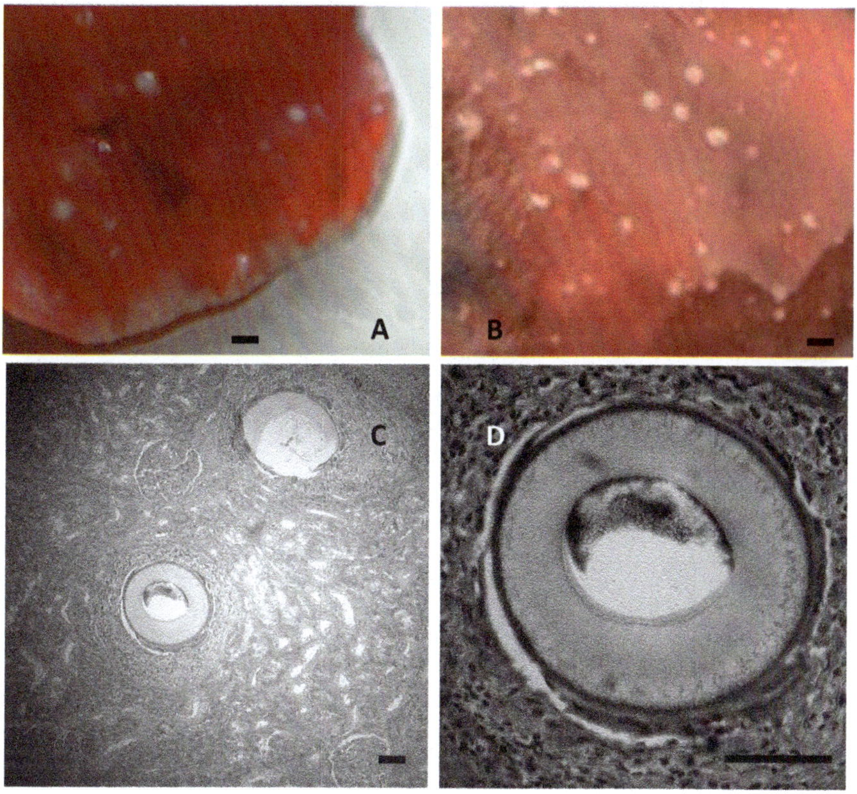

Fig. 7.1 Adiaspiromycosis in mammalian lungs. Gross lesions of adiaspiromycosis in the lung of a mole (**a**), and an otter (**b**). (**c**, **d**) Histopathological section of experimentally infected monkey kidney (inoculation directly into kidney) showing adiaspores surrounded by granulomata. H&E stain. Scale bar = 1 mm (Panels A and B) or 100 μm (Panels C and D)

7.2 Biology and Epidemiology of Epizootic *Emmonsia*, *Blastomyces* and *Emergomyces* Species

In their mycelial form, many members of Ajellomycetaceae resemble each other in the production of spherical or disc-shaped conidia borne on narrow, cylindrical or swollen stalks (Dukik et al. 2017; Fig. 7.2). However, in mammalian hosts, the genera *Emmonsia*, *Blastomyces* and *Emergomyces* can be differentiated by the appearance of their thermodependent growth phase. *Blastomyces parvus* produces uninucleate, typically 10–40 μm wide, thin-walled cells (Sigler 2005), whereas cells of *A. crescens* (adiaspores) are multinucleate and up to 400 μm in size (Dvořák et al. 1973; Boisseau-Lebreuil 1975; Sigler 2005). Moreover, there are differences between these species in the temperatures at which thermal dimorphism occurs, with *A. crescens* and *B. parvus* switching at temperatures approaching 37 and 40 °C,

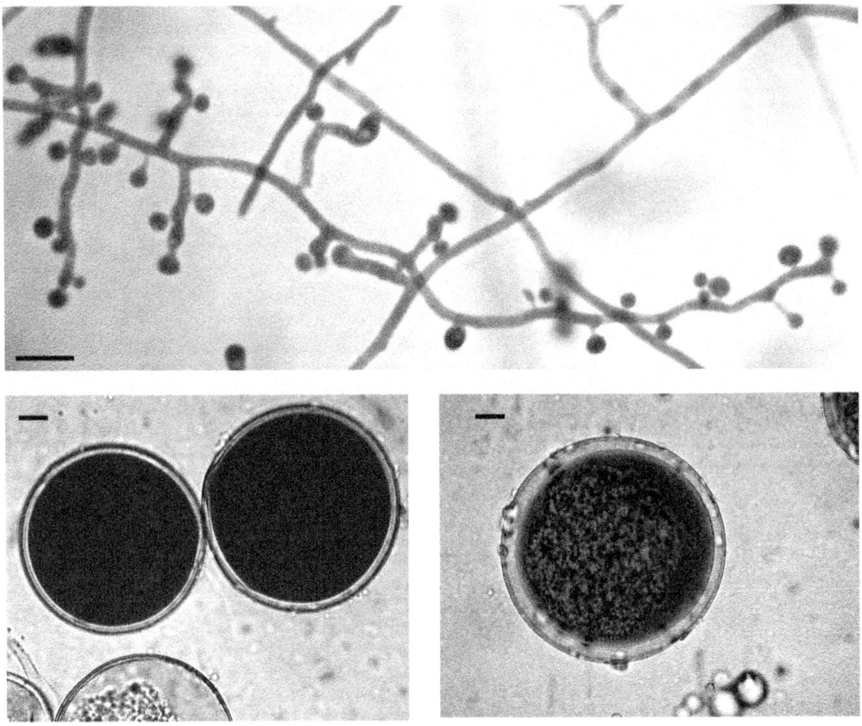

Fig. 7.2 Microscopic appearance of *Emmonsia crescens*, NCPF 4268 on Sabouraud's agar after 3 weeks at 30 °C (Top), 3 weeks at 37 °C (bottom left) or blood agar after 3 weeks at 37 °C (bottom right). Scale bar = 10 μm

respectively (Borman et al. 2009; Jiang et al. 2018). Similar adiaspores can easily be produced from mycelial cultures of *A. crescens* in vitro if cultures are moved to and maintained at 37 °C, with typical thick-walled multinucleate cells developing after several weeks (Fig. 7.2).

Judging from historical reports in which adiaspore size was accurately recorded, *A. crescens* was encountered worldwide, and apparently has a broad host range, although it has a focus with small terrestrial animals. Infections have been reported in foxes, Eurasian otters, stoats, weasels, and various species of mice, voles and shrews in Central and Eastern Europe (Sharapov 1969; Krivanec 1977; Krivanec and Otcenasek 1977; Hubalek 1999; Borman et al. 2009), ferrets in New Zealand (Lugton et al. 1997), hairy nosed wombats in Australia (Mason and Gauhwin 1982), striped skunks and ground squirrels in Canada (Leighton and Wobeser 1978; Albassam et al. 1986), hedgehogs, moles, voles and shrews in France (Doby et al. 1971), otters, voles, muskrats and lemmings in Finland (Jellison et al. 1960), voles, shrews and otters in Sweden (Jellison 1969), mole-rats in Zambia and Israel (Hubalek et al. 2005) and various free-living mammals in the UK (McDiarmid and Austwick 1954; Tevis 1956; Austwick 1968, Chantrey et al. 2006; Borman et al.

2009). Indeed, previous reviews have enumerated in excess of 100 host animal species for *A. crescens* (Sigler 2005). Conversely, *B. parvus* has a lower prevalence, with reports from the USA, Kenya, Zambia, Israel, Australia and Eastern Europe, and with a narrower host range (Jellison 1969; Krivanec et al. 1976; Krivanec 1977; Mason and Gauhwin 1982; Hubalek et al. 2005; Sigler 2005).

Although the natural habitat of *Emmonsia*-like species has not been precisely defined, there is strong evidence that infection is acquired by inhalation of conidia from soil and/or nesting materials of burrowing mammals. *Emergomyces africanus* has been detected in soil and air samples from South Africa (Schwartz et al. 2018a, b), but natural infections of animals have not been proven (Cronje et al. 2017). Experimental infections have successfully been induced in a variety of mammalian species after intranasal inoculation of conidial suspensions prepared from the mycelial phase of *A. crescens* (Jellison 1969), with mature adiaspores measuring up to 100 μm developing within 15 days of inoculation. Moreover, several reports have detailed the direct isolation of *Emmonsia*-like fungi from soil and nest materials with or without passage through rodents (Jellison 1969; Hubalek et al. 1995; Peterson and Sigler 1998; Sigler and Flis 1998 and later editions), and viable adiaspores have been recovered from the digestive tracts of rodents and also from the dung of larger carnivores that prey on infected rodents (Krivanec et al. 1975; reviewed in Sigler 2005). Finally, studies from Czechia reported high prevalence in nesting rodents in wooded areas (Hubalek et al. 1995), but failed to find evidence of adiaspiromycosis in rodents that do not nest in soil (Hubalek 1999). However, although the above data are good evidence that *A. crescens* may be vectored by animals (via predation/excretion/decomposition), there exists no evidence that the pathogen can be transmitted directly from animals to man.

Several studies have evaluated prevalence of infection. Studies from Czechia, France and UK estimated prevalence rates between 30 and 70% (e.g. Dvořák et al. 1973; Hubalek 1999; Borman et al. 2009), with higher rates in rodents than in carnivores and a predilection for mustelid carnivores (Doby et al. 1971; Krivanec et al. 1975; Borman et al. 2009). Several reports also suggested seasonal variation in infection rates (Dvořák et al. 1969; Hubalek et al. 1993), and presented evidence for wide heterogeneity of adiaspore sizes within the lungs of single affected mammals, suggestive of multiple exposures, possibly by several species. Similarly, the analysis of infectious burdens in several studies revealed large variations in mean adiaspore numbers amongst infected animals, with heavy infections associated with poor general condition, emaciation and even death (Jellison 1969; Hubalek et al. 1995; Simpson and Gavier-Wilden 2000; Borman et al. 2009). Finally, isolated cases of adiaspiromycosis have been sporadically reported in domestic mammals, including farm animals, dogs, goats and horses (reviewed in Sigler 2005).

7.3 Adiaspiromycosis and *Emmonsia*-Like Infections in Humans

To date, almost 70 human cases of pulmonary adiaspiromycosis have been reported, mostly ascribed to *A. crescens* (Sigler 2005, Anstead et al. 2012). The first was reported in France in 1960 (Doby-Dubois et al. 1964), presenting as a solitary human pulmonary nodule. A decade later, the first case of disseminated pulmonary disease was documented (Kodousek et al. 1971). The pathological effects of adiaspiromycosis in humans appear to depend upon adiaspore burden and host immune status, and range from asymptomatic infection which is usually self-resolving (Buyuksirin et al. 2011), through necrogranulomatous pneumonia, respiratory failure (Barbas Filho et al. 1990) and death (Peres et al. 1992). Many infections have been discovered incidentally during histopathological evaluations for other conditions (see, for example, Denson et al. 2009). The typical radiological appearance of adiaspiromycosis is of bilateral reticulonodular infiltrates (Barbas Filho et al. 1990; England and Hochholzer 1993; Denson et al. 2009) that can be mistaken for sarcoidosis, cryptococcosis and other fungal pneumonias, miliary tuberculosis or malignancy. In symptomatic patients, the disease may have a protracted, indolent course typified by fatigue, fever, weight loss, cough and dyspnoea, or a rapid acute course if initial fungal exposure was high (Barbas Filho et al. 1990; de Almeida Barbosa et al. 1997). In many symptomatic cases, infection can be linked to recent or repeated exposure to soil, nesting material or animal carcases (Barbas Filho et al. 1990; de Almeida Barbosa et al. 1997; Nuorva et al. 1997).

A small number of extrapulmonary cases of human adiaspiromycosis-like infections have been reported, but in the light of recent taxonomic developments recognising several novel *Emergomyces* species as agents of infection in immunocompromised humans (Kenyon et al. 2014; Schwartz et al. 2015b; Dukik et al. 2017; Jiang et al. 2018), the identity of published agents needs critical re-evaluation. A case involving the appendix presumably following ingestion of the fungus was reported (Kodousek 1972). Two cases in AIDS patients ascribed to *B. parvus* (*Emmonsia parva*) (Echavarria et al. 1993; Turner et al. 1999), including a case of disseminated disease with skeletal and bone marrow involvement, likely represent cases of disseminated disease caused by an *Emergomyces* species. A report of an outbreak of ocular adiaspiromycosis in Brazilian children after immersion in a river (Mendes et al. 2009) almost certainly concerned an unrelated fungus.

Over the last several decades, multiple reports have appeared of mostly immunocompromised patients with disseminated infections with widespread cutaneous manifestations, caused by *Emmonsia*-like fungi now recognised as *Emergomyces* species. The tissue form consists of small yeast cells rather than adiaspores. *Emergomyces pasteurianus* was described in 1998 from an HIV-positive Italian woman with cutaneous lesions (Drouhet et al. 1998; Gori and Drouhet 1998), with additional cases reported from Spain in a liver transplant recipient with HIV (Pelegrin et al. 2011), India in an HIV-infected man (Malik et al. 2016), China in a renal transplant patient and a steroid-treated patient (Feng et al. 2015; Tang et al. 2015), and South Africa in an HIV-infected woman (Dukik et al. 2017).

Emergomyces africanus has since emerged as a leading cause of disseminated mycoses in AIDS patients in South Africa, in which pulmonary disease was common and cutaneous involvement universal (Kenyon et al. 2014; Dukik et al. 2017). By 2015, at least 55 cases of *E. africanus* have been diagnosed, all occurring in patients with immuncompromising conditions (mostly advanced HIV infection) (Schwartz et al. 2015a).

7.4 Histopathology and Diagnosis

In the absence of specific antigen tests for *A. crescens* and relatives, and since the fungus cannot easily be cultured from respiratory secretions, the diagnosis of virtually all human and animal cases of adiaspiromycosis has relied on histopathological examination of biopsy or necropsy tissues (see, for example, Austwick 1968; Kodousek 1972; Nuorva et al. 1997; Simpson and Gavier-Wilden 2000; Moraes and Gomes 2004; Borman et al. 2009; Denson et al. 2009). Visualisation of adiaspores can be facilitated in fresh intact tissue by KOH digestion (Borman et al. 2009; Chantrey et al. 2006) and the formal identification of the etiological agent can be confirmed by PCR amplification and sequencing of fungal genomic DNA extracted from infected lung tissue (Borman et al. 2009; Dot et al. 2009). In severe animal infections in which adiaspore burdens are high, firm whitish nodules measuring several millimetres in diameter may be visible in the external lung parenchyma (Peres et al. 1992; England and Hochholzer 1993; Chantrey et al. 2006; Borman et al. 2009) (Fig. 7.1a, b). In fixed histopathological sections, individual giant adiaspores can be visualised as round to oval cells with a multi-laminar thick cell wall, and granular cytoplasmic contents (Fig. 7.1c). In human infections caused by *A. crescens*, reported adiaspore sizes vary quite widely (40–500 μm). Whether this is a function of the age of the infection (Anstead et al. 2012) or of different aetiology (Schwartz et al. 2015a) remains to be established. Typically, pulmonary adiaspores in immunocompetent hosts are surrounded by a dense granulomatous inflammatory infiltrate comprising macrophages, lymphocytes, scanty neutrophils together with epithelioid cells, multinucleate giant cells and fibromatous matter (Fig. 7.1c, d). Depending on section thickness and tissue integrity, it is not uncommon to see empty granulomata and granulomata with no central adiaspore in heavily infected lungs (Fig. 7.1c). Pulmonary functional compromise results from compression of smaller airways via tissue disruption due to the granulomata (Watts and Chandler 1975).

Emergomycosis is diagnosed by culture of *Emergomyces* species from blood or biopsy material from skin or other affected tissues (Kenyon et al. 2014). The fungus usually grows on Sabouraud's agar within 2 weeks, though incubation up to 6 weeks at 30 °C is recommended. Yield from blood culture can be improved by use of mycobacterial/fungal blood culture bottles. Histopathological examination of skin tissue can secure the diagnosis of deep fungal infection with shorter turnaround time. *Emergomyces* species appear as small yeasts measuring 3–7 μm and cannot be distinguished from other dimorphic fungi with small yeasts (i.e. *Histoplasma*, *Sporothrix*, etc.) (Schwartz et al. 2015a).

7.5 Control

Due to the rarity of the disease, no official guidelines exist for the management of human adiaspiromycosis. Many mild infections are asymptomatic or self-limiting and spontaneously resolve. Patients reportedly receive antifungal treatment even for relatively mild disease (Anstead et al. 2012) and the benefits of therapy may be anecdotal. Only one study of in vitro antifungal susceptibilities of *E. crescens* exists, with only a single isolate reported (Borman et al. 2009). In vitro, both the adiaspore and mycelial forms of *A. crescens* appeared susceptible to several conventional antifungal agents, including amphotericin B, itraconazole, voriconazole and caspofungin; there were no discernible differences in susceptibility between the two growth forms (Borman et al. 2009). Similarly, low MICs have been reported for these antifungals with many of the other members of the Ajellomycetaceae, including *Blastomyces* and *Histoplasma* spp. (Li et al. 2000). Clinical improvement has been reported in patients treated with itraconazole (de Almeida Barbosa et al. 1997; Dot et al. 2009; Anstead et al. 2012), ketoconazole (Severo et al. 1989; Martins et al. 1997; Santos et al. 2000), voriconazole (Denson et al. 2009) and amphotericin B (Nuorva et al. 1997; Anstead et al. 2012), although in the absence of controls, whether antifungals contributed to recovery is unknown. Since adiaspores are non-replicative structures and pulmonary damage results primarily from the host immune response, some investigators have advocated for the use of corticosteroids, either in combination with antifungals (Anstead et al. 2012; de Almeida Barbosa et al. 1997) or alone (Silva et al. 2010). In severe human disease, a multipronged approach that includes an antifungal agent and corticosteroids to kill the organism and dampen the host inflammatory response that it provokes, respectively, seems reasonable.

Treatment of emergomycosis should follow guidelines for the management of other endemic mycoses in immunocompromised hosts such as progressive disseminated histoplasmosis (Wheat et al. 2007). This should consist of amphotericin B for 10–14 days, followed by itraconazole for a year or longer, pending immune reconstitution.

7.6 Conclusions

According to the medical literature, adiaspiromycosis remains a relatively rare disease of otherwise healthy human hosts. However, reports from many countries worldwide have detailed high prevalence rates in indigenous wild mammals, which suggests that this fungal infection may be substantially under-diagnosed in the human population, especially since infections with low fungal burdens are usually asymptomatic, or at worst mild and self-limiting. Moreover, recent descriptions of novel pathogenic *Emergomyces* species with a yeast tissue form rather than an adiaspore form, and which affect predominantly immunocompromised patients underscore the morphological, taxonomic and pathogenic complexities of this interesting group of *Emmonsia*-like fungi.

References

Albassam MA, Bhatnagar R, Lillie LE, Roy L (1986) Adiaspiromycosis in striped skunks in Alberta, Canada. J Wildl Dis 22:13–18

Anstead GM, Sutton DA, Graybill JR (2012) Adiaspiromycosis causing respiratory failure and a review of human infections due to *Emmonsia* and *Chrysosporium* spp. J Clin Microbiol 50:1346–1354

Austwick PKC (1968) Mycotic infections. Symp Zool Soc Lond 24:249–271

Barbas Filho JV, Amato MB, Deheinzelin D, Saldiva PH, de Carvalho CR (1990) Respiratory failure caused by adiaspiromycosis. Chest 97:1171–1175

Boisseau-Lebreuil HT (1975) *In vitro* formation of adiaspores in 10 strains of *Emmonsia crescens*, the fungal agent of adiaspiromycosis. CR Séances Soc Biol Fil 169:1057–1061

Borman AM, Simpson VR, Palmer MD, Linton CJ, Johnson EM (2009) Adiaspiromycosis due to *Emmonsia crescens* is widespread in native British mammals. Mycopathologia 68:153–163

Buyuksirin M, Ozkayaa S, Yucel N, Guldaval F, Ceylan K, Erbay Polat G (2011) Pulmonary adiaspiromycosis: the first reported case in Turkey. Respir Med CME 4:166–169

Chantrey JC, Borman AM, Johnson EM, Kipar A (2006) *Emmonsia crescens* infection in a British water vole (*Arvicola terrestris*). Med Mycol 44:375–378

Ciferri R, Montemartini A (1959) Taxonomy of *Haplosporangium parvum*. Mycopathologia 10:303–316

Cronje N, Schwartz IS, Retief L et al (2017) Attempted molecular detection of the dimorphic fungal pathogen, *Emergomyces africanus*, in small terrestrial mammals in South Africa. Med Mycol. https://doi.org/10.1093/mmy/myx065

de Almeida Barbosa A, Moreira Lemos AC, Severo LC (1997) Acute pulmonary adiaspiromycosis. Report of three cases and a review of 16 other cases collected from the literature. Rev Iberoam Micol 14:177–180

de Hoog GS, Redhead SA, Feng P, Jiang Y, Dukik K, Sigler L (2017) Proposals to conserve *Blastomyces* Gilchrist & W.R. Stokes against *Blastomyces* Costantin & Rolland and *Ajellomycetaceae* against *Paracoccidioidaceae* (*Ascomycota*: *Onygenales*). Taxon 65: 1167–1169

Denson JL, Keen CE, Froeschle PO, Toy EW, Borman AM (2009) Adiaspiromycosis mimicking widespread malignancy in a patient with pulmonary adenocarcinoma. J Clin Pathol 62:837–839

Doby JM, Boisseau-Lebreuil MT, Rault B (1971) L'adiaspiromycose par *Emmonsia crescens* chez les petits mammiferes sauvages en France. Mycopathologia 44:107–115

Doby-Dubois M, Chevrel ML, Doby JM, Louvet M (1964) Premier cas humain d'adiaspiromycose par *Emmonsia crescens*, Emmons et Jellison 1960. Bull Soc Pathol Exot 57:240–244

Dot J-M, Debourgogne A, Champigneulle J, Salles Y, Brizion M, Puyhardy JM, Collomb J, Plénat F, Machouart M (2009) Molecular diagnosis of disseminated adiaspiromycosis due to *Emmonsia crescens*. J Clin Microbiol 47:1269–1273

Drouhet E, Guého E, Gori S, Huerre M, Provost F, Borgers M, Dupont B (1998) Mycological, ultrastructural and experimental aspects of a new dimorphic fungus *Emmonsia pasteuriana* sp. nov. isolated from a cutaneous disseminated mycosis in AIDS. J Mycol Med 8:64–77

Dukik K, Muñoz JF, Jiang Y, Feng P, Sigler L, Stielow JB et al (2017) Novel taxa of thermally dimorphic systemic pathogens in the *Ajellomycetaceae* (*Onygenales*). Mycoses 60:296–309

Dvořák J, Otcenášek M, Prokopič J (1969) The spring peak of adiaspiromycosis due to *Emmonsia crescens* Emmons and Jellison 1960. Sabouraudia 7:12–14

Dvořák J, Otcenášek M, Rosicky B (1973) Adiaspiromycosis caused by *Emmonsia crescens* Emmons and Jellison 1960. Studie CSAV c14. Academia Press, Prague

Echavarria E, Cano EL, Restrepo A (1993) Disseminated adiaspiromycosis in a patient with AIDS. J Med Vet Mycol 31:91–97

Emmons CW, Jellison WL (1960) *Emmonsia crescens* sp. nov. and adiaspiromycosis (haplomycosis) in mammals. Ann N Y Acad Sci 89:91–101

England DM, Hochholzer L (1993) Adiaspiromycosis: an unusual fungal infection of the lung. Report of 11 cases. Am J Surg Pathol 17:876–886

Feng P, Yin S, Zhu G et al (2015) Disseminated infection caused by *Emmonsia pasteuriana* in a renal transplant recipient. J Dermatol 42:1179–1182

Gori S, Drouhet E (1998) Cutaneous disseminated mycosis in a patient with AIDS due to a new dimorphic fungus. J Mycol Med 8:57–63

Hubalek Z (1999) Emmonsiosis of wild rodents and insectivores in Czechland. J Wildl Dis 35: 243–249

Hubalek Z, Nesvadboda J, Rychnovsky B (1995) A heterogeneous distribution of *Emmonsia parva* var. *crescens* in an agro-system. J Med Vet Mycol 33:197–200

Hubalek Z, Zejda J, Svobodová Š, Kučera J (1993) Seasonality of rodent adiaspiromycosis in a lowland forest. J Med Vet Mycol 31:359–366

Hubalek Z, Burda H, Scharff A, Heth G, Nevo E, Sumbera R, Pesko J, Zima J (2005) Emmonsiosis of subterranean rodents (Bathyergidae, Spalacidae) in Africa and Israel. Med Mycol 43(8): 691–697

Jellison WL (1969) Adiaspiromycosis (=haplomycosis). Mountain Press, Montana

Jellison WL, Helminen M, Vinson JW (1960) Presence of a pulmonary fungus in rodents in Finland. Ann Med Exp Fenn 38:3–8

Jiang Y, Dukik K, Muñoz J, Sigler L, Schwartz IS et al (2018) Phylogeny, ecology and taxonomy of systemic pathogens in *Ajellomycetaceae* (*Onygenales*). Fungal Divers (in press)

Kenyon C, Corcoran C, Govender NP (2014) An *Emmonsia* species causing disseminated infection in South Africa. N Engl J Med 370:283–284

Krivanec K (1977) Adiaspiromycosis in Czechoslovakian mammals. Sabouraudia 15:221–223

Krivanec K, Otcenasek M (1977) Importance of free living mustelid carnivores in circulation of adiaspiromycosis. Mycopathologia 60:139–144

Krivanec K, Octenasek M, Rosicky B (1975) The role of polecats of the genus *Putorius* Cuvier, 1987 in natural foci of adiaspiromycosis. Folia Parasitol 22:245–249

Krivanec K, Otcenasek M, Slais J (1976) Adiaspiromycosis in large free-living carnivores. Mycopathologia 58:21–25

Kodousek R (1972) Finding of isolated spherules of the fungus *Emmonsia crescens* in surgical specimen of the appendix in a 7-year-old boy. Cesk Patol 8:160–162

Kodousek R, Vortel V, Fingerland A (1971) Pulmonary adiaspiromycosis in man caused by *Emmonsia crescens*: report of a unique case. Am J Clin Pathol 56:394–399

Leighton FA, Wobeser G (1978) The prevalence of adiaspiromycosis in three sympatric species of ground squirrels. J Wildl Dis 14:362–365

Li RK, Ciblak MA, Nordoff N, Pasarell L et al (2000) *In vitro* activities of voriconazole, itraconazole, and amphotericin B against *Blastomyces dermatitidis*, *Coccidioides immitis*, and *Histoplasma capsulatum*. Antimicrob Agents Chemother 44(6):1734–1736

Lugton IW, Wobeser G, Morris RS, Caley P (1997) Epidemiology of *Mycobacterium bovis* in feral ferrets (*Mustela furo*) in New Zealand. N Z Vet J 45:140–150

Malik R, Capoor MR, Vanidassane I et al (2016) Disseminated *Emmonsia pasteuriana* infection in India: a case report and a review. Mycoses 59:127–132

Martins RL, Santos CG, França FR, Moraes MA (1997) Human adiaspiromycosis. A report of a case treated with ketoconazole. Rev Soc Bras Med Trop 30:507–509

Mason R, Gauhwin M (1982) Adiaspiromycosis in south Australian hairy-nosed wombats (*Lasiorhinus latifrons*). J Wildl Dis 18:3–8

McDiarmid A, Austwick PK (1954) Occurrence of *Haplosporangium parvum* in the lungs of the mole (*Talpa europaea*). Nature 174:843–844

Mendes MO, Moraes MA, Renoiner EI, Dantas MH, Lanzieri TM, Fonseca CF, Luna EJ, Hatch DL (2009) Acute conjunctivitis with episcleritis and anterior uveitis linked to adiaspiromycosis and freshwater sponges, Amazon region, Brazil, 2005. Emerg Infect Dis 15:633–639

Moraes MA, Gomes MI (2004) Human adiaspiromycosis: cicatricial lesions in mediastinal lymph nodes. Rev Soc Bras Med Trop 37:177–178

Nuorva K, Pitkänen R, Issakainen J, Huttunen NP, Juhola M (1997) Pulmonary adiaspiromycosis in a two year old girl. J Clin Pathol 50:82–85

Pelegrin I, Ayats J, Xiol X, Cuenca-Estrella M et al (2011) Disseminated adiaspiromycosis: case report of a liver transplant patient with human immunodeficiency infection, and literature review. Transpl Infect Dis 13(5):507–514

Peres LC, Figueiredo F, Peinado M, Soares FA (1992) Fulminant disseminated pulmonary adiaspiromycosis in humans. Am J Trop Med Hyg 46:146–150

Peterson SW, Sigler L (1998) Molecular genetic variation in *Emmonsia crescens* and *Emmonsia parva*, etiologic agents of adiaspiromycosis, and their phylogenetic relationship to *Blastomyces dermatitidis* (*Ajellomyces dermatitidis*) and other systemic fungal pathogens. J Clin Microbiol 36:2918–2925

dos Santos VM, Fatureto MC, Saldanha JC, Adad SJ (2000) Pulmonary adiaspiromycosis: report of two cases. Rev Soc Bras Med Trop 33:483–488

Schwartz IS, Govender NP, Corcoran C et al (2015a) Clinical characteristics, diagnosis, management and outcomes of disseminated emmonsiosis: a retrospective case series. Clin Infect Dis 61:1004–1012

Schwartz IS, Kenyon C, Feng P et al (2015b) 50 years of *Emmonsia* disease in humans: the dramatic emergence of a cluster of novel fungal pathogens. PLoS Pathog 11:e100e5198

Schwartz IS, Lerm B, Hoving JC et al (2018a) Emergomyces africanus in soil, South Africa. Emerg Infect Dis 24(2):377–380. https://doi.org/10.3201/eid2402.171351

Schwartz IS, McLoud JD, Berman D et al (2018b) Molecular detection of airborne Emergomyces africanus, a thermally dimorphic fungal pathogen, in Cape Town, South Africa. PLoS Negl Trop Dis 12(1):e0006174. https://doi.org/10.1371/journal.pntd.0006174

Sharapov VM (1969) Adiaspiromycosis in the U.S.S.R. Izv Sibirsk Otdel AN SSSR Ser Biol-Med Nauk 1:86–95

Severo LC, Geyer GR, Camargo JJ, Porto NS (1989) Adiaspiromycosis treated successfully with ketoconazole. J Med Vet Mycol 27:265–268

Sigler L (1996) *Ajellomyces crescens* sp. nov.: taxonomy of *Emmonsia* species and relatedness with *Blastomyces dermatitidis* (teleomorph *Ajellomyces dermatitidis*). J Med Vet Mycol 34: 303–314

Sigler L (2005) Adiaspiromycosis and other infections caused by *Emmonsia* species. In: Merz WG, Hay RJ (eds) Medical mycology, Topley and Wilson's microbiology and microbial infections, 10th edn. Hodder Arnold Press, London, pp 809–824

Sigler L, Flis A (1998) Catalogue of the University of Alberta Microfungus Collection and Herbarium. University of Alberta Devonian Botanic Garden, Edmonton, pp 1–213

Silva RM, Liporoni GA, Botto CC, Rodrigues BC, Scudeler D, Cunha Junior W (2010) Pulmonary adiaspiromycosis treated without antifungal drugs. Rev Soc Bras Med Trop 43:95–97

Simpson VR, Gavier-Wilden D (2000) Fatal adiaspiromycosis in a wild Eurasian otter (*Lutra lutra*). Vet Rec 147:239–241

Tang XH, Zhou H, Zhang XQ, Han JD, Gao Q (2015) Cutaneous disseminated emmonsiosis due to *Emmonsia pasteuriana* in a patient with cytomegalovirus enteritis. JAMA Dermatol 151:1026–1028

Tevis L (1956) Additional records of *Haplosporangium parvum* in mammals in Britain. Nature 177:437

Turner D, Burke M, Bashe E, Blinder S, Yust I (1999) Pulmonary adiaspiromycosis in a patient with acquired immunodeficiency syndrome. Eur J Clin Microbiol Infect Dis 18:893–895

Wang P, Kenyon C, de Hoog S, Guo L, Fan H, Liu H, Li Z, Sheng R, Yang Y, Jiang Y, Zhang L, Xu Y (2017) A novel dimorphic pathogen, *Emergomyces orientalis* (*Onygenales*), agent of disseminated infection. Mycoses 60:310–319

Watts JC, Chandler FW (1975) Human pulmonary adiaspiromycosis. Arch Pathol 99:11–15

Wheat LJ, Freifeld AG, Kleiman MB, Baddley JW, McKinsey DS, Loyd JE, Kauffman CA, Infectious Diseases Society of America (2007) Clinical practice guidelines for the management of patients with histoplasmosis: 2007 update by the Infectious Diseases Society of America. Clin Infect Dis 45:807–825

Blastomycosis in Mammals

8

Ilan S. Schwartz

Abstract

Blastomycosis is a serious fungal disease of dogs, humans, and occasionally other mammals caused by geographically restricted, thermally dimorphic *Blastomyces* species. Blastomycosis is primarily a canine disease, with approximately ten dogs diagnosed for every human case. Dogs also develop disease more rapidly, thus becoming sentinels for possible human disease. Human and canine blastomycosis may differ according to epidemiology/epizoology, clinical features, performance and use of diagnostics, and management.

8.1 Introduction

Blastomycosis is a disease of mammals caused by the geographically restricted, thermally dimorphic fungi *Blastomyces* species. Human blastomycosis was first reported by Gilchrist in 1894 in a case of cutaneous disease first mistakenly attributed to protozoan disease (Gilchrist 1894). Four years later, Gilchrist and Stokes isolated the causative agent, a fungus they called *Blastomyces dermatitidis* (Gilchrist and Stokes 1898). The first case of canine blastomycosis was reported by Meyer in 1912 (Meyer 1912). Since then, blastomycosis has been recognized as a common and serious disease of people and animals in endemic/enzootic areas.

I. S. Schwartz (✉)
Division of Infectious Diseases, Department of Medicine, University of Alberta, Edmonton, AB, Canada
e-mail: ilan.steven.schwartz@gmail.com

S. Seyedmousavi et al. (eds.), *Emerging and Epizootic Fungal Infections in Animals*,
https://doi.org/10.1007/978-3-319-72093-7_8

8.2 Ecology and Distribution

Our understanding of the ecology of *Blastomyces* species is incomplete due to the difficulty in isolating the fungus from the environment (Restrepo et al. 2000). Blastomycosis is acquired primarily through inhalation of airborne conidia of *Blastomyces* species. These are liberated from the mold phase, which is associated with moist, acidic, sandy soils enriched with decaying organic matter and animal droppings (Restrepo et al. 2000). Aerosolization of conidia is promoted by disturbances to soil that may be caused by natural phenomena or due to human or animal activities such as excavation (Bradsher 2014b). Upon inhalation, conidia undergo a temperature-dependent transformation to yeast-like cells, capable of causing local and disseminated disease (Bradsher 2014b).

The geographic range of endemicity for canine and human blastomycosis includes North America, where it primarily occurs in states and provinces along the Great Lakes, and Ohio, Mississippi, Missouri, and St. Lawrence rivers (Bradsher 2014b). Autochthonous blastomycosis has also been reported from most of Africa (Broc and Haddad 1952; Carman et al. 1989), parts of India (Randhawa et al. 1983), and the Middle East (Kuttin et al. 1978; Kingston et al. 1980). However, the etiological agent may not always be the same species between and within geographical regions. Genetic studies of large collections of *B. dermatitidis* isolates have identified the presence of two distinct genetic populations (Meece et al. 2011; Brown et al. 2013), leading Brown et al. (2013) to conclude the presence of a cryptic species which they called *B. gilchristii*. These species are indistinguishable in morphology, physiology, and in most currently applied molecular bar codes (Dukik et al. 2017). Although clinical and demographic phenotypic differences have been suggested (Meece et al. 2013), the clinical significance of distinguishing *B. dermatitidis* from *B. gilchristii* is not yet established. Geographic differences exist, and epidemiological differences (such as outbreak potential) are surmised by the fact that isolates identified as *B. gilchristii* predominate in northern Ontario and Wisconsin, areas with the highest reported incidence of blastomycosis (Brown et al. 2013).

On the other hand, differences have long been noted in isolates from Africa compared to those implicated in disease in North America. In fact, isolates from Africa have been observed to be slightly smaller (Kaufman et al. 1983), more difficult to convert to yeast phase (Lombardi et al. 1988), and have different antigenic expression (Kaufman et al. 1983) than isolates from North America. Moreover, clinical differences in human disease have been suggested (Vandepitte and Gatti 1972). Strikingly, no cases of animal blastomycosis have been reported from Africa (Carman et al. 1989). Recently, a new species of *Blastomyces*, *B. percursus*, was described from Israel and South Africa (Dukik et al. 2017); *B. dermatitidis* and *B. gilchristii* have also been confirmed in isolates from sub-Saharan Africa (Brown et al. 2013), and so the extent that disease is attributable to each species there and elsewhere outside of North America has not yet been defined.

Most human and animal cases of blastomycosis are sporadic or endemic/enzootic (Bradsher 2014b), although occasionally outbreaks have occurred which have informed our understanding of the ecology, attack rate, and natural history of outbreak-related *B. dermatitidis* infection (Klein et al. 1986; Armstrong et al. 1987; Baumgardner and Burdick 1991; Smith and Gauthier 2015). Outbreaks have involved both rural and urban exposures (Smith and Gauthier 2015). Recreational outdoor activities and especially water activities as well as exposure to excavation and construction are frequently implicated in human and canine blastomycosis (Baumgardner et al. 1995; Smith and Gauthier 2015). Proximity to waterways has been identified as a risk factor for sporadic blastomycosis in people (Baumgardner et al. 1992) and in dogs (Archer et al. 1987; Baumgardner et al. 1995; Arceneaux et al. 1998). For instance, case-control studies of canine blastomycosis in Louisiana and Wisconsin have identified residence within 400 m of a body of water to be a significant risk factor for the disease in dogs (Baumgardner et al. 1995; Arceneaux et al. 1998). In Louisiana, Arceneaux et al. found the odds of living near water was tenfold higher for dogs with blastomycosis than controls (Arceneaux et al. 1998).

Most people who develop blastomycosis are immunocompetent. Persons with immunodeficiencies who develop blastomycosis are reported to have more severe forms of the disease (Pappas et al. 1993), but the numbers of cases reported to date have been small. Persons treated with tumor necrosis factor (TNF)-α inhibitors may represent a growing cohort at risk of blastomycosis (Smith and Kauffman 2009). Individuals with diabetes mellitus appear to be at higher risk of blastomycosis (Lemos et al. 2002) and of requiring management in an intensive care unit (Kralt et al. 2009). Most animals with blastomycosis were previously healthy, although Davies and Troy reported 10% of infected cats in a small series had feline leukemia virus (Davies and Troy 1996).

Anderson et al. (2016) recently showed that people can become reinfected with *Blastomyces* spp.; previously, whether a second episode of blastomycosis represented reinfection and not relapse was inconclusive. These authors reported on two persons in whom blastomycosis was diagnosed and treated, only to later develop the disease again (Anderson et al. 2016). By genotyping isolates from the initial and subsequent episodes in each respective patient using 27 polymorphic microsatellite markers, they demonstrated that relapse occurred in one case (concordance between the two isolates at 27/27 loci) and reinfection occurred in the other (concordance at just 15/27 loci) (Anderson et al. 2016).

Blastomycosis is most common in dogs residing in or visiting enzootic areas (Baumgardner et al. 1995). The incidence of blastomycosis in dogs is eight to ten times that of humans (Baumgardner et al. 1995; Herrmann et al. 2011), presumably related to time spent outdoors, proximity to soil, and activities, such as digging, that may result in soil disturbances and conidial exposure. Most affected dogs are immunocompetent (Sykes and Merkel 2014). The incidence appears to be highest in young, large sporting dogs and hounds, including coonhounds, pointers, Weimaraners (Rudmann et al. 1992), golden retrievers, Labrador retrievers, and Doberman pinschers (Arceneaux et al. 1998). Sporting dogs may be more likely to be exposed due to selective use in hunting (Rudmann et al. 1992). Some but not all studies have found the disease is more common in intact males (Rudmann et al. 1992; Arceneaux et al. 1998). Blastomycosis has also been described in wild canids.

For example, Nemeth et al. (2016) reviewed the database of wild animals sent to the Canadian Wildlife Health Cooperative from 1991 to 2014. Blastomycosis was diagnosed in 14 wild canids, including 11 of 149 (7.6%) red foxes (*Vulpes vulpes*) and 3 of 185 (1.6%) gray wolves (*Canis lupus*).

Feline blastomycosis is encountered 28–100 times less frequently than canine blastomycosis (Legendre 2012; Davies et al. 2013) and has been reported even among indoor-only cats (Blondin et al. 2007; Houseright et al. 2015). Blastomycosis has also been reported in captive wild felids, including lions (*Panthera leo*), Siberian tiger (*Panthera tigris*), cheetah (*Acinonyx jubatus*), and snow leopard (*Panthera uncia*) (Storms et al. 2003).

Blastomycosis has been reported in a range of domestic and captive animals including kinkajou (*Potos flavus*) (Harris et al. 2011), ferret (Nemeth et al. 2016), red ruffed lemur (*Varecia rubra*) (Rosser et al. 2016), and rhesus monkey (*Macaca mulatta*) (Wilkinson et al. 1999). Marine mammals reported with blastomycosis include sea lion (*Zalophus californianus*) (Zwick et al. 2000) and Atlantic bottlenose dolphin (*Tursiops truncatus*) (Cates et al. 1986). Wild, free-roaming animals (other than canids) that have developed blastomycosis include an American black bear (*Ursus americanus*) (Dykstra et al. 2012).

Among livestock, some cases of blastomycosis have been described in horses living in endemic areas (Wilson et al. 2006; Stewart and Cuming 2015). Blastomycosis has also been reported in an alpaca (Imai et al. 2014).

Animals do not play a role in transmission of *Blastomyces* spp., aside from rare cases of inoculation blastomycosis reported due to a bite (Gray and Baddour 2002; Harris et al. 2011) or percutaneous injury during autopsy of an infected animal (Gray and Baddour 2002). The concurrent or sequential infection of a person and his or her dog is common (Baumgardner et al. 1992) and likely due to a common exposure (Sarosi et al. 1979; Armstrong et al. 1987; Baumgardner et al. 1992). Even so, dogs appear to develop disease earlier (Sarosi et al. 1979). In experimental murine blastomycosis, larger inocula lead to earlier disease (Williams and Moser 1987), and it has been inferred that the shorter prepatent period in dogs reflects increased inocula from being closer to the ground (Legendre 2012). In any case, a history of blastomycosis in one's dog should raise suspicion for the disease in a person with a compatible syndrome (Sarosi et al. 1979).

8.3 Clinical Signs

8.3.1 In Humans

Clinical signs encountered in persons with blastomycosis will depend on the organ systems involved, but clinicians should be aware of the protean nature of the disease (Bradsher 2014a). Pulmonary infection can be subclinical or can result in an acute or chronic pneumonia (Sarosi et al. 1974). Acute pulmonary blastomycosis can present with fevers, sepsis, and hypoxia, with clinical examination and radiographs consistent with focal airspace disease (Sarosi et al. 1974; Lemos et al. 2002). In other words, the disease can be indistinguishable from community-acquired (bacterial) pneumonia (Lemos et al. 2002; Bradsher 2014a; Alpern et al. 2016), and it is common for patients to receive multiple courses of antibiotics before the correct diagnosis is established (Alpern et al. 2016). Acute respiratory distress syndrome (ARDS) occurs in 8–15% of cases of symptomatic blastomycosis (Meyer et al. 1993; Vasquez et al. 1998; Lemos et al. 2001; Azar et al. 2015) and is associated with mortality rates of at least 40% (Meyer et al. 1993; Vasquez et al. 1998; Lemos et al. 2001; Azar et al. 2015; Schwartz et al. 2016). Patients with chronic pulmonary involvement may present with chronic dyspnea, cough and hemoptysis, often accompanied by constitutional symptoms. The radiographic appearance is like acute disease but with a third of patients having mass-like lesions (Patel et al. 1999). Not surprisingly, chronic blastomycosis is frequently mistaken for pulmonary malignancies or tuberculosis (Lemos et al. 2002; Bradsher 2014a).

Extrapulmonary disease occurs in ~25–40% of cases (Baumgardner et al. 1992; Lemos et al. 2002). The most common extrapulmonary site of blastomycosis is the skin, usually manifesting as ulcerative or verrucous lesions (Bradsher 2014b). Osteoarticular disease is the next most common form (Kralt et al. 2009). In a series of persons with osteoarticular blastomycosis, disease of the axial skeleton was most common, followed by long bones of the lower limb (Oppenheimer et al. 2007). Vertebral disease may rarely present with spinal cord compression syndromes (Saccente et al. 1998). As with bacterial osteomyelitis, disease of long bones most often localizes to the metaphyses (Oppenheimer et al. 2007). Arthritis is less common than osteomyelitis. It is generally monoarticular and mimics pyogenic bacterial septic arthritis (Oppenheimer et al. 2007). In males, the third most common site of extrapulmonary blastomycosis is the prostate and genitourinary system (Saccente and Woods 2010). Central nervous system disease is less common, occurring in approximately 5% of cases with extrapulmonary dissemination (Bariola et al. 2010). Patients may present with meningitis, encephalitis, or signs of a space-occupying lesion (Bush et al. 2013). Cerebrospinal fluid pleocytosis occurs and can have either lymphocytic or neutrophilic predominance (Bariola et al. 2010). Ocular involvement has been reported but it is uncommon (Lopez et al. 1994).

8.3.2 In Animals

Canine disease is more commonly disseminated beyond the lungs at the time of diagnosis (Legendre 2012). Chest radiographs may show focal airspace disease or a "snowstorm" pattern of reticulonodular disease (Crews et al. 2008b) (Fig. 8.1). Ocular disease is much more common in dogs than in humans, occurring in up to half of all cases (Bloom et al. 1996; Arceneaux et al. 1998). Among these, bilateral disease occurs in half (Bloom et al. 1996). Ocular disease can be localized to anterior segments, posterior segments, or, most commonly, both (i.e., endophthalmitis) (Bloom et al. 1996). Like in human disease, cutaneous involvement is common in canine blastomycosis, occurring in approximately half of infected dogs (Arceneaux et al. 1998). An example of a dog with bilateral ocular and cutaneous disease is shown in Fig. 8.2. Other common sites of extrapulmonary disease include lymphatic

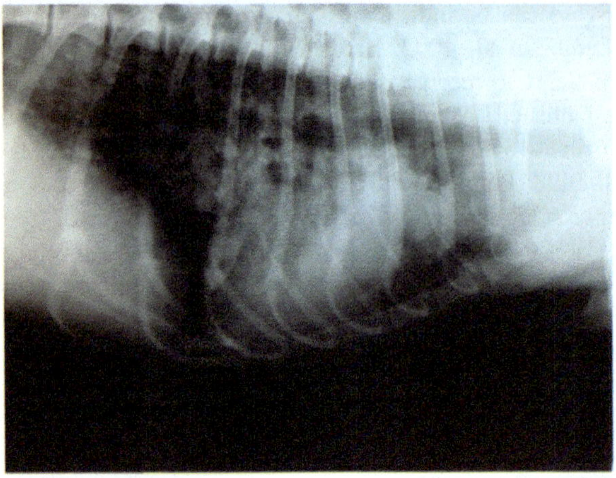

Fig. 8.1 Lateral thoracic X-ray of a dog with pulmonary blastomycosis demonstrating diffuse reticulonodular disease (Courtesy of Dr. Peter Schwartz, DVM, Assiniboine Animal Hospital, Winnipeg, Manitoba)

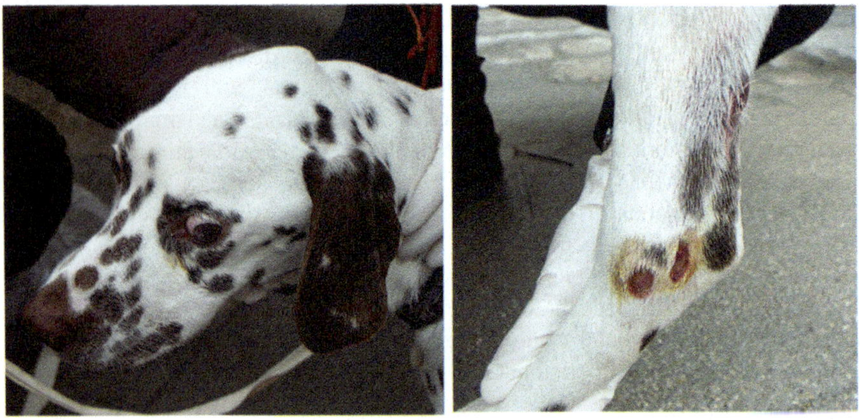

Fig. 8.2 Ocular involvement and ulcerative cutaneous lesions in a dog infected in Quebec, Canada (Courtesy of Dr. René Chermette, DVM, dipl. EVPC, Parasitology-Mycology, EnvA, Maisons-Alfort, France)

and osteoarticular structures (Arceneaux et al. 1998). Central nervous system disease is less common, occurring in 6% of dogs in a series (Arceneaux et al. 1998). Cardiovascular blastomycosis has been reported in nine dogs (Langlois et al. 2013; Schmiedt et al. 2015); it is extraordinarily uncommon in people.

Like the disease in humans, clinical signs commonly encountered in canine blastomycosis are nonspecific. In decreasing frequency of occurrence reported by a large study, these include fever, lymphadenopathy, harsh lung sounds, cutaneous lesions, chorioretinitis, anterior uveitis, cough, emaciation, and retinal detachment (Arceneaux et al. 1998).

Cats are often diagnosed only upon autopsy and with widespread disease (Davies and Troy 1996). In one series, clinical signs of respiratory, neurologic, and cutaneous disease were present in 59, 41, and 18% of cats, respectively (Davies and Troy 1996).

In horses, clinical signs generally include pneumonia and with reports of respiratory infection or cutaneous and subcutaneous lesions (Wilson et al. 2006; Stewart and Cuming 2015).

8.4 Diagnosis

8.4.1 Culture

Culture of *Blastomyces* spp. from a clinical specimen is the gold standard diagnostic for blastomycosis. *Blastomyces dermatitidis/gilchristii* grow on general fungal media such as Sabouraud's dextrose agar or potato dextrose agar, incubated at 25–30 °C (Bradsher 2014b). Growth of white to buff colonies usually appears within 10–14 days but may require up to 6 weeks incubation (Bradsher 2014b). Microscopically, *B. dermatitidis/gilchristii* characteristically have conidiophores of varying lengths that run perpendicular to hyphae and terminate with single conidia that resemble "lollipops" (Bradsher 2014b). In contradistinction, the conidia of *B. percursus* appear in clusters at the end of conidiophores (Dukik et al. 2017). Traditionally, conversion of the mold to a yeast phase at 37°C was performed for confirmation of the identification, but this is rarely done today (Saccente and Woods 2010). Most microbiology laboratories confirm identification of *B. dermatitidis/gilchristii* with a DNA probe (AccuProbe; GenProbe Inc., San Diego, CA), but cross-reactions can occur, including with other dimorphic fungi (Saccente and Woods 2010).

The diagnostic yield of culture is surmised from a retrospective review of cases of human pulmonary blastomycosis (Martynowicz and Prakash 2002). *Blastomyces dermatitidis* was isolated from the first sputum sample in 75% of cases, which increased to 81% after a mean of 2.3 samples (Martynowicz and Prakash 2002); the diagnostic yield of bronchial washings was even higher. Nonetheless, the true sensitivities are likely lower since only diagnosed cases (i.e., those with at least one positive test) are included in this type of study.

Despite favorable operating characteristics, culture has important limitations. Slow turnaround times limit reliance on this test for management decision (Bradsher 2014b). In addition, occupational exposure of laboratory workers to highly infectious mycelia is a real concern (Denton et al. 1967). While culture is a standard investigation of humans suspected of blastomycosis (Bradsher 2014b), it is rarely used in veterinary practice (Legendre 2012; Sykes and Merkel 2014). In a survey of small-animal veterinary practices in Wisconsin, 80% of respondents reported that they never used culture for the diagnosis of blastomycosis (Anderson et al. 2014). Similarly, large retrospective case series of dogs with pulmonary blastomycosis from

Louisiana (Arceneaux et al. 1998) and Minnesota (Crews et al. 2008a) found that fungal cultures were sent for only 17 of 115 dogs (12%) and 6 of 125 dogs (5%), respectively.

8.4.2 Microscopy, Cytology, and Histopathology

Microscopic examination of clinical specimens using wet smears, cytopathology, and histopathology are cornerstone diagnostic tools for blastomycosis in humans and animals (Saccente and Woods 2010). The classic appearance is of large (8–15 μm), round, multinucleate yeast-like cells with thick, double refractile walls that may have a single bud with broad bases (Guarner and Brandt 2011) (Fig. 8.3). Small yeast forms are also known to occur and may be difficult to distinguish histologically from the yeast-like cells of other dimorphic fungi (Guarner and Brandt 2011).

The most expedient test is a wet preparation using potassium hydroxide (KOH) solution, with or without staining with calcofluor white or lactophenol blue (Saccente and Woods 2010). The sensitivity of this test is poor: a study found the yield of sputum for KOH preparation in human pulmonary blastomycosis to be 25% for a single specimen (Martynowicz and Prakash 2002). Nonetheless, examination of multiple specimens often leads to the correct diagnosis. For example, in a series of 100 persons with pulmonary blastomycosis, KOH preparation of respiratory samples (invasively and noninvasively collected) led to the immediate diagnosis in 66% of cases (Patel et al. 1999).

Visualization of *Blastomyces* yeast-like cells is enhanced with cytologic and histopathologic preparations. Cytology stains like Papanicolaou and Wright stains can be used to identify the organism in respiratory specimens (usually sputum and bronchoalveolar lavage in humans and transtracheal lavage or transthoracic fine-needle aspirates in dogs), lymph node aspirates, and impression smears or discharge from cutaneous lesions (Martynowicz and Prakash 2002). In tissue, the yeast-like

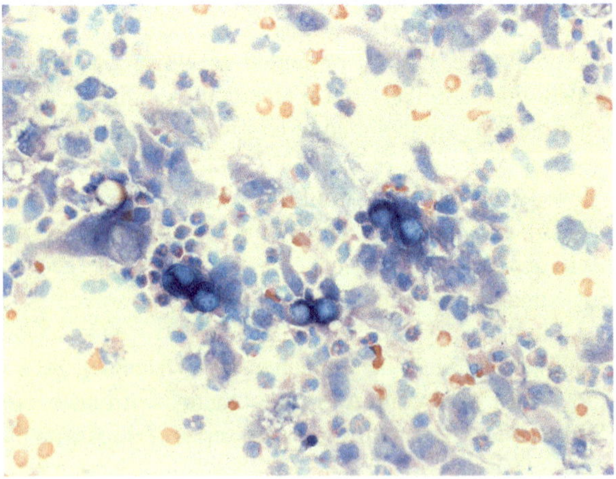

Fig. 8.3 Modified Wright's stain of impression smears collected from a hilar mass during necropsy of a 4-year-old mixed breed male dog, viewed at 60×. Note the large, round yeast-like cells with thick, double refractile walls and broad-based budding (Courtesy of Dr. Angelica Galezowski, DVM, MVetSc, DACVP, Faculty of Veterinary Medicine, University of Calgary, Canada)

cells may first be visualized with hematoxylin and eosin, although sensitivity and specificity are greatly improved by use of fungal stains like periodic acid-Schiff (PAS) or Grocott-Gomori methenamine-silver nitrate (GMS) (Saccente and Woods 2010). The host response is characterized by mixed inflammatory reaction, predominated first by neutrophils and eventually by noncaseating granulomas (Saccente and Woods 2010).

Cytologic and histopathologic examinations are frequently used in medical and veterinary practice. These tests perform well in diagnosing blastomycosis compared to culture (Patel et al. 2010), and with faster turnaround time. Cytologic examination led to the diagnosis in 71% of cases of canine disease from Louisiana, and histopathologic examination diagnosed an additional 9% (Arceneaux et al. 1998).

8.4.3 Antigen and Antibody Detection

An antigen enzyme immunoassay (EIA) for *B. dermatitidis* galactomannan is commercially available for the diagnosis of blastomycosis in humans and animals (MiraVista Diagnostics, Indianapolis, IN, USA). In a study of 89 people with blastomycosis proven by histopathology or culture, the quantitative detection of antigenuria had a sensitivity of 90% (Connolly et al. 2012). Sensitivity was higher in patients with pulmonary disease (with or without extrapulmonary dissemination) compared to isolated extrapulmonary disease. The specificity was 99% in controls without fungal infections, but cross-reactivity occurred in 96% of controls with histoplasmosis (Connolly et al. 2012). However, another study from Marshfield, Wisconsin, evaluated antigen tests in persons with blastomycosis over a course of 10 years and found the sensitivity of antigenuria to be 76% (Frost and Novicki 2015).

The use of the antigen EIA has also been studied for the diagnosis of canine blastomycosis (Spector et al. 2008). In a study of 46 dogs with blastomycosis, the sensitivities of antigen detection in urine and blood were 93 and 87%, respectively; false-positive results occurred in 2% of controls (Spector et al. 2008). Antigen detection is commonly used by veterinarians for the diagnosis of blastomycosis: Wisconsin veterinarians reported relying on this test more than any other to diagnose the disease (Anderson et al. 2014).

There is not currently a role for serological testing for *Blastomyces*-specific antibodies due to poor sensitivity (Smith and Gauthier 2015). An investigational EIA for antibodies against *Blastomyces adhesion-1* (BAD-1) was reported to have sensitivities and specificities of 88% and 95–99%, respectively (Richer et al. 2014). Confirmatory studies are warranted.

8.4.4 Nucleic Acid Detection

Nucleic acid detection is not yet a part of routine diagnostics for blastomycosis, and the role for this tool in clinical laboratories has not been defined. Theoretically, nucleic acid detection could reduce turnaround time compared with culture while obviating laboratory-associated hazard. It could also improve sensitivity compared to cytology or histopathology (Babady et al. 2011).

8.5 Treatment

8.5.1 In Humans

Clinical guidelines have been published for the management of human blastomycosis by the Infectious Diseases Society of America (IDSA) updated in 2008 (Chapman et al. 2008) and of human fungal pneumonias including blastomycosis by the American Thoracic Society (ATS) in 2011 (Limper et al. 2011). In humans, the recommended treatment for mild to moderate pulmonary or extrapulmonary blastomycosis (other than osteoarticular or CNS disease) is itraconazole 200 mg once or twice daily for 6–12 months; treatment of osteoarticular disease should be for 12 months. Moderately severe to severe pulmonary or extrapulmonary disease should be treated preferably with lipid formulation amphotericin B (3–5 mg/kg) or alternatively amphotericin deoxycholate (0.7–1.0 mg/kg) until clinical improvement, followed by de-escalation to a triazole for 6–12 months (Chapman et al. 2008). Central nervous system disease should be treated with a lipid formulation of amphotericin B (3–5 mg/kg) (for 4–6 weeks or until clinical improvement) followed by step-down to a triazole for 12 months (Chapman et al. 2008; Limper et al. 2011). Recent interest in the use of voriconazole for CNS blastomycosis has been driven by improved CSF and brain tissue penetration (Ta et al. 2009); accumulating published clinical experience provides anecdotal support for its use for this indication (Bakleh et al. 2005; Borgia et al. 2006; Bariola et al. 2010; Bush et al. 2013). Therapeutic drug monitoring is recommended for itraconazole and possibly voriconazole to ensure adequate levels are maintained (Chapman et al. 2008; Limper et al. 2011).

Acute respiratory distress syndrome due to blastomycosis occurs in 8–15% of persons with symptomatic blastomycosis (Meyer et al. 1993; Vasquez et al. 1998; Lemos et al. 2001; Azar et al. 2015), but it portends a grave prognosis with case fatality rates of at least 40% (Meyer et al. 1993; Vasquez et al. 1998; Lemos et al. 2001; Azar et al. 2015; Schwartz et al. 2016). Consequently, the roles of adjunctive, rescue therapies for ARDS caused by blastomycosis are an area of great interest, but little data. Some investigators have suggested a role for adjunctive corticosteroids considering the inflammatory cascade involved in the pathogenesis ARDS (Hough 2014). Anecdotal support has come from case reports and series (Lahm et al. 2008; Plamondon et al. 2010; Azar et al. 2015; Schwartz et al. 2016), although sufficiently powered prospective or retrospective studies are unlikely to be forthcoming (Schwartz et al. 2016). Extracorporeal membrane oxygenation (ECMO) is another promising adjunctive rescue therapy for ARDS due to blastomycosis (Dalton et al. 1999; Resch et al. 2009; Bednarczyk et al. 2015; Schwartz et al. 2016) and should be considered in centers where the capacity exists.

8.5.2 In Animals

Guidelines for the management of veterinary blastomycosis are currently available for cats as part of guidelines for the prevention and management of rare systemic mycoses published in 2013 by the European Advisory Board on Cat Diseases (Lloret et al. 2013). These suggest that itraconazole (10 mg/kg once daily) should be the preferred therapy for most cases, usually given for >3 months and that amphotericin B (0.25 mg/kg every 48 h to a total dose of 4–16 mg/kg) or fluconazole (2.5–10 mg/kg twice daily) are preferred for severe cases or those with CNS involvement (Lloret et al. 2013).

No guidelines currently exist for the diagnosis or management of canine blastomycosis. Randomized controlled trials do not exist to guide management decisions, but several prospective and retrospective studies of canine blastomycosis are instructive.

Legendre et al. prospectively treated 112 dogs with blastomycosis with itraconazole at doses of 5 or 10 mg/kg daily for 60 days and compared outcomes to historical controls treated with amphotericin B (at a cumulative dose of 8–9 mg/kg) in a study setting (Legendre et al. 1996). No differences were observed in outcomes between dogs treated with either of the itraconazole doses and amphotericin B: cure was achieved in 54–57% of all dogs, with disease recurring in 20–21% and death in 23–26% (Legendre et al. 1996). These observations are for the most part congruent with other reports (Arceneaux et al. 1998).

Fluconazole has been suggested by some authors as a reasonable (and cheaper) alternative to itraconazole in the treatment of canine blastomycosis (Mazepa et al. 2011). In vitro susceptibility testing on a small number of human *B. dermatitidis* isolates suggests fluconazole has less activity than itraconazole (Li et al. 2000; González et al. 2005). While no trial has compared these head-to-head, small single-arm trials in humans are informative. In a prospective, open-label trial of 48 persons with non-life-threatening, non-CNS blastomycosis, itraconazole (at a dose of 200–400 mg daily, for a mean of 6 months) resulted in 90% success (response with no relapse by 1 year follow-up) (Dismukes et al. 1992). In contrast, a trial of fluconazole at a dose of 400–800 mg daily (for a mean of 8.9 months) in 39 patients with non-life-threatening, non-CNS blastomycosis resulted in successful outcome in 87% (Pappas et al. 1997). Prospective trials are lacking in dogs, but Mazepa et al. (2011) retrospectively compared 36 and 31 dogs treated with fluconazole and itraconazole, respectively. The study was not powered to detect a difference in outcomes, but dogs treated with fluconazole required longer courses of therapy than those treated with itraconazole (median 183 vs 138 days, respectively). Nonetheless, costs for fluconazole were much less than for itraconazole, with median costs of $1223 and $3717, respectively (Mazepa et al. 2011). Taken together, this data suggests that fluconazole at higher doses and for longer courses may be reasonable alternatives to itraconazole for blastomycosis in animals.

For CNS and ocular disease, fluconazole achieves better penetration than itraconazole into these structures (Perfect et al. 1986; Savani et al. 1987), although both drugs were effective in animal models of non-blastomycotic fungal meningitis (Perfect et al. 1986) and endophthalmitis (Savani et al. 1987).

Adjunctive systemic corticosteroids have been suggested for ocular blastomycosis in dogs. In a retrospective study of 12 dogs with ocular involvement in 19 eyes, systemic corticosteroids were administered in addition to triazoles (Finn et al. 2007). The success rate in this series was 74% across all eyes involved, including 67% of eyes with endophthalmitis, the most common complication that generally carries a poor prognosis (Finn et al. 2007). For comparison, another retrospective study of itraconazole monotherapy for ocular blastomycosis reported favorable outcomes in just 13% of eyes with endophthalmitis (Brooks et al. 1992). In another study, amphotericin B plus ketoconazole resulted in favorable outcomes for 20% of eyes with endophthalmitis, although this improved to 43% if only non-severely affected eyes were considered (Bloom et al. 1996).

8.5.3 Monitoring Response to Therapy

Radiographic worsening is commonly observed soon after initiation of effective antifungal therapy for canine blastomycosis, occurring in almost a quarter of dogs (Crews et al. 2008b). However, this does not portend a worse outcome (Crews et al. 2008b) and—in the absence of other signs of clinical failure—should not alter management. For this reason, follow-up chest radiography is recommended no sooner than 4–6 weeks after therapy initiation in stable patients (Crews et al. 2008b).

Quantitative *Blastomyces* antigen detection by EIA has been evaluated for monitoring of remission in dogs during and after treatment for blastomycosis (Foy et al. 2014). Foy and colleagues prospectively studied 27 dogs with blastomycosis who were monitored clinically, radiographically, and with detection and quantification of *Blastomyces* antigen in urine and serum following discontinuation of antifungal therapy; among these, 12 dogs were also monitored from time of therapy initiation (Foy et al. 2014). The investigators found that urine antigen levels dropped dramatically within several months of initiation of antifungal therapy. Seven of 27 dogs (26%) relapsed at a median of 4 months following treatment discontinuation. Five of these had detectable antigenuria at the time of clinical relapse, but only one had rising levels of antigenuria preceding clinically detectable relapse. Moreover, persistence of positive urinary antigen at treatment discontinuation did not predict relapse: only two of seven dogs that relapsed had detectable antigenuria at that treatment discontinuation. On the other hand, five of eight of dogs with antigenuria at the end of therapy did not relapse (Foy et al. 2014). In summary, monitoring urine antigen levels during and/or after discontinuation of therapy is unlikely to add significant value to serial clinical and radiographic evaluations for most dogs with blastomycosis.

8.6 Prevention

A commercially available vaccine against blastomycosis does not currently exist, although the science is advancing. Wüthrich et al. (2000, 2011) developed a recombinant, live-attenuated vaccine using an avirulent, genetically engineered strain of *B. dermatitidis* lacking *Blastomyces* adhesin-1 (*BAD-1*, previously called *WI-1*), an essential virulence factor. This vaccine has been demonstrated to be protective against experimental blastomycosis in mice (Wüthrich et al. 2000), although efficacy in dogs has not yet been established. Experimental infection and a field study of beagles and foxhounds, respectively, demonstrated acceptable safety and immunogenicity (Wüthrich et al. 2011). Adverse reactions included fever, lymphadenopathy, and draining cutaneous lesions at the site of inoculation (Wüthrich et al. 2011); these may limit use in all but highly enzootic regions.

References

Alpern JD, Bahr NC, Vazquez-Benitez G et al (2016) Diagnostic delay and antibiotic overuse in acute pulmonary blastomycosis. Open Forum Infect Dis 3:ofw078. https://doi.org/10.1093/ofid/ofw078

Anderson JL, Dieckman JL, Reed KD, Meece JK (2014) Canine blastomycosis in Wisconsin: a survey of small-animal veterinary practices. Med Mycol 52:774–779. https://doi.org/10.1093/mmy/myu051

Anderson JL, Meece JK, Hall MC, Frost HM (2016) Evidence of delayed dissemination or re-infection with Blastomyces in two immunocompetent hosts. Med Mycol Case Rep 13:9–11. https://doi.org/10.1016/j.mmcr.2016.09.002

Arceneaux KA, Taboada J, Hosgood G (1998) Blastomycosis in dogs: 115 cases (1980–1995). J Am Vet Med Assoc 213:658–664

Archer JR, Trainer DO, Schell RF (1987) Epidemiologic study of canine blastomycosis in Wisconsin. J Am Vet Med Assoc 190:1292–1295

Armstrong CW, Jenkins SR, Kaufman L et al (1987) Common-source outbreak of blastomycosis in hunters and their dogs. J Infect Dis 155:568–570

Azar MM, Assi R, Relich RF et al (2015) Blastomycosis in Indiana: clinical and epidemiologic patterns of disease gleaned from a multicenter retrospective study. Chest 148:1276–1284. https://doi.org/10.1378/chest.15-0289

Babady NE, Buckwalter SP, Hall L et al (2011) Detection of Blastomyces dermatitidis and Histoplasma capsulatum from culture isolates and clinical specimens by use of real-time PCR. J Clin Microbiol 49:3204–3208. https://doi.org/10.1128/JCM.00673-11

Bakleh M, Aksamit AJ, Tleyjeh IM, Marshall WF (2005) Successful treatment of cerebral blastomycosis with voriconazole. Clin Infect Dis 40:e69–e71. https://doi.org/10.1086/429319

Bariola JR, Perry P, Pappas PG et al (2010) Blastomycosis of the central nervous system: a multicenter review of diagnosis and treatment in the modern era. Clin Infect Dis 50:797–804. https://doi.org/10.1086/650579

Baumgardner DJ, Buggy BP, Mattson BJ et al (1992) Epidemiology of blastomycosis in a region of high endemicity in north central Wisconsin. Clin Infect Dis 15:629–635

Baumgardner DJ, Burdick JS (1991) An outbreak of human and canine blastomycosis. Rev Infect Dis 13:898–905

Baumgardner DJ, Paretsky DP, Yopp AC (1995) The epidemiology of blastomycosis in dogs: north central Wisconsin, USA. J Med Vet Mycol 33:171–176

Bednarczyk JM, Kethireddy S, White CW et al (2015) Extracorporeal membrane oxygenation for blastomycosis related acute respiratory distress syndrome: a case series. Can J Anaesth 62:807–815

Blondin N, Baumgardner DJ, Moore GE, Glickman LT (2007) Blastomycosis in indoor cats: suburban Chicago, Illinois, USA. Mycopathologia 163:59–66. https://doi.org/10.1007/s11046-006-0090-1

Bloom JD, Hamor RE, Gerding PA (1996) Ocular blastomycosis in dogs: 73 cases, 108 eyes (1985–1993). J Am Vet Med Assoc 209:1271–1274

Borgia SM, Fuller JD, Sarabia A, El-Helou P (2006) Cerebral blastomycosis: a case series incorporating voriconazole in the treatment regimen. Med Mycol 44:659–664. https://doi.org/10.1080/13693780600803870

Bradsher RW (2014a) The endemic mimic: blastomycosis an illness often misdiagnosed. Trans Am Clin Climatol Assoc 125:188–202

Bradsher RW (2014b) Blastomycosis. In: Bennett JE, Dolin R, Blaser MJ (eds) Mandell, Douglas, and Bennett's principles and practice of infectious diseases, 8th edn. Elsevier, Amsterdam, pp 2963–2973

Broc R, Haddad N (1952) Tumeur bronchique a Scopulariopsis americana determination precoce d'une maladie de Gilchrist. Bull Mem Soc Med Hop Paris 68:678–682

Brooks DE, Legendre AM, Gum GG et al (1992) The treatment of canine ocular blastomycosis with systemically administered itraconazole. Prog Vet Comp Ophthalmol 4:263–268

Brown EM, McTaggart LR, Zhang SX et al (2013) Phylogenetic analysis reveals a cryptic species Blastomyces gilchristii, sp. nov. within the human pathogenic fungus Blastomyces dermatitidis. PLoS One 8:e59237

Bush JW, Wuerz T, Embil JM et al (2013) Outcomes of persons with blastomycosis involving the central nervous system. Diagn Microbiol Infect Dis 76:175–181. https://doi.org/10.1016/j.diagmicrobio.2013.03.002

Carman WF, Frean JA, Crewe-Brown HH et al (1989) Blastomycosis in Africa. A review of known cases diagnosed between 1951 and 1987. Mycopathologia 107:25–32

Cates MB, Kaufman L, Grabau JH et al (1986) Blastomycosis in an Atlantic bottlenose dolphin. J Am Vet Med Assoc 189:1148–1150

Chapman SW, Dismukes WE, Proia LA et al (2008) Clinical practice guidelines for the management of blastomycosis: 2008 update by the Infectious Diseases Society of America. Clin Infect Dis 46:1801–1812

Connolly P, Hage CA, Bariola JR et al (2012) Blastomyces dermatitidis antigen detection by quantitative enzyme immunoassay. Clin Vaccine Immunol 19:53–56. https://doi.org/10.1128/CVI.05248-11

Crews LJ, Feeney DA, Jessen CR et al (2008a) Utility of diagnostic tests for and medical treatment of pulmonary blastomycosis in dogs: 125 cases (1989–2006). J Am Vet Med Assoc 232: 222–227. https://doi.org/10.2460/javma.232.2.222

Crews LJ, Feeney DA, Jessen CR, Newman AB (2008b) Radiographic findings in dogs with pulmonary blastomycosis: 125 cases (1989–2006). J Am Vet Med Assoc 232:215–221. https://doi.org/10.2460/javma.232.2.215

Dalton HJ, Hertzog JH, Hannan RL et al (1999) Extracorporeal membrane oxygenation for overwhelming Blastomyces dermatitidis pneumonia. Crit Care 3:91–94

Davies C, Troy G (1996) Deep mycotic infections in cats. J Am Anim Hosp Assoc 32:380–391. https://doi.org/10.5326/15473317-32-5-380

Davies JL, Epp T, Burgess HJ (2013) Prevalence and geographic distribution of canine and feline blastomycosis in the Canadian prairies. Can Vet J 54:753–760

Denton JF, Di Salvo AF, Hirsch ML et al (1967) Laboratory-acquired North American blastomycosis. JAMA 199:935. https://doi.org/10.1001/jama.1967.03120120123030

Dismukes WE, Bradsher RW, Cloud GC et al (1992) Itraconazole therapy for blastomycosis and histoplasmosis. NIAID Mycoses Study Group. Am J Med 93:489–497

Dukik K, Muñoz JF, Jiang Y et al (2017) Novel taxa of thermally dimorphic systemic pathogens in the Ajellomycetaceae (Onygenales). Mycoses 60:297–309. https://doi.org/10.1111/myc.12601

Dykstra JA, Rogers LL, Mansfield SA et al (2012) Fatal disseminated blastomycosis in a free-ranging American black bear (Ursus americanus). J Vet Diagn Investig 24:1125–1128. https://doi.org/10.1177/1040638712461788

Finn MJ, Stiles J, Krohne SG (2007) Visual outcome in a group of dogs with ocular blastomycosis treated with systemic antifungals and systemic corticosteroids. Vet Ophthalmol 10:299–303. https://doi.org/10.1111/j.1463-5224.2007.00554.x

Foy DS, Trepanier LA, Kirsch EJ, Wheat LJ (2014) Serum and urine *Blastomyces* antigen concentrations as markers of clinical remission in dogs treated for systemic blastomycosis. J Vet Intern Med 28:305–310. https://doi.org/10.1111/jvim.12306

Frost HM, Novicki TJ (2015) Blastomyces antigen detection for the diagnosis and management of blastomycosis. J Clin Microbiol 53:02352–02315. https://doi.org/10.1128/JCM.02352-15

Gilchrist TC (1894) Protozoan dermatitis. J Cutan Gen Dis 12:496–499

Gilchrist TC, Stokes WR (1898) A case of pseudo-lupus vulgaris caused by a blastomyces. J Exp Med 3:53–78

González GM, Fothergill AW, Sutton DA et al (2005) In vitro activities of new and established triazoles against opportunistic filamentous and dimorphic fungi. Med Mycol 43:281–284. https://doi.org/10.1080/13693780500088416

Gray NA, Baddour LM (2002) Cutaneous inoculation blastomycosis. Clin Infect Dis 34:E44–E49. https://doi.org/10.1086/339957

Guarner J, Brandt ME (2011) Histopathologic diagnosis of fungal infections in the 21st century. Clin Microbiol Rev 24:247–280. https://doi.org/10.1128/CMR.00053-10

Harris JR, Blaney DD, Lindsley MD et al (2011) Blastomycosis in man after kinkajou bite. Emerg Infect Dis 17:268–270. https://doi.org/10.3201/eid1702.101046

Herrmann JA, Kostiuk SL, Dworkin MS, Johnson YJ (2011) Temporal and spatial distribution of blastomycosis cases among humans and dogs in Illinois (2001–2007). J Am Vet Med Assoc 239:335–343. https://doi.org/10.2460/javma.239.3.335

Hough CL (2014) Steroids for acute respiratory distress syndrome? Clin Chest Med 35:781–795

Houseright RA, Webb JL, Claus KN (2015) Pathology in practice. Blastomycosis in an indoor-only cat. J Am Vet Med Assoc 247:357–359. https://doi.org/10.2460/javma.247.4.357

Imai DM, McGreevey N, Anderson JL, Meece JK (2014) Disseminated *Blastomyces dermatitidis*, genetic group 2, infection in an alpaca (*Vicugna pacos*). J Vet Diagn Invest 26:442–447. https://doi.org/10.1177/1040638714523773

Kaufman L, Standard PG, Weeks RJ, Padhye AA (1983) Detection of two Blastomyces dermatitidis serotypes by exoantigen analysis. J Clin Microbiol 18:110–114

Kingston M, El-Mishad MM, Ali MA (1980) Blastomycosis in Saudi Arabia. Am J Trop Med Hyg 29:464–466

Klein BS, Vergeront JM, Weeks RJ et al (1986) Isolation of Blastomyces dermatitidis in soil associated with a large outbreak of blastomycosis in Wisconsin. N Engl J Med 314:529–534

Kralt D, Light B, Cheang M et al (2009) Clinical characteristics and outcomes in patients with pulmonary blastomycosis. Mycopathologia 167:115–124

Kuttin ES, Beemer AM, Levij J et al (1978) Occurrence of Blastomyces dermatitidis in Israel. First autochthonous Middle Eastern case. Am J Trop Med Hyg 27:1203–1205

Lahm T, Neese S, Thornburg AT et al (2008) Corticosteroids for blastomycosis-induced ARDS: a report of two patients and review of the literature. Chest 133:1478–1480

Langlois DK, Pelosi A, Kruger JM (2013) Successful treatment of intracardiac and intraocular blastomycosis in a dog with combination azole therapy. J Am Anim Hosp Assoc 49:273–280. https://doi.org/10.5326/JAAHA-MS-5874

Legendre AM (2012) Blastomycosis. In: Greene CE (ed) Infectious diseases of the dog and cat, 4th edn. Elsevier, Amsterdam, pp 606–614

Legendre AM, Rohrbach BW, Toal RL et al (1996) Treatment of blastomycosis with itraconazole in 112 dogs. J Vet Intern Med 10:365–371

Lemos LB, Baliga M, Guo M (2002) Blastomycosis: the great pretender can also be an opportunist. Initial clinical diagnosis and underlying diseases in 123 patients. Ann Diagn Pathol 6:194–203

Lemos LB, Baliga M, Guo M (2001) Acute respiratory distress syndrome and blastomycosis: presentation of nine cases and review of the literature. Ann Diagn Pathol 5:1–9

Li RK, Ciblak MA, Nordoff N et al (2000) In vitro activities of voriconazole, itraconazole, and amphotericin B against Blastomyces dermatitidis, Coccidioides immitis, and Histoplasma capsulatum. Antimicrob Agents Chemother 44(6):1734. https://doi.org/10.1128/AAC.44.6.1734-1736.2000

Limper AH, Knox KS, Sarosi GA et al (2011) An official American Thoracic Society statement: treatment of fungal infections in adult pulmonary and critical care patients. Am J Respir Crit Care Med 183:96–128

Lloret A, Hartmann K, Pennisi MG et al (2013) Rare systemic mycoses in cats: blastomycosis, histoplasmosis and coccidioidomycosis: ABCD guidelines on prevention and management. J Feline Med Surg 15(7):624. https://doi.org/10.1177/1098612X13489226

Lombardi G, Padhye AA, Ajello L (1988) In vitro conversion of African isolates of Blastomyces dermatitidis to their yeast form. Mycoses 31:447–450

Lopez R, Mason JO, Parker JS, Pappas PG (1994) Intraocular blastomycosis: case report and review. Clin Infect Dis 18:805–807

Martynowicz MA, Prakash UBS (2002) Pulmonary blastomycosis: an appraisal of diagnostic techniques. Chest 121:768–773. https://doi.org/10.1378/chest.121.3.768

Mazepa ASW, Trepanier LA, Foy DS (2011) Retrospective comparison of the efficacy of fluconazole or itraconazole for the treatment of systemic blastomycosis in dogs. J Vet Intern Med 25: 440–445. https://doi.org/10.1111/j.1939-1676.2011.0710.x

Meece JK, Anderson JL, Fisher MC et al (2011) Population genetic structure of clinical and environmental isolates of Blastomyces dermatitidis, based on 27 polymorphic microsatellite markers. Appl Environ Microbiol 77:5123–5131. https://doi.org/10.1128/AEM.00258-11

Meece JK, Anderson JL, Gruszka S et al (2013) Variation in clinical phenotype of human infection among genetic groups of Blastomyces dermatitidis. J Infect Dis 207:814–822. https://doi.org/10.1093/infdis/jis756

Meyer K (1912) Blastomycosis in dogs. Proc Path Soc Philadelphia 15:10

Meyer KC, McManus EJ, Maki DG (1993) Overwhelming pulmonary blastomycosis associated with the adult respiratory distress syndrome. N Engl J Med 329:1231–1236

Nemeth NM, Campbell GD, Oesterle PT et al (2016) Red fox as sentinel for Blastomyces dermatitidis, Ontario, Canada. Emerg Infect Dis 22(7):1275. https://doi.org/10.3201/eid2207.151789

Oppenheimer M, Embil JM, Black B et al (2007) Blastomycosis of bones and joints. South Med J 100:570–578. https://doi.org/10.1097/SMJ.0b013e3180487a92

Pappas PG, Bradsher RW, Kauffman CA et al (1997) Treatment of blastomycosis with higher doses of fluconazole. The National Institute of Allergy and Infectious Diseases Mycoses Study Group. Clin Infect Dis 25:200–205

Pappas PG, Threlkeld MG, Bedsole GD et al (1993) Blastomycosis in immunocompromised patients. Medicine (Baltimore) 72:311–325

Patel AJ, Gattuso P, Reddy VB (2010) Diagnosis of blastomycosis in surgical pathology and cytopathology: correlation with microbiologic culture. Am J Surg Pathol 34:256–261. https://doi.org/10.1097/PAS.0b013e3181ca48a5

Patel RG, Patel B, Petrini MF et al (1999) Clinical presentation, radiographic findings, and diagnostic methods of pulmonary blastomycosis: a review of 100 consecutive cases. South Med J 92:289–295

Perfect JR, Savani DV, Durack DT (1986) Comparison of itraconazole and fluconazole in treatment of cryptococcal meningitis and candida pyelonephritis in rabbits. Antimicrob Agents Chemother 29:579–583

Plamondon M, Lamontagne F, Allard C, Pépin J (2010) Corticosteroids as adjunctive therapy in severe blastomycosis-induced acute respiratory distress syndrome in an immunosuppressed patient. Clin Infect Dis 51:e1–e3

Randhawa HS, Khan ZU, Gaur SN (1983) Blastomyces dermatitidis in India: first report of its isolation from clinical material. Sabouraudia 21:215–221

Resch M, Kurz K, Schneider-Brachert W et al (2009) Extracorporeal membrane oxygenation (ECMO) for severe acute respiratory distress syndrome (ARDS) in fulminant blastomycosis in Germany. BMJ Case Rep 2009:bcr0720080392. https://doi.org/10.1136/bcr.07.2008.0392
Restrepo A, Baumgardner DJ, Bagagli E et al (2000) Clues to the presence of pathogenic fungi in certain environments. Med Mycol 38(Suppl 1):67–77
Richer SM, Smedema ML, Durkin MM et al (2014) Development of a highly sensitive and specific blastomycosis antibody enzyme immunoassay using Blastomyces dermatitidis surface protein BAD-1. Clin Vaccine Immunol 21:143–146. https://doi.org/10.1128/CVI.00597-13
Rosser MF, Lindemann DM, Barger AM et al (2016) Systemic blastomycosis in a captive red ruffed lemur (Varecia rubra). J Zoo Wildl Med 47:912–916. https://doi.org/10.1638/2016-0019.1
Rudmann DG, Coolman BR, Perez CM, Glickman LT (1992) Evaluation of risk factors for blastomycosis in dogs: 857 cases (1980–1990). J Am Vet Med Assoc 201:1754–1759
Saccente M, Abernathy RS, Pappas PG et al (1998) Vertebral blastomycosis with paravertebral abscess: report of eight cases and review of the literature. Clin Infect Dis 26:413–418
Saccente M, Woods GL (2010) Clinical and laboratory update on blastomycosis. Clin Microbiol Rev 23:367–381
Sarosi GA, Eckman MR, Davies SF, Laskey WK (1979) Canine blastomycosis as a harbinger of human disease. Ann Intern Med 91:733. https://doi.org/10.7326/0003-4819-91-5-733
Sarosi GA, Hammerman KJ, Tosh FE, Kronenberg RS (1974) Clinical features of acute pulmonary blastomycosis. N Engl J Med 290:540–543. https://doi.org/10.1056/NEJM197403072901004
Savani DV, Perfect JR, Cobo LM, Durack DT (1987) Penetration of new azole compounds into the eye and efficacy in experimental Candida endophthalmitis. Antimicrob Agents Chemother 31:6–10
Schmiedt C, Kellum H, Legendre AM et al (2015) Cardiovascular involvement in 8 dogs with blastomyces dermatitidis infection. J Vet Intern Med 20:1351–1354. https://doi.org/10.1892/0891-6640(2006)20[1351:CIIDWB]2.0.CO;2
Schwartz IS, Embil JM, Sharma A et al (2016) Management and outcomes of acute respiratory distress syndrome caused by blastomycosis: a retrospective case series. Medicine (Baltimore) 95:e3538. https://doi.org/10.1097/MD.0000000000003538
Smith JA, Gauthier G (2015) New developments in blastomycosis. Semin Respir Crit Care Med 36: 715–728. https://doi.org/10.1055/s-0035-1562898
Smith JA, Kauffman CA (2009) Endemic fungal infections in patients receiving tumour necrosis factor-alpha inhibitor therapy. Drugs 69:1403–1415. https://doi.org/10.2165/00003495-200969110-00001
Spector D, Legendre AM, Wheat J et al (2008) Antigen and antibody testing for the diagnosis of blastomycosis in dogs. J Vet Intern Med 22:839–843. https://doi.org/10.1111/j.1939-1676.2008.0107.x
Stewart AJ, Cuming RS (2015) Update on fungal respiratory disease in horses. Vet Clin North Am Equine Pract 31:43–62. https://doi.org/10.1016/j.cveq.2014.11.005
Storms TN, Clyde VL, Munson L, Ramsay EC (2003) Blastomycosis in nondomestic felids. J Zoo Wildl Med 34:231–238. https://doi.org/10.1638/1042-7260(2003)034[0231:BINF]2.0.CO;2
Sykes JE, Merkel LK (2014) Blastomycosis. In: Sykes JE (ed) Canine and feline infectious diseases. Elsevier, St. Louis, pp 574–586
Ta M, Flowers SA, Rogers PD (2009) The role of voriconazole in the treatment of central nervous system blastomycosis. Ann Pharmacother 43:1696–1700. https://doi.org/10.1345/aph.1M010
Vandepitte J, Gatti F (1972) A case of North American blastomycosis in Africa. Its existence in Republic of Zaire. Ann la Société belge médecine Trop 52:467–479
Vasquez JE, Mehta JB, Agrawal R, Sarubbi FA (1998) Blastomycosis in northeast Tennessee. Chest 114:436–443
Wilkinson LM, Wallace JM, Cline JM (1999) Disseminated blastomycosis in a rhesus monkey (*Macaca mulatta*). Vet Pathol 36:460–462. https://doi.org/10.1354/vp.36-5-460

Williams JE, Moser SA (1987) Chronic murine pulmonary blastomycosis induced by intratracheally inoculated Blastomyces dermatitidis conidia. Am Rev Respir Dis 135:17–25. https://doi.org/10.1164/arrd.1987.135.1.17
Wilson JH, Olson EJ, Haugen EW et al (2006) Systemic blastomycosis in a horse. J Vet Diagn Investig 18:615–619. https://doi.org/10.1177/104063870601800619
Wüthrich M, Filutowicz HI, Klein BS (2000) Mutation of the WI-1 gene yields an attenuated blastomyces dermatitidis strain that induces host resistance. J Clin Investig 106:1381–1389. https://doi.org/10.1172/JCI11037
Wüthrich M, Krajaejun T, Shearn-Bochsler V et al (2011) Safety, tolerability, and immunogenicity of a recombinant, genetically engineered, live-attenuated vaccine against canine blastomycosis. Clin Vaccine Immunol 18:783–789. https://doi.org/10.1128/CVI.00560-10
Zwick LS, Briggs MB, Tunev SS et al (2000) Disseminated blastomycosis in two California sea lions (Zalophus californianus). J Zoo Wildl Med 31:211–214. https://doi.org/10.1638/1042-7260(2000)031[0211:DBITCS]2.0.CO;2

Paracoccidioidomycosis ceti (Lacaziosis/Lobomycosis) in Dolphins

9

Raquel Vilela and Leonel Mendoza

Abstract

Infections caused by the fungal pathogen *Lacazia loboi* were first reported in 1931 by Jorge de Oliveira Lobo in a human with granulomatous skin lesions in Pernambuco, Brazil. Early histopathological and serological analyses found morphological similarities and cross-reactive antigens with *Paracoccidioides brasiliensis*. In 1971, veterinarians working with dolphins in Florida, USA, reported granulomatous skin lesions in a dolphin, similar to that in human lacaziosis. Based on histopathological findings, *L. loboi* was initially believed to be also the etiologic agent of cutaneous disease in dolphins. Ever since, cutaneous granulomas have been reported in different dolphin species around the coast of Asia, Europe, and North and South America. Recently, using molecular biology approaches, some investigators stated that the DNA sequences extracted from cases of cutaneous granulomas in dolphins were closely related to those of *P. brasiliensis*. This chapter deals with the history, taxonomy, and other features of *L. loboi* in humans and the unculturable *P. brasiliensis* var. *ceti* type affecting the skin of dolphins.

R. Vilela
Faculty of Pharmacy, Federal University of Minas Gerais, Belo Horizonte, Minas Gerais, Brazil

Biomedical Laboratory Diagnostics, Michigan State University, East Lansing, MI, USA
e-mail: vilelar@msu.edu

L. Mendoza (✉)
Microbiology and Molecular Genetics, Biomedical Laboratory Diagnostics, Michigan State University, East Lansing, MI, USA
e-mail: mendoza9@msu.edu

S. Seyedmousavi et al. (eds.), *Emerging and Epizootic Fungal Infections in Animals*,
https://doi.org/10.1007/978-3-319-72093-7_9

9.1 History of *Lacazia loboi* in Humans and *P. brasiliensis* var. *ceti* in Dolphins

The first case of lacaziosis (lobomycosis) in humans was reported by Jorge de Oliveira Lobo in a male patient from the Amazon basin with chronic (19 years) nodular lesions on his sacral anatomical area, in Recife, Pernambuco, Brazil (Lobo 1931). Forty years later, Migaki et al. (1971) reported skin granulomas in dolphins (*Tursiops truncatus*) with the development of yeast-like cells similar to those initially reported by Jorge Lobo.

In the following years, it was evident that, despite the numerous yeast-like cells present in the infected tissue of humans with lacaziosis, *L. loboi* resisted culture on most mycological media (Almeida and Lacaz 1948–49; Borelli 1968; Fonseca and Lacaz 1971; Furtado et al. 1967; Lacaz 1996). This unique feature of the pathogen led to false claims on its isolation in pure culture (Fonseca and Area 1940; Lacaz et al. 1986). However, those claims were challenged, and their isolates later identified as contaminating fungi (Fonseca and Lacaz 1971; Lacaz 1996; Vilela et al. 2007). Even today, allegations on the isolation of contaminating fungi from cases of human lacaziosis persist (Costa 2015).

The pathogen was known under names such as *Blastomyces brasiliensis*, *B. loboi*, *Glenosporella loboi*, *G. amazonica*, *Lobomyces loboi*, *Loboa loboi*, and *Paracoccidioides loboi* (Camargo et al. 1998; Fonseca and Lacaz 1971; Furtado et al. 1967; Lacaz 1996). Due to many taxonomic uncertainties surrounding the etiologic agent of cutaneous granulomas in humans, Taborda et al. (1999a, b) proposed the genus *Lacazia* (genus name dedicated to Dr. Carlos da Silva Lacaz, for his contribution on *L. loboi*), ending 70 years of taxonomic uncertainties. Likewise, the disease name periodically changed according to the proposed taxonomy. Examples of these are lobomycosis (*Lobomyces*, Borelli 1968) and lacaziosis (*Lacazia*, Vilela et al. 2005). In addition, the following regional disease names were also used for both human or dolphin keloid blastomycosis (Lobo 1931), Jorge Lobo disease (Lacaz et al. 1986), blastomycosis Jorge Lobo type (Almeida and Lacaz 1948–49), and others (Almeida and Lacaz 1948–49; Arju et al. 2014; Azulay et al. 1976; Baruzzi et al. 1973; Lacaz et al. 1986; Xavier et al. 2008).

The first report of a common bottlenose dolphin (*Tursiops truncatus*) displaying cutaneous lesions similar to human lacaziosis took place in the coastal areas of Florida, USA (Migaki et al. 1971). The following year, Woodard (1972) reported the second case on the same dolphin species and again in Florida. In 1973, De Vries and Laarman reported cutaneous lesions in a Guiana dolphin (*Sotalia guianensis*). As many others, the latter authors isolated fungal contaminants (*Scedosporium apiospermum* and *Candida haemulonis*) that they believed were the etiologic agents of the granulomas in dolphins. These cases were soon followed by reports of severe skin infections in several dolphin species. Most of them diagnosed around the US coastal areas, South America, and in other oceans (Bermudez et al. 2009; Dudok van Heel 1977; Kiszka et al. 2009; Lane et al. 2014; Symmers 1983; Tajima et al. 2015; Van Bressem et al. 2009). More recently, Minakawa et al. (2016) reported the infection in a new dolphin species (the Pacific white-sided dolphin, *Lagenorhynchus*

Table 9.1 Names applied to uncultivated types of humans and dolphins during the past 70 years

Nomenclature and synonyms	Human	Dolphin
Glenosporella loboi (1940)	√	
Blastomyces brasiliensis (1941)	√	
Glenosporopsis amazonica (1943)	√	
Paracoccidioides loboi (1948)	√	√
Blastomyces loboi (1952)	√	
Loboa loboi (1956)	√	
Lobomyces loboi (1958)	√	√
Lacazia loboi (1999)	+	√
Paracoccidioides brasiliensis var. *ceti* (2017)		+

√ = misapplied or synonymous name; + = correct name

obliquidens) suggesting that other cetacean species should also be investigated. Reports of infections are consistently diagnosed every year in different geographical areas, confirming the importance of the disease in this protected species (Table 9.1).

There are several reports of the unculturable nature of *L. loboi* in humans (Azulay et al. 1976; Borelli 1962; Fonseca and Area 1940; Furtado et al. 1967; Lacaz et al. 1986), but only two in dolphins. The first study indicated culture failure from an infected dolphin (Caldwell et al. 1975); the second is a recent well-documented study confirming its unculturable nature (Schaefer et al. 2016). In that study, Schaefer et al. (2016) tested fresh samples from common bottlenose dolphins (*T. truncatus*) collected at the Indian River Lagoon, Florida, USA. This study concluded that the etiologic agent was an unculturable version of *P. brasiliensis*. Based on phylogenetic analyses, in the interim, we use the name *P. brasiliensis* var. *ceti* to identify properly this unique type within the culturable *P. brasiliensis* as proposed by Vilela et al. (2016).

9.2 Taxonomy of *Lacazia loboi* and *Paracoccidioides brasiliensis* var. *ceti*

Paracoccidioides brasiliensis (Splendore) de Almeida **var.** *ceti* Vilela, St. Leger, Bossart, and Mendoza, **var. nov**.

Holotype B92-932, H&E histopathological slide collected from a US dolphin, deposited at the Michigan State University Herbarium, East Lansing.

Etymology To differentiate this novel uncultivated variety of *P. brasiliensis* affecting dolphins from the cultivated type causing human systemic paracoccidioidomycosis, the variety *ceti* (Cetacea) is proposed.

Description Uncultivated fungus causing cutaneous disease in dolphins. In vivo, numerous branching chains of globose to subglobose, yeast-like cells (5–10 μm) present, some connected by slender bridges (2–3 μm). Adjacent older cells detach by increasing cell wall thickness.

Disease nomenclature We propose the name "paracoccidioidomycosis ceti" to differentiate this cutaneous disease of dolphins from human systemic paracoccidioidomycosis (Vilela et al. 2016).

Due to the unculturable nature of *L. loboi* and *P. brasiliensis* var. *ceti*, the taxonomy of these two etiologic agents of cutaneous granulomas has been contentious (see above). *Lacazia loboi* from humans and *P. brasiliensis* var. *ceti* from dolphins share several features: (1) they resist culture (Lacaz et al. 1986; Schaefer et al. 2016); (2) they are restricted to cutaneous granulomas in mammalian hosts (Baruzzi et al. 1973; Bossart et al. 2015); (3) they develop similar yeast-like cells in chains (Haubold et al. 2000; Vilela et al. 2016); and (4) they possess cross-reactive antigens (Mendoza et al. 2008). Using traditional approaches, these two mammalian pathogens were believed to be the same organism. Only recently, with the use of molecular methodologies, their true position in the tree of life was unveiled (Herr et al. 2001; Rotstein et al. 2009; Vilela et al. 2016).

Initially Herr et al. (2001) extracted total *L. loboi* DNA from a biopsied tissue collected from a Brazilian man with cutaneous lacaziosis. They found that ITS, chitin synthase 4 (*CHS4*), and 18S SSU rDNA sequences placed this pathogen as a sister group to *P. brasiliensis*, the etiologic agent of systemic paracoccidioidomycosis. The authors argued that their phylogenetic data confirmed Lacaz (1996) position on the similarities shared by these two pathogens. The main problem studying the phylogenetics of *L. loboi*, however, was that the total DNA recovered from biopsied tissues contained DNA from the host and from normal skin and environmental microbiota. Vilela et al. (2005) proposed a molecular model to study this unculturable pathogen using specific primers. They argued that due to their phylogenetic proximity, *L. loboi* might share common DNA sequences with *P. brasiliensis*. Using this approach, they were able to amplify the *gp43*-like gene in *L. loboi* using well known DNA sequences of this gene in *P. brasiliensis*. This approach was later validated when Vilela et al. (2009) amplified the ITS rDNA and chitin synthase 4, ADP-ribosylation factor, and *gp43* coding genes. The phylogenetic data showed that indeed *L. loboi* clustered with strong support in its own genus, confirming previous analysis (Vilela et al. 2009). These analyses showed that *P. brasiliensis* and *P. lutzii* shared the same ancestor with *L. loboi* and *P. brasiliensis* var. *ceti*; thus the latter unculturable microbes may possess putative dimorphic capabilities.

When Vilela et al. (2009) concluded their studies, *L. loboi* was still considered the same etiology affecting both humans and dolphins. However, Rotstein et al. (2009) found that the DNA sequences (26S LSrDNA) recovered from an offshore dolphin

(*T. truncatus*) with cutaneous granulomas displaying yeast-like cells shared 97% identity with *P. brasiliensis*. Unfortunately, their DNA sequences are not available. Subsequently, two teams in Japan (Minakawa et al. 2016; Ueda et al. 2013) and one in Spain (Esperon et al. 2012) reported that the DNA sequences, using at least two types of sequences (ITS and *gp43*), placed the etiologic agent of dolphin granulomas within the DNA sequences of the culturable *P. brasiliensis* isolates causing human systemic paracoccidioidomycosis. These findings came as a surprise since the clinical morphological features of the yeast-like cells in the dolphin-infected tissues look similar to that in human lacaziosis (Haubold et al. 2000; Lacaz et al. 1986). To validate these findings, Vilela et al. (2016) amplified the coding *kex* gene from six dolphins (*T. truncatus*) with cutaneous granulomas collected and two additional *CHS4* DNA sequences from different dolphins at the Indian River Lagoon, Florida, USA (Fig. 9.1). They confirmed previous studies (Esperon et al. 2012; Minakawa et al. 2016; Rotstein et al. 2009; Ueda et al. 2013) and concluded that an unculturable type of *P. brasiliensis*, different from the culturable type causing human systemic infections, is the etiologic agent of cutaneous granulomas in dolphins. They named the disease paracoccidioidomycosis ceti, epithet used throughout this chapter.

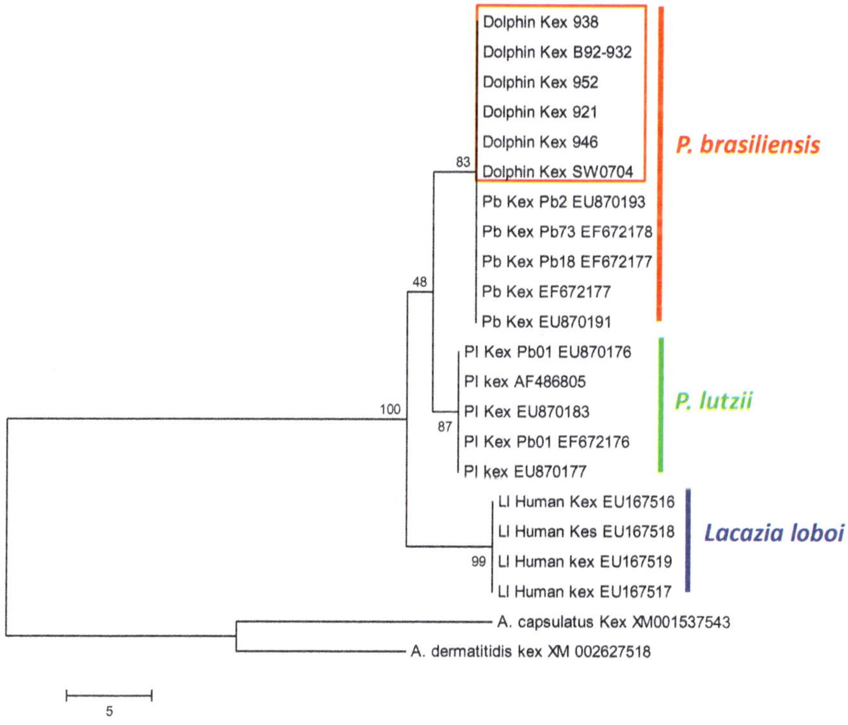

Fig. 9.1 Maximum parsimony tree of the exon partial *kex* DNA sequences PCR amplified from six US dolphins. The phylogenetic tree depicts the dolphin *kex* DNA sequences clustered among the culturable *P. brasiliensis* homologs. The species *P. lutzii* grouped as the sister taxon to *Lacazia loboi* (Vilela et al. 2016). Note the low bootstrap support placing *L. loboi* as an independent genus. The scale bar indicates nucleotide substitutions per site

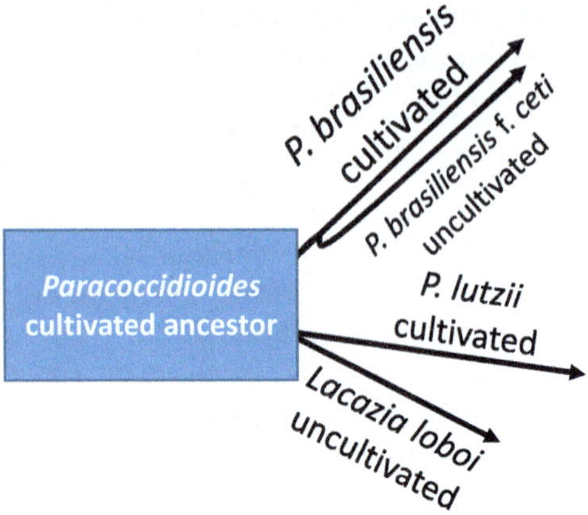

Fig. 9.2 The putative evolutionary paths of the culturable *P. brasiliensis* and *P. lutzii* recovered from humans with systemic paracoccidioidomycosis and the unculturable *P. brasiliensis* var. *ceti* and *Lacazia loboi*, restricted to subcutaneous tissues in dolphins and humans, respectively. The diagram shows the phylogenetic placement of these pathogens according to current phylogenetic analyses with one or more genes (Vilela et al. 2009, 2016) (Fig. 9.1)

One of the striking findings of the latter study (Vilela et al. 2016) was that after the addition of *P. brasiliensis* var. *ceti*, the DNA sequences of *L. loboi* showed lower bootstrap support to be in its own genus. This was best illustrated using the partial DNA sequences of the exons *CHS4*, *gp43*-like, and *kex* (Fig. 9.1) (Vilela et al. 2016). In previous analysis (Vilela et al. 2009), these exons placed the DNA sequences of human *L. loboi* with strong support as the sister taxon of two *Paracoccidioides* species (98–100% bootstrap support). However, with the addition of the dolphin DNA sequences, the bootstrap values plummeted to 83 (*CHS4*) and 48 (*kex*) (Fig. 9.1), and the *gp43*-like exon formed a strong supported sister taxon with *P. lutzii*, but did not affect bootstrap support using ITS DNA sequences. These analyses suggested that *L. loboi* maybe just another species in the genus *Paracoccidioides* (Vilela et al. 2016). The placement of the unculturable *P. brasiliensis* var. *ceti* among well-known DNA sequences of culturable isolates from human is intriguing. However, the fact that *L. loboi* and *P. brasiliensis* var. *ceti* shared many features strongly supports their common origin (Fig. 9.2).

9.3 Ecology and Epidemiology

Infections caused by *L. loboi* in humans are restricted to Mexico, Central America, and some countries in South America (Bermudez et al. 2009; Francesconi et al. 2014; Paniz-Mondolfi et al. 2012; Paniz-Mondolfi and Sander-Hoffmann 2009; Talhari and Talhari 2012; Vilela and Mendoza 2015). In these areas, the disease is more prevalent in dense forests with large rivers, geographical areas usually having elevated annual rainfall, high humidity, and hot temperatures. Most cases are diagnosed around the

Brazilian Amazon basin, but several cases are also reported in the rainforest areas of Colombia and Venezuela and less frequently in Bolivia, Ecuador, Guyana, Peru, Surinam, Central America (Costa Rica and Panama), and Mexico. In addition, imported cases outside these areas such as North America (Canada and the USA), Europe (France, Germany, Greece, the Netherlands), and South Africa have been reported (Arju et al. 2014; Burns et al. 2000; Elsayed et al. 2004; Fischer et al. 2002; Papadavid et al. 2012; Saint-Blancard et al. 2000; Symmers 1983; Vilela and Mendoza 2015). These cases were mainly diagnosed in individuals that had visited endemic areas in South America or were exposed, through direct contact with infected dolphins.

In contrast, the majority of reports in dolphins occurred in the coastal areas of Florida, USA (Bossart et al. 2015; Murdoch et al. 2010; Reif et al. 2006, 2009). The disease has been also diagnosed in other geographical areas such as Brazil (Daura-Jorge and Simões-Lopes 2011; Sacristan et al. 2016), Costa Rica (Bessesen et al. 2014), France (Symmers 1983), Spain (Esperon et al. 2012), Japan (Minakawa et al. 2016; Tajima et al. 2015; Ueda et al. 2013), Madagascar (*T. aduncus*, Kiszka et al. 2009), South Africa, Surinam (de Moura et al. 2014), and Venezuela (Bermudez et al. 2009).

Because human infections by *L. loboi* and the culturable *P. brasiliensis* are restricted to Central and South America, the finding of *P. brasiliensis* var. *ceti* affecting dolphins in Spain (Esperon et al. 2012), Japan (Minakawa et al. 2016; Tajima et al. 2015; Ueda et al. 2013), and the USA (Rotstein et al. 2009; Vilela et al. 2016) is significant. The phylogenetic data suggest that earlier in the life history of the pathogen, some dolphins swimming along South America coastal areas could have been infected around the river estuaries of the above endemic countries, with propagules from an ancestor of *P. brasiliensis*. Newly infected dolphins probably stayed around South America coastal areas, and others could have migrated to North America and other oceans. The fact that *P. brasiliensis* var. *ceti* cannot be cultured suggests that the pathogen muted from its original form or that this was indeed a unique strain derived from culturable *P. brasiliensis* ancestor. Two hypotheses are possible: (1) infected animals could transmit the pathogen by direct contact with non-infected dolphins or (2) dolphins are constantly in contact with propagules of the pathogen located around South America river estuaries. One clue supporting the first hypothesis is that the majority of bottlenose dolphins around the US coastal regions stay in those areas for long periods before developing skin granulomas (Bossart et al. 2015; Murdoch et al. 2010; Reif et al. 2006, 2009). More importantly, phylogenetic analysis showed that *P. brasiliensis* var. *ceti* affecting dolphins probably originated from an ancestral strain located in South America.

Since *L. loboi* and *P. brasiliensis* var. *ceti* resist culture, the epidemiology of the infection caused by these two pathogens remains an enigma. In humans, however, it is believed that *L. loboi* is acquired after contact with environmental propagules. A support for this theory came after a tribe of Caiabi Indians in Brazil was relocated from the Tapajos River (a hyperendemic area of lacaziosis) to the Xingu National Park. Before relocation, numerous cases of the disease were annually diagnosed, whereas new cases have not been reported after relocation (Baruzzi et al. 1973;

Talhari and Talhari 2012; Lacaz et al. 1986). Since phylogenetic analysis placed *L. loboi* with the dimorphic *Onygenales*, it is quite possible that conidia (yet to be found) of this unique pathogen may be present in the environment. In humans, the disease is considered occupational or related to recreational activities. *Paracoccidioides brasiliensis* var. *ceti* affecting dolphins can be considered a zoonotic pathogen because transmission between dolphins and humans has been documented (Symmers 1983). However, the infection rate may be low (Norton 2006; Reif et al. 2013). Accidental laboratory transmission (Rosa et al. 2009) and successful experimental infections in animals (Belone et al. 2001; Madeira et al. 2000; Sampaio and Dias 1970) and humans (Borelli 1962; Lacaz et al. 1986) have been reported.

9.4 Host Response and Pathogenesis

In contrast with the epidemiological features of *P. brasiliensis* infecting their hosts by aerial propagules (Lacaz et al. 1986) (Chap. 6), *L. loboi* in humans and *P. brasiliensis* var. *ceti* in dolphins are restricted to the subcutaneous tissues. This fact suggests that environmental propagules of both pathogens are more likely introduced by traumatic implantation of such elements in the subcutaneous tissues of the infected hosts. Accounts of humans with lacaziosis after traumatic lesions involving contaminated plants, snake bites, insect bites, stingray trauma, and others support this hypothesis (Almeida and Lacaz 1948–49; Azulay et al. 1976; Baruzzi et al. 1973; Lacaz et al. 1986; Rodriguez-Toro 1993; Talhari and Talhari 2012). Once these environmental propagules reach the subcutaneous tissues, *L. loboi* and *P. brasiliensis* var. *ceti* switch to a yeast-like form and successfully establish in the host subcutaneous tissues.

According to many observations (Bossart et al. 2015; Daura-Jorge and Simões-Lopes 2011; De Vries and Laarman 1973; Silva and Brito 1994; Talhari and Talhari 2012; Woods et al. 2010), both pathogens increase in size very slowly and could take months or even years to produce large granulomatous parakeloidal lesions. The disease rarely disseminates to other organs (Azulay et al. 1976; Opromolla et al. 2003). It has been found that yeast-like cells of both pathogens actively proliferate inside inflammatory cells (macrophages, mainly giant cells) activating the release of transforming growth factor β1 (TGF-β1), a powerful cytokine involved in immunosuppressive events (Francesconi et al. 2014; Reif et al. 2009; Vilani-Moreno et al. 2004a; Xavier et al. 2008), and can block the release of nitric oxide in giant cells and macrophages, also inhibiting the production of interferon gamma (IFN-γ) and thus locking the immune response in a Th2 subset, a typical reaction observed in both infected humans and dolphins (Goihman-Yahr et al. 1989; Vilani-Moreno et al. 2004b, 2011). The subcutaneous granulomas in infected humans and dolphins showed numerous branching yeast-like cells uniform in size and arranged in chains, surrounded by inflammatory cells and heavy fibrosis (Francesconi et al. 2014; Goihman-Yahr et al. 1989; Pecher et al. 1979; Reif et al. 2009; Vilani-Moreno and Opromolla 1997). It is believed that the proliferation of CD8 T cells promoted by TGF-β1 is responsible also for the production of immunoglobulins and other factors that favor the process of fibrosis giving the external parakeloidal appearance of the cutaneous lesions.

The presence of IL-10, IL-4 and IL-6 has been found in infected humans, an evidence that the immune response is locked into a Th2 subset (Lacaz et al. 1986; Vilani-Moreno et al. 2007). It is likely that the Th2 events are triggered by metabolites released by the fungus during its in vivo reproduction, but those factors are yet to be investigated. Mendoza et al. (2008) detected several antigens using sera from experimentally inoculated mice and infected humans and dolphins. The study found a ~193kD *gp43*-like antigen in infected humans and dolphins. The presence of melanin in the cell wall of *L. loboi* and *P. brasiliensis* var. *ceti* is believed to protect the pathogens from the host immune response (Taborda et al. 1999b). Vilani-Moreno and Opromolla (1997) reported that the viability of *L. loboi* yeast-like cells in humans is reduced, and probably only ~40% of the cells are still viable in the infected host tissues. A similar finding was reported in infected dolphins (Lane et al. 2014; Reif et al. 2006). A dramatic decrease in the number of circulating B- and T-helper cells was observed, which might contribute to impairment of adaptive immunity. Antibodies against *P. brasiliensis* var. *ceti* were found to cross-react with the antigens of the culturable *P. brasiliensis* isolates (Landman et al. 1988; Mendes et al. 1986; Puccia and Travassos 1991; Silva et al. 1968; Vidal et al. 1997).

9.5 Clinical Signs and Lesions

The disease in humans and dolphins is usually chronic and takes time to develop into large lesions (Bossart et al. 2015; Reif et al. 2006; Rodriguez-Toro 1993; Tapia et al. 1978; Talhari and Talhari 2012). Typical lesions are monomorphic or multimorphic and painless but may develop mild pruritus. The most affected anatomical sites in humans are ears, shoulders, limbs, back, and abdominal areas. Human lacaziosis clinical features have been subjected to extensive reviews since the first case was reported (Cardoso de Brito and Quaresma 2007; Francesconi et al. 2014; Talhari and Talhari 2012). The first effort to describe the polymorphic clinical characteristics of the disease in humans came from Silva and Brito (1994). These authors described five clinical forms including the typical parakeloidal granuloma and the gummatous, infiltrate, ulcerated, and verrucous form. The same year Machado (1972) described two basic forms. The first one is a hyperergic state that includes the macular, gummatous, and nodular forms described by Silva and Brito (1994). The second form is a hypoergic state including the parakeloidal and verrucous forms (Fig. 9.3). In the polymorphic form, the host immune system may react to the antigens presented during infections in a similar way as in the paucibacillary and multibacillary forms of *M. leprae* infection (Eichelmann et al. 2013). In a revision of 40 cases in Acre, Brazil, Opromolla et al. (2000) also described the above forms and indicated that the most frequent anatomical area was the ear (Fig. 9.3a). Differential diagnosis in humans includes chromoblastomycosis, leishmaniasis, leprosy, neoplasia, and paracoccidioidomycosis (Eichelmann et al. 2013; Lacaz et al. 1986).

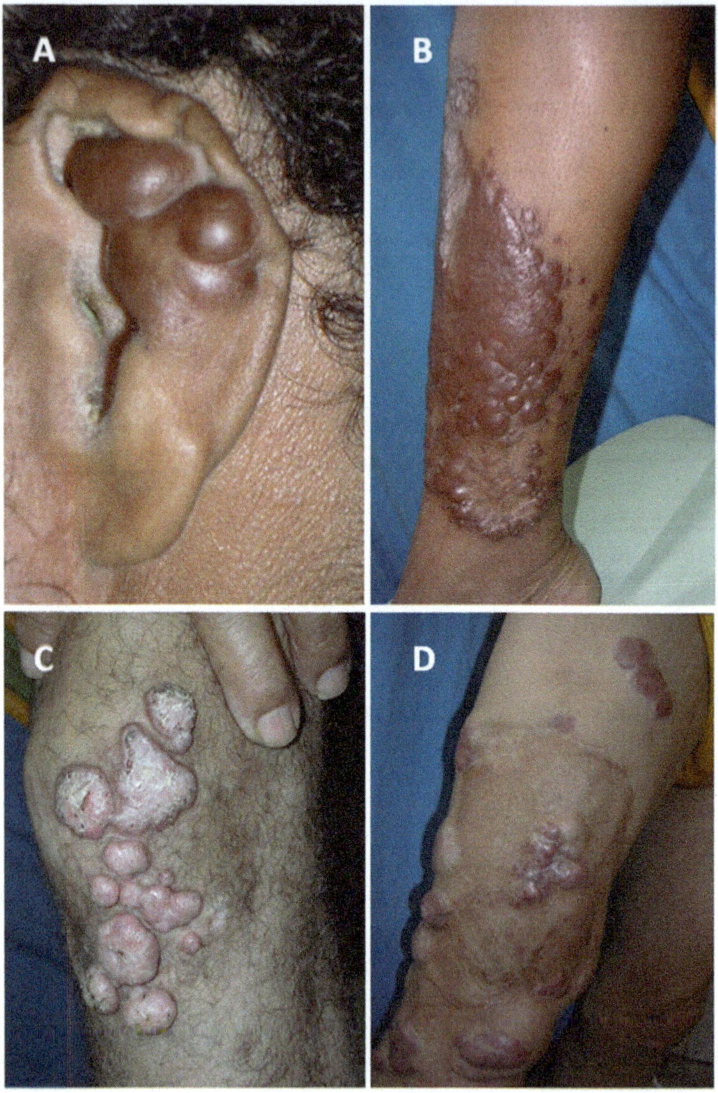

Fig. 9.3 Cases of human lacaziosis from Acre, Brazil. (**a** and **d**) Some of the clinical manifestation of the infection described by Silva and Brito (1994). (**a**) The most common anatomical site followed by different type of granulomatous lesions on the limbs. (**d**) An example of recurrence after surgical removal of old lesions (Courtesy of Dr. P. Rosa)

The infection has been diagnosed in several dolphin species including *Tursiops truncatus*, *T. aduncus*, *Sotalia guianensis*, and *Lagenorhynchus obliquidens* (Bossart et al. 2015; De Vries and Laarman 1973; Kiszka et al. 2009; Minakawa et al. 2016; Van Bressem et al. 2009). The terms "lacaziosis-like" and "lobomycosis-like" are currently used to identify the clinical features of putative cases of the disease based only on gross anatomical observations, but lacking histopathological confirmation (Daura-Jorge and Simões-Lopes 2011; Kiszka et al. 2009; Reif et al. 2006; Sacristan et al. 2016; Tajima et al. 2015). However, due to the new proposed names for the disease, the use of paracoccidioidomycosis ceti-like is recommended. The clinical features of the disease in dolphins are characterized by the formation of white, gray to reddish nodular or verrucous lesions (cauliflower-like), sometimes with prominent elevation over non-infected skin areas (Fig. 9.4) (Bossart et al. 2015; Minakawa et al. 2016; Rotstein et al. 2009; Reif et al. 2006; Ueda et al. 2013). These areas could ulcerate and form papillary nodules becoming large plaques. The lesions bleed easily after small traumas (Fig. 9.4a). The most frequently affected anatomical areas are anterior dorsum, dorsal and pectoral fins, flukes, rostrum, dorsal cranial surface, and the mid body (Bossart et al. 2015; Reif et al. 2006). Photographic

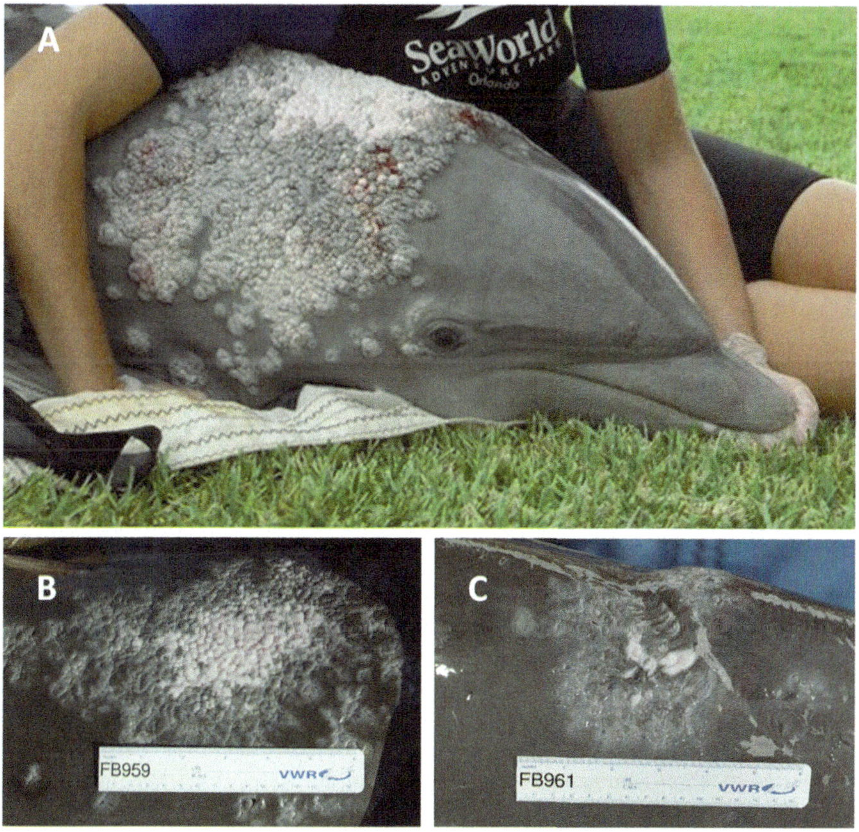

Fig. 9.4 (**a**) A common bottlenose dolphin (*Tursiops truncates*) with extensive plaques over the frontal anterior section of the dorsal fin caused by *Paracoccidioides brasiliensis* var. *ceti*. Note numerous elevated verrucous gray and white plaques some of them bleeding. New nodular small satellite lesions adjacent to the large plaque are observed (Courtesy of Dr. J. St. Leger). (**b** and **c**) Two different dolphins with large nodular white plaques typical of paracoccidioidomycosis ceti (Courtesy of Drs. G.D. Bossart, P.A. Fair and J.S. Reif)

observations for long periods revealed that small lesions in dolphins slowly progressed to form extensive granulomatous plaques. Most veterinarians agreed that environmental factors might play a key role in some epizootic reports of the disease in the coastal areas of Florida and North Carolina, USA (de Moura et al. 2014; Lane et al. 2014; Reif et al. 2006, 2009; Tajima et al. 2015).

9.6 Pathological Findings

The diagnosis of the lacaziosis in humans is confirmed after biopsy collection from the infected areas. Hematoxylin-eosin (H&E)-stained tissue samples show the presence of atrophy of the epidermis with extensive fibrosis interspersed with microabscesses containing numerous lymphocytes, plasma cells, foamy histiocytes, and multinucleated giant cells, some enclosing the pathogens (Fig. 9.5) (Baruzzi et al. 1981; Carneiro et al. 2009; Francesconi et al. 2014; Lacaz et al. 1986; Talhari and Talhari 2012). Acanthosis with hyperkeratosis is commonly observed. The Splendore/Hoeppli phenomenon enclosing one or more *L. loboi* yeast-like cells has been occasionally reported (Opromolla et al. 2000). In H&E *L. loboi* yeast-like cells appear as unstained hyaline structures resembling ghost-like bodies (Fig. 9.5a). In Gomori methenamine silver (GMS) stain, *L. loboi* cells appear uniform in shape (4–12 μm in diameter) developing one or more chains of three or more yeast-like cells connected by slender tubes (Fig. 9.5b). In some instances, few chains of yeast-like cells are present forming individual or budding spherical structures. The yeast-like cells shape varied between spherical to oval (lemon shape).

Dolphins are protected mammalian species; thus the collection of material for histopathology is strictly regulated. Biopsies are collected after the diseased dolphin is restrained and precautions are taken to avoid traumas that may have future repercussions. In H&E, the inflammatory process and the pathogen seem very similar to that in cases of human lacaziosis (Esperon et al. 2012; Minakawa et al. 2016; Rotstein et al. 2009; Ueda et al. 2013). Minor differences in the pathogen cell size in humans and dolphins have been mentioned (Haubold et al. 2000; Minakawa et al. 2016). Acanthosis, hyperkeratosis, hyperpigmentation, and extensive progressive fibrosis are the main pathological features in dolphins (Bossart et al. 2015; Tajima et al. 2015; Ueda et al. 2013). Microabscesses containing numerous unstained thick-walled hyaline yeast-like cells are often present (Fig. 9.5c, d). Histiocytes, giant cells, few lymphocytes, and plasma cells can also be observed. In GMS, the presence of numerous dark, 2–10 μm wide, oval yeast-like cells connected by short isthmuses is diagnostic of dolphin cases (Fig. 9.5e, f).

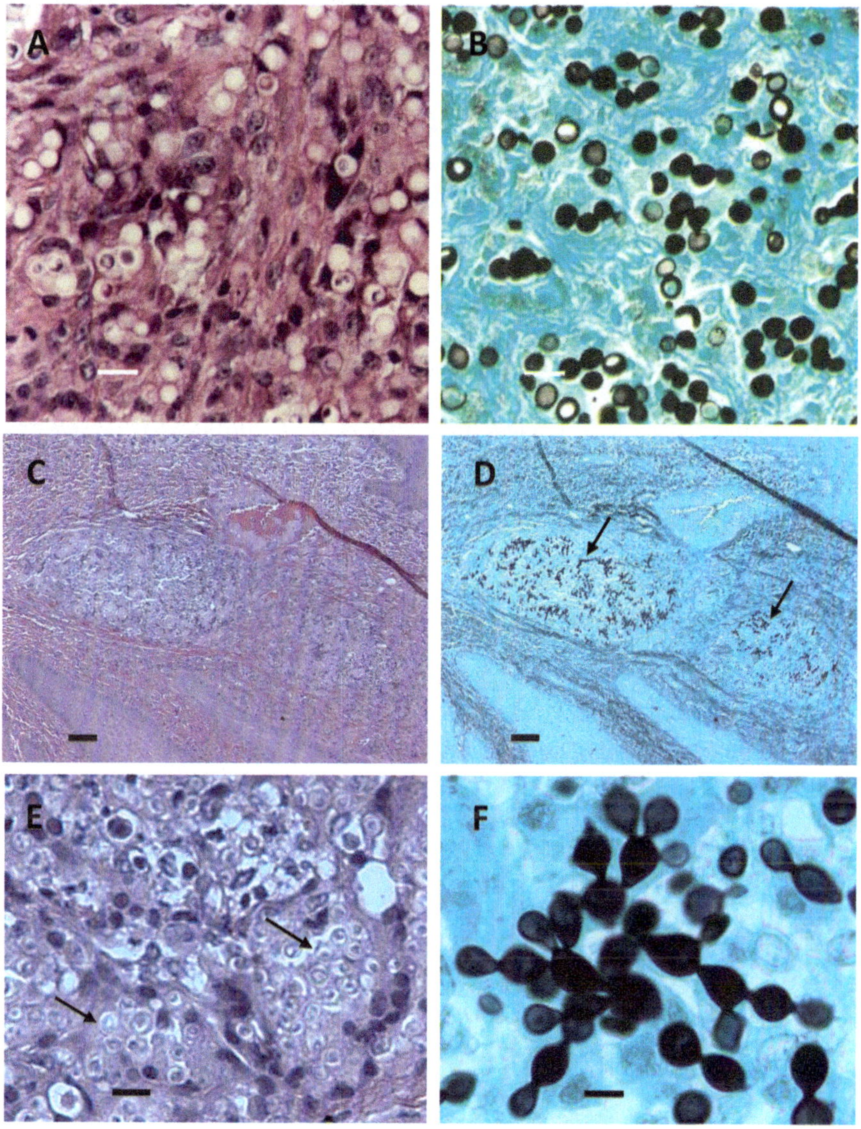

Fig. 9.5 (**a**) Unstained *Lacazia loboi* yeast-like cells (H&E) (arrows) from a case of human lacaziosis Acre, Brazil (bar = 18 μm). (**b**) A Gomori methenamine silver stain sample of panel A (bar = 18 μm). Small yeast-like cells chain of more than two cells connected by slender isthmuses are visible (Courtesy of Dr. Patricia Rosa). (**c** and **e**) H&E stained histological sections of a dolphin with paracoccidioidomycosis ceti. Presence of microabscesses containing inflammatory infiltrate, extensive fibrosis, acanthosis (Panel C; bar = 65 μm), and numerous unstained *Paracoccidioides brasiliensis* var. *ceti* yeast-like cells (Panel E; bar = 20 μm) (arrows). (**d**) A low magnification of panel C stained with Gomori methenamine silver. Presence of numerous black chains of yeast-like cells (arrows) (bar = 65 μm). (**f**) A close-up of Panel E. Presence of the yeast-like cells uniform in size and formation of branching chain connected by small bridges (bar = 10 μm)

9.7 Laboratory Diagnosis

The ideal clinical specimen to be sent to the laboratory is a biopsy in both humans and dolphins (Vilela and Mendoza 2015). However, skin smears of the granulomas collected with a scalp blade (Talhari et al. 2009) or the Scotch tape technique (Miranda and Silva 2005) are also valuable tools for the microscopic visualization of the pathogens. Both *L. loboi* and *P. brasiliensis* var. *ceti* resist culture; thus the use of traditional laboratory media has to be performed with the purpose only of ruling out other pathogens.

To perform the wet mount technique, the biopsy is divided into 2–3 mm cubes and placed onto a slide containing two drops of 10 or 20% KOH (Vilela and Mendoza 2015). The addition of calcofluor enhances sensitivity when observed under fluorescence. The samples are heated for 5 min and then left on the counter. In wet mount preparations, *L. loboi* from humans and *P. brasiliensis* var. *ceti* from dolphins appear as single spherical or oval yeast-cells developing branching chains of cells (Fig. 9.6). The yeast-like cells in dolphins measure ~2–10 μm, whereas in humans ~4–14 μm, according to Haubold et al. (2000). In 10% KOH, the cells appear hyaline with a thick cell wall (Fig. 9.6).

Experimental inoculation of *L. loboi* has been successfully obtained in several mammalian species including humans (Belone et al. 2001; Borelli 1962; Madeira et al. 2000). The mouse model proposed by Madeira et al. (2000) showed better results using propagules from human lacaziosis. Inoculated mice on the injected pads developed small granulomas that slowly enlarged. In histopathology, diffuse infiltrate-containing macrophages, lymphocytes, plasma cells, fibrosis, and numerous yeast-like cells can be observed. There are no official reports of *P. brasiliensis* var. *ceti* experimental infection in mice. Incidentally, few years ago, one of the authors (LM) was informed of a successful experimental inoculation in a mouse injected with yeast-like cells from an infected dolphin (Dr. Libero Ajello, personal communication).

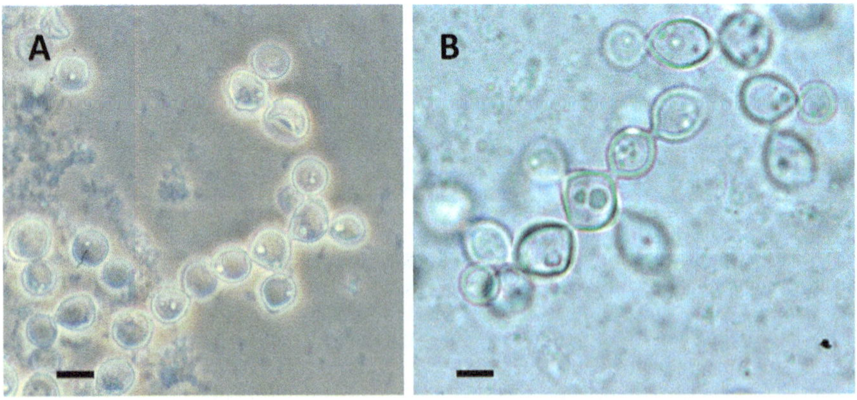

Fig. 9.6 (**a**) *Paracoccidioides brasiliensis* var. *ceti* yeast-like cells in 10% KOH wet mount preparation from a case of dolphin with cutaneous granulomas (bar = 10 μm). Presence of uniform single cells and branching chains of yeast-like cells (Nomarski microscopy). (**b**) A wet mount (10% KOH) preparation from a human lacaziosis case (bar = 10 μm). Presence of yeast-like cells in chain

9.8 Treatment

Currently, there is no effective drug or therapeutic protocol for the management of the infections caused by *L. loboi* in humans or *P. brasiliensis* var. *ceti* in dolphins. Surgery in the early stages of the disease remains the treatment of choice in both conditions (Baruzzi et al. 1981; Bossart et al. 2015; Lacaz et al. 1986). In advanced cases, surgery can be also performed, but recurrence rate is high (Lacaz et al. 1986; Opromolla et al. 2000; Rodriguez-Toro 1993; Talhari and Talhari 2012). These two atypical microorganisms were always classified within the fungi, thus treated with antifungal drugs (Bustamante et al. 2013; Cucé et al. 1980; Dudok van Heel 1977; Lawrence and Ajello 1986). The response of infected hosts to antifungal therapy has been unsatisfactory (Cucé et al. 1980; Lacaz et al. 1986; Woods et al. 2010). For instance, amphotericin B, itraconazole, 5-fluorocytosine, ketoconazole, posaconazole, terbinafine, and others resulted in the reduction of the granuloma size, but cure was achieved only in few instances (Francesconi et al. 2014). On the other hand, antibacterial drugs such as sulfadimethoxine and sulfamethoxypyridazine were found to improve the nodular lesions, but the presence of yeast-like cells persisted in the infected areas (Woods et al. 2010). Because leprosy is the differential diagnosis of lacaziosis, drugs such as clofazimine have been used in patients with dual infection. According to studies conducted in Acre, Brazil (Opromolla et al. 2000; Silva 1972; Talhari and Talhari 2012), the use of dapsone and clofazimine dramatically improved lesion size and itching, usually associate to active lacaziosis, and in some cases the lesions decreased or disappeared (Woods et al. 2010). Moreover, when surgery and these two drugs were combined, no recurrence was reported (Woods et al. 2010).

Only a handful of wild infected dolphins have been kept in captivity for treatment; thus there is limited information on the management strategies in animals. Dolphins treated with the strategies used in humans showed similar responses (Minakawa et al. 2016). Surgical removal of small lesions is the best option. However, surgery in large or multicentric lesions is not recommended. In 1977, a dolphin captured in Florida, USA, was successfully treated with miconazole (Dudok van Heel 1977). But, this antifungal drug has not been used ever since. More recently, a dolphin treated with topical itraconazole and ketoconazole did not improve (Esperon et al. 2012). The same animal was then treated with oral 2.5 mg/kg itraconazole and 2.0 mg/kg terbinafine. With this treatment protocol, cutaneous lesions reduced to small nodules that later disappeared, and no relapse was reported. Currently, there is little information on the best antifungal approach to treat dolphins. This is probably in part due to the price of antifungals and the difficulties to keep wild animals for long periods outside their natural ecological habitat.

References

Almeida FP, Lacaz CDS (1948–1949) Blastomicose "tipo Jorge Lobo". An Fac Med Sao Paulo 24: 5–37

Arju R, Kothadia JP, Kaminski M, Abraham S, Giashuddin S (2014) Jorge Lobo's disease: a case of keloidal blastomycosis (lobomycosis) in a nonendemic area. Ther Adv Infect Dis 2:91–96

Azulay RD, Carneiro JA, Da Graça M, Cunha S, Reis LT (1976) Keloidal blastomycosis (Lobo's disease) with lymphatic involvement: a case report. Int J Dermatol 15:40–42

Baruzzi RG, Castro RM, D'Andretta C Jr, Carvalhal S, Ramos OL, Pontes PL (1973) Occurrence of Lobo's blastomycosis among "Caiabi," Brazilian Indians. Int J Dermatol 12:95–99

Baruzzi RG, Marcopito LF, Michalany NS, Livianu J, Pintoe NR (1981) Early diagnosis and prompt treatment by surgery in Jorge Lobo's disease (keloidal blastomycosis). Mycopathologia 74:51–54

Belone AFF, Madeira S, Rosa PS, Opromolla DVA (2001) Experimental reproduction of the Jorge Lobo's disease in BALB/c mice inoculated with *Lacazia loboi* obtained from a previously infected mouse. Mycopathologia 155:191–194

Bermudez L, Van Bressem MF, Reyes-Jaimes O, Sayegh AJ, Paniz-Mondolfi AE (2009) Lobomycosis in man and lobomycosis-like disease in bottlenose dolphin, Venezuela. Emerg Infect Dis 15:1301–1303

Bessesen BL, Oviedo L, Burdett Hart L, Herra-Miranda D, Pacheco-Polanco JD, Baker L, Saborío-Rodriguez G, Bermúdez-Villapol L, Acevedo-Gutiérrez A (2014) Lacaziosis-like disease among bottlenose dolphins *Tursiops truncatus* photographed in Golfo Dulce, Costa Rica. Dis Aquat Org 107:173–180

Borelli D (1962) Lobomicosis experimental. Derm Venez 3:72–82

Borelli D (1968) Lobomicosis: nomenclatura de agente (revisión critica). Med Cutanea 3(2): 151–156

Bossart GD, Schaefer AM, McCulloch S, Goldstein J, Fair PSA, Reif JS (2015) Mucocutaneous lesions in free-ranging Atlantic bottlenose dolphins *Tursiops truncates* from the southeastern USA. Dis Aquat Org 115:175–184

Burns RA, Roy JS, Woods C, Padhye AA, Warnock DW (2000) Report of the first human case of lobomycosis in the United States. J Clin Microbiol 38:1283–1285

Bustamante B, Seas C, Salomon M, Bravo F (2013) Lobomycosis successfully treated with posaconazole. Am J Trop Med Hyg 88:1207–1208

Caldwell DK, Caldwell MC, Woodard JC, Ajello L, Kaplan W, McClure HM (1975) Lobomycosis as a disease of the Atlantic bottle-nosed dolphin (*Tursiops truncatus* Montagu, 1821). Am J Trop Med Hyg 24:105–114

Camargo ZP, Baruzzi RG, Maeda SM, Floriano MC (1998) Antigenic relationship between *Loboa loboi* and *Paracoccidioides brasiliensis* as shown by serological methods. Med Mycol 36: 413–417

Cardoso de Brito A, Quaresma JAS (2007) Lacaziosis (Jorge Lobo's disease): review and update. An Bras Dermatol 82:461–474

Carneiro FP, Maia LB, Moraes MA, deMagalhaes AV, Vianna LM (2009) Lobomycosis: diagnosis and management of relapsed and multifocal lesions. Diagn Microbiol Infect Dis 65:62–64

Costa PF (2015) Isolation and in vitro cultivation of the etiologic agent of Jorge Lobo: morphology physiology and complete genome of *Candida loboi* sp. nov. PhD. Thesis, Universidade Federal do Para

Cucé LC, Wroclawski EL, Sampaio SA (1980) Treatment of paracoccidioidomycosis, candidiasis, chromomycosis, lobomycosis, and mycetoma with ketoconazole. Int J Dermatol 19:405–408

Daura-Jorge FG, Simões-Lopes PC (2011) Lobomycosis-like disease in wild bottlenose dolphins *Tursiops truncatus* of Laguna, southern Brazil: monitoring of a progressive case. Dis Aquat Org 93:163–170

De Moura JF, Hauser-Davis RA, Lemos L, Emin-Lima R, Siciliano S (2014) Guiana dolphins (*Sotalia guianensis*) as marine ecosystem sentinels: ecotoxicology and emerging diseases. Rev Environ Contam Toxicol 228:1–29

De Vries GA, Laarman JJ (1973) A case of Lobo's disease in the dolphin *Sotalia guianensis*. Aquat Mamm 1:26–33

Dudok van Heel WH (1977) Successful treatment in a case of lobomycosis (Lobo's disease) in *Tursiops truncatus* (Mont) at Dolfinarium, Harderwijk. Aquat Mamm 5:8–15

Eichelmann K, González González SE, Salas-Alanis JC, Ocampo-Candiani J (2013) Leprosy. An update: definition, pathogenesis, classification, diagnosis, and treatment. Actas Dermosifiliogr 104:554–563

Elsayed S, Kuhn SM, Barber D, Church DL, Adams S, Kasper R (2004) Human case of lobomycosis. Emerg Infect Dis 10:715–718

Esperon F, García-Párraga D, Bellière EN, Sánchez-Vizcaíno JM (2012) Molecular diagnosis of lobomycosis-like disease in a bottlenose dolphin in captivity. Med Mycol 50:106–109

Fischer M, Chrusciak-Talhari A, Reinel D, Talhari S (2002) Successful treatment with clofazimine and itraconazole in a 46 year old patient after 32 years duration of disease. Hautarzt 53:677–681

Fonseca FO, Area LAE (1940) Contribuição para o conhecimento das granulomatoses blastomycoides: o agente etiológico da doença de Jorge Lobo. Rev Med Cirurg Brasil 48:147–158

Fonseca OJM, Lacaz CDS (1971) Estudo de culturas isoladas de blastomicose queloidiforme (doença de Jorge Lobo). Denominação ao seu agente etiològico. Rev Inst Med Trop São Paulo 13:225–251

Francesconi VA, Klein AP, Santos APBG, Ramasawmy R, Francesconi F (2014) Lobomycosis: epidemiology, clinical presentation, and management options. Ther Clin Risk Manag 10: 851–860

Furtado JS, De Brito T, Freymuller E (1967) Structure and reproduction of *Paracoccidioides loboi*. Mycologia 59:286–294

Goihman-Yahr M, Cabello de Brito I, Bastardo de Albornoz MC, Pereira MHDGJ, de Roman A, Martin BS, Molina T (1989) Functions of polymorphonuclear leukocytes and individuality of Jorge Lobo's disease: absence of the specific leukocyte digestive defect against *Paracoccidioides brasiliensis*. Mycoses 32:603–608

Haubold EM, Cooper CR Jr, Wen JW, McGinnis MR, Cowan DF (2000) Comparative morphology of *Lacazia loboi* (syn. *Loboa loboi*) in dolphins and humans. Med Mycol 38:9–14

Herr RA, Tarcha EJ, Taborda PR, Taylor JW, Ajello L, Mendoza L (2001) Phylogenetic analysis of *Lacazia loboi* places this previously uncharacterized pathogen within the dimorphic Onygenales. J Clin Microbiol 39(1):309–314

Kiszka J, Van Bressem MF, Pusineri C (2009) Lobomycosis-like disease and other skin conditions in indo-Pacific bottlenose dolphins *Tursiops aduncus* from the Indian Ocean. Dis Aquat Org 84:151–157

Lacaz CDS (1996) *Paracoccidioides loboi* (Fonseca Filho et Arêa Leão, 1940) Almeida et Lacaz, 1948–1949. Description of the fungus in Latin. Rev Inst Med Trop São Paulo 38(3).229–231

Lacaz CDS, Baruzzi RG, Rosa MDCB (1986) Doença de Jorge Lobo. IPSIS Editora SA, São Paulo, pp 1–92

Landman G, Velludo MA, Lopes JA, Mendes E (1988) Crossed-antigenicity between the etiologic agents of lobomycosis and paraccocidioidomycosis evidenced by an immunoenzymatic method (PAP). Allergol Immunopathol 16:215–218

Lane EP, de Wet M, Thompson P, Siebert U, Wohlsein P, Plön S (2014) A systemic health assessment of Indian Ocean bottlenose (*Tursiops aduncus*) and Indo-Pacific humpback (*Sousa plumbea*) dolphins accidentally caught in shark nets off the KwaZulu-Natal coast, South Africa. PLoS One 9:1–13

Lawrence DN, Ajello L (1986) Lobomycosis in western Brazil: report of a clinical trial with ketoconazole. Am J Trop Med Hyg 35:162–166

Lobo JO (1931) Um caso de blastomicose produzido por uma espécie nova, encontrada em Recife. Rev Med Pernamb 1:763–765

Machado PA (1972) Polimorfismo das lesões dermatológicas na blastomicose de Jorge Lobo entre os índios Caiabi. Acta Amaz 2:93–97

Madeira S, Opromolla DVA, Belone ADFF (2000) Inoculation of BALB/C mice with *Lacazia loboi*. Rev Inst Med Trop São Paulo 42:239–243

Mendes E, Michalany N, Mendes NF (1986) Comparison of the cell walls of *Paracoccidioides loboi* and *Paracoccidioides brasiliensis* by using polysaccharide-binding dyes. Int J Tissue React 6:229–231

Mendoza L, Belone AF, Vilela R, Rehtanz M, Bossart GD, Reif JS, Fair PA, Durden WN, St. Leger J, Travassos LR, Rosa PS (2008) Use of sera from humans and dolphins with lacaziosis and sera from experimentally infected mice for western blot analyses of *Lacazia loboi* antigens. Clin Vaccine Immunol 15:164–167

Migaki G, Valerio MG, Irvine B, Garner FM (1971) Lobo's disease in an Atlantic bottle-nosed dolphin. J Am Vet Med Assoc 159:578–582

Minakawa T, Ueda K, Tanaka M, Tanaka N, Kuwamura M, Izawa T, Konno T, Yamate J, Itano EN, Sano A, Wada S (2016) Detection of multiple budding yeast cells and partial sequence of 43-Kda glycoprotein coding gene of *Paracoccidioides brasiliensis* from a case of lacaziosis in a female Pacific white-sided dolphin (*Lagenorhyncus obliquidens*). Mycopathologia 181: 523–529

Miranda MF, Silva AJ (2005) Vinyl adhesive tape also effective for direct microscopy diagnosis of chromomycosis, lobomycosis, and paracoccidioidomycosis. Diagn Microbiol Infect Dis 52: 39–43

Murdoch ME, Mazzoil M, McCulloch S, Bechdel S, O'Corry-Crowe G, Bossart GD, Reif JS (2010) Lacaziosis in bottlenose dolphins *Tursiops truncatus* along the coastal Atlantic Ocean, Florida, USA. Dis Aquat Organ 92:69–73

Norton SA (2006) Dolphin-to-human transmission of lobomycosis? J Am Acad Dermatol 55: 723–724

Opromolla DVA, Belone AFF, Taborda PR, Taborda VBA (2000) Clinic-pathological correlation in 40 cases of lobomycosis. An Bras Dermatol 75:425–434

Opromolla DV, Belone AF, Taborda PR, Rosa PS (2003) Lymph node involvement in Jorge Lobo's disease: report of two cases. Int J Dermatol 42:938–941

Paniz-Mondolfi AE, Sander-Hoffmann L (2009) Lobomycosis in inshore and estuarine dolphins. Emerg Infect Dis 15:672–673

Paniz-Mondolfi A, Talhari C, Sander Hoffmann L, Connor DL, Talhari S, Bermudez-Villapol L, Hernadez-Perez M, VanBressem MF (2012) Lobomycosis: an emerging disease in humans and delphinidae. Mycoses 55:298–309

Papadavid E, Dalamaga M, Kapniari I, Pantelidaki E, Papageorgiou S, Vassiliki P, Tsirigotis P, Dervenoulas I, Stavrianeas N, Rigopoulos D (2012) Lobomycosis: a case from southeastern Europe and review of the literature. J Dermatol Case Rep 6:65–69

Pecher SA, Croce J, Ferri RG (1979) Study of humoral and cellular immunity in lobomycosis. Allergol Immunopathol (Madr) 7(6):439–444

Puccia R, Travassos LR (1991) 43-kilodalton glycoprotein from *Paracoccidioides brasiliensis*: immunochemical reactions with sera from patients with paracoccidioidomycosis, histoplasmosis, or Jorge Lobo's disease. J Clin Microbiol 29:1610–1615

Reif JS, Mazzoil MS, McCulloch SD, Varela RA, Doldstein JD, Fair PA, Bossart GD (2006) Lobomycosis in Atlantic bottlenose dolphins form the Indian River lagoon, Florida. J Am Vet Med Assoc 228:104–108

Reif JS, Peden-Adams MM, Romano TA, Rice CD, Fair PA, Bossart GD (2009) Immune dysfunction in Atlantic bottlenose dolphins (*Tursiops truncatus*) with lobomycosis. Med Mycol 47:125–135

Reif JS, Schaefer AM, Bossart GD (2013) Lobomycosis: risk of zoonotic transmission from dolphins to humans. Vector Borne Zoonotic Dis 13:689–693

Rodriguez-Toro G (1993) Lobomycosis. Int J Dermatol 32:324–332
Rosa PS, Soares CT, Belone ADFF, Vilela R, Ura S, Filho MC, Mendoza L (2009) Accidental Jorge Lobo's disease in a worker dealing with *Lacazia loboi* infected mice: a case report. J Med Case Rep 3:67–71
Rotstein DS, Burdett LG, McLellan W, Schwacke L, Rwles T, Terio KA, Schultz S, Pabst A (2009) Lobomycosis in offshore bottlenose dolphins (*Tursiops truncatus*), North Carolina. Emerg Infect Dis 15:588–590
Sacristan C, Réssio RA, Castilho P, Fernandes N, Costa-Silva S, Esperón F, Daura-Jorge FG, Groch KR, Kolesnikovas CK, Marigo J, Ott PH, Oliveira LR, Sánchez-Sarmiento AM, Simões-Lopes PC, Catão-Dias JL (2016) Lacaziosis-like disease in *Tursiops truncatus* from Brazil: a histopathological and immunohistochemical approach. Dis Aquat Org 117:229–235
Saint-Blancard P, Maccari F, Le Guyadec T, Lanternier G, Le Vagueresse R (2000) Lobomycosis: a mycosis seldom observed in metropolitan France. Ann Pathol 20(3):241–244
Sampaio MM, Dias LB (1970) Experimental infection of Jorge Lôbo's disease in the cheek-pouch of the golden hamster. Rev Inst Med Trop Sao Paulo 12:115–120
Schaefer AM, Reif JS, Guzmán EA, Bossart GD, Ottuso P, Snyder J, Medalie N, Rosato R, Han S, Fair PA, McCarthy PJ (2016) Toward the identification, characterization and experimental culture of *Lacazia loboi* from Atlantic bottlenose dolphin (*Tursiops truncatus*). Med Mycol 54:659–565
Silva D (1972) Micose de Lobo. Rev Soc Bras Med Trop 6:85–98
Silva D, Brito A (1994) Formas clínicas não usuais da micose de Lobo. An Bras Dermatol 69: 133–136
Silva ME, Kaplan W, Miranda JL (1968) Antigenic relationships between *Paracoccidioides loboi* and other pathogenic fungi determined by immunofluorescence. Mycopathol Mycol Appl 36: 97–106
Symmers WS (1983) A possible case of Lôbo's disease acquired in Europe from a bottle-nosed dolphin (*Tursiops truncatus*). Bull Soc Pathol Exot Filiales 76:777–784
Taborda PR, Taborda VA, McGinnis MR (1999a) *Lacazia loboi* gen. nov, comb. nov, the etiologic agent of lobomycosis. J Clin Microbiol 37:2031–2033
Taborda VB, Taborda PR, McGinnis MR (1999b) Constitutive melanin in the cell wall of the etiologic agent of Lobo's disease. Rev Inst Med Trop Sao Paulo 41:9–12
Tajima Y, Sasaki K, Kashiwagi N, Yamada TK (2015) A case of stranded Indo-Pacific bottlenose dolphin (Tursiops aduncus) with lobomycosis-like skin lesions in Kinko-wan, Kagoshima, Japan. J Vet Med Sci 77:989–992
Talhari S, Talhari C (2012) Lobomycosis. Clin Dermatol 30:420–424
Talhari C, Chrusciak-Talhari A, de Souza JV, Araújo JR, Talhari S (2009) Exfoliative cytology as a rapid diagnostic tool for lobomycosis. Mycoses 52:187–189
Tapia A, Torres-Calcindo A, Arosemena R (1978) Keloidal blastomycosis (Lobo's disease) in Panama. Int J Dermatol 17:572–574
Ueda K, Sano A, Yamate J, Nakagawa EI, Kuwamura M, Izawa T, Tanaka M, Hasegawa Y, Chibana H, Izumisawa Y, Miyahara H, Senzo Uchida S (2013) Two cases of Lacaziosis in bottlenose dolphins (*Tursiops truncatus*) in Japan. Case Rep Vet Med 2013:1–9. https://doi.org/10.1155/2013/318548
Van Bressem MF, Santos MC, Oshima JE (2009) Skin diseases in Guiana dolphins (*Sotalia guianensis*) from the Paranaguá estuary, Brazil: a possible indicator of a compromised marine environment. Mar Environ Res 67:63–68
Vidal MS, Palacios SA, de Melo NT, Lacaz CDS (1997) Reactivity of anti-gp43 antibodies from *Paracoccidioides brasiliensis* antiserum with extracts from cutaneous lesions of Lobo's disease. Preliminary note. Rev Inst Med Trop Sao Paulo 39:35–37
Vilani-Moreno FR, Opromolla DV (1997) Determinação da viabilidade do *Paracoccidioides loboi* em biópsias de pacientes portadores de doença de Jorge Lobo. Ann Bras Dermatol 72:433–437

Vilani-Moreno FR, Silva LM, Opromolla DV (2004a) Evaluation of the phagocytic activity of peripheral blood monocytes of patients with Jorge Lobo's disease. Rev Soc Bras Med Trop 37: 165–168

Vilani-Moreno FR, Lauris JR, Opromolla DV (2004b) Cytokine quantification in the supernatant of mononuclear cell cultures and in blood serum from patients with Jorge Lobo's disease. Mycopathologia 158:17–24

Vilani-Moreno FR, Mozer E, de Sene AM, Ferasçoli MDO, Pereira TC, Miras MG, Souza GHDP, ADFF B (2007) In vitro and in situ activation of the complement system by the fungus *Lacazia loboi*. Rev Inst Med Trop Sao Paulo 49:97–101

Vilani-Moreno FR, Belone Ade F, Lara VS, Venturini J, Lauris JR, Soares CT (2011) Detection of cytokines and nitric oxide synthase in skin lesions of Jorge Lobo's disease patients. Med Mycol 49(6):643–648

Vilela R, Mendoza L (2015) Lacazia, lagenidium, pythium, and rhinosporidium. In: Jorgenses JH, Pfaller MA, Carrol KC, Funke G, Landry ML, Richerter SS, Warnock DW (eds) Manual of clinical microbiology, 11th edn. ASM Press, Washington, pp 2196–2208

Vilela R, Mendoza L, Rosa PS, Belone AFF, Madeira S, Opromolla DVA, de Resende MA (2005) Molecular model for studying the uncultivated fungal pathogen *Lacazia loboi*. J Clin Microbiol 43:3657–3661

Vilela R, Martins JEC, Pereira CN, Melo N, Mendoza L (2007) Molecular study of archival fungal strains isolated from cases of lacaziosis (Jorge Lobo's disease). Mycoses 50:470–474

Vilela R, Rosa PS, Belone AF, Taylor JW, Diório SM, Mendoza L (2009) Molecular phylogeny of animal pathogen *Lacazia loboi* inferred from rDNA and DNA coding sequences. Mycol Res 113:851–857

Vilela R, Bossart GD, St. Leger JA, Dalton LM, Reif JS, Schaefer AM, McCarthy PJ, Fair PA, Mendoza L (2016) Cutaneous granulomas in dolphins caused by novel uncultivated *Paracoccidioides brasiliensis*. Emerg Infect Dis 22:2097–2103

Woodard JC (1972) Electron microscope study of lobomycosis (*Loboa loboi*). Lab Investig 27: 606–612

Woods WJ, Belone Ade F, Carneiro LB, Rosa PS (2010) Ten years experience with Jorge Lobo's disease in the state of acre, Amazon region, Brazil. Rev Inst Med Trop Sao Paulo 52:273–278

Xavier MB, Libonati RM, Unger D, Oliveira C, Corbett CE, de Brito AC, Quaresma JA (2008) Macrophage and TGF-beta immunohistochemical expression in Jorge Lobo's disease. Hum Pathol 39(2):269–274

Part III

Emerging Mycoses in Animals

Feline Sporotrichosis

10

Anderson Messias Rodrigues, G. Sybren de Hoog,
and Zoilo Pires de Camargo

Abstract

A wide diversity of emerging fungal diseases has affected humans and animals in recent decades. Several diseases are caused by pathogens of animal origin, which sometimes affect the human host due to close contact between humans and companion animals. Sporotrichosis is an example of how cat-human interactions can lead to zoonotic transmission of a disease and increases the incidence of the mycosis to epidemic levels observed in South and Southeast Brazil. Human and animal sporotrichosis is an infection that is classically acquired by inoculation of contaminated materials into the host's cutaneous and subcutaneous tissues. Feline transmission (cat-cat and cat-human) through scratching and biting is an alternative transmission route that is highly effective, putting a larger number of individuals at risk of acquiring the infection. The main etiological agent of feline sporotrichosis is *Sporothrix brasiliensis*, a highly pathogenic organism to the mammalian host. In this chapter, we discuss the main advances in taxonomy, ecology, epidemiology, diagnostics, host-parasite interactions, and treatments related to this disease.

A. M. Rodrigues (✉) · Z. P. de Camargo
Department of Microbiology, Immunology and Parasitology, Cell Biology Division, Federal University of São Paulo, São Paulo, Brazil
e-mail: rodrigues_bio@yahoo.com.br; zpcamargo1@gmail.com

G. S. de Hoog
Center of Expertise in Mycology RadboudUMC/CWZ, Nijmegen, The Netherlands

Westerdijk Fungal Biodiversity Institute, Utrecht, The Netherlands
e-mail: s.hoog@westerdijkinstitute.nl

S. Seyedmousavi et al. (eds.), *Emerging and Epizootic Fungal Infections in Animals*,
https://doi.org/10.1007/978-3-319-72093-7_10

10.1 Brief History of the Causative Agents of Sporotrichosis

Sporothrix species are the etiological agents of human and animal sporotrichosis, a chronic, subcutaneous mycosis with frequent lymphatic involvement (Rodrigues et al. 2016b). On November 30, 1896, the American physician Benjamin Robinson Schenck, working at Johns Hopkins Hospital in Baltimore, Maryland (USA), examined a patient with an infection in his right arm. The patient had a primary infection in the index finger that had spread to his arm via the lymphatic system. Schenck isolated the causative agent and sent a sample to the mycologist Erwin F. Smith, who classified the fungus in the genus *Sporotrichum* (Schenck 1898) (Fig. 10.1).

In 1900, Hektoen and Perkins, also in the USA, reported the second case of the disease when they isolated the pathogen from a skin lesion of a child who had injured her index finger with a hammer (Hektoen and Perkins 1900). Based on experimental studies with laboratory animals, they described the pathogen morphology in detail and named the fungus *Sporothrix schenckii.*

In 1910, in France, Matruchot also encountered this species. Several similar cases began to occur in the USA and Europe, first in France (Beurmann and Ramond 1903), where it became a rather common disease, and later in England (Walker and Ritchie 1911; Adamson 1911). Several reports of cutaneous mycosis, and even of the rare, disseminated form of the disease, were published by experts of that time (Beurmann and Ramond 1903; Beurmann and Gougerot 1912).

Animal sporotrichosis was first described in 1907 in São Paulo (Brazil) by Lutz and Splendore, who isolated pathogenic microorganisms of the genus *Sporothrix* from the buccal mucosa of rats (Lutz and Splendore 1907). Several attempts to classify and name the pathogen were made until 1921, when Davis, studying the samples isolated in France and the Americas, concluded that they were identical and classified the etiological agent found in Brazil as *Sporotrichum schenckii* (Davis 1921). However, *Sporotrichum* is of basidiomycetous nature, and therefore Carmichael (1962) recommended *Sporothrix schenckii* as the correct name for the causative agent of sporotrichosis.

After these initial findings, cases of human sporotrichosis were described in the Americas (Dixon et al. 1991), Asia (Song et al. 2013), Europe (Bachmeyer et al. 2006), and Africa (Vismer and Hull 1997), with Brazil, South Africa, and China considered highly endemic areas. The first sporotrichosis epidemic in South Africa was described by Pijper and Pullinger (1927) and affected 14 native miners, but the source of infection was not identified. The largest epidemic in humans occurred in the early 1940s, with approximately 3000 cases identified in 3 years, at the same place in South Africa where the first cases were diagnosed. The infection occurred when miners touched untreated wood beams supporting the mine shafts; transmission ceased after the wood was treated with antifungals (Helm and Berman 1947).

In all the scenarios described above, sporotrichosis is characterized as a disease acquired through direct contact between the human host and the environment. *Sporothrix* was first described in cats in the mid-1950s (Singer and Muncie 1952), and three decades later, the first case of zoonotic transmission was reported (Read and Sperling 1982). In recent years (1998–2016), large epizootics have occurred in domestic cats in South and Southeast Brazil, with an increasing number of zoonotic

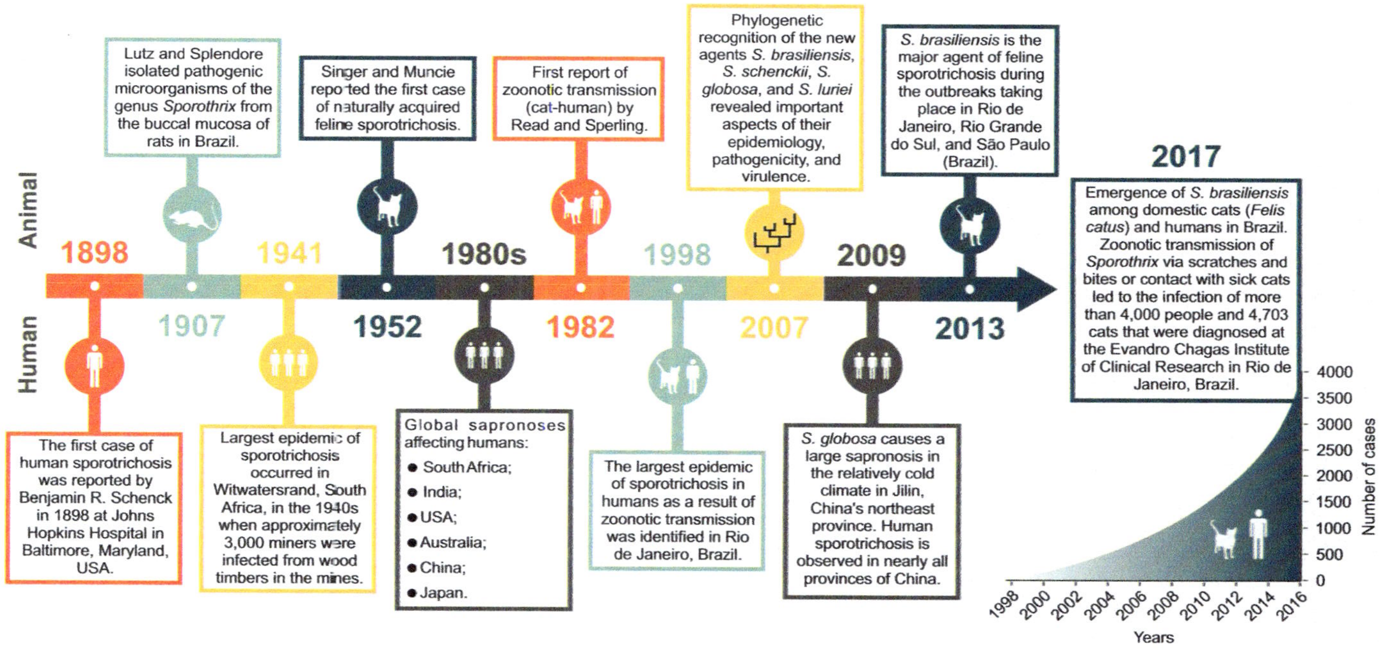

Fig. 10.1 Chronological order of the main facts and pioneering events that contributed to the epidemiological studies on human and animal sporotrichosis. Despite a century of history, the emergence of *Sporothrix brasiliensis* that cause large outbreaks in humans and animals in South and Southeast Brazil has only been observed in the last few decades

transmissions (Montenegro et al. 2014; Gremião et al. 2015; Sanchotene et al. 2015). Due to the negligence of human and animal cases, it is difficult to estimate the true magnitude of this epizootic/epidemic (Fig. 10.1).

10.2 Taxonomic Developments in *Sporothrix*

Sporothrix has great phenotypic plasticity. Phenotypic variations include changes in colony morphology, growth on different culture media, thermal tolerance, virulence in a murine model, susceptibility to antifungal drugs, protein secretion, genome architecture, and melanin synthesis (Marimon et al. 2008b; Fernandes et al. 2009a; Arrillaga-Moncrieff et al. 2009; Sasaki et al. 2014; Almeida-Paes et al. 2015; Della Terra et al. 2017). Recent phylogenetic studies have shown that *S. schenckii* is actually a complex of distinct species, with more than one taxon capable of causing disease in humans and animals (Zhang et al. 2015).

The first molecular evidence that the classical agent *S. schenckii* is represented by two or more species came from the studies of de Beer et al. (2003) through sequencing and phylogenetic analysis of the ITS (internal transcribed spacer) region located in the ribosomal operon (rDNA). Strains of patients from the Americas and South Africa were classified in group I. Interestingly, the strains isolated directly from environmental sources such as soil and decaying plant material were classified in group II, with very few representatives of clinical origin. Thus, the phylogenetic splitting observed was consistent with the niche of these groups, separating clinical from environmental isolates (Fig. 10.2). In fact, *Sporothrix* samples from soil and decaying organic matter tend to be less pathogenic to murine models than clinical isolates (Howard and Orr 1963; Arrillaga-Moncrieff et al. 2009; Rodrigues et al. 2016a).

A few years later, Marimon et al. (2006) proposed three sibling species of medical interest within *S. schenckii*: *Sporothrix brasiliensis*, *S. mexicana*, and *S. globosa*. Marimon et al. (2007) also studied growth at different temperatures (30, 35, and 37 °C), assimilation of sugars (sucrose, raffinose, and ribitol), morphology of sessile conidia, and partial sequence of the calmodulin gene. The analysis based on the calmodulin gene grouped the *Sporothrix* species in different clades: clade I comprised only Brazilian isolates (later named *S. brasiliensis*); clade II comprised most of the isolates from the USA and isolates from South America (later named *S. schenckii s. str.*); clade III included isolates from China, India, Italy, Japan, Spain, and the USA (later named *S. globosa*); and clade IV consisted of environmental isolates from Mexico (later named *S. mexicana*).

Isolated in 1956, *S. schenckii* var. *luriei* was described by Ajello and Kaplan (1969) from a human in South Africa. This isolate exhibited morphological characteristics different from other *S. schenckii* isolates, including the presence of large, round cells with thick walls and typically with one septum, and the production of sclerotia in vitro when grown on oatmeal agar. Phylogenetic data demonstrated a clear separation between *S. schenckii* var. *luriei* (clade VI) and the other species in the pathogenic clade. Moreover, morphological and physiological characteristics

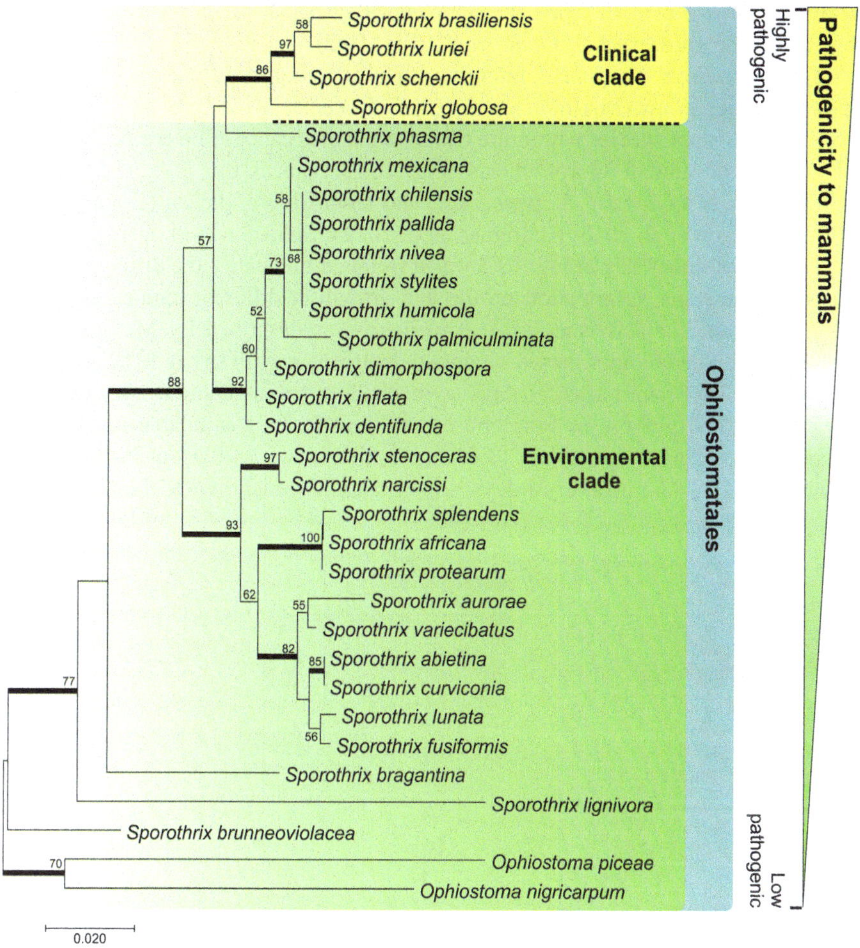

Fig. 10.2 Phylogenetic relations between members of clinical and environmental relevance in *Sporothrix*, based on sequences from the ribosomal operon rDNA (IT1 + 5.8 s + ITS2). Analysis method: neighbor joining (model: T93 + G). The numbers next to the branches refer to the percentages of resampling (1000 bootstraps)

showed differences between the species, including the absence of pigmented sessile conidia, presence of long conidia, good growth at 35–37 °C, and the inability to assimilate sucrose and raffinose; consequently, *S. schenckii* var. *luriei* was named as *S. luriei* (Marimon et al. 2008a).

Today, the "clinical clade" of *Sporothrix* is composed of *S. brasiliensis* (clade I), *S. schenckii s. str.* (clades IIa and IIb), *S. globosa* (clade III), and *S. luriei* (clade VI) (Zhou et al. 2014). The strains from human and animal sources are located in the clades *S. brasiliensis*, *S. schenckii s. str.*, *S. globosa*, and *S. luriei*. Located at a relatively large genetic distance, the environmental species of *Sporothrix* are flanked by environmental species of *Ophiostoma* (Fig. 10.2).

The *Sporothrix* of clinical interest, often infecting vertebrate hosts, form a monophyletic group (Marimon et al. 2007; Roets et al. 2008); however, other genetically distant species within the Ophiostomatales can also be accidental agents of human and animal sporotrichosis (Fig. 10.2). This is the case of *S. pallida*, *S. mexicana*, *S. chilensis*, *S. stenoceras*, and *Ophiostoma piceae* (Bommer et al. 2009; Morrison et al. 2013; Rodrigues et al. 2013a; Scheufen et al. 2015).

The dichotomous key in Fig. 10.3 was developed to identify the different species based on the sugar assimilation profile and growth at different temperatures after 21 days. Main characteristics of *S. brasiliensis*, as described by Marimon et al. (2007), were that colonies grown on potato dextrose agar (PDA) at 30 °C attained a diameter of 15–38 mm after 21 days of incubation; sessile conidia were brown to dark brown and globose to subglobose; and the species was unable to assimilate sucrose and raffinose. *Sporothrix globosa* colonies grown on PDA at 30 °C attained a diameter of 18–40 mm after 21 days; sessile conidia were brown to dark brown and

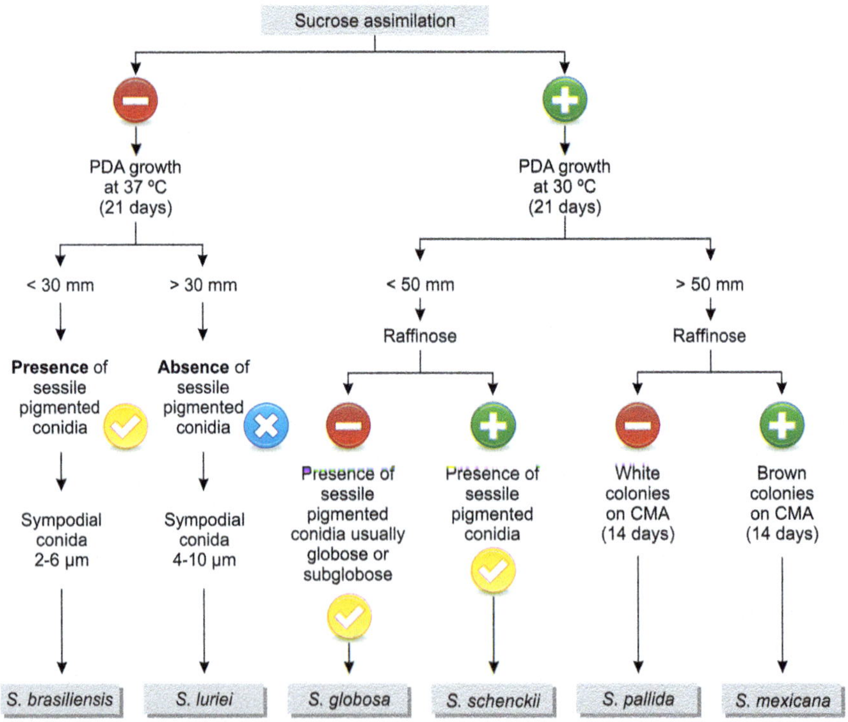

Fig. 10.3 Dichotomous key for identification of species of medical relevance in the genus *Sporothrix*. The criteria used for identification are sucrose and raffinose assimilation and colony size after 21 days of growth on potato dextrose agar (PDA), as well as micromorphological characteristics. *CMA* Corn Meal Agar

globose to subglobose; and the fungus was unable to assimilate raffinose, a key difference with *S. schenckii*. In turn, the *S. mexicana* colonies grown on PDA at 30 °C attained a diameter of 66–69 mm after 21 days; the sessile conidia were brown to dark brown and were predominantly globose, ovoidal, or ellipsoidal; and absence of assimilation of sucrose, raffinose, and ribitol. However, much phenotypic overlap has been reported in the literature (Oliveira et al. 2011; Rodrigues et al. 2013a; Camacho et al. 2015), and the dichotomous key should be used with caution to identify clinical *Sporothrix* species.

10.3 Ecology

Sporothrix schenckii s. str., the classical etiological agent of sporotrichosis, is a saprophyte widely distributed in nature. It is associated with decaying organic matter in soil, water, and rotten bark (Lacaz et al. 2002). Ideal conditions for fungal development in the environment include temperatures between 25 and 28 °C and relative humidity between 92 and 100% (Kenyon et al. 1984; Findlay et al. 1984; Conti Diaz 1989). From propagules in the environment, *Sporothrix* is able to cause disease in humans and other animals such as cats, dogs (Schubach et al. 2001; Govender et al. 2015), rats, mice, pigs, camels, chimpanzees, armadillos, cattle (Rodrigues et al. 2013a,b), donkeys, and horses (Crothers et al. 2009). Due to those climatic conditions, sporotrichosis is common in temperate and subtropical regions, being rarer in arid, semiarid, and cold zones (Dixon et al. 1991). Brazilian *Sporothrix brasiliensis* is mainly cat-transmitted, whereas the mainly East Asian species *S. globosa* practically never infects cats. It might be expected that the habitat determinants of these two species are fundamentally different, but as yet no data are available.

Few ecological studies have evaluated the natural reservoirs and the factors that affect the population growth of *Sporothrix* in nature (Mackinnon et al. 1969; Criseo and Romeo 2010; Rodrigues et al. 2014a). Studies on physiological characteristics of *S. schenckii* allow development of strategies for selective isolation (Fernandes et al. 2009b). Ghosh et al. (2002) reported that the mycelial phase tolerates osmotic pressure of up to 20% glycerol. The yeast form is slightly more tolerant to osmotic pressure, tolerating 6–7% salt concentration, but all are inhibited at 8% salt. *Sporothrix* species grow well between pH 3.0 and 11.5, but few can tolerate pH 12.5 (Ghosh et al. 2002; Fernandes et al. 2009b).

Sporothrix species exhibit thermal dimorphism, growing in the mycelial phase at 25 °C but as yeast at 37 °C (Howard 1961). In addition to temperature, pH seems to have an important role in the transformation of *S. schenckii*. At pH 4.0–5.0 at 25 °C in rich medium, only hyphae develop; however, yeast cells grow in the same medium at pH 6.5–8.0 and at 35 °C (Rodriguez-Del Valle et al. 1983).

At room temperature, colonies are initially white to cream and, over time, usually become dark, with colors ranging from brown to black. Some isolates may take months to exhibit these characteristics, and some may never become black (Rippon 1988) (Fig. 10.4). Microscopically, the hyphae are thin and septate, and the diameter does not exceed 3 μm; ovoid conidia are 2–3 × 3–6 μm in size and are arranged sympodially, in the shape of a daisy at the conidiophore apex. Sessile brown conidia are inserted along undifferentiated hyphae (Fig. 10.4).

At 37 °C the colony morphology changes from pasty white to grayish yellow, resembling a bacterial colony (Fig. 10.4). Mother yeast cells are spherical and daughter cells are oval to cigar-shaped, budding on a narrow base. Yeast cell size ranges from 1–3 to 3–10 μm; multiple buds may be observed, but a single bud is typical (Howard 1961; Rodrigues et al. 2015d) (Fig. 10.5).

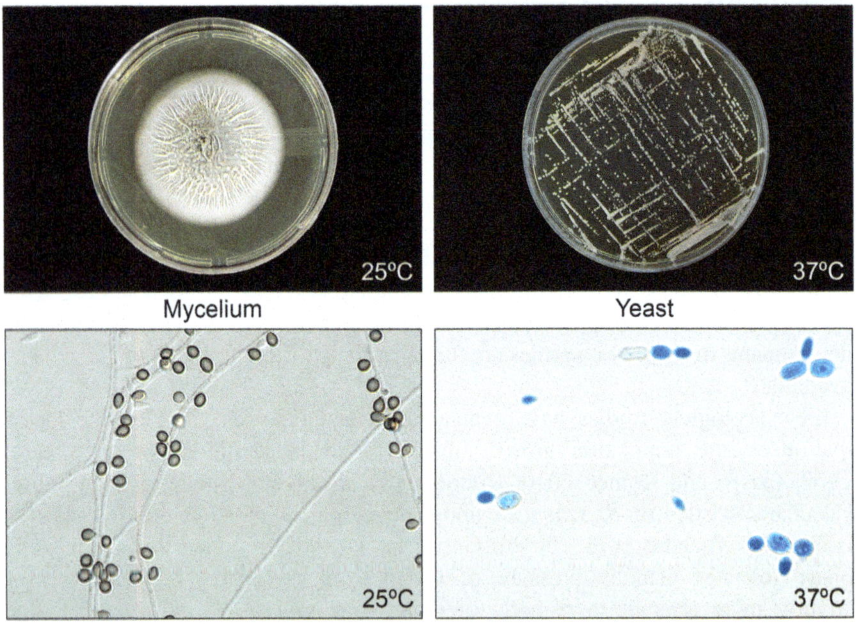

Fig. 10.4 Thermal dimorphism in *Sporothrix* spp. is a morphophysiological adaptation for infection of warm-blooded hosts. The saprophytic phase occurs at room temperature (25 °C). When grown on Sabouraud agar, the colonies are at first white to cream and leathery with a grooved surface. Microscopically (25 °C), many formations of sessile dematiaceous conidia along undifferentiated hyphae are present. When the isolates are grown at 37 °C in brain heart infusion (BHI) agar, the yeast grows, forming small, creamy colonies. Microscopically (37 °C) yeast cells are spherical to oval, commonly known as cigar-shaped. Petri dish: 9 cm in diameter

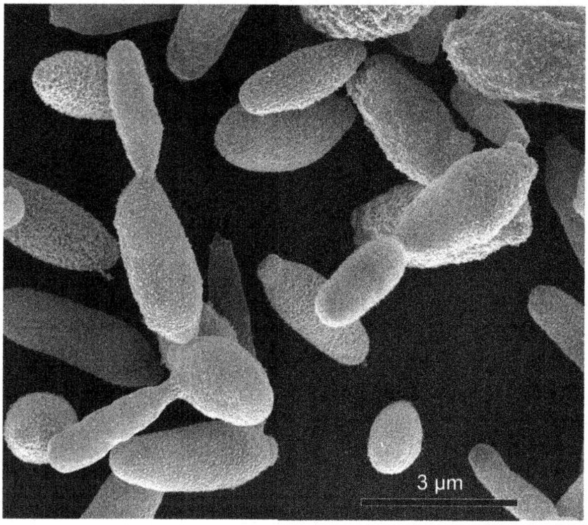

Fig. 10.5 Morphology of *Sporothrix brasiliensis* yeast cells. The parasitic phase is represented by oval or spherical cells (1–3 µm to 3–10 µm), sometimes elongated and with buds

10.4 Epidemiology

Sporotrichosis is not a registered disease; therefore its real prevalence in a given region is difficult to determine. Also inadequate diagnostics hinder understanding of epidemiology. Therefore reliable data on disease incidence in endemic regions are not always available and vary between countries (Pappas et al. 2000, 2003).

Sporotrichosis is considered a cosmopolitan mycosis, but it occurs more often in regions with humid (sub)tropical and temperate climates. Countries with highly endemic areas include South and Central America, South Africa, China, Japan, and India (Mayorga et al. 1978; Fukushiro 1984; Conti Diaz 1989; Verma et al. 2012; Schubach et al. 2005b; Camacho et al. 2015; Chakrabarti et al. 2015). In Brazil, feline sporotrichosis is more common in the South and Southeast regions, being considered endemic in the state of Rio de Janeiro (Oliveira et al. 2011), especially in areas with high population density and low sanitation rates (Barros et al. 2008). In other regions of Brazil, basal transmission levels have been reported (de Araujo et al. 2015), and human infections are usually associated with direct contact with soil and decaying organic matter (Rodrigues et al. 2014d). *Sporothrix globosa* causes a large sapronosis in the relatively cold climate in Jilin, China's northeast province (Zhang et al. 2015). Species-specific differences in biogeography of the disease are of clinical interest (Rodrigues et al. 2016b). For example, in the USA and Australia, major outbreaks were described in patients who reported previous contact with *Sphagnum* moss or hay (Dixon et al. 1991; Coles et al. 1992; Feeney et al. 2007). In these cases, *S. schenckii s. str.* was the etiological agent. However, in the large outbreaks in China, India, and Japan (Fukushiro 1984;

Verma et al. 2012; Song et al. 2013), humans were probably infected through soil and plant debris.

The classical transmission route occurs via traumatic inoculation of *Sporothrix* propagules from soil and plant material (Fig. 10.6). In many cases, lesions caused by thorns or splinters, or scratches against tree bark or contaminated wood, can result in the development of the disease (Powell et al. 1978; de Araujo et al. 2015).

The disease can also be acquired through animal bites or scratches, especially from cats (Schubach et al. 2004; Schubach et al. 2005a; Barros et al. 2008). Sporotrichosis has also been reported in other animals such as dogs, rats, horses, mules, armadillos, goats, bats, squirrels, camels, dolphins, and foxes (Fig. 10.7). Among the animals able to develop sporotrichosis, the domestic cat (*Felis catus*) is the animal host more susceptible to infection by *Sporothrix*, and due to the proximity to humans, it presents the greatest potential for zoonotic transmission. When the cat transmits the fungus by scratching or biting, the contamination is caused by the yeast inoculum and not by conidia (Fig. 10.6). According to Fernandes et al. (2000), the yeast is more virulent in the murine model than the conidia. This may enhance the alternative, mammal-borne transmission route (Almeida-Paes et al. 2014).

The presence of the fungus in the cat's skin and fur, the habit of foraying away from home, and the involvement in fights with other animals likely favor fungal transmission among cats (Barbee et al. 1977). In cats, mycosis follows a severe course in which the fungus spreads quickly from initial skin lesions. Schubach et al. (2002) isolated *Sporothrix* from 100% of skin lesion samples, 66.2% of nasal cavity samples, 41.8% of oral cavity samples, and 39.5% of nail samples of cats with sporotrichosis. Transmission of the fungus to humans occurs through scratches or bites as well as through contamination of preexisting skin wounds (Dunstan et al. 1986). The role of cats in the epidemiology of human sporotrichosis has been emphasized since the 1980s; however, little was known about feline sporotrichosis and the role of cats in the epidemiological chain (Larsson et al. 1989).

The first reports of human cases due to transmission by cats were described by Read and Sperling (1982). In the state of Rio de Janeiro, the epicenter of areas of cat-cat and cat-human transmission, several outbreaks have been reported (de Lima Barros et al. 2003; Oliveira et al. 2011). Currently a large epidemic takes place. Between the 1980s and 1990s, a low number of human sporotrichosis cases were registered by the Evandro Chagas Institute of Clinical Research (Fiocruz, Rio de Janeiro, Brazil). In the late 1990s, 759 cases of human sporotrichosis were diagnosed, with a direct correlation between human and cat cases (de Lima Barros et al. 2001). From 1998 to 2004, Fiocruz identified 1503 cats and 64 dogs with sporotrichosis in Rio de Janeiro (Schubach et al. 2005b). The epidemiological data show an unprecedented spread of the epidemic, with more than 4000 human and feline cases (Gremião et al. 2015, 2017). Such data underline the successful transmission via cats. Currently, the feces of contaminated animals are identified as possible sources of *S. brasiliensis* in nature (Montenegro et al. 2014; Rodrigues et al. 2015a). In the Brazilian states of Rio Grande do Sul and São Paulo, epizootics in cats exist with similar characteristics to those observed in Rio de Janeiro (Sanchotene et al. 2015; Montenegro et al. 2014). Early diagnosis of the disease

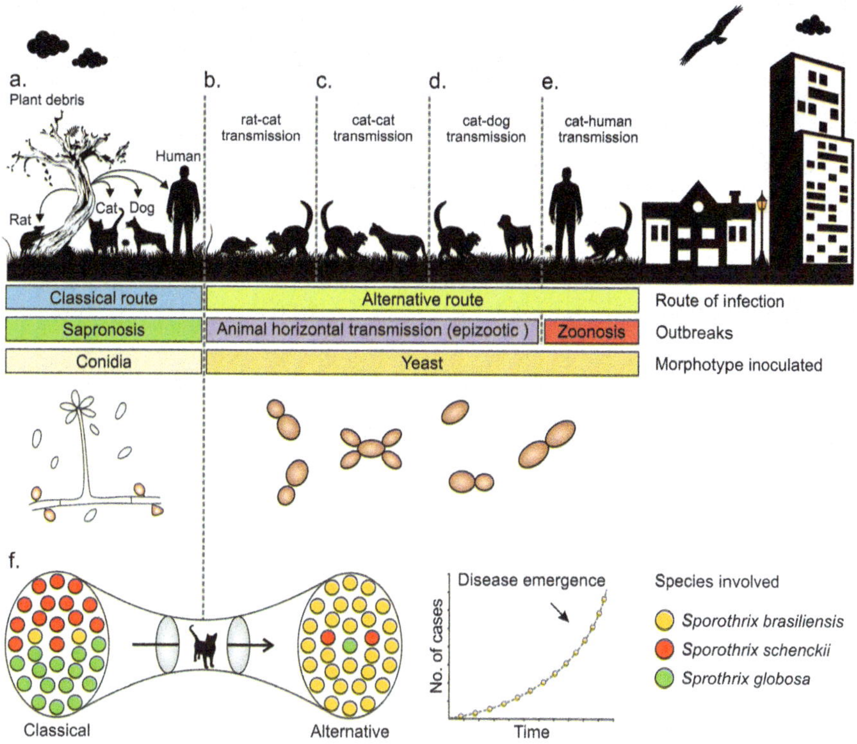

Fig. 10.6 Epidemiological aspects of the transmission chain of human and animal sporotrichosis. (**a**) Classically, sporotrichosis is a sapronosis, in which the mammalian host (humans, rats, cats, and dogs) can acquire the disease through direct contact with the environment (classical transmission route). (**b**) Horizontal animal transmission may involve the rat-cat contact (rat-rat, not shown). (**c**) The cat-cat transmission causes epizootics in populations of susceptible cats, especially in densely populated urban areas such as the metropolitan regions of Rio de Janeiro, São Paulo, and Rio Grande do Sul (Brazil). (**d**) Cats can transmit *Sporothrix* spp. to dogs by traumatic inoculation. (**e**) Due to the proximity to humans, cats have great potential for zoonotic transmission in urban areas. Note that in scenarios (**b–e**) (alternative transmission route), traumatic inoculation of *Sporothrix* yeast, a more virulent morphotype, occurs. Furthermore, a greater fungal load is transmitted in the alternative transmission route. Such factors (inoculum size, yeast morphotype, and *S. brasiliensis*) lead to the onset of more severe forms of the disease in the population. (**f**) In the classical transmission route, transmission occurs mainly by traumatic inoculation of *S. schenckii* and *S. globosa* from soil and plant debris. However, the *S. brasiliensis*-cat association causes horizontal transmission of *S. brasiliensis* to other animals (rats, cats, dogs, etc.) and humans. In areas with epizootics of feline sporotrichosis, *S. brasiliensis* is prevalent in cats and, consequently, in humans because of zoonotic transmission. In areas where the classical infection route is prevalent, the main etiological agent is *S. schenckii s. str.* or *S. globosa* (depending on the region). Therefore, reservoirs, vectors, microbes, and hosts may vary in the disease transmission chain

and control and prevention measures are extremely important (Madrid et al. 2010, 2012). Dogs do not seem to play an important role in disease transmission (Barros et al. 2010).

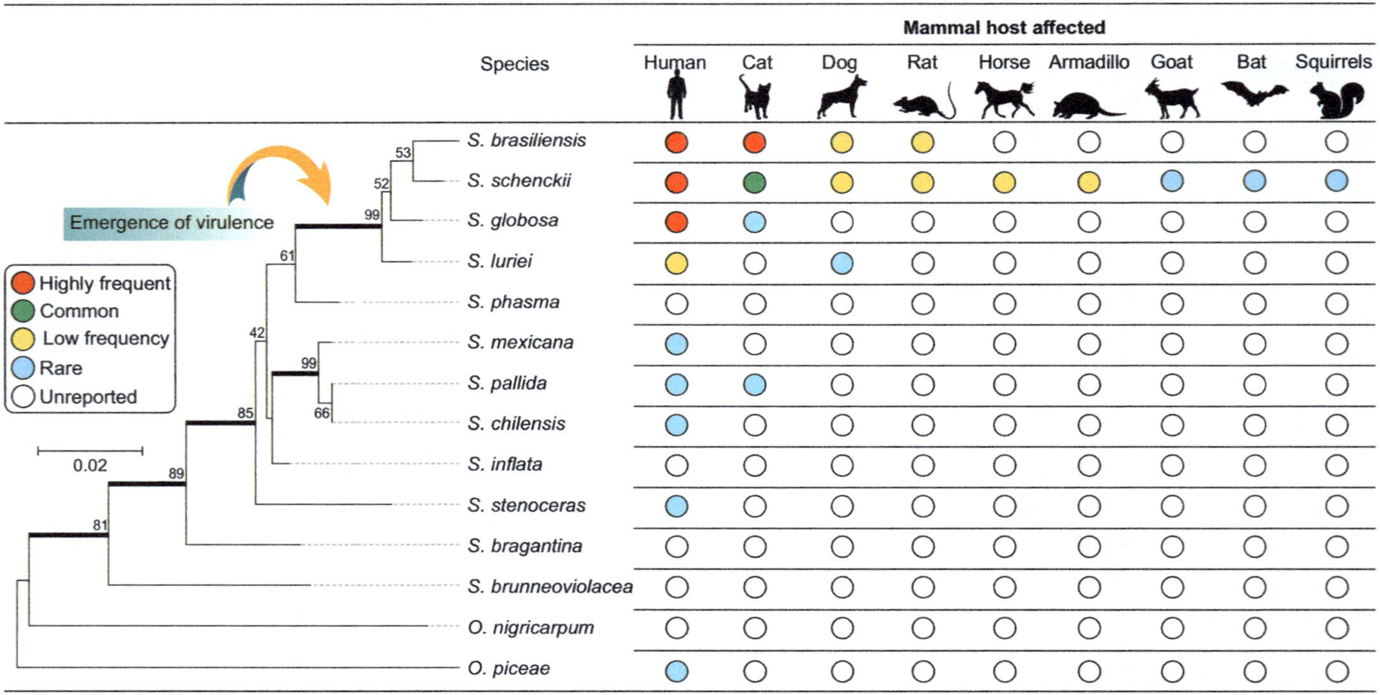

Fig. 10.7 Mammalian hosts of sporotrichosis. Among the species of clinical relevance, *Sporothrix brasiliensis*, *S. schenckii*, and *S. globosa* are highly frequent in humans. A few cases of infection by *S. luriei* have been described in the literature. In Brazil, large epizootics are caused by *S. brasiliensis* in cats. In turn, *S. schenckii* is an important etiological agent of feline sporotrichosis in other regions of the world (e.g., Peru, the USA, and Malaysia). Although diagnosed in other animals such as dogs, rats, armadillos, goats, bats, and squirrels, *Sporothrix* spp. infection in these animals is less frequent than that observed in cats and humans. Rare cases of human infection have been reported for members of the environmental clade such as *S. mexicana*, *S. pallida*, *S. chilensis*, *S. stenoceras*, and *Ophiostoma piceae*. Note that the frequency of infection caused by members of the *S. schenckii* complex is higher due to the emergence of pathogenicity in this group

The prevalent agent in epizootics of feline sporotrichosis is *S. brasiliensis* (96.9%), an agent endemic to Brazil. Phylogenetic and population genetic analyses confirm zoonotic transmission (cat-human) of *S. brasiliensis* and the low genetic diversity of this species. A positive correlation exists between the number of feline cases and the marked increase in human cases. Over time, *S. brasiliensis* will expand its biogeographical occurrence.

On rare occasions, sporotrichosis has been associated with scratches and bites during the hunting of wild animals. In the state of Rio Grande do Sul (Brazil), ten cases of sporotrichosis associated with armadillo hunting were recorded from 2005 to 2009 in hunters scratched by the animals (Alves et al. 2010). The etiological agent was *S. schenckii* (Rodrigues et al. 2014a), the same species found in 18 clinical isolates from cats in Malaysia (Kano et al. 2015) (Fig. 10.7).

10.5 Clinical Signs

Classically, human and animal sporotrichosis is acquired by traumatic inoculation of the fungus in the subcutaneous tissue (Barros et al. 2011; Orofino-Costa et al. 2017) and is usually associated with lymphangitis. The inoculum load, the immune status of the host, the virulence of the inoculated strain, the depth of traumatic inoculation, and the fungus species influence the development of sporotrichosis (Rodrigues et al. 2016b; Della Terra et al. 2017).

After traumatic inoculation in cats, the incubation period ranges from days to months. The domestic cat is highly susceptible to infection by *S. brasiliensis*. Several clinical manifestations have been reported, characterizing sporotrichosis as a polymorphic disease, which can be mistaken for a number of other dermatologic manifestations such as atypical mycobacteriosis, cryptococcosis, histoplasmosis, neoplasias, and cutaneous leishmaniasis. Feline sporotrichosis ranges from an infection with no apparent cutaneous manifestations to dissemination with or without extracutaneous signs. The lesions are ulcerative and crusty, with alopecia, and generally exhibit central ulceration (Fig. 10.8).

In cats, the disease occurs more often among males during reproductive age. This finding suggests an important relation with forays and animal fights (Schubach et al. 2001, 2002). Most skin lesions are observed in the head, the most affected area during fights. However, mucosal involvement may also occur due to cat habits such as licking, which can transfer a considerable amount of yeast cells from the skin lesions to the oral cavity, as well as to other distant body parts.

A considerable amount of yeast is present in skin lesions and exudates of cats, boosting horizontal animal transmission or zoonotic transmission. Owners of a sick animal commonly acquire sporotrichosis during oral administration of antifungals because the animal may bite or scratch, consequently, inoculating the yeast. Therefore, care must be taken, such as using gloves and safety equipment during animal handling, to prevent the progression of the sporotrichosis transmission chain.

Clinical manifestations in humans range from localized nodules and ulcers to ulcers along the lymphatic pathway, dissemination through the skin with mucosal

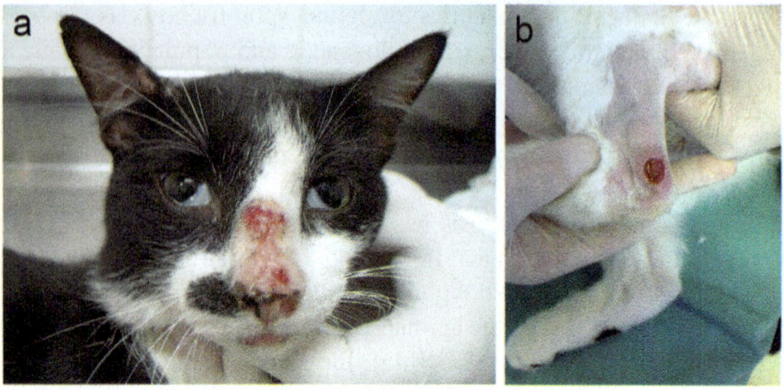

Fig. 10.8 Clinical aspects of feline sporotrichosis caused by *Sporothrix brasiliensis* in South and Southeast Brazil. Images of feline sporotrichosis courtesy of Dr. Sandro Antonio Pereira (INI/Fiocruz, Brazil) (right) and Dr. Flávio de Queiroz Telles Filho (Federal University of Paraná, Brazil) (left)

tissue involvement, and even lung involvement. When acquired via the classical transmission route, the mycosis is more common in workers such as gardeners, vegetable growers, and farmers (Powell et al. 1978). Conidial inhalation is uncommon in sporotrichosis, but it may result in pulmonary and systemic forms of the disease, with spread to the bones, eyes, central nervous system, and internal organs; these forms mainly occur in immunocompromised patients (Kauffman 1999; Kauffman et al. 2007; Aung et al. 2013). When acquired through the alternative transmission route, the mycosis is more frequent in cat owners and veterinarians, ceasing to be an occupational hazard and exposing more individuals to the disease due to intimate contact between domestic cats and humans.

Human sporotrichosis is mostly a benign infection limited to the skin and the subcutaneous tissue, with spread to the bones and internal organs possible. The initial lesion that develops as granulomatous nodules may progress to a necrotic or ulcerative lesion. During this period, the adjacent lymphatic vessels become thick and similar to a rope. Along the lymphatic vessels, multiple subcutaneous nodules and abscesses are observed. The disease has subacute to chronic evolution in immunocompetent patients and may spread in immunocompromised patients (Silva-Vergara et al. 2012). Cases in which the fungus has reached the joints causing osteoarthritis have frequently been reported (Al-Tawfiq and Wools 1998; Koeter and Jackson 2006; Orofino-Costa et al. 2010). The mechanism and tropism of the fungus to the synovial fluid, bones, and cartilage are still unknown. Sporotrichosis lesions are often misdiagnosed as other diseases such as cutaneous leishmaniasis (Castrejon et al. 1995).

Sporotrichosis has often been described among HIV patients, alcoholics, diabetics, and patients who used corticoids for a long period (Al-Tawfiq and Wools 1998; Silva-Vergara et al. 2012; Paixao et al. 2015). AIDS patients have a very poor prognosis because sporotrichosis spreads very easily. However, less commonly, immunocompetent individuals may have hematogenous spread and develop osteoarticular injuries (Almeida-Paes et al. 2014).

The most common clinical form in 80% of human cases is the lymphocutaneous form. Seven to 30 days after the traumatic inoculation, a nodule or ulcer develops at the inoculation site and adjacent lymphatic vessels (Bonifaz and Vázquez-González 2013). The fixed form, which is more prevalent in *Sporothrix globosa*, consists of a single non-lymphatic nodule, which is limited and less progressive. This clinical manifestation is characterized by erythematous and occasionally ulcerated plaques, usually located on the face, neck, trunk, and legs. The lesions may undergo spontaneous remission but may relapse and persist for years if not treated (Fischman Gompertz et al. 2016).

The disseminated form is usually associated with some type of immunosuppression: HIV/AIDS, chemotherapy patients, advanced age, chronic alcoholism, diabetes, Cushing's syndrome, prolonged corticosteroid therapy, kidney diseases, and other conditions. After traumatic inoculation, the fungus spreads hematogenically (Al-Tawfiq and Wools 1998; Koeter and Jackson 2006; Orofino-Costa et al. 2010). Extracutaneous forms are also often associated with immunosuppression, but this relation is not mandatory (Kauffman et al. 2000, 2007). The extracutaneous form is common in immunocompromised patients such as diabetics, alcoholics, bone narrow transplant recipients, and patients with chronic lung disease or blood cancer. If *Sporothrix* spreads to the organs, the disease may present various symptoms such as sinusitis, meningitis, laryngitis, endophthalmitis, osteoarticular and pulmonary involvement, and fungemia (Silva-Vergara et al. 2012).

10.6 Host Response and Pathogenesis

Temperature plays an important role in the *Sporothrix* mycelium-yeast transition. The thermal dimorphism is a morphophysiological adaptation to infection of warm-blooded hosts. This transition to an invasive phase is a unique characteristic among Ophiostomatales and is shared with phylogenetically distant human pathogens in *Onygenales* and *Eurotiales* (Rodrigues et al. 2016b). In members of the environmental *Sporothrix* clades such as *S. chilensis*, *S. mexicana*, and *S. pallida*, the mycelium-yeast transition occurs poorly, with development of few yeast cells, which can be related to low virulence to mammalians (Rodrigues et al. 2016a).

Once *Sporothrix* comes into contact with the host, it can stimulate or trigger the natural (innate) and acquired (adaptive) defense mechanisms of the host. Immunity to *Sporothrix* is conferred by neutrophils and monocytes (Davis 1996). The hypothesis that suggests immunity mediated by T cells as a limiting factor in the progression of infection is because sporotrichosis is more severe in athymic mice and AIDS patients (Donabedian et al. 1994; Tachibana et al. 1999).

The host defense mechanisms affect the manifestation and severity of fungal infections, thereby generating different clinical forms of the disease. Isolates from skin lesions activate dendritic cells, thereby inducing Th1 immune response (Uenotsuchi et al. 2006). *S. schenckii* is capable of inducing a mixed Th1/Th2 response in experimental sporotrichosis (Uenotsuchi et al. 2006; Maia et al. 2006). Sassa et al. (2009) suggested that the toll-like receptors (TLR4) play an important role in macrophages in *S. schenckii* experimental infection.

The humoral response involves potent antigens able to elicit immune responses in the host during colonization and tissue invasion. After infection by *S. brasiliensis*, B lymphocytes of cats produce an IgG-mediated immune response against *Sporothrix* antigens, similar to the humoral response observed in murines (Fernandes et al. 2013) and human sporotrichosis (Rodrigues et al. 2015c). Significant cross-reactivity exists among antigens of *S. brasiliensis*, *S. schenckii*, and *S. globosa*, supporting the hypothesis that the antigens share epitopes highly conserved among the members of the clinical clade. This also suggests that infection in warm-blooded hosts (humans, cats, and rats) is the result of evolutionary convergence among *S. brasiliensis*, *S. schenckii*, and *S. globosa* (Rodrigues et al. 2015b, c).

A protein identified as 3-carboxymuconate cyclase (*gp60-70*) is considered the main antigen expressed by pathogenic *Sporothrix* species and recognized by circulating IgG antibodies in the serum of hosts (Rodrigues et al. 2015c). This highly polymorphic glycoprotein, with a molecular weight ranging from 60 to 70 kDa (*gp60* and *gp70*, respectively) and an isoelectric point between 4.3 and 4.8, is present in at least six isoforms and glycoforms. The *gp60-70* protein is located in the cell wall of *Sporothrix* (Castro et al. 2013), where it functions as an adhesin for extracellular matrix molecules such as fibronectin and laminin (Teixeira et al. 2009; Nascimento et al. 2008). Note that anti-*gp60-70* antibodies reduce adherence of *Sporothrix* to the host tissue in a concentration-dependent manner. They facilitate yeast opsonization by phagocytes, reducing the fungal load in the host (Ruiz-Baca et al. 2009; Nascimento et al. 2008). Thus, it is suggested that anti-3-carboxymuconate cyclase antibodies are important effectors in the host defense against the establishment and spread of infection. Rodrigues et al. (2015b) demonstrated the presence of anti-*gp60-70* antibodies in felines during epizootics by *S. brasiliensis*. Preimmunization with 3-carboxymuconate cyclase in experimental sporotrichosis has a negative regulatory effect on the Th1 immune response, with a reduction in cytokine levels (Alba-Fierro et al. 2016). Knowledge of the humoral immune response in feline sporotrichosis is essential to the advance of serological diagnostic techniques, to vaccine production, and to improve our understanding of the evolution of the interaction between *Sporothrix* and its hosts (Rodrigues et al. 2015b).

In experimental sporotrichosis, the study of the host-pathogen interaction focuses on pathogenicity and virulence using several species of *Sporothrix*. The strains most frequently used are BALB/c, C57BL/6, and OF1 mice in subcutaneous or disseminated inoculation models (Castro et al. 2013; Fernandes et al. 2013; Almeida-Paes et al. 2015; Arrillaga-Moncrieff et al. 2009; Della Terra et al. 2017). The amount of inoculum and the morphotype inoculated, i.e., conidium or yeast, vary among research groups. However, a clear difference exists in the infection ability of species belonging to clinical and environmental clades, respectively. Despite methodological divergences, *S. brasiliensis* is by far the most virulent species of the clinical clade, followed by *S. schenckii*, *S. luriei*, and *S. globosa*

(Fernandes et al. 2013; Arrillaga-Moncrieff et al. 2009; Fernandez-Silva et al. 2012). At the other extreme, *S. chilensis*, *S. mexicana*, and *S. pallida* are not pathogenic, and inoculated animals are able to resolve the infection a few days after the challenge (Arrillaga-Moncrieff et al. 2009; Rodrigues et al. 2016a). Alternative methods such as the use of invertebrate models are available for assessing the virulence of *Sporothrix*, especially the larvae of *Galleria mellonella*, with results similar to those observed in traditional animal models (Freitas et al. 2015; Clavijo-Giraldo et al. 2016).

10.7 Diagnosis

Clinical evaluation and patient history are the first tools used for diagnosis. Sporotrichosis can be diagnosed by direct examination of tissue biopsies or lesion pus. *Sporothrix* from pus presents as gram-positive bacilliform corpuscles, phagocytized by giant and polymorphonuclear cells. Typically, 10% potassium hydroxide is used for the observation of yeast in human tissue; however, the cells are small and sparse, making detection in direct examination difficult. By contrast, in samples from infected felines, yeast can be easily found due to the large fungal load in the lesions (Barros et al. 2011). Feline sporotrichosis is often diagnosed by cytological evaluation of samples collected from abscess aspirates. Advantages include convenience, low cost, and high sensitivity of the technique (Pereira et al. 2011; Sanchotene et al. 2015). In disseminated sporotrichosis, sputum, urine, blood, and cerebrospinal and synovial fluids samples can be used in the diagnosis.

Isolation of the fungus is obtained after spreading clinical specimens on Sabouraud agar with chloramphenicol or cycloheximide. For species identification, dimorphism must be confirmed by the mycelium-yeast transition by subculturing the fungus on enriched media such as BHI agar at 37 °C for 5–7 days (Morris-Jones 2002). In 89% of cases, the fungus can be isolated in 8 days, but it may take up to 4 weeks.

In the typical histopathology of the disease, a mixed, granulomatous, and pyogenic inflammatory process can be observed, in which the characteristic structures are cigar-shaped or oval yeast cells (Fig. 10.9). Mycelium or asteroid bodies are not commonly observed in the tissue (Zhang et al. 2011). Asteroid bodies occur due to intense deposit formation in vivo of eosinophilic material around the microorganism, known as the Splendore-Hoeppli phenomenon. In sporotrichosis, asteroid bodies are characterized by the presence of radial, star-shaped deposition and are more frequent in lesions of humans with the lymphocutaneous form (Hussein 2008; Zhang et al. 2011). The origin of the events leading to the formation of asteroid bodies is not clear, but the deposition of antibodies and the incorporation of calcite and apatite crystals, which form the crystallized spicules, are known to occur. Moreover, the asteroid body serves as a resistance structure and plays an important role in fungus survival in the host (da Rosa et al. 2008).

Noninvasive methods are available for diagnosis of sporotrichosis. Techniques using fluorescent antibodies or immunohistochemistry are quick but are not

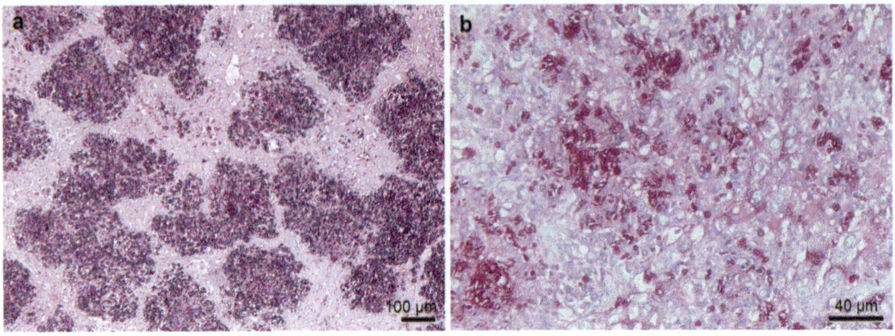

Fig. 10.9 (**a**) Histopathological aspects of *Sporothrix brasiliensis* during infection in a warm-blooded host (murine model of infection; magnification of 10×). (**b**) A mixed inflammatory process, granulomatous, with many cigar-shaped or oval yeast cells, is observed in the tissue (magnification of 40×). Staining: Schiff periodic acid

available in most laboratories (Davis 1996). Blumer et al. (1973) compared several serological methods, including latex agglutination, tube agglutination, complement fixation, immunodiffusion, and immunofluorescence, using crude fungal culture filtrate as the antigen. The latex agglutination technique was the most sensitive and specific, and immunodiffusion was also considered specific. However, Welsh and Dolan (1973) observed nonspecific reactions with the sera of normal individuals and heterologous sera using the agglutination technique.

The ELISA test has been extensively studied as a tool for sporotrichosis diagnosis and serologic follow-up of the patient. The antigens used include mycelium or yeast-culture filtrates, as well as partially purified molecules. The literature shows that the tests have the same sensitivity and specificity, which range from 80 to 100% in humans (Almeida-Paes et al. 2007a; Almeida-Paes et al. 2007b; Bernardes-Engemann et al. 2005, 2009) and felines (Fernandes et al. 2011; Rodrigues et al. 2015b).

Regarding the humoral immune response of patients with sporotrichosis, Scott and Muchmore (1989), using Western blot, observed that 100% of patient sera recognize the 40 and 70 kDa molecules and mostly also those of 22 and 36 kDa. However, Mendoza et al. (2002) reported that positive sera recognize additional molecules such as 40, 55, 74, 90, and 147 kDa, depending on the culture medium used for the production of the protein/glycoprotein extract. In contrast, in experimental murine sporotrichosis, antibodies recognize 67 kDa (Carlos et al. 1998) and 70 kDa molecules (Nascimento and Almeida 2005).

A well-known component of the *Sporothrix* cell wall is peptido-rhamnomannan. This glycopeptide or glycoconjugate is a complex of molecules with different molecular weights that are difficult to purify. The fraction of the peptido-rhamnomannan extract able to bind to concanavalin-A (ConA) has antigenic characteristics, and molecules of 84, 70, and 58 kDa were observed when probed with anti-*S. schenckii* rabbit serum (Lima and Bezerra 1997). This fraction was called *Ss*CBF (*S. schenckii* ConA-binding fraction), and *O*-glycosylated chains of peptido-rhamnomannan from the cell wall are responsible for binding lectin ConA. In addition, this fraction was recognized by sera from patients with sporotrichosis when tested by ELISA, suggesting that the fraction *Ss*CBF may be an important tool for the complementary and/or differential diagnosis of the mycosis (Bernardes-Engemann et al. 2005).

Another important antigen is *gp60-70* (3-carboxymuconate cyclase), which is present in the culture filtrate. All BALB/c mice infected with *S. schenckii* were able to produce antibodies (IgG1 and IgG3) against this molecule in ELISA (Nascimento and Almeida 2005). This glycoprotein is also recognized by circulating IgG in 100% of cats with sporotrichosis in the main endemic areas of Brazil (Rodrigues et al. 2015b).

The study of antigenic molecules for diagnosis is important, and the use of antigens such as the cell-free antigen (CFA) preparation is simple and requires no special equipment. Mycelial and yeast CFA preparations of *S. brasiliensis* have 14 and 23 reactive bands, respectively, against sera from immunized rabbits (Almeida-Paes et al. 2012). Antigens from yeast can be successfully used in the diagnosis of human sporotrichosis.

The use of one type of antigen or another depends on the structure of each laboratory. The use of *Ss*CBF allows for better test standardization because it is more specific; however, the culture filtrate is easier to produce because it does not require expensive equipment. The disadvantage lies in the production, which takes time and requires that strains, culture media, cultivation time, and preparation methods be standardized among laboratories.

Methods based on molecular biology are still scarce in the diagnosis of sporotrichosis. The regions of the ribosomal RNA gene can be used to establish phylogenetic relations among the different species. The ITS regions, which are the regions between the genes encoding the rRNA regions 18S, 5.8S, and 28S, are particularly useful for identifying *S. brasiliensis*, *S. schenckii*, *S. globosa*, and *S. luriei* (Zhou et al. 2014) using the primers ITS1 and ITS4 (White et al. 1990). Phylogenetic analysis based on protein-encoding genes such as beta-tubulin (de Meyer et al. 2008), calmodulin (Marimon et al. 2006, 2007), and elongation factor 1α (Rodrigues et al. 2013b) is also important for the molecular identification of etiological agents of human and animal sporotrichosis.

New techniques to identify the newly described species are in progress. Such methods include polymerase chain reaction (PCR) fingerprinting with the T3B oligonucleotide (de Oliveira et al. 2012) and PCR-restriction fragment length polymorphism (RFLP) with calmodulin amplification followed by digestion with the enzyme *Hha*I (Rodrigues et al. 2014b). A comparison of these techniques with analysis of the partial sequence of the calmodulin gene showed that both were able to identify the majority of the isolates studied and were very useful in the identification of isolates with atypical phenotype.

Only a few methods are available for the direct molecular diagnosis of sporotrichosis. Unlike the methods described above, these methods are sensitive and specific for the detection and identification of all *Sporothrix* species of clinical interest (Rodrigues et al. 2015a). For direct molecular diagnosis of complex samples

Table 10.1 Comparison of the different molecular methods for identification and/or detection of etiological agents of human and animal sporotrichosis

Method	Samples					*In silico* analysis[f]	Time ID (h)[g]	Equipment[h]	Reference
	Culture[a]	Biopsy[b]	Soil[c]	Plant[d]	Feces[e]				
DNA sequencing	x	x	x	x	x	Yes	10–12	Thermocycler	Marimon et al. (2007)
								Electrophoresis system	
								Sequencer	
Amplified fragment length polymorphism (AFLP)	x					Yes	10–12	Thermocycler	Neyra et al. (2005); Zhang et al. (2015)
								Sequencer	
Random amplified polymorphic DNA (RAPD)	x					Yes	6–7	Thermocycler	de Oliveira et al. (2012)
								Electrophoresis system	
PCR-restriction fragment length polymorphism (PCR-RFLP)	x					No	6–20	Thermocycler	Rodrigues et al. (2014b)
								Electrophoresis system	
Species-specific PCR	x	x	x	x	x	No	6	Thermocycler	Rodrigues et al. (2015a)
								Electrophoresis system	
Rolling circle amplification (RCA)	x	x	x	x	x	No	5	Thermocycler	Rodrigues et al. (2015e)
								Electrophoresis system	
MALDI-ToF	x					Yes	2	Mass spectrometer	Oliveira et al. (2015)

[a]Material isolated in pure culture
[b]Clinical sample, fresh tissue
[c]Environmental soil sample
[d]Plant debris or decaying organic matter
[e]Animal feces
[f]Bioinformatics of the generated data
[g]Average execution time in hours, *ID* identification
[h]Equipment necessary to perform the method

(target plus exogenous DNA, e.g., host DNA), species-specific PCR is used for the selective detection of *Sporothrix* DNA using primer pairs for *S. brasiliensis*, *S. schenckii*, *S. globosa*, *S. mexicana*, *S. pallida*, and *S. stenoceras* (Rodrigues et al. 2015a). Molecular diagnosis using species-specific PCR was validated using a murine model (BALB/c) of disseminated sporotrichosis. The infections caused by *S. brasiliensis* and *S. schenckii* were successfully detected in different organs such as the spleen, lungs, liver, kidneys, heart, brain, and tail using species-specific PCR (Rodrigues et al. 2015a). PCR was also useful for the detection of *Sporothrix* DNA in feces of infected animals, suggesting the use of species-specific PCR in ecoepidemiological studies (Rodrigues et al. 2015a).

The identification and specific detection of *Sporothrix* using rolling circle amplification (RCA) were recently proposed (Rodrigues et al. 2015e). Bioinformatics analysis revealed six candidate probes specific for *S. brasiliensis*, *S. schenckii*, *S. globosa*, *S. luriei*, *S. mexicana*, and *S. pallida*, with no significant homology to sequences of the genomes of humans, mice, or microorganisms outside members of *Sporothrix*. The accuracy of RCA was demonstrated in vitro by the binding specificity (probe DNA) in various isolates with no cross-reactivity and high detection sensitivity (up to 3×10^6 target copies). RCA-based identification is consistent with the phylogenetic analysis results of the major loci used in the *S. schenckii* complex, reinforcing the use of RCA as a reliable identification method alternative to DNA sequencing. The method also provides a powerful tool for the rapid and specific detection of clinically relevant *Sporothrix* in biopsies, soil, feces, and plant debris. Due to the robustness of RCA, the technique has potential for ecological studies (Rodrigues et al. 2015e).

In general, molecular methods may be useful for the diagnosis of sporotrichosis. A small amount of sample is typically required, and the method is sensitive and specific, taking less time than the traditional methods. Table 10.1 shows a comparative summary of the main methods available for the identification and/or detection of *Sporothrix* spp. in different biological samples.

10.8 Control

Treatment of feline sporotrichosis is a major challenge to public health because the number of oral antifungal agents is limited; these have numerous adverse effects and high cost (Rodrigues et al. 2016c). The drug choice should take into account the clinical condition of the patient, the extent of the skin lesions, the drug interactions, the adverse effects, and the systemic involvement (Orofino-Costa et al. 2015, 2017). The treatment of feline sporotrichosis includes local hyperthermia, potassium iodide, azoles (ketoconazole and itraconazole), amphotericin B, and the allylamine terbinafine. Currently, itraconazole and potassium iodide are the most used drugs in the treatment of feline sporotrichosis (Pereira et al. 2010; Reis et al. 2012; Gremião et al. 2015, 2017), and clinical cure is observed regardless of the initial clinical conditions or coinfection with FIV and/or FeLV. The treatment of feline sporotrichosis may last several weeks to several months (average time 4–9 months) and must be continued for at least 1 month after clinical cure. Recently, the combination of itraconazole and potassium iodide was successfully reported as a new therapeutic option (itraconazole 100 mg/day and KI 2.5–20 mg/kg/day), with a high cure rate (96.15%) and 14 weeks of treatment on average (Reis et al. 2016). Recurrence may occur, demonstrating the possibility of lesion reactivation, despite the completion of treatment (Pereira et al. 2010; Gremião et al. 2011, 2015; Chaves et al. 2013).

Sporothrix brasiliensis is the main etiological agent of feline sporotrichosis in Brazil, and little variation exists in the susceptibility profile to azoles in this species (Rodrigues et al. 2014c; Espinel-Ingroff et al. 2017). Significant activity of itraconazole (minimum inhibitory concentration (MIC), 0.125–2 μg/mL), amphotericin B (MIC, 0.125–4 μg/mL), and ketoconazole (0.0312–2 μg/mL) against *S. brasiliensis* in feline sporotrichosis has been reported (Brilhante et al. 2016). Fluconazole, caspofungin, and 5-fluorocytosine have no antifungal in vitro activity against *S. brasiliensis* or other species of the clinical clade (Rodrigues et al. 2014c; Marimon et al. 2008b). Significant differences exist between the minimum concentrations necessary to inhibit the growth of *Sporothrix* in vitro and the concentration necessary to reduce the number of colony-forming units, demonstrating the fungistatic effect of most commercially available antifungals (Rodrigues et al. 2014c). It is important to note that isolates of *S. brasiliensis* recovered recently (2011–2012) from humans during the epidemic in Rio de Janeiro were less susceptible to amphotericin B, itraconazole, posaconazole, voriconazole, and terbinafine compared to isolates collected before 2004 (1998–2004), suggesting that this epidemiological profile might be changing over time (Borba-Santos et al. 2015). In this scenario, terbinafine may be a promising alternative against the mycelium and yeast forms of *S. brasiliensis*, followed by posaconazole (Borba-Santos et al. 2015).

In the last decades, human sporotrichosis has been treated with itraconazole, the drug of choice for cutaneous and lymphocutaneous forms of the mycosis in humans due to its good tolerability and lower incidence of adverse effects (Kauffman et al. 2007). This drug is also used in immunocompromised patients with clinical forms with systemic involvement (Bolao et al. 1994). However, for some researchers, the drug of choice is potassium iodide (de Macedo et al. 2015a, b; Costa et al. 2013), which provides an effective and affordable treatment. The treatment with antifungal drugs can take from weeks to months to restore a favorable clinical profile and is continued for at least 1 month after the complete disappearance of symptoms (Bonifaz and Vázquez-González 2013). The prognosis of sporotrichosis varies from good to regular even in immunocompromised patients, although relapse may occur (Kauffman et al. 2007).

The drug of choice for the treatment of fixed cutaneous and lymphocutaneous sporotrichosis is 200 mg itraconazole, orally for 3–6 months (Kauffman et al. 2007). This drug belongs to the azole class and acts by inhibiting the ergosterol biosynthesis pathway, the primary sterol of fungal membranes. As ergosterol is depleted, it is replaced by unusual sterols, changing the normal permeability and fluidity of the membrane and, consequently, the binding of enzymes to the membrane such as those involved in cell wall synthesis. Other drugs of the same class are used, such as fluconazole, which are less effective than itraconazole but are recommended to patients who do not tolerate or use drugs that interact with itraconazole (Kauffman et al. 2007).

Amphotericin B binds to ergosterol and disturbs the normal functioning of the membrane, causing leakage of the intracellular content. One of the problems associated with this drug is its toxicity to mammalian cells, particularly nephrotoxicity. The treatment of choice when patients have severe sporotrichosis or are at risk of death is amphotericin B (Kauffman et al. 2007; Silva-Vergara et al. 2012).

In developing countries, the first-choice drug is potassium iodide because it is more economically viable. The mechanism of action of potassium iodide is unknown; whether it acts against the fungus via a fungicidal mechanism or by increasing the body's defense mechanism through the immune response is not known (Rex and Bennett 1990; Sterling and Heymann 2000).

Alternatively, thermotherapy has been used for treatment of sporotrichosis, with the advantage of being low cost and having no side effects. The temperature of the human body is higher than the optimum temperature for fungal growth, and thus, the fungus does not grow well at 37 °C. Judging from this information, the growth of a *Sporothrix* isolate was assessed under different temperatures. This study showed that temperatures higher than 39 °C inhibit fungal growth (Mackinnon and Conti-Díaz 1963). Indeed, reports exist of successful treatment of human sporotrichosis using thermotherapy (Hiruma et al. 1987). Thermotherapy has also been successfully used in an infected cat (Honse et al. 2010).

10.9 Conclusion and Perspectives

In the last two decades, we have observed the return of infectious diseases as one of the major causes of human death and a reduced quality of life in patients (Fisher et al. 2012). Factors that contribute to this situation include the presence of organisms more resistant to drugs, the emergence of new pathogens and/or reemerging pathogens, and the increased number of immunocompromised patients such as those with AIDS, cancer, or organ transplantation. For a long time, sporotrichosis was described as a disease of low incidence in Brazil; however, recent reports show that the number of cases has been increasing, and the incidence of more severe or atypical clinical forms of the disease is occurring more frequently (Rodrigues et al. 2014d).

In general, zoonotic pathogens are twice as likely to be associated with emerging diseases than non-zoonotic pathogens (Taylor et al. 2001). *Sporothrix brasiliensis* appears as a new emerging pathogen in a highly susceptible feline population and has developed an effective transmission route to humans in densely populated urban areas (Rodrigues et al. 2013b, 2014d; Montenegro et al. 2014). Furthermore, *S. brasiliensis* is more virulent than other members of the clinical clade (Fernandes et al. 2013), a factor that intensifies the atypical clinical manifestations in humans and animals.

The diagnosis of sporotrichosis must be quick and precise, thus avoiding disease complications in animals and interrupting the transmission to humans, minimizing the effects on the population (Rodrigues et al. 2015a). The treatment should be carried out effectively with the administration of antifungals such as itraconazole, potassium iodide, or a combination of therapies (Gremião et al. 2015).

Notably, outbreaks caused by the classical transmission route, in which *S. schenckii* and *S. globosa* are the typical etiological agents, require the removal of the fungus sources in nature such as decaying plant debris (Rodrigues et al. 2016b). The alternative transmission route, which involves epizootics caused by *S. brasiliensis* during horizontal animal transmission as well as during zoonotic transmission, requires different strategies for epidemic containment. Prevention measures, appropriate treatment, and castration of cats are extremely important in the management of sick animals and education of pet owners on the main aspects of fungal transmission, especially in hyperendemic areas. Infected dead animals should be cremated but never buried, thus preventing new *S. brasiliensis* foci in the soil and the perpetuation of the pathogen in nature.

References

Adamson HG (1911) Case of sporotrichosis. Proc R Soc Med 4(Dermatol Sect):113–121

Ajello L, Kaplan W (1969) A new variant of *Sporothrix schenckii*. Mycoses 12(11):633–644. https://doi.org/10.1111/j.1439-0507.1969.tb03423.x

Alba-Fierro CA, Pérez-Torres A, Toriello C, Pulido-Camarillo E, López-Romero E, Romo-Lozano-Y, Gutiérrez-Sánchez G, Ruiz-Baca E (2016) Immune response induced by an immuno-dominant 60 kDa glycoprotein of the cell wall of *Sporothrix schenckii* in two mice strains with experimental sporotrichosis. J Immunol Res 2016:15. https://doi.org/10.1155/2016/6525831

Almeida-Paes R, Pimenta MA, Monteiro PC, Nosanchuk JD, Zancope-Oliveira RM (2007a) Immunoglobulins G, M, and A against *Sporothrix schenckii* exoantigens in patients with sporotrichosis before and during treatment with itraconazole. Clin Vaccine Immunol 14(9): 1149–1157. https://doi.org/10.1128/cvi.00149-07

Almeida-Paes R, Pimenta MA, Pizzini CV, Monteiro PC, Peralta JM, Nosanchuk JD, Zancope-Oliveira RM (2007b) Use of mycelial-phase *Sporothrix schenckii* exoantigens in an enzyme-linked immunosorbent assay for diagnosis of sporotrichosis by antibody detection. Clin Vaccine Immunol 14(3):244–249. https://doi.org/10.1128/cvi.00430-06

Almeida-Paes R, Bailao AM, Pizzini CV, Reis RS, Soares CM, Peralta JM, Gutierrez-Galhardo MC, Zancope-Oliveira RM (2012) Cell-free antigens of *Sporothrix brasiliensis*: antigenic diversity and application in an immunoblot assay. Mycoses 55(6):467–475. https://doi.org/10.1111/j.1439-0507.2012.02175.x

Almeida-Paes R, de Oliveira MM, Freitas DF, do Valle AC, Zancope-Oliveira RM, Gutierrez-Galhardo MC (2014) Sporotrichosis in Rio de Janeiro, Brazil: *Sporothrix brasiliensis* is associated with atypical clinical presentations. PLoS Negl Trop Dis 8(9):e3094. https://doi.org/10.1371/journal.pntd.0003094

Almeida-Paes R, de Oliveira LC, Oliveira MME, Gutierrez-Galhardo MC, Nosanchuk JD, Zancopé Oliveira RM (2015) Phenotypic characteristics associated with virulence of clinical isolates from the *Sporothrix* complex. Biomed Res Int 2015:1–10

Al-Tawfiq JA, Wools KK (1998) Disseminated sporotrichosis and *Sporothrix schenckii* fungemia as the initial presentation of human immunodeficiency virus infection. Clin Infect Dis 26(6): 1403–1406. https://doi.org/10.1086/516356

Alves SH, Boettcher CS, Oliveira DC, Tronco-Alves GR, Sgaria MA, Thadeu P, Oliveira LT, Santurio JM (2010) *Sporothrix schenckii* associated with armadillo hunting in Southern Brazil: epidemiological and antifungal susceptibility profiles. Rev Soc Bras Med Trop 43(5):523–525

Arrillaga-Moncrieff I, Capilla J, Mayayo E, Marimon R, Mariné M, Gené J, Cano J, Guarro J (2009) Different virulence levels of the species of *Sporothrix* in a murine model. Clin Microbiol Infect 15(7):651–655. https://doi.org/10.1111/j.1469-0691.2009.02824.x

Aung AK, Teh BM, MCgrath C, Thompson PJ (2013) Pulmonary sporotrichosis: case series and systematic analysis of literature on clinico-radiological patterns and management outcomes. Med Mycol 51(5):534–544. https://doi.org/10.3109/13693786.2012.751643

Bachmeyer C, Buot G, Binet O, Beltzer-Garelly E, Avram A (2006) Fixed cutaneous sporotrichosis: an unusual diagnosis in West Europe. Clin Exp Dermatol 31(3):479–481

Barbee WC, Ewert A, Davidson EM (1977) Animal model of human disease: sporotrichosis. Am J Pathol 86(1):281–284

Barros MBL, Schubach AO, Schubach TMP, Wanke B, Lambert-Passos SR (2008) An epidemic of sporotrichosis in Rio de Janeiro, Brazil: epidemiological aspects of a series of cases. Epidemiol Infect 136(09):1192–1196. https://doi.org/10.1017/S0950268807009727

Barros MBL, Schubach TP, Coll JO, Gremião ID, Wanke B, Schubach A (2010) Sporotrichosis: development and challenges of an epidemic. Rev Panam Salud Publica 27(6):455–460. https://doi.org/10.1590/S1020-49892010000600007

Barros MB, de Almeida Paes R, Schubach AO (2011) *Sporothrix schenckii* and sporotrichosis. Clin Microbiol Rev 24(4):633–654. https://doi.org/10.1128/cmr.00007-11

Bernardes-Engemann AR, Orofino Costa RC, Miguens BP, Penha CVL, Neves E, Pereira BAS, Dias CMP, Mattos M, Gutierrez MC, Schubach A, Oliveira Neto MP, Lazéra M, Lopes-Bezerra LM (2005) Development of an enzyme-linked immunosorbent assay for the serodiagnosis of several clinical forms of sporotrichosis. Med Mycol 43(6):487–493. https://doi.org/10.1080/13693780400019909

Bernardes-Engemann AR, Loureiro y Penha CV, Benvenuto F, Braga JU, Barros ML, Orofino-Costa R, Lopes-Bezerra LM (2009) A comparative serological study of the SsCBF antigenic fraction isolated from three *Sporothrix schenckii* strains. Med Mycol 47(8):874–878. https://doi.org/10.3109/13693780802695520

Beurmann L, Gougerot H (1912) Les Sporotrichose. Librairie Felix Alcan, Paris

Beurmann L, Ramond L (1903) Abcès sous-cutanés multiples d'origine mycosique. Ann Dermatol Syph 4:678–685

Blumer SO, Kaufman L, Kaplan W, McLaughlin DW, Kraft DE (1973) Comparative evaluation of five serological methods for the diagnosis of sporotrichosis. Appl Microbiol 26(1):4–8

Bolao F, Podzamczer D, Ventin M, Gudiol F (1994) Efficacy of acute phase and maintenance therapy with itraconazole in an AIDS patient with sporotrichosis. Eur J Clin Microbiol Infect Dis 13(7):609–612

Bommer M, Hütter M-L, Stilgenbauer S, de Hoog GS, de Beer ZW, Wellinghausen N (2009) Fatal *Ophiostoma piceae* infection in a patient with acute lymphoblastic leukaemia. J Med Microbiol 58(3):381–385. https://doi.org/10.1099/jmm.0.005280-0

Bonifaz A, Vázquez-González D (2013) Diagnosis and treatment of lymphocutaneous sporotrichosis: what are the options? Curr Fungal Infect Rep 7(3):252–259. https://doi.org/10.1007/s12281-013-0140-3

Borba-Santos LP, Rodrigues AM, Gagini TB, Fernandes GF, Castro R, de Camargo ZP, Nucci M, Lopes-Bezerra LM, Ishida K, Rozental S (2015) Susceptibility of *Sporothrix brasiliensis* isolates to amphotericin B, azoles, and terbinafine. Med Mycol 53(2):178–188. https://doi.org/10.1093/mmy/myu056

Brilhante RS, Rodrigues AM, Sidrim JJ, Rocha MF, Pereira SA, Gremião ID, Schubach TM, de Camargo ZP (2016) In vitro susceptibility of antifungal drugs against *Sporothrix brasiliensis* recovered from cats with sporotrichosis in Brazil. Med Mycol 54(3):275–279. https://doi.org/10.1093/mmy/myv039

Camacho E, León-Navarro I, Rodríguez-Brito S, Mendoza M, Niño-Vega GA (2015) Molecular epidemiology of human sporotrichosis in Venezuela reveals high frequency of *Sporothrix globosa*. BMC Infect Dis 15(1):94. https://doi.org/10.1186/s12879-015-0839-6

Carlos IZ, Sgarbi DB, Placeres MC (1998) Host organism defense by a peptide-polysaccharide extracted from the fungus *Sporothrix schenckii*. Mycopathologia 144(1):9–14

Carmichael JW (1962) Chrysosporium and some other aleuriosporic hyphomycetes. Can J Bot 40(8):1137–1173. https://doi.org/10.1139/b62-104

Castrejon OV, Robles M, Zubieta Arroyo OE (1995) Fatal fungaemia due to *Sporothrix schenckii*. Mycoses 38(9–10):373–376

Castro RA, Kubitschek-Barreira PH, Teixeira PAC, Sanches GF, Teixeira MM, Quintella LP, Almeida SR, Costa RO, Camargo ZP, Felipe MSS, de Souza W, Lopes-Bezerra LM (2013) Differences in cell morphometry, cell wall topography and Gp70 expression correlate with the virulence of *Sporothrix brasiliensis* clinical isolates. PLoS One 8(10):e75656. https://doi.org/10.1371/journal.pone.0075656

Chakrabarti A, Bonifaz A, Gutierrez-Galhardo MC, Mochizuki T, Li S (2015) Global epidemiology of sporotrichosis. Med Mycol 53(1):3–14. https://doi.org/10.1093/mmy/myu062

Chaves AR, de Campos MP, Barros MBL, do Carmo CN, Gremião IDF, Pereira SA, Schubach TMP (2013) Treatment abandonment in feline sporotrichosis – study of 147 cases. Zoonoses Public Health 60(2):149–153. https://doi.org/10.1111/j.1863-2378.2012.01506.x

Clavijo-Giraldo DM, Matinez-Alvarez JA, Lopes-Bezerra LM, Ponce-Noyola P, Franco B, Almeida RS, Mora-Montes HM (2016) Analysis of *Sporothrix schenckii sensu stricto* and *Sporothrix brasiliensis* virulence in *Galleria mellonella*. J Microbiol Methods 122:73–77. https://doi.org/10.1016/j.mimet.2016.01.014

Coles FB, Schuchat A, Hibbs JR, Kondracki SF, Salkin IF, Dixon DM, Chang HG, Duncan RA, Hurd NJ, Morse DL (1992) A multistate outbreak of sporotrichosis associated with *Sphagnum* moss. Am J Epidemiol 136(4):475–487

Conti Diaz IA (1989) Epidemiology of sporotrichosis in Latin America. Mycopathologia 108(2): 113–116

Costa RO, Macedo PM, Carvalhal A, Bernardes-Engemann AR (2013) Use of potassium iodide in dermatology: updates on an old drug. An Bras Dermatol 88(3):396–402. https://doi.org/10.1590/abd1806-4841.20132377

Criseo G, Romeo O (2010) Ribosomal DNA sequencing and phylogenetic analysis of environmental *Sporothrix schenckii* strains: comparison with clinical isolates. Mycopathologia 169(5): 351–358. https://doi.org/10.1007/s11046-010-9274-9

Crothers SL, White SD, Ihrke PJ, Affolter VK (2009) Sporotrichosis: a retrospective evaluation of 23 cases seen in northern California (1987–2007). Vet Dermatol 20(4):249–259. https://doi.org/10.1111/j.1365-3164.2009.00763.x

da Rosa WD, Gezuele E, Calegari L, Goni F (2008) Asteroid body in sporotrichosis. Yeast viability and biological significance within the host immune response. Med Mycol 46(5):443–448. https://doi.org/10.1080/13693780801914898

Davis DJ (1921) The identity of American and French sporotrichosis. U Wis Stud Sci 2:104–130

Davis BA (1996) Sporotrichosis. Dermatol Clin 14(1):69–76

Della Terra PP, Rodrigues AM, Fernandes GF, Nishikaku AS, Burger E, de Camargo ZP (2017) Exploring virulence and immunogenicity in the emerging pathogen *Sporothrix brasiliensis*. PLoS Negl Trop Dis 11(8):e0005903. https://doi.org/10.1371/journal.pntd.0005903

de Araujo ML, Rodrigues AM, Fernandes GF, de Camargo ZP, de Hoog GS (2015) Human sporotrichosis beyond the epidemic front reveals classical transmission types in Espírito Santo, Brazil. Mycoses 58(8):485–490. https://doi.org/10.1111/myc.12346

de Beer ZW, Harrington TC, Vismer HF, Wingfield BD, Wingfield MJ (2003) Phylogeny of the *Ophiostoma stenoceras–Sporothrix schenckii* complex. Mycologia 95:434–441

de Lima Barros MB, Schubach TM, Galhardo MC, de Oliviera Schubach A, Monteiro PC, Reis RS, Zancope-Oliveira RM, dos Santos Lazera M, Cuzzi-Maya T, Blanco TC, Marzochi KB,

Wanke B, do Valle AC (2001) Sporotrichosis: an emergent zoonosis in Rio de Janeiro. Mem Inst Oswaldo Cruz 96(6):777–779
de Lima Barros MB, de Oliveira Schubach A, Galhardo MC, Schubach TM, dos Reis RS, Conceicao MJ, do Valle AC (2003) Sporotrichosis with widespread cutaneous lesions: report of 24 cases related to transmission by domestic cats in Rio de Janeiro, Brazil. Int J Dermatol 42(9):677–681. https://doi.org/10.1046/j.1365-4362.2003.01813.x
de Macedo PM, Lopes-Bezerra LM, Bernardes-Engemann AR, Orofino-Costa R (2015a) New posology of potassium iodide for the treatment of cutaneous sporotrichosis: study of efficacy and safety in 102 patients. J Eur Acad Dermatol Venereol 29(4):719–724. https://doi.org/10.1111/jdv.12667
de Macedo PM, Sztajnbok DC, Camargo ZP, Rodrigues AM, Lopes-Bezerra LM, Bernardes-Engemann AR, Orofino-Costa R (2015b) Dacryocystitis due to *Sporothrix brasiliensis*: a case report of a successful clinical and serological outcome with low-dose potassium iodide treatment and oculoplastic surgery. Br J Dermatol 172(4):1116–1119. https://doi.org/10.1111/bjd.13378
de Meyer EM, de Beer ZW, Summerbell RC, Moharram AM, de Hoog GS, Vismer HF, Wingfield MJ (2008) Taxonomy and phylogeny of new wood- and soil-inhabiting *Sporothrix* species in the *Ophiostoma stenoceras-Sporothrix schenckii* complex. Mycologia 100(4):647–661. https://doi.org/10.3852/07-157r
de Oliveira MME, Sampaio P, Almeida-Paes R, Pais C, Gutierrez-Galhardo MC, Zancope-Oliveira RM (2012) Rapid identification of *Sporothrix* species by T3B fingerprinting. J Clin Microbiol 50(6):2159–2162. https://doi.org/10.1128/JCM.00450-12
Dixon DM, Salkin IF, Duncan RA, Hurd NJ, Haines JH, Kemna ME, Coles FB (1991) Isolation and characterization of *Sporothrix schenckii* from clinical and environmental sources associated with the largest U.S. epidemic of sporotrichosis. J Clin Microbiol 29(6):1106–1113
Donabedian H, O'Donnell E, Olszewski C, MacArthur RD, Budd N (1994) Disseminated cutaneous and meningeal sporotrichosis in an AIDS patient. Diagn Microbiol Infect Dis 18(2): 111–115
Dunstan RW, Reimann KA, Langham RF (1986) Feline sporotrichosis. J Am Vet Med Assoc 189(8):880–883
Espinel-Ingroff A, Abreu DPB, Almeida-Paes R, Brilhante RSN, Chakrabarti A, Chowdhary A, Hagen F, Cordoba S, Gonzalez GM, Govender NP, Guarro J, Johnson EM, Kidd SE, Pereira SA, Rodrigues AM, Rozental S, Szeszs MW, Balleste Alaniz R, Bonifaz A, Bonfietti LX, Borba-Santos LP, Capilla J, Colombo AL, Dolande M, Isla MG, Melhem MSC, Mesa-Arango AC, Oliveira MME, Panizo MM, Pires de Camargo Z, Zancope-Oliveira RM, Meis JF, Turnidge J (2017) Multicenter and international study of MIC/MEC distributions for definition of epidemiological cutoff values (ECVs) for species of *Sporothrix* identified by molecular methods. Antimicrob Agents Chemother 61(10):e01057–e01017. https://doi.org/10.1128/aac.01057-17
Feeney KT, Arthur IH, Whittle AJ, Altman SA, Speers DJ (2007) Outbreak of sporotrichosis, Western Australia. Emerg Infect Dis 13(8):1228–1231. https://doi.org/10.3201/eid1308.061462
Fernandes KSS, Coelho ALJ, Bezerra LML, Barja-Fidalgo C (2000) Virulence of *Sporothrix schenckii* conidia and yeast cells, and their susceptibility to nitric oxide. Immunology 101(4): 563–569. https://doi.org/10.1046/j.1365-2567.2000.00125.x
Fernandes GF, Amaral CCD, Sasaki A, Godoy PM, De Camargo ZP (2009a) Heterogeneity of proteins expressed by Brazilian *Sporothrix schenckii* isolates. Med Mycol 47(8):855–861. https://doi.org/10.3109/13693780802713216
Fernandes GF, dos Santos PO, Amaral CC, Sasaki AA, Godoy-Martinez P, Camargo ZP (2009b) Characteristics of 151 Brazilian *Sporothrix schenckii* isolates from 5 different geographic regions of Brazil: a forgotten and re-emergent pathogen. Open Mycol J 3(1):48–58. https://doi.org/10.2174/1874437000903010048
Fernandes GF, Lopes-Bezerra LM, Bernardes-Engemann AR, Schubach TM, Dias MA, Pereira SA, de Camargo ZP (2011) Serodiagnosis of sporotrichosis infection in cats by enzyme-linked

immunosorbent assay using a specific antigen, SsCBF, and crude exoantigens. Vet Microbiol 147(3–4):445–449. https://doi.org/10.1016/j.vetmic.2010.07.007

Fernandes GF, dos Santos PO, Rodrigues AM, Sasaki AA, Burger E, de Camargo ZP (2013) Characterization of virulence profile, protein secretion and immunogenicity of different *Sporothrix schenckii sensu stricto* isolates compared with *S. globosa* and *S. brasiliensis* species. Virulence 4(3):241–249. https://doi.org/10.4161/viru.23112

Fernandez-Silva F, Capilla J, Mayayo E, Guarro J (2012) Virulence of *Sporothrix luriei* in a murine model of disseminated infection. Mycopathologia 173(4):245–249. https://doi.org/10.1007/s11046-011-9506-7

Findlay GH, Vismer HF, Dreyer L (1984) Studies on sporotrichosis. Pathogenicity and morphogenesis in the Transvaal strains of *Sporothrix schenckii*. Mycopathologia 87(1–2):85–93

Fischman Gompertz O, Rodrigues AM, Fernandes GF, Bentubo HD, Pires de Camargo Z, Petri V (2016) Atypical clinical presentation of sporotrichosis caused by *Sporothrix globosa* resistant to itraconazole. Am J Trop Med Hyg 94(6):1218–1222. https://doi.org/10.4269/ajtmh.15-0267

Fisher MC, Henk DA, Briggs CJ, Brownstein JS, Madoff LC, McCraw SL, Gurr SJ (2012) Emerging fungal threats to animal, plant and ecosystem health. Nature 484(7393):186–194. https://doi.org/10.1038/nature10947

Freitas DF, Santos SS, Almeida-Paes R, de Oliveira MM, do Valle AC, Gutierrez-Galhardo MC, Zancope-Oliveira RM, Nosanchuk JD (2015) Increase in virulence of *Sporothrix brasiliensis* over five years in a patient with chronic disseminated sporotrichosis. Virulence 6(2):112–120. https://doi.org/10.1080/21505594.2015.1014274

Fukushiro R (1984) Epidemiology and ecology of sporotrichosis in Japan. Zentralbl Bakteriol Mikrobiol Hyg A 257(2):228–233

Ghosh A, Maity PK, Hemashettar BM, Sharma VK, Chakrabarti A (2002) Physiological characters of *Sporothrix schenckii* isolates. Mycoses 45(11–12):449–554

Govender NP, Maphanga TG, Zulu TG, Patel J, Walaza S, Jacobs C, Ebonwu JI, Ntuli S, Naicker SD, Thomas J (2015) An outbreak of lymphocutaneous sporotrichosis among mine-workers in South Africa. PLoS Negl Trop Dis 9(9):e0004096. https://doi.org/10.1371/journal.pntd.0004096

Gremião I, Schubach T, Pereira S, Rodrigues A, Honse C, Barros M (2011) Treatment of refractory feline sporotrichosis with a combination of intralesional amphotericin B and oral itraconazole. Aust Vet J 89(9):346–351. https://doi.org/10.1111/j.1751-0813.2011.00804.x

Gremião ID, Menezes RC, Schubach TM, Figueiredo AB, Cavalcanti MC, Pereira SA (2015) Feline sporotrichosis: epidemiological and clinical aspects. Med Mycol 53(1):15–21. https://doi.org/10.1093/mmy/myu061

Gremião ID, Miranda LH, Reis EG, Rodrigues AM, Pereira SA (2017) Zoonotic epidemic of sporotrichosis: cat to human transmission. PLoS Pathog 13(1):e1006077. https://doi.org/10.1371/journal.ppat.1006077

Hektoen L, Perkins CF (1900) Refractory subcutaneous abscesses caused by *Sporothrix schenckii*: a new pathogenic fungus. J Exp Med 5(1):77–89

Helm M, Berman C (1947) The clinical, therapeutic and epidemiological features of the sporotrichosis infection on the mines. In: Proceedings of the Transvaal Mine Medical Officers' Association. Sporotrichosis infection on mines of the Witwatersrand. Johannesburg, South Africa, pp 59–67

Hiruma M, Katoh T, Yamamoto I, Kagawa S (1987) Local hyperthermia in the treatment of sporotrichosis. Mycoses 30(7):315–321. https://doi.org/10.1111/j.1439-0507.1987.tb04396.x

Honse CO, Rodrigues AM, Gremião ID, Pereira SA, Schubach TM (2010) Use of local hyperthermia to treat sporotrichosis in a cat. Vet Rec 166(7):208–209. https://doi.org/10.1136/vr.b4768

Howard DH (1961) Dimorphism of *Sporotrichum schenckii*. J Bacteriol 81:464–469

Howard DH, Orr GF (1963) Comparison of strains of *Sporotrichum schenckii* isolated from nature. J Bacteriol 85:816–821

Hussein MR (2008) Mucocutaneous Splendore-Hoeppli phenomenon. J Cutan Pathol 35(11):979–988. https://doi.org/10.1111/j.1600-0560.2008.01045.x

Kano R, Okubo M, Siew HH, Kamata H, Hasegawa A (2015) Molecular typing of *Sporothrix schenckii* isolates from cats in Malaysia. Mycoses 58(4):220–224. https://doi.org/10.1111/myc.12302
Kauffman CA (1999) Sporotrichosis. Clin Infect Dis 29(2):231–237. https://doi.org/10.1086/520190
Kauffman CA, Hajjeh R, Chapman SW, Group MS (2000) Practice guidelines for the management of patients with sporotrichosis. Clin Infect Dis 30(4):684–687. https://doi.org/10.1086/313751
Kauffman CA, Bustamante B, Chapman SW, Pappas PG (2007) Clinical practice guidelines for the management of sporotrichosis: 2007 update by the Infectious Diseases Society of America. Clin Infect Dis 45(10):1255–1265. https://doi.org/10.1086/522765
Kenyon EM, Russell LH, McMurray DN (1984) Isolation of *Sporothrix schenckii* from potting soil. Mycopathologia 87(1–2):128
Koeter S, Jackson RW (2006) Successful total knee arthroplasty in the presence of sporotrichal arthritis. Knee 13(3):236–237. https://doi.org/10.1016/j.knee.2006.02.004
Lacaz CS, Porto E, Martins JEC, Heins-Vaccari EM, de Melo NT (2002) Tratado de Micologia Médica, 9th edn. Sarvier, São Paulo
Larsson CE, Goncalves Mde A, Araujo VC, Dagli ML, Correa B, Fava Neto C (1989) Feline sporotrichosis: clinical and zoonotic aspects. Rev Inst Med Trop Sao Paulo 31(5):351–358
Lima OC, Bezerra LML (1997) Identification of a concanavalin A-binding antigen of the cell surface of *Sporothrix schenckii*. J Med Vet Mycol 35(3):167–172. https://doi.org/10.1080/02681219780001101
Lutz A, Splendore A (1907) On a mycosis observed in men and mice: contribution to the knowledge of the so-called sporotrichosis. Revista Médica de São Paulo 21:443–450
Mackinnon JE, Conti-Díaz IA (1963) The effect of temperature on sporotrichosis. Sabouraudia 2(2):56–59. https://doi.org/10.1080/00362176385190141
Mackinnon JE, Conti-Díaz IA, Gezuele E, Civila E, Da Luz S (1969) Isolation of *Sporothrix schenckii* from nature and considerations on its pathogenicity and ecology. Med Mycol 7(1):38–45. https://doi.org/10.1080/00362177085190071
Madrid IM, Mattei A, Martins A, Nobre M, Meireles M (2010) Feline sporotrichosis in the southern region of Rio Grande do Sul, Brazil: clinical, zoonotic and therapeutic aspects. Zoonoses Public Health 57(2):151–154. https://doi.org/10.1111/j.1863-2378.2008.01227.x
Madrid IM, Mattei AS, Fernandes CG, Oliveira Nobre M, Meireles MCA (2012) Epidemiological findings and laboratory evaluation of sporotrichosis: a description of 103 cases in cats and dogs in Southern Brazil. Mycopathologia 173(4):265–273. https://doi.org/10.1007/s11046-011-9509-4
Maia DC, Sassa MF, Placeres MC, Carlos IZ (2006) Influence of Th1/Th2 cytokines and nitric oxide in murine systemic infection induced by *Sporothrix schenckii*. Mycopathologia 161(1):11–19. https://doi.org/10.1007/s11046-005-0142-y
Marimon R, Gené J, Cano J, Trilles L, Dos Santos Lazéra M, Guarro J (2006) Molecular phylogeny of *Sporothrix schenckii*. J Clin Microbiol 44(9):3251–3256. https://doi.org/10.1128/JCM.00081 06
Marimon R, Cano J, Gené J, Sutton DA, Kawasaki M, Guarro J (2007) *Sporothrix brasiliensis*, *S. globosa*, and *S. mexicana*, three new *Sporothrix* species of clinical interest. J Clin Microbiol 45(10):3198–3206. https://doi.org/10.1128/JCM.00808-07
Marimon R, Gené J, Cano J, Guarro J (2008a) *Sporothrix luriei*: a rare fungus from clinical origin. Med Mycol 46(6):621–625. https://doi.org/10.1080/13693780801992837
Marimon R, Serena C, Gené J, Cano J, Guarro J (2008b) In vitro antifungal susceptibilities of five species of *Sporothrix*. Antimicrob Agents Chemother 52(2):732–734. https://doi.org/10.1128/AAC.01012-07
Mayorga R, Caceres A, Toriello C, Gutierrez G, Alvarez O, Ramirez ME, Mariat F (1978) An endemic area of sporotrichosis in Guatemala. Sabouraudia 16(3):185–198
Mendoza M, Diaz AM, Hung MB, Zambrano EA, Diaz E, De Albornoz MC (2002) Production of culture filtrates of *Sporothrix schenckii* in diverse culture media. Med Mycol 40(5):447–454

Montenegro H, Rodrigues AM, Galvão Dias MA, da Silva EA, Bernardi F, Camargo ZP (2014) Feline sporotrichosis due to *Sporothrix brasiliensis*: an emerging animal infection in São Paulo, Brazil. BMC Vet Res 10(1):269. https://doi.org/10.1186/s12917-014-0269-5

Morris-Jones R (2002) Sporotrichosis. Clin Exp Dermatol 27(6):427–431

Morrison AS, Lockhart SR, Bromley JG, Kim JY, Burd EM (2013) An environmental *Sporothrix* as a cause of corneal ulcer. Med Mycol Case Rep 2(0):88–90. https://doi.org/10.1016/j.mmcr.2013.03.002

Nascimento RC, Almeida SR (2005) Humoral immune response against soluble and fractionate antigens in experimental sporotrichosis. FEMS Immunol Med Microbiol 43(2):241–247. https://doi.org/10.1016/j.femsim.2004.08.004

Nascimento RC, Espíndola NM, Castro RA, Teixeira PAC, Loureiro y Penha CV, Lopes-Bezerra LM, Almeida SR (2008) Passive immunization with monoclonal antibody against a 70-kDa putative adhesin of *Sporothrix schenckii* induces protection in murine sporotrichosis. Eur J Immunol 38(11):3080–3089. https://doi.org/10.1002/eji.200838513

Neyra E, Fonteyne P-A, Swinne D, Fauche F, Bustamante B, Nolard N (2005) Epidemiology of human sporotrichosis investigated by amplified fragment length polymorphism. J Clin Microbiol 43(3):1348–1352. https://doi.org/10.1128/jcm.43.3.1348-1352.2005

Oliveira MM, Almeida-Paes R, Muniz MM, Gutierrez-Galhardo MC, Zancope-Oliveira RM (2011) Phenotypic and molecular identification of *Sporothrix* isolates from an epidemic area of sporotrichosis in Brazil. Mycopathologia 172(4):257–267. https://doi.org/10.1007/s11046-011-9437-3

Oliveira MM, Santos C, Sampaio P, Romeo O, Almeida-Paes R, Pais C, Lima N, Zancope-Oliveira RM (2015) Development and optimization of a new MALDI-TOF protocol for identification of the *Sporothrix* species complex. Res Microbiol 166(2):102–110. https://doi.org/10.1016/j.resmic.2014.12.008

Orofino-Costa R, Boia MN, Magalhaes GA, Damasco PS, Bernardes-Engemann AR, Benvenuto F, Silva IC, Lopes-Bezerra LM (2010) Arthritis as a hypersensitivity reaction in a case of sporotrichosis transmitted by a sick cat: clinical and serological follow up of 13 months. Mycoses 53(1):81–83. https://doi.org/10.1111/j.1439-0507.2008.01661.x

Orofino-Costa R, de Macedo PM, Bernardes-Engemann AR (2015) Hyperendemia of sporotrichosis in the Brazilian Southeast: learning from clinics and therapeutics. Curr Fungal Infect Rep 9(4):220–228. https://doi.org/10.1007/s12281-015-0235-0

Orofino-Costa R, Macedo PM, Rodrigues AM, Bernardes-Engemann AR (2017) Sporotrichosis: an update on epidemiology, etiopathogenesis, laboratory and clinical therapeutics. An Bras Dermatol 92(5):606–620. https://doi.org/10.1590/abd1806-4841.2017279

Paixao AG, Galhardo MC, Almeida-Paes R, Nunes EP, Goncalves ML, Chequer GL, Lamas Cda C (2015) The difficult management of disseminated *Sporothrix brasiliensis* in a patient with advanced AIDS. AIDS Res Ther 12:16. https://doi.org/10.1186/s12981-015-0051-1

Pappas PG (2003) Sporotrichosis. Clinical mycology. Oxford University Press, New York

Pappas PG, Tellez I, Deep AE, Nolasco D, Holgado W, Bustamante B (2000) Sporotrichosis in Peru: description of an area of hyperendemicity. Clin Infect Dis 30(1):65–70. https://doi.org/10.1086/313607

Pereira SA, Passos SR, Silva JN, Gremião ID, Figueiredo FB, Teixeira JL, Monteiro PC, Schubach TM (2010) Response to azolic antifungal agents for treating feline sporotrichosis. Vet Rec 166(10):290–294. https://doi.org/10.1136/vr.b4752

Pereira SA, Menezes RC, Gremião IDF, Silva JN, Honse Cde O, Figueiredo FB, da Silva DT, Kitada AAB, dos Reis ÉG, Schubach TMP (2011) Sensitivity of cytopathological examination in the diagnosis of feline sporotrichosis. J Feline Med Surg 13(4):220–223. https://doi.org/10.1016/j.jfms.2010.10.007

Pijper A, Pullinger BD (1927) An outbreak of sporotrichosis among South African native miners. Lancet 210(5435):914–916. https://doi.org/10.1016/S0140-6736(01)35176-0

Powell KE, Taylor A, Phillips BJ, Blakey DL, Campbell GD, Kaufman L, Kaplan W (1978) Cutaneous sporotrichosis in forestry workers. Epidemic due to contaminated *Sphagnum* moss. JAMA 240(3):232–235

Read SI, Sperling LC (1982) Feline sporotrichosis: transmission to man. Arch Dermatol 118(6): 429–431

Reis EG, Gremião ID, Kitada AA, Rocha RF, Castro VS, Barros MB, Menezes RC, Pereira SA, Schubach TM (2012) Potassium iodide capsule treatment of feline sporotrichosis. J Feline Med Surg 14(6):399–404. https://doi.org/10.1177/1098612x12441317

Reis EG, Schubach TM, Pereira SA, Silva JN, Carvalho BW, Quintana MS, Gremião ID (2016) Association of itraconazole and potassium iodide in the treatment of feline sporotrichosis: a prospective study. Med Mycol 54(7):684–690. https://doi.org/10.1093/mmy/myw027

Rex JH, Bennett JE (1990) Administration of potassium iodide to normal volunteers does not increase killing of *Sporothrix schenckii* by their neutrophils or monocytes. J Med Vet Mycol 28(3):185–189

Rippon JW (1988) Medical mycology–the pathogenic fungi and the pathogenic actinomycetes, 3rd edn. W. B. Saunders Company, Philadelphia

Rodrigues AM, de Hoog S, de Camargo ZP (2013a) Emergence of pathogenicity in the *Sporothrix schenckii* complex. Med Mycol 51(4):405–412. https://doi.org/10.3109/13693786.2012.719648

Rodrigues AM, de Melo Teixeira M, de Hoog GS, Schubach TMP, Pereira SA, Fernandes GF, Bezerra LML, Felipe MS, de Camargo ZP (2013b) Phylogenetic analysis reveals a high prevalence of *Sporothrix brasiliensis* in feline sporotrichosis outbreaks. PLoS Negl Trop Dis 7(6): e2281. https://doi.org/10.1371/journal.pntd.0002281

Rodrigues AM, Bagagli E, de Camargo ZP, Bosco SMG (2014a) *Sporothrix schenckii sensu stricto* isolated from soil in an armadillo's burrow. Mycopathologia 177:199–206. https://doi.org/10.1007/s11046-014-9734-8

Rodrigues AM, de Hoog GS, Camargo ZP (2014b) Genotyping species of the *Sporothrix schenckii* complex by PCR-RFLP of calmodulin. Diagn Microbiol Infect Dis 78(4):383–387. https://doi.org/10.1016/j.diagmicrobio.2014.01.004

Rodrigues AM, de Hoog GS, de Cassia Pires D, Brihante RSN, da Costa Sidrim JJ, Gadelha MF, Colombo AL, de Camargo ZP (2014c) Genetic diversity and antifungal susceptibility profiles in causative agents of sporotrichosis. BMC Infect Dis 14(1):219. https://doi.org/10.1186/1471-2334-14-219

Rodrigues AM, de Hoog GS, Zhang Y, Camargo ZP (2014d) Emerging sporotrichosis is driven by clonal and recombinant *Sporothrix* species. Emerg Microbes Infect 3(5):e32. https://doi.org/10.1038/emi.2014.33

Rodrigues AM, de Hoog GS, de Camargo ZP (2015a) Molecular diagnosis of pathogenic *Sporothrix* species. PLoS Negl Trop Dis 9(12):e0004190. https://doi.org/10.1371/journal.pntd.0004190

Rodrigues AM, Fernandes GF, Araujo LM, Della Terra PP, Dos Santos PO, Pereira SA, Schubach TM, Burger E, Lopes-Bezerra LM, de Camargo ZP (2015b) Proteomics based characterization of the humoral immune response in sporotrichosis: toward discovery of potential diagnostic and vaccine antigens. PLoS Negl Trop Dis 9(8):e0004016. https://doi.org/10.1371/journal.pntd.0004016

Rodrigues AM, Kubitschek-Barreira PH, Fernandes GF, de Almeida SR, Lopes-Bezerra LM, de Camargo ZP (2015c) Immunoproteomic analysis reveals a convergent humoral response signature in the *Sporothrix schenckii* complex. J Proteome 115:8–22. https://doi.org/10.1016/j.jprot.2014.11.013

Rodrigues AM, Kubitschek-Barreira PH, Fernandes GF, de Almeida SR, Lopes-Bezerra LM, de Camargo ZP (2015d) Two-dimensional gel electrophoresis data for proteomic profiling of *Sporothrix* yeast cells. Data Brief 2:32–38. https://doi.org/10.1016/j.dib.2014.11.004

Rodrigues AM, Najafzadeh MJ, de Hoog GS, de Camargo ZP (2015e) Rapid identification of emerging human-pathogenic *Sporothrix* species with rolling circle amplification. Front Microbiol 6:1385. https://doi.org/10.3389/fmicb.2015.01385

Rodrigues AM, Cruz Choappa R, Fernandes GF, De Hoog GS, Camargo ZP (2016a) *Sporothrix chilensis* sp. nov. (Ascomycota: Ophiostomatales), a soil-borne agent of human sporotrichosis with mild-pathogenic potential to mammals. Fungal Biol 120(2):246–264. https://doi.org/10.1016/j.funbio.2015.05.006

Rodrigues AM, de Hoog GS, Camargo ZP (2016b) *Sporothrix* species causing outbreaks in animals and humans driven by animal-animal transmission. PLoS Pathog 12(7):e1005638. https://doi.org/10.1371/journal.ppat.1005638

Rodrigues AM, Fernandes GF, de Camargo ZP (2016c) Sporotrichosis. In: Bayry J (ed) Emerging and re-emerging infectious diseases of livestock. Springer, Berlin

Rodriguez-Del Valle N, Rosario M, Torres-Blasini G (1983) Effects of pH, temperature, aeration and carbon source on the development of the mycelial or yeast forms of *Sporothrix schenckii* from conidia. Mycopathologia 82(2):83–88

Roets F, de Beer ZW, Wingfield MJ, Crous PW, Dreyer LL (2008) *Ophiostoma gemellus* and *Sporothrix variecibatus* from mites infesting *Protea* infructescences in South Africa. Mycologia 100(3):496–510

Ruiz-Baca E, Toriello C, Perez-Torres A, Sabanero-Lopez M, Villagomez-Castro JC, Lopez-Romero E (2009) Isolation and some properties of a glycoprotein of 70 kDa (Gp70) from the cell wall of *Sporothrix schenckii* involved in fungal adherence to dermal extracellular matrix. Med Mycol 47(2):185–196. https://doi.org/10.1080/13693780802165789

Sanchotene KO, Madrid IM, Klafke GB, Bergamashi M, Terra PPD, Rodrigues AM, de Camargo ZP, Xavier MO (2015) *Sporothrix brasiliensis* outbreaks and the rapid emergence of feline sporotrichosis. Mycoses 58(11):652–658. https://doi.org/10.1111/myc.12414

Sasaki AA, Fernandes GF, Rodrigues AM, Lima FM, Marini MM, dos Feitosa SL, de Melo Teixeira M, Felipe MSS, da Silveira JF, de Camargo ZP (2014) Chromosomal polymorphism in the *Sporothrix schenckii* complex. PLoS One 9(1):e86819. https://doi.org/10.1371/journal.pone.0086819

Sassa MF, Saturi AE, Souza LF, Ribeiro LC, Sgarbi DB, Carlos IZ (2009) Response of macrophage toll-like receptor 4 to a *Sporothrix schenckii* lipid extract during experimental sporotrichosis. Immunology 128(2):301–309. https://doi.org/10.1111/j.1365-2567.2009.03118.x

Schenck BR (1898) On refractory subcutaneous abscesses caused by a fungus possibly related to the *Sporotricha*. Bull Johns Hopkins Hosp 9:286–290

Scheufen S, Strommer S, Weisenborn J, Prenger-Berninghoff E, Thom N, Bauer N, Köhler K, Ewers C (2015) Clinical manifestation of an amelanotic *Sporothrix schenckii* complex isolate in a cat in Germany. JMM Case Rep 2(4). https://doi.org/10.1099/jmmcr.0.000039

Schubach TM, Valle AC, Gutierrez-Galhardo MC, Monteiro PC, Reis RS, Zancope-Oliveira RM, Marzochi KB, Schubach A (2001) Isolation of *Sporothrix schenckii* from the nails of domestic cats (*Felis catus*). Med Mycol 39(1):147–149

Schubach TM, de Oliveira Schubach A, dos Reis RS, Cuzzi-Maya T, Blanco TC, Monteiro DF, Barros BM, Brustein R, Zancope-Oliveira RM, Fialho Monteiro PC, Wanke B (2002) *Sporothrix schenckii* isolated from domestic cats with and without sporotrichosis in Rio de Janeiro, Brazil. Mycopathologia 153(2):83–86

Schubach TM, Schubach A, Okamoto T, Barros MB, Figueiredo FB, Cuzzi T, Fialho-Monteiro PC, Reis RS, Perez MA, Wanke B (2004) Evaluation of an epidemic of sporotrichosis in cats: 347 cases (1998–2001). J Am Vet Med Assoc 224(10):1623–1629. https://doi.org/10.2460/javma.2004.224.1623

Schubach A, Schubach TM, Barros MB, Wanke B (2005a) Cat-transmitted sporotrichosis, Rio de Janeiro, Brazil. Emerg Infect Dis 11(12):1952–1954. https://doi.org/10.3201/eid1112.040891

Schubach AO, Schubach TM, Barros MB (2005b) Epidemic cat-transmitted sporotrichosis. N Engl J Med 353(11):1185–1186. https://doi.org/10.1056/NEJMc051680

Scott EN, Muchmore HG (1989) Immunoblot analysis of antibody responses to *Sporothrix schenckii*. J Clin Microbiol 27(2):300–304

Silva-Vergara ML, de Camargo ZP, Silva PF, Abdalla MR, Sgarbieri RN, Rodrigues AM, dos Santos KC, Barata CH, Ferreira-Paim K (2012) Disseminated *Sporothrix brasiliensis* infection with endocardial and ocular involvement in an HIV-infected patient. Am J Trop Med Hyg 86(3):477–480. https://doi.org/10.4269/ajtmh.2012.11-0441

Singer JI, Muncie JE (1952) Sporotrichosis: etiologic considerations and report of additional cases from New York. N Y State J Med 52(17:1):2147–2153

Song Y, Li SS, Zhong SX, Liu YY, Yao L, Huo SS (2013) Report of 457 sporotrichosis cases from Jilin province, northeast China, a serious endemic region. J Eur Acad Dermatol Venereol 27(3): 313–318. https://doi.org/10.1111/j.1468-3083.2011.04389.x

Sterling JB, Heymann WR (2000) Potassium iodide in dermatology: a 19th century drug for the 21st century-uses, pharmacology, adverse effects, and contraindications. J Am Acad Dermatol 43(4): 691–697. https://doi.org/10.1067/mjd.2000.107247

Tachibana T, Matsuyama T, Mitsuyama M (1999) Involvement of CD4+ T cells and macrophages in acquired protection against infection with *Sporothrix schenckii* in mice. Med Mycol 37(6): 397–404

Taylor LH, Latham SM, woolhouse MEJ (2001) Risk factors for human disease emergence. Philos Trans R Soc Lond Ser B Biol Sci 356(1411):983–989. https://doi.org/10.1098/rstb.2001.0888

Teixeira PAC, de Castro RA, Nascimento RC, Tronchin G, Pérez Torres A, Lazéra M, de Almeida SR, Bouchara J-P, Loureiro y Penha CV, Lopes-Bezerra LM (2009) Cell surface expression of adhesins for fibronectin correlates with virulence in *Sporothrix schenckii*. Microbiology 155(11):3730–3738. https://doi.org/10.1099/mic.0.029439-0

Uenotsuchi T, Takeuchi S, Matsuda T, Urabe K, Koga T, Uchi H, Nakahara T, Fukagawa S, Kawasaki M, Kajiwara H, Yoshida S, Moroi Y, Furue M (2006) Differential induction of Th1-prone immunity by human dendritic cells activated with *Sporothrix schenckii* of cutaneous and visceral origins to determine their different virulence. Int Immunol 18(12):1637–1646. https://doi.org/10.1093/intimm/dxl097

Verma S, Verma GK, Singh G, Kanga A, Shanker V, Singh D, Gupta P, Mokta K, Sharma V (2012) Sporotrichosis in sub-himalayan India. PLoS Negl Trop Dis 6(6):e1673. https://doi.org/10.1371/journal.pntd.0001673

Vismer HF, Hull PR (1997) Prevalence, epidemiology and geographical distribution of *Sporothrix schenckii* infections in Gauteng, South Africa. Mycopathologia 137(3):137–143. https://doi.org/10.1023/A:1006830131173

Walker N, Ritchie J (1911) Remarks on a case of sporotrichosis. Br Med J 2:1–5

Welsh RD, Dolan CT (1973) *Sporothrix* whole yeast agglutination test – low-titer reactions of sera of subjects not known to have sporotrichosis. Am J Clin Pathol 59(1):82–85

White TJ, Bruns T, Lee S, Taylor J (1990) Amplification and direct sequencing of fungal ribosomal RNA genes for phylogenetics. In: Innis M, Gelfand D, Shinsky J, White T (eds) PCR protocols: a guide to methods and applications. Academic Press, New York, pp 315–322

Zhang YQ, Xu XG, Zhang M, Jiang P, Zhou XY, Li ZZ, Zhang MF (2011) Sporotrichosis: clinical and histopathological manifestations. Am J Dermatopathol 33(3):296–302. https://doi.org/10.1097/DAD.0b013e3181f5b622

Zhang Y, Hagen F, Stielow B, Rodrigues AM, Samerpitak K, Zhou X, Feng P, Yang L, Chen M, Deng S, Li S, Liao W, Li R, Li F, Meis JF, Guarro J, Teixeira M, Al-Zahrani HS, Camargo ZP, Zhang L, de Hoog GS (2015) Phylogeography and evolutionary patterns in *Sporothrix* spanning more than 14,000 human and animal case reports. Persoonia 35:1–20. https://doi.org/10.3767/003158515x687416

Zhou X, Rodrigues AM, Feng P, Hoog GS (2014) Global ITS diversity in the *Sporothrix schenckii* complex. Fungal Divers 66(1):153–165. https://doi.org/10.1007/s13225-013-0220-2

Lethargic Crab Disease: Now You See, Now You Don't

11

Vania A. Vicente, Raphael Orélis-Ribeiro, G. Sybren de Hoog, and Walter A. Boeger

Abstract

An infectious disease has caused high mortalities in the Brazilian mangrove-land crab, *Ucides cordatus*. The lethargic crab disease spread northward and southward in waves from Pernambuco State in Brazil. Primary causative agent was the black yeast *Exophiala cancerae*, with *Fonsecaea brasiliensis* as a secondary, opportunistic invader. The reasons for coming and going of the disease may be of intrinsic as well as of environmental nature.

V. A. Vicente
Laboratory of Molecular Microbiology—LABMICRO, Department of Basic Pathology, Federal University of Paraná, Curitiba, Paraná, Brazil
e-mail: vaniava63@gmail.com

R. Orélis-Ribeiro
Aquatic Parasitology Laboratory, School of Fisheries, Aquaculture and Aquatic Sciences, College of Agriculture, Auburn University, Auburn, AL, USA
e-mail: raphael.orelis@gmail.com

G. S. de Hoog
Center of Expertise in Mycology RadboudUMC/CWZ, Nijmegen, The Netherlands

Westerdijk Fungal Biodiversity Institute, Utrecht, The Netherlands
e-mail: s.hoog@westerdijkinstitute.nl

W. A. Boeger (✉)
Laboratory of Molecular Ecology and Evolutionary Parasitology, Department of Zoology, Federal University of Paraná, Curitiba, Paraná, Brazil
e-mail: wboeger@gmail.com

S. Seyedmousavi et al. (eds.), *Emerging and Epizootic Fungal Infections in Animals*,
https://doi.org/10.1007/978-3-319-72093-7_11

11.1 The Sudden Disaster

In 1997, crab collectors and biologists first reported an unusually large-scale mortality in a population of the Brazilian mangrove-land crab *Ucides cordatus* (Brachyura: Ocypodidae) near the city of Goiana (state of Pernambuco, Brazil). Subsequently, epizootic events spread northward and southward, affecting populations of *U. cordatus* in 8 of the 17 coastal states, particularly in the northeast (states of Piauí, Ceará, Rio Grande do Norte, Paraíba, Pernambuco, Sergipe, Bahia) and southeast regions (state of Espírito Santo), spanning over 3500 km of the Brazilian Atlantic coast (Fig. 11.1). Mortalities occurred extensively at least until 2006, but anecdotal reports of mortalities in the mangroves of the state of Espírito Santo (southeast Brazil) prevailed for the next 4–5 years.

Crabs in areas of high-profile mortality shared several common clinical signs, such as lethargy, poor motor control, and inability to return to the upright position when turned upside down, and the name lethargic crab disease (LCD) was coined (Boeger et al. 2005). The disease caused extensive population depressions in the mangrove-land crab, which represents an important fishery resource for artisanal exploitation by local communities along the Brazilian coast (Alves and Nishida 2003; Schmidt 2006). Some mangroves in the states of Paraíba and Bahia have experienced reductions in the fishing yields of 84 and 97.6%, respectively (Nóbrega and Nishida 2003; Schmidt 2006), causing severe socioeconomic problems in the affected regions. The overall impact of this infirmity, although not directly evaluated, was likely severe (Glaser 2003) since the crab species affected has unquestionable ecological importance to West Atlantic mangroves. The species is responsible for the accelerated decomposition of mangrove litter and, thus, for nutrient remineralization and energy transfer into the sediment (Nordhaus 2003).

Greatly ignored by the authorities, despite its extent and conspicuous damage, the mortalities remained unstudied for almost a decade, with only anecdotal hypotheses about its origin and causative agent. Newspapers and even scientists, in unpublished accounts (i.e., abstracts in meetings and conferences), provided some reasonable but others not so serious hypotheses to explain the ongoing mortalities. The popular opinion speculated that the mortalities were caused by sewage, in contaminated areas close to large cities; by oil spills, where petroleum was being explored; and by fecal "chloroform" (this is the actual spelling used in some newspapers, but we can imagine what it meant) near large cities. Naive hypotheses apart, some scientists suggested that deforestation of mangroves, over-capture, and even pathogens of invasive species could be associated with the mortalities. One hypothesis, however, was preferred by most: that viruses associated with the Pacific shrimp (*Litopenaeus vannamei*), a species widely cultured in the Brazilian northeastern coast, were causing the mortalities (Schaeffer-Novelli et al. 2004).

However, it was not until 2003 that a systematic scientific research was conducted to define the origin and the causative agent of these extensive mortalities. The study resulted in a series of publications that revealed the causative agent and the dynamics of the disease (Boeger et al. 2005, 2007; Ferreira et al. 2009; Orélis-Ribeiro et al. 2011; de Hoog et al. 2011; Pie et al. 2011; Ávila et al. 2012; Vicente et al. 2012; Guerra et al. 2013) (Fig. 11.2).

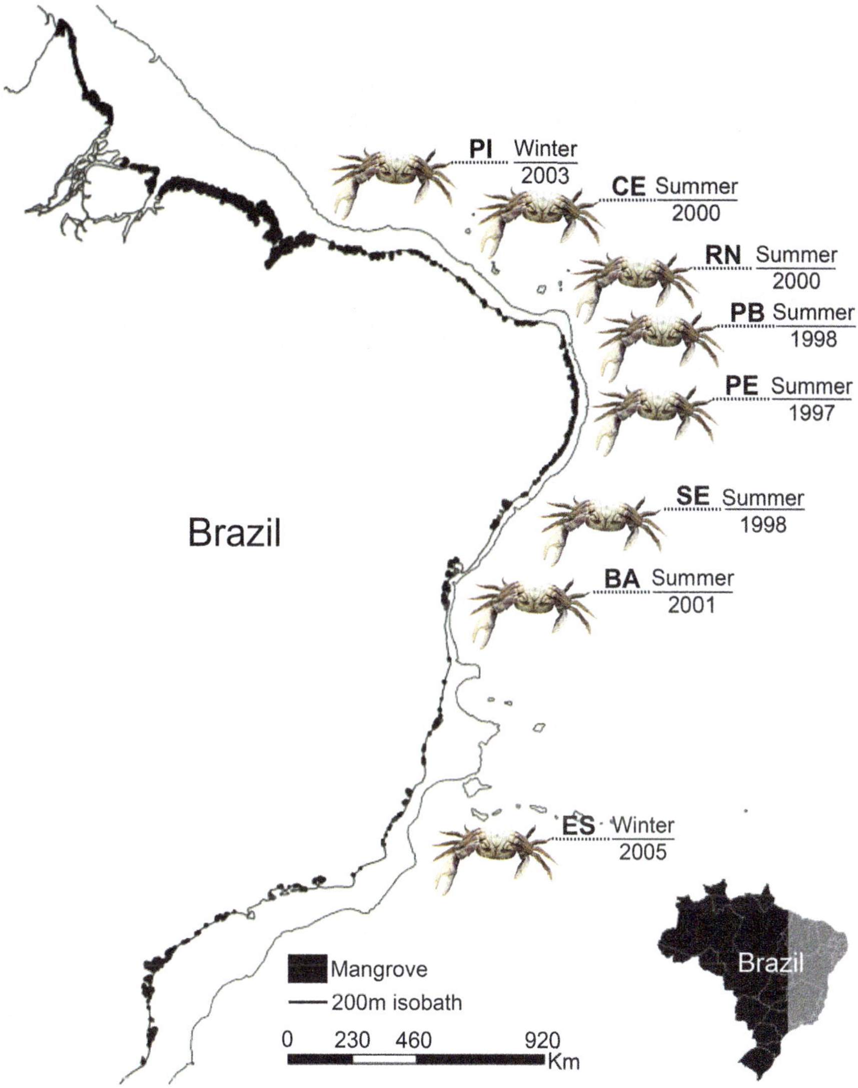

Fig. 11.1 Spatiotemporal distribution of lethargic crab disease outbreaks in Brazil. Season and year of the first epizootic events are indicated alongside crab pictures to the right of each Brazilian state affected. *PI* Piauí, *CE* Ceará, *RN* Rio Grande do Norte, *PB* Paraíba, *BA* Bahia, *ES* Espírito Santo. Adapted from Vicente et al. (2012)

11.2 The Lethargic Crab Disease (LCD)

LCD is known solely from the crab species *U. cordatus*. The great majority of the mortality outbreaks associated with LCD occurred during the summer of the southern hemisphere, but some anecdotal accounts reported mortalities during the winter

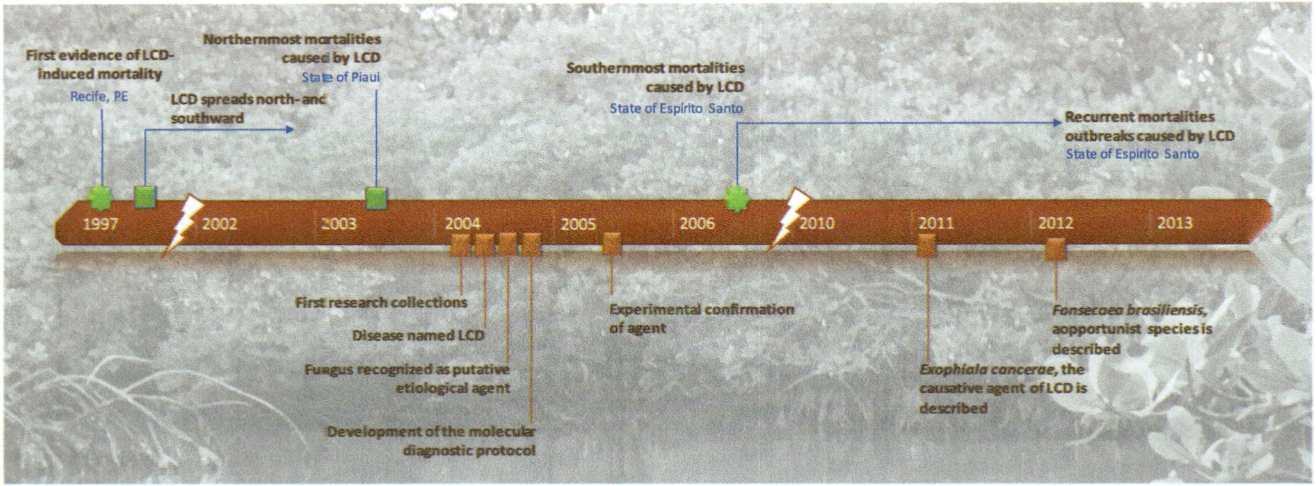

Fig. 11.2 Chronological order of the main facts and pioneering events that contributed to the epidemiological studies on lethargic crab disease

months as well. The infected crabs become increasingly lethargic and irresponsive as the disease progresses, resulting in the inability to feed or escape from predators and fishermen. In the mangrove, crabs apparently crawl out of their burrows and die outside. When captured, sick crabs die quickly, usually during the transfer from the mangrove and the point of commercialization. Tetany was observed in many moribund animals. These signs strongly suggest injury to the nervous and respiratory system, as revealed in subsequent studies (Boeger et al. 2007; Orélis-Ribeiro et al. 2011). The rapid death of sick crabs, associated with capture and transport, appears to be correlated with their inability to respond adequately to corresponding stress.

These clinical signs are explained by the histopathology of the disease (Boeger et al. 2007). Free or encapsulated (hemocytic encapsulation) yeast-like cells of a fungal agent are abundant in epidermal and connective tissues, gills, intestinal wall, thoracic ganglion, hepatopancreas, hemolymph, and heart of moribund animals showing clinical signs of LCD (Fig. 11.3). Gonads, somatic muscles, and digestive system are less affected by the fungus. In advanced stages of the disease, hyphae may also be present. Necrosis, tissue degeneration, and congestion of hemal sinuses and vessels are present in heavily infected organs (Fig. 11.3). Nerve fibers, especially those associated with the ventral thoracic ganglion, may be compressed by accumulation of yeast-like cells. Cardiac musculature is greatly compromised. In heavy infections, the tissue of gill lamellae, including epithelium and pillar cells, is destroyed with subsequent dilation or compression of the lacunae. Cellular immune responses include hemocytic infiltration, agglutination and encapsulation, and phagocytosis. Phagocytosis of yeast-like cells is abundant in the connective tissue associated with the exoskeleton.

11.3 The Etiological Agent

LCD was diagnosed as a systemic infection associated to black yeasts, supporting preliminary identification of the causative agent as a species of *Exophiala* (Boeger et al. 2005). The hypothesis of a viral origin, proposed by Schaeffer-Novelli et al. (2004), was refuted due to the lack of evidence from histological analyses, by molecular diagnostic tests for a number of known shrimp viruses (white spot syndrome virus [WSSV], infectious hypodermal hematopoietic necrosis virus [IHHNV], hematopancreatic parso-like virus [HPV], mesodermal baculovirus [MBV], baculovirus penaei [BP], Taura syndrome virus [TSV], gill-associated virus [GAV]), and by experiments of artificial infection which failed to detect any evidence of viral infection (Orélis-Ribeiro et al. unpublished).

Putative causative agents of LCD in the mangrove coast of Brazil were isolated from direct culturing of tissues (hepatopancreas and heart) from moribund mangrove crabs (*U. cordatus*) showing signs of the disease (Boeger et al. 2007). A black yeast-like species was isolated from crabs in all stages of the disease. The agent was described as a new species, *Exophiala cancerae* (de Hoog et al. 2011) (Fig. 11.4). In two cases, a coinfection by a second melanized, fonsecaea-like fungus was detected. Sequences of the ribosomal operon were obtained for isolates originating from

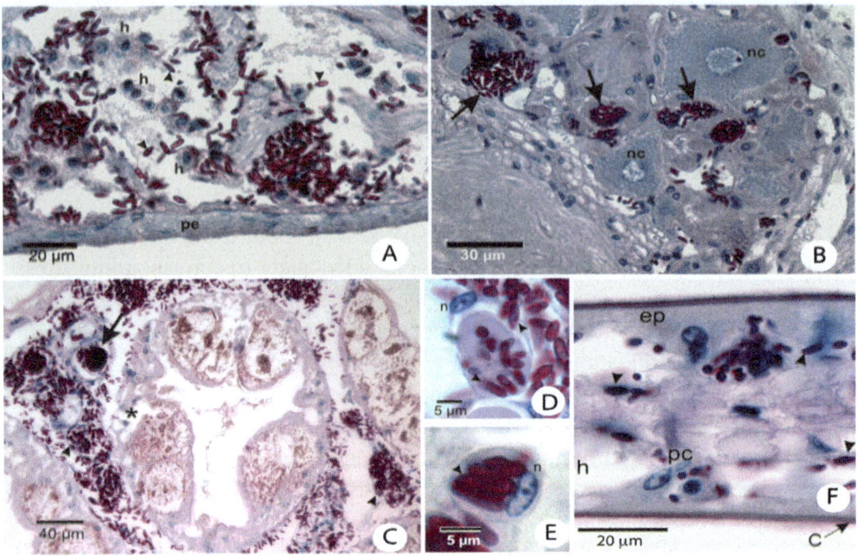

Fig. 11.3 Histopathology of *Ucides cordatus* infected showing signs of lethargic crab disease: (**a**) Heart of a moribund mangrove-land crab with LCD. Disruption and disorganization of cardiac-muscle fibers associated with the yeast-like cells (arrowheads) with extensive hemocytic infiltration (h). Note that pericardia (pe) are free of yeast-like cells. PAS and HE stains. (**b**) Neurosecretory area of the thoracic ganglion of a moribund mangrove-land crab with LCD with hemal sinuses congested by yeast-like cells (arrows). *nc* neurosecretory cells. PAS and HE stains. (**c**) Large hemocytic agglutination congesting the hemal sinus between tubules of the hepatopancreas, with both yeast cells (asterisk) and hyphae (arrows). PAS and HE stains. Yeast-like cells (arrowhead) and hemocytic encapsulations (arrow) occlude hemal sinuses of the hepatopancreas. The hepatopancreas caecum shows signs of necrosis (asterisk). PAS and HE stains. (**d**) RI cell with numerous yeast-like cells (arrowheads) in the connective tissue associated with the exoskeleton. *n* nucleus of the cell. Periodic acid Schiff (PAS) and HE stain. (**e**) Phagocytic cell (probably a hyalinocyte) of the connective tissue associated with the exoskeleton with numerous yeast-like cells (arrowhead). *n* nucleus of the cell. PAS and HE stains. (**f**) Gills of mangrove-land crab with LCD in intermediate stage of infection. Although many yeast-like cells (arrowheads) are visible, most of its components are relatively intact. *c* cuticle, *pc* pillar cell, *ep* epithelium, *h* hemal lacuna. PAS and HE stains

different crab tissues, and also this species appeared to represent an unknown taxon, which was described as *Fonsecaea brasiliensis* by Vicente et al. (2012) (Fig. 11.4).

LCD was associated with the presence of a black yeast-like fungus, *E. cancerae*, a member of the ascomycete order *Chaetothyriales* comprising the black yeasts and their filamentous relatives, as was proven studies using light and electron microscopy, behavioral and experimental tests, and molecular phylogenetics. Species of this order are regularly encountered as causative agents of cutaneous and systemic disorders in humans and cold-blooded animals (Nyaoke et al. 2009; de Hoog et al. 2011; Kano et al. 2000). Human infections range from commensalism or mild cutaneous infection to fatal neurotropism or osteotropism, sometimes with extracutaneous dissemination with severe mutilation. Waterborne vertebrates such as toads, frogs, and turtles are particularly susceptible to infection by *Exophiala* or fonsecaea-like species (de Hoog et al. 2011). Some species are host-specific to

A

Exophiala cancerae
de Hoog, Vicente,
Najafzadeh, Harrak, et al.,
Persoonia 27: 58 (2011)

B

Fonsecaea brasiliensis
Vicente, Najafzadeh,
Klaassen and de Hoog.,
Vet Microbiol 158: (2012)

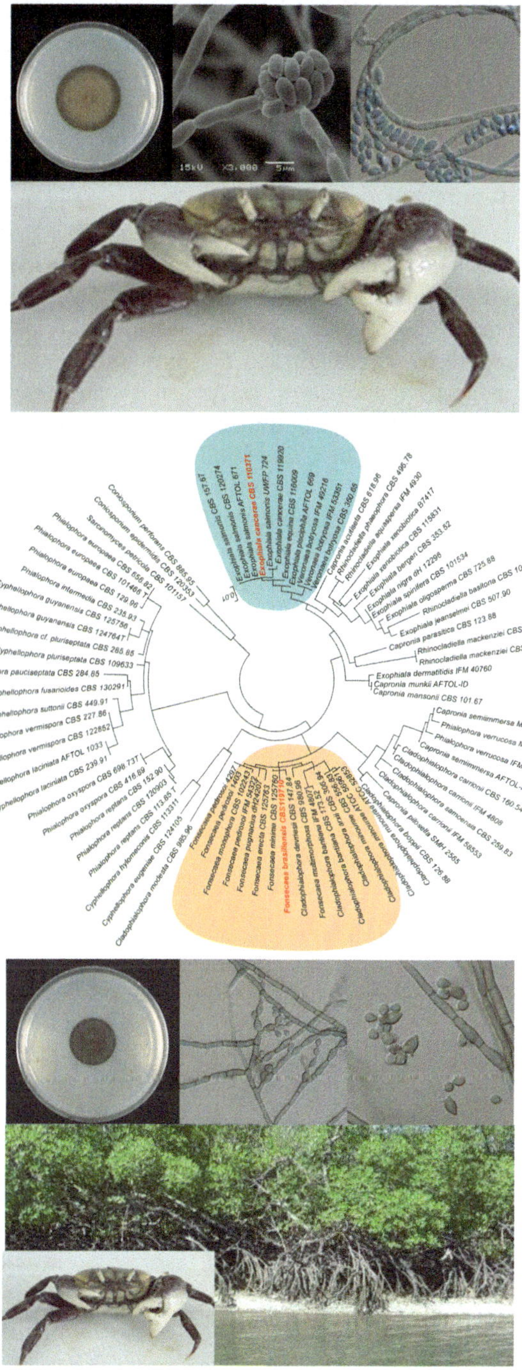

Fig. 11.4 Phylogenetic tree based on confidently aligned of LSU sequences constructed with maximum likelihood implemented in RAxML. Colored rectangles indicate main clades recognized within by Vicente et al. (2012) and de Hoog et al. (2011) which grouped the two agents of lethargic

particular fish taxa (Nyaoke et al. 2009). Virulence factors are largely unknown, but the ability to assimilate alkylbenzenes that are present in sweat and in nervous tissues of mammals as well as in the toxic skin of amphibians, has been suggested degradation of monoaromates is a rare property in the fungal Kingdom (Prenafeta-Boldú et al. 2006; de Hoog et al. 2011). Older literature data demonstrate that black yeast infections taking epizootic dimensions are relatively common in fish and frogs (e.g. Fijan 1969; Richards et al. 1978; Reuter et al. 2003). Several opportunistic species are involved (e.g. Manharth et al. 2005; Nyaoke et al. 2009; de Hoog et al. 2011) which are morphologically very similar. Invertebrate hosts in moist habitats may also be infected. In addition to crabs, *Exophiala* species have been reported from earthworms (Vakili 1993) and from mussels (Dover et al. 2007).

Fonsecaea brasiliensis and *E. cancerae* are members of rather distant clades within the *Chaetothyriales*, as proven by phylogenetic analysis of partial large subunit (LSU) rDNA data (Fig. 11.3). Isolates of *E. cancerae* belong to a clade of waterborne species of *Exophiala*, which are potentially able to cause infectious diseases in cold-blooded animals (de Hoog et al. 2011). *Fonsecaea brasiliensis* from crab clustered with *Cladophialophora devriesii* and *C. bantiana*. The former species was originally described from a fatal disseminated infection in a healthy appearing Caribbean woman (Gonzalez et al. 1984), although an inherited immune disorder such as a CARD9 defect cannot be excluded, while *C. bantiana* is an important agent with neurotropic invasion in human (de Hoog et al. 2009).

A multilocus analysis based on sequences of ITS (rDNA internal transcribed spacer) and partial sequences of translation elongation factor 1-a (*TEF1*), β-tubulin (*BT2*), as well as fingerprint profiles generated by amplification of restriction length polymorphism (AFLP) were used to confirm the identity of *F. brasiliensis* from crab and its relationship to *C. devriesii* (Vicente et al. 2012).

Exophiala cancerae was present in all crabs exhibiting clinical signs of LCD investigated and is considered to be the main agent of the disease. The natural niches of both species remain unclear. *Fonsecaea brasiliensis* was encountered coinfecting mangrove crab, but Vicente et al. (2012) could not establish whether severity of the disease increased with double infections. Environmental isolation studies showed that *F. brasiliensis* was abundantly present in the environmental niches known as crab habitats but also in plant debris in the Brazilian rainforest (Vicente et al. 2012, 2014; Guerra et al. 2013) suggesting that *F. brasiliensis* was a general opportunist of moribund crabs. In contrast, *E. cancerae* could not be found in the local environment (Vicente et al. 2012), not even from focus areas of the disease, suggesting a certain degree of host specificity. Remarkably, however the species was found elsewhere in water, once causing an outbreak in toads (J. Cunningham, pers. comm.), and was also repeatedly isolated from human nails (de Hoog et al. 2011).

Fig. 11.4 (continued) crab disease (LCD), *Fonsecaea brasiliensis* and *Exophiala cancaerae* showed in red letter in the tree. The macro/micromorphology and ecological aspects of the two agents of LCD are presented: (**A**) colony, conidiophores apparatus, and crab host; (**B**) colony, conidiophores apparatus, and environmental niche/crab host

Experimental infections of *E. cancerae* (Orélis-Ribeiro et al. 2011) and of *F. brasiliensis* (Vicente et al. 2012) provided strong evidence that the former species is likely the major causative agent of LCD (Fig. 11.5). When infected with *E. cancerae*, crab dying is beginning to be observed between the 7th and the 10th

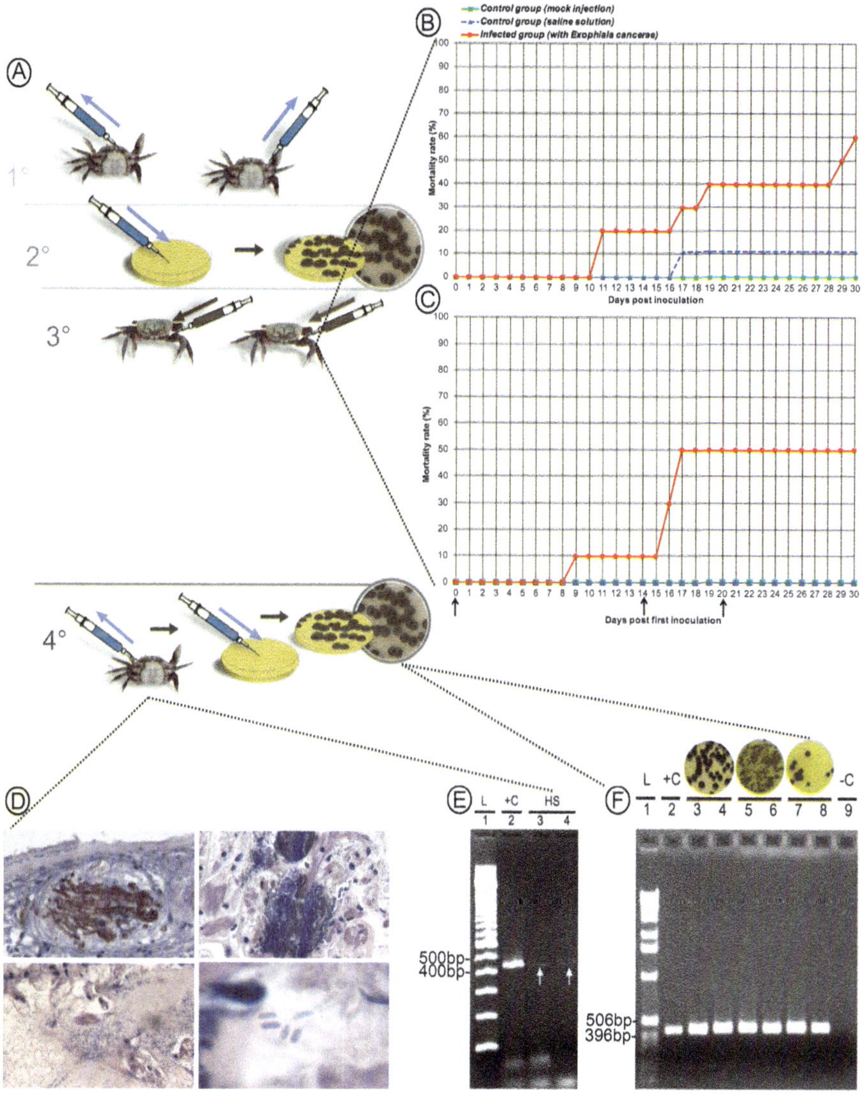

Fig. 11.5 Diagram illustrating the fulfillment of Koch's postulates on the context of LCD. (A) Koch's postulates; (B and C) Mortality curves for crabs that were subjected to a single and multiple experimental infection;(D) Histology of experimentally infected crabs; (E) Molecular diagnosis of infected tissues; (F) Molecular identification of colonies recovered from the tissue of artificially infected crabs. Adapted from Orélis-Ribeiro et al. (2011)

day post infection and reaches up to 60% mortality between days 17 and 33, when the experiment was terminated. The diseased crabs, inoculated with *E. cancerae*, reproduced the clinical signs of LCD observed in the mangroves. None or very limited mortalities were observed in the control groups (mock or saline-solution injections). Mortality of crabs injected with a conidial suspension of *F. brasiliensis* was significantly less pronounced and began 27 days after infection, reaching 40% by the end of the experiment. Two weeks after plating the collected tissues from crabs that died during the experiment on Mycosel medium, resulting colonies were identified down to genus level by morphology and to species level by the use of specific PCR primers. The lower virulence of *F. brasiliensis* supports its nature as a secondary invader in natural infections, probably taking advantage of weakened crabs initially infected by *E. cancerae*.

11.4 The Development of Tools for Identification and Diagnosis

Mortality due to LCD was initially determined from anecdotal observations or inference from clinical signs. To improve estimations, a set of species-specific PCR primers was developed in order to allow correct diagnosis, fungal identification, and environmental prospection of the species involved in LCD outbreaks (Pie et al. 2011; Guerra et al. 2013). All primers sets were based on sequences of rDNA internal transcribed spacer (ITS) regions. PCR products generated from extracts of samples (environment, colonies, crab tissue) differed between the two species and remained negative if the respective DNA targets were not present, allowing unambiguous identification of the agent. Amplicons measured 450 or 396 bp, corresponding to the expected product sizes of *E. cancerae* (strain CBS 120420) and *F. brasiliensis* (strain CBS 119710), respectively.

11.5 The Dynamics of the Epizootic

The global distribution of *E. cancerae* (de Hoog et al. 2011) suggests that it may have been present in Brazil prior to the beginning of the epizootic in the late 1990s. LCD outbreaks, with local crab mortality rates of up to 90%, may have taken place before. Indeed elder fishermen from regions along the Brazilian coast more than 1500 km apart indicated that extensive mortalities of mangrove-land crabs took place more than 50 years ago. Assuming that the species has a global distribution, changes in host or in environmental conditions, rather than emergence of virulent fungal genotypes, are likely causes of the epizootic.

The original habitat of the etiologic species of LCD, despite many attempts, remains unknown. Search efforts to identify habitats and eventual other hosts of *E. cancerae* and *F. brasiliensis* endemic areas of crab disease in the recent past were partially successful. Guerra et al. (2013), using techniques of fungus isolation and molecular detection on mangrove plant substrates and mud associated with crab burrows yielded positive results with *F. brasiliensis* in samples from plant branches and roots. However, *E. cancerae* was again not recovered from any environmental sample. Perhaps this species has a hitherto uncovered reservoir. During periods of the absence of LCD, *E. cancerae* was detected in specimens of *U. cordatus* and, in a single specimen of *Cardisoma guanhumi* ("guaiamum"), another Ocypodidae crab species (Boeger and Pie 2006). Although animals were positive for the agent of LCD, they were free of clinical signs of the disease. It was therefore concluded that *E. cancerae* was able to reside asymptomatically in the mangrove-land crab and eventually in its close relative *C. guanhumi*. Perhaps this specific lineage of *E. cancerae* is a true host-specific pathogen, requiring as yet unknown conditions to become fulminant. In contrast, *F. brasiliensis* represents an opportunist that is capable of colonizing weakened, diseased crabs.

Starting in 1997 from the first reported locality with high mortality in the mangrove land of Goiana in Pernambuco, putative LCD spread to northern and southern mangroves in a wave-like fashion (Fig. 11.2). Greater mortalities were usually seen during the first events (first wave) in each direction, with lower mortalities occurring in populations where the disease had occurred previously. Unsuitable environmental conditions are known to cause stress and subsequent weakening of the immune response of the host and making the crabs susceptible to infection (Stentiford and Feist 2005). Several studies reported that that relative immune competence may vary among individual mangrove-land crabs. Some individuals were apparently capable of impeding (at least temporarily) the invasion and multiplication of the fungus within their tissues. In some crabs that looked healthy, hyphae and yeast-like cells were found within hemocytic encapsulations and/or agglutinations only (Boeger and Pie 2006). In other crabs, however, rapid multiplication of fungal cells appeared to overwhelm their immune response, resulting in tissue invasion and dissemination leading to LCD. A small percentage of animals managed to survive events of LCD of mangrove-land crabs (Nóbrega and Nishida 2003; Schmidt 2006). Part of these survivors may not have been exposed to the causative agents, but others may have been capable of clearing fungal invasion of their tissues. Genetic, physiological, and/or environmental factors may determine differences in strength of the immunological response to *E. cancerae* within a population of the mangrove-land crab; the relative contribution of these factors warrants further exploration.

The wave pattern observed in the mortality associated with LCD suggests that there may be an interplay between demographic and epidemiologic processes within each population. Schmidt (2006) suggested that the most likely dispersion pathway for the fungus of LCD is the marine environment, since he found crabs to present higher mortality rates in the lower intertidal zones. Environmental processes interacted with each other; general pattern of mortalities and dispersion were reproduced in computational models by Ferreira et al. (2009) and Ávila et al. (2012). However, Boeger and Pie (2006) suggested that the wave pattern of dispersion could be associated with a variation of the resistance of crabs along the year. As mortalities often occurred following the "andada" (period when crab leave their burrows to copulate), these authors suggested that the resistance of the crabs would be reduced following this period of apparent stress. The dispersion would promote extensive mortalities of mangrove-land crab population in the limits of distribution during the first year, but subsequent outbreaks would be less extensive due to the larger percentage of resistant individuals in the populations—since susceptible hosts had largely disappeared from the population during previous outbreaks.

Curiously, the outbreaks spread north and southward, but in both directions reached a limit in distribution. The southernmost limit of dispersion of mortalities associated with LCD occurred in the mangroves of the state of Espírito Santo in the year 2006. Despite recurrent mortalities during the next 2 years in the region, LCD did not disperse toward mangrove populations located south of this state. Similarly, mortalities were reported in populations of mangrove-land crab in the northern states of Rio Grande do Norte and Ceará in 2000. The northernmost report occurred in the state of Piauí, in 2003. No mortalities were reported in the extensive mangroves located in the northern states of Maranhão and Pará—the latter is the state that presents that largest concentration of mangroves in Brazil.

An in vitro experiment (Orélis-Ribeiro et al. 2017) showed that the agent of LCD, the black yeast *E. cancerae* was remarkably tolerant to variable salinity levels, supporting an extensive period of exposure (1 week) to the osmotic stress of 38 ppt salinity, which in nature is recorded only during dry periods of the Brazilian northeastern coast (Mémery et al. 2000, Ffield 2005). Thus, the experimental results support the hypothesis of dispersion of the etiological agent through the marine environment in most Brazilian coastal areas. However, higher temperature (30 °C) caused a reduction of about 1.2-fold in colony-forming units. The maximum growth temperature of *E. cancerae* is 33 °C and optimal growth at 24–27 °C (de Hoog et al. 2011) (Fig. 11.6). Hence, it seems plausible to suggest that sea surface temperatures either above or below this optimum growth range apparently represent a key factor influencing the dispersion of LCD-related mortalities to the north-south limits of distribution of *U. cordatus*.

Still, the factors associated particularly with the southern limits of dispersion of *E. cancerae* are not clear. It appears that the rarefaction in mangrove distribution in the southern coast of Espírito Santo and the northern coast of the state of Rio de Janeiro could represent a buffer area precluding the stepping-stone dispersion of the etiologic agent of LCD. The change in direction of the coast from a roughly north-south to a northeast-southwestern, observed immediately southern of the limit of distribution of the outbreaks, increases the distance between the main coastal current from the coast which may further reduce the possibility for the dispersing fungal forms to reach the coastal mangroves.

Mathematical modeling has proven to be a valuable tool to providing insights on the spatiotemporal dynamics of LCD (Ferreira et al. 2009, Ávila et al. 2012). In a first attempt to elaborate a model that reflects the periodicity and the observed fluctuations in the incidence of the mortality events within a mangrove area, Ferreira et al. (2009) separated the adult mangrove-land crab population into infected and susceptible groups. Noteworthy is that simulations over the range of parameter values successfully reproduced the cyclic pattern of the epizootic events observed in nature. Essentially, the model predicts that in areas where the mangrove-land crab population is more susceptible, or *E. cancerae* is highly virulent, the incidence of high-profile mortality events is higher, and the cycles occur because the susceptible crabs are driven to low densities at which the black yeast population can no longer

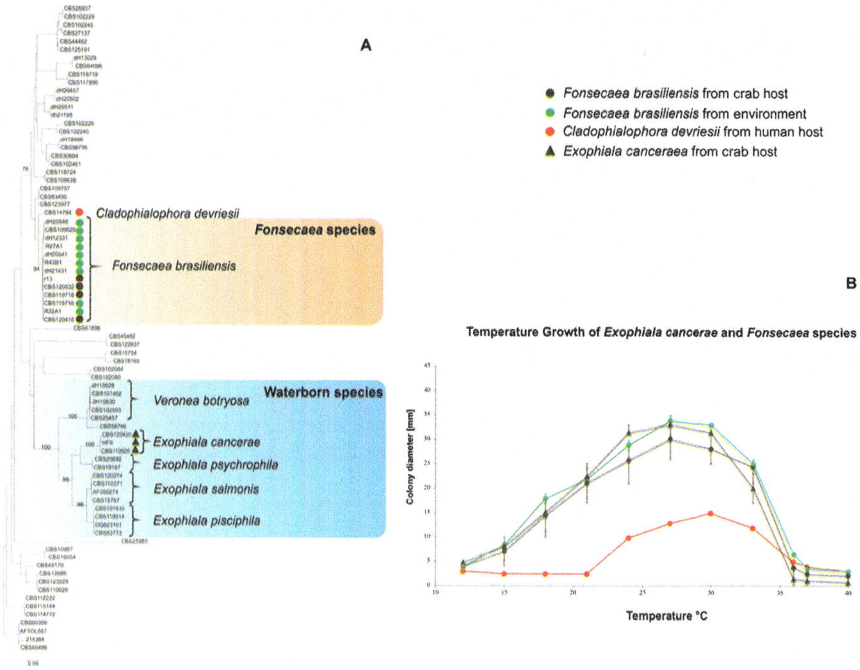

Fig. 11.6 Molecular ecology of the lethargic crab disease agents: (**A**) phylogenetic tree based on confidently aligned ITS sequences constructed with maximum likelihood implemented in RAxML. Colored rectangles indicate main clades recognized within by Vicente et al. (2012) and de Hoog et al. (2011) which grouped the two disease agents *Fonsecaea brasiliensis* and *Exophiala cancaerae*, respectively; (**B**) Temperature relations of *F. brasiliensis*, *E. cancerae* and *Cladophialophora devriesii*. For interpretation environmental strains (in green), strains isolated from crabs (in brown), clinical strain from human patient (in red)

spread, therefore also decreasing. When the susceptible crab population recovers to a threshold, another epidemic peak emerges, and so on. In summary, although stress associated with mating season has been previously suggested as a possible factor influencing the dynamic of the outbreaks, the developed model shows that the oscillations may be a consequence of the relationship between demographic and epidemiological parameters. Furthermore, considering the implications of this model, the authors hypothesized that the introduction of resistant mangrove-land crabs or the sustainable management of its population through capture could represent effective steps toward preventing the spread of LCD. Adding another layer of complexity to the previous model, Ávila et al. (2012) formulated a model to study the propagation of LCD between mangroves as a result of dispersion of *E. cancerae* in sea (either by diffusion or ocean currents). The study assessed different conditions for the existence of epidemic waves, thus determining the minimum speed at which they travel. This model pointed out the existence of traveling wave solutions, which strengthened the previous hypotheses regarding strategies for disease control. Simulations on the study of the speed of the epidemic wave indicated that ceasing the collection of mangrove-land crabs in estuaries close to the LCD-affected areas will increase disease dispersion throughout the Brazilian coast. In addition, the results also suggested that spread of the epizootic events mainly from the north to south could be owing to the influence of ocean currents in the dispersion of the fungus.

The reports of mortalities of populations of the mangrove-land crab are presently scarce and limited to mangroves located in the southernmost limit known for the outbreaks, in the state of Espírito Santo and southern Bahia.

References

Alves RRN, Nishida AK (2003) Socio-economical aspects, environmental perception of 'Caranguejo-uçá', *U. cordatus* (L.1763) (Decapoda, Brachyura) gatherers in the Mamanguape river estuary, northeast Brazil. Interciencia 28(1):36–43

Ávila RP, Mancera PFA, Esteva L et al (2012) Traveling waves in the lethargic crab disease. Appl Math Comput 218(19):9898–9910

Boeger WA, Pie MR (2006) Montando o quebra-cabeça. Revista do GIA 2:34–36

Boeger WA, Pie MR, Ostrensky A et al (2005) Lethargic crab disease: multidisciplinary evidence supports a mycotic etiology. Mem Inst Oswaldo Cruz 100(2):161–167

Boeger WA, Pie MR, Vicente VA et al (2007) Histopathology of the mangrove land crab *Ucides cordatus* (Ocypodidae) affected by lethargic crab disease. Dis Aquat Org 78:73–81

de Hoog GS, Guarro J, Figueras MJ et al (2009) Atlas of clinical fungi, 3rd CD-ROM edn. CBS-KNAW Fungal Biodiversity Centre. Universitat Rovira i Virgili Reus, Utrecht

de Hoog GS, Vicente VA, Najafzadeh MJ et al (2011) Waterborne *Exophiala* species causing disease in cold-blooded animals. Persoonia 27:46–72

Dover CLV, Ward ME, Scott JL et al (2007) A fungal epizootic in mussels at a deep-sea hydrothermal vent. Mar Ecol 28:54–62

Ferreira CP, Pie MR, Esteva L et al (2009) Modelling the lethargic crab disease. J Biol Dyn 3 (6):620–634

Ffield A (2005) North Brazil current rings viewed by TRMM microwave imager SST and the influence of the Amazon plume. Deep Sea Res Part 1 Oceanogr Res Pap 52(1):137–160

Fijan N (1969) Systemic mycosis in channel catfish. Bull Wildl Dis Assoc 5:109–110

Glaser M (2003) Interrelations between mangrove ecosystem, local economy and social sustainability in Caeté Estuary, North Brazil. Wetl Ecol Manag 11:265–272

Gonzalez MS, Alfonso B, Seckinger D et al (1984) Subcutaneous phaeohyphomycosis caused by *Cladosporium devriesii*, sp. nov. Sabouraudia 22:427–432

Guerra RS, do Nascimento MMF, Miesch S et al (2013) Black yeast biota in the mangrove, in search of the origin of the lethargic crab disease (LCD). Mycopathologia 175(5–6):421–430

Kano R, Kusuda M, Nakamura Y (2000) First isolation of *Exophiala dermatitidis* from a dog: identification by molecular analysis. Vet Microbiol 76:201–205

Manharth A, Lemberger K, Mylniczenko N et al (2005) Disseminated phaeohyphomycosis due to *Exophiala* species in a Galapagos tortoise, *Geochelone nigra*. J Herpetol Med Surg 15:20–26

Mémery L, Arhan M, Alvarez-Salgado XA et al (2000) The water masses along the western boundary of the south and equatorial Atlantic. Prog Oceanogr 47:69–98

Nóbrega RR, Nishida AK (2003) Aspectos socioeconômicos e percepção ambiental dos catadores de caranguejo-uçá, *Ucides cordatus* (L. 1763) (Decapoda, Brachyura) do estuário do rio Mamanguape, Nordeste do Brasil. Interciencia 28:36–43

Nordhaus I (2003) Feeding ecology of the semi-terrestrial crab *U. cordatus* (Decapoda: Brachyura) in a mangrove forest in northern Brazil. Dissertation, Universität Bremen

Nyaoke A, Weber ES, Innis C et al (2009) Disseminated phaeohyphomycosis in weedy, *Phyllopteryx taeniolatus*, and leafy, *Phycodurus eques*, seadragons caused by species of *Exophiala*, including a novel species. J Vet Diagn Investig 21:69–79

Orélis-Ribeiro R, Boeger WA, Vicente VA et al (2011) Fulfilling Koch's postulates confirms the mycotic origin of lethargic crab disease. Antonie Van Leeuwenhoek 99:601–608

Orélis-Ribeiro R, Vicente VA, Ostrensky A et al (2017) Is Marine Dispersion of the Lethargic Crab Disease Possible? Assessing the Tolerance of Exophiala cancerae to a Broad Combination of Salinities, Temperatures, and Exposure Times. Mycopathologia 182(11–12): 997–1004

Pie MR, Boeger W, Patella L et al (2011) Specific primers for the detection of the black-yeast fungus associated with lethargic crab disease (LCD). Dis Aquat Org 94(1):73–75

Prenafeta-Boldú FX, Summerbell RC, de Hoog GS (2006) Fungi growing on aromatic hydrocarbons: biotechnology's unexpected encounter with biohazard. FEMS Microbiol Rev 30:109–130

Reuter RE, Hutchinson W, Ham J et al (2003) *Exophiala* sp. infection in captured king George whiting (*Sillaginodes punctata*). Bull Eur Assoc Fish Pathol 23:128–134

Richards RH, Holliman A, Helgason S (1978) *Exophiala salmonis* infection in Atlantic salmon *Salmo salar* L. J Fish Dis 1:357–368

Schaeffer-Novelli Y, Cintrón-Molero G, Coelho-Jr C et al (2004) The mangrove mud crab die-offs in northeastern Brazil: circumstantial evidence for an epizootic origin related to marine penaeid shrimp production. In: Resumos do 2° Simpósio Brasileiro de Oceanografia, São Paulo

Schmidt AJ (2006) Estudo da dinâmica populacional do caranguejo-uçá, *Ucides cordatus* cordatus e dos efeitos de uma mortalidade em massa desta espécie em manguezais do Sul da Bahia. Thesis, Universidade de São Paulo

Stentiford GD, Feist SW (2005) A histopathological survey of shore crab (*Carcinus maenas*) and brown shrimp (*Crangon crangon*) from six estuaries in the United Kingdom. J Invertebr Pathol 88:136–146

Vakili NG (1993) *Exophiala jeanselmei*, a pathogen of earthworm species. Med Mycol 31:343–346

Vicente VA, Orélis-Ribeiro R, Najafzadeh MJ et al (2012) Black yeast-like fungi associated with lethargic crab disease (LCD) in the mangrove-land crab, *Ucides cordatus* (Ocypodidae). Vet Microbiol 158:109–122

Vicente VA, Najafzadeh MJ, Sun J et al (2014) Environmental siblings of black agents of human chromoblastomycosis. Fungal Divers 65:47–63

Cryptococcosis: Emergence of *Cryptococcus gattii* in Animals and Zoonotic Potential 12

Karuna Singh, Macit Ilkit, Tahereh Shokohi, Ali Tolooe, Richard Malik, and Seyedmojtaba Seyedmousavi

Abstract

Cryptococcosis is one of the most serious fungal diseases of animals worldwide, affecting a wide variety of mammals (including humans) and, occasionally, birds, reptiles, and amphibians. The disease is caused by pathogenic members of the encapsulated, melanin-forming, basidiomycetous yeast genus *Cryptococcus*, namely, *Cryptococcus neoformans* and *Cryptococcus gattii* species complexes. These two species have different ecological niches across climate zones: *C. neoformans* has been isolated primarily from soil and avian excrement, whereas *C. gattii* is mainly

K. Singh
Department of Zoology, Mahila Mahavidyalaya, Banaras Hindu University, Varanasi, India
e-mail: karunasingh5@gmail.com

M. Ilkit
Division of Mycology, Department of Microbiology, Faculty of Medicine, University of Çukurova, Adana, Turkey
e-mail: macitilkit@gmail.com

T. Shokohi
Invasive Fungi Research Center (IFRC), and Department of Medical Mycology and Parasitology, School of Medicine, Mazandaran University of Medical Sciences, Sari, Iran
e-mail: shokohi.tahereh@gmail.com

A. Tolooe
Department of Microbiology and Immunology, Center of Expertise in Microbiology, Infection Biology and Antimicrobial Pharmacology, Tehran, Iran
e-mail: atolooe@gmail.com

R. Malik
Centre for Veterinary Education, University of Sydney, Sydney, NSW, Australia
e-mail: richard.malik@sydney.edu.au

S. Seyedmousavi (✉)
Laboratory of Clinical Immunology and Microbiology (LCIM), National Institute of Allergy and Infectious Diseases (NIAID), National Institutes of Health (NIH), Bethesda, MD, USA
e-mail: Seyedmousavi@nih.gov; S.Seyedmousavi@gmail.com

S. Seyedmousavi et al. (eds.), *Emerging and Epizootic Fungal Infections in Animals*,
https://doi.org/10.1007/978-3-319-72093-7_12

associated with decaying wood and other plant materials, particularly in and around various species of trees. Cryptococcosis, which appears to be acquired by the inhalation of yeasts from environmental niches and penetration into the sinonasal cavity (animals) or pulmonary alveoli (humans) of the host, followed by hematogenous dissemination (humans) or penetration of the cribriform plate of the ethmoid bones (many animals), often manifests as skin and soft tissue infections, rhinosinusitis, pneumonia, and meningoencephalitis. Animals and people may become infected via the same environmental source; however, no convincing mammal-to-mammal transmission has been documented to date. This chapter highlights the diseases and complications that *Cryptococcus* species may cause in invertebrates, cold- and warm-blooded animals, marine mammals, and nonhuman primates. The potential role of animal hosts as sentinels of human cryptococcosis is discussed.

12.1 Introduction

Cryptococcosis is a primary or opportunistic, life-threatening fungal infection of worldwide distribution caused by members of the genus *Cryptococcus*, viz., *Cryptococcus neoformans* and *Cryptococcus gattii* species complexes (Kwon-Chung et al. 2017). The major environmental sources of *Cryptococcus* comprise soil contaminated with avian (most often pigeon) guano (*C. neoformans*) or plant debris and decaying wood, especially in tree hollows protected from light (*C. gattii*) (Sorrell 2001; Spickler 2013). The presence of virulent isolates of *C. neoformans* in pigeon excrement was reported by Passoni (1999).

In humans, *C. neoformans* is responsible for 98% of cryptococcal infections in patients with AIDS (Litvintseva et al. 2005). Most cryptococcal infections are caused by *C. neoformans* var. *grubii*, while *C. neoformans* var. *neoformans* causes disease exclusively in immunocompromised individuals (Litvintseva et al. 2005). In contrast, the sibling species *C. gattii* causes cryptococcosis predominantly but not exclusively in immunocompetent individuals (Casadevall and Perfect 1998; Byrnes et al. 2010).

Both species of *Cryptococcus* may additionally cause cryptococcosis in a wide range of animals, from lower invertebrates such as soil-dwelling amoebae, nematodes, cockroaches, and mites to higher mammals (Steenbergen et al. 2001; Mylonakis et al. 2002). Naturally occurring cryptococcosis has also been documented in amphibians, reptiles, birds, and mammals (Refai et al. 2014). Cryptococcosis has been recorded in cattle (Emmons 1952), horses and dogs (Barclay 1979; Sutton 1981), cats (Trivedi et al. 2011), foxes and ferrets (Lewington 1982; Morera et al. 2011), goats (Chapman et al. 1990), monkeys (Pal et al. 1984), cheetahs and pigs (Ajello 1958), llamas and alpacas (Stephen et al. 2002), koalas and other marsupials (Krockenberger et al. 2002), and various marine mammals (Duncan et al. 2006; MacDougall et al. 2007; Rotstein et al. 2010). Interestingly, the clinical manifestations of cryptococcosis in various animals may differ from those in humans according to animal species and site and extent of colonization. For example, in outbreaks of mastitis in ruminants, the clinical signs may include anorexia, decreased milk production, and enlargement of the supramammary lymph nodes (Lin and Heitman 2006).

Moreover, some animals act as reservoirs or passengers for *Cryptococcus* spp., representing a constant source of infection in the environment. *Cryptococcus* spp. have been isolated from asymptomatic animals as well as clinical cases. *C. neoformans* var. *grubii* has been shown to be a transient colonizer of the nasal mucus of cats, dogs, koalas, and ferrets (Malik et al. 1995, 1997, 2002; Malik et al. 2006a; Sorrell et al. 1996; Lester et al. 2004; Morera et al. 2014; Danesi et al. 2014a). Furthermore, nasal colonization of wild and domestic animals has been suggested to be a good indicator of the environmental presence of *C. gattii* (Duncan et al. 2006).

Among animals, cryptococcosis is most commonly observed in cats (Pennisi et al. 2013), with the involvement of the upper and/or lower respiratory tract, subcutaneous granulomata (typically contiguous with the sinonasal cavity), and disseminated infections (Sorrell 2001; Duncan et al. 2006; McGill et al. 2009; Headley et al. 2016). Clinically, multifocal skin involvement is a typical marker for widespread hematogenous dissemination to multiple tissues. Animals, including wild and domesticated cats and wildlife, share urban and rural habitats with people, and often one or more species serve as potential sentinel hosts for human exposure (Danesi et al. 2014b).

The main epidemiological findings from human and animal studies of cryptococcosis are summarized in Fig. 12.1.

12.2 Etiology and Taxonomy

The genus *Cryptococcus* (teleomorph *Filobasidiella*) comprises basidiomycetous yeast species, most of which are environmental saprophytes that do not cause infections in human or animals (Kwon-Chung et al. 2017). The pathogenic agents of cryptococcosis are classified into two species, *C. neoformans* and *C. gattii* (Table 12.1). The species *C. neoformans* comprises two varieties, *C. neoformans* var. *grubii* and *C. neoformans* var. *neoformans* (Meyer et al. 2009). The species *C. neoformans* consists of the VNI–VNIV and VNB molecular genotypes, comprising var. *grubii* (serotype A or VNI, VNII, and VNB strains), var. *neoformans* (serotype D or VNIV strains), and serotype AD strains (VNIII), which represents hybrids of the two varieties (Kwon-Chung and Varma 2006). The species *C. gattii* is subdivided into two serotypes (B and C) and four molecular types (VGI, VGII, VGIII, and VGIV) (Boekhout et al. 2001). Diseases caused by other *Cryptococcus* species, such as *Cryptococcus laurentii* and *Cryptococcus albidus*, have been reported infrequently and generally in immunocompromised hosts (Harris et al. 2012).

Recently, by combining various molecular genotyping methods, a new taxonomical classification system, based on a seven species/four hybrid scheme, has been proposed for pathogenic cryptococcal species (Hagen et al. 2015). This scheme includes the following species: *C. neoformans* [*C. neoformans Cryptococcus neoformans* var. *grubii* (serotype A) with three genotypes (VNI, VNII, and

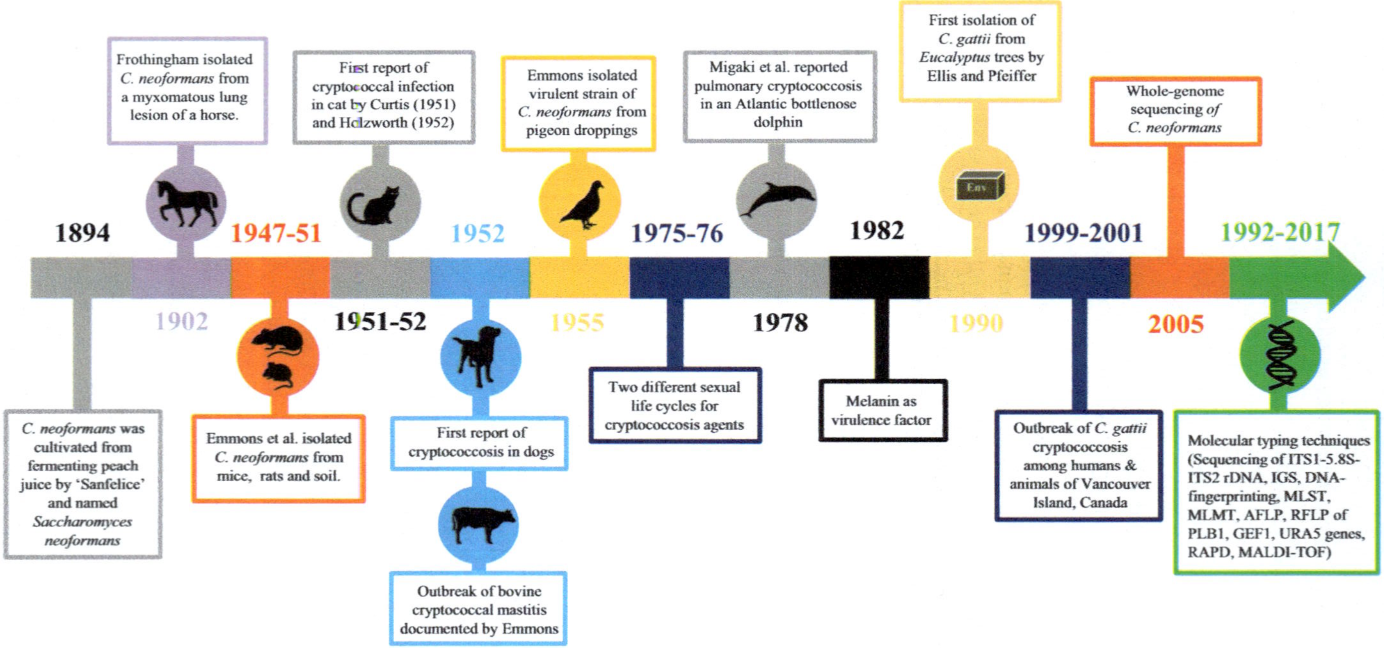

Fig. 12.1 The main facts and pioneering events that contributed to the epidemiological studies on human and animal cryptococcosis. *Env* environment, *MLEE* multilocus enzyme electrophoresis, *RAPD* random amplification of polymorphic DNA, *AFLP* PCR-fingerprinting amplified fragment length polymorphism, *RFLP* restriction fragment length polymorphism, *rDNA* ribosomal DNA region, *IGS* intergenic spacer region, *MLST* multilocus sequence typing, *MLMT* multilocus microsatellite typing, *MALDI-TOF* matrix-assisted laser desorption ionization–time-of-flight mass spectrometry-based method analysis

Table 12.1 Classification of *Cryptococcus neoformans* and *Cryptococcus gattii* species complexes

Species complexes		Molecular type
Cryptococcus neoformans	*Cryptococcus neoformans* var. *grubii*	VNI/VNII/VNB (AFLP1,1A/1B, VNB)
	Cryptococcus neoformans var. *neoformans*	VNIV (AFLP2)
	Cryptococcus neoformans AD hybrid	VNIII (AFLP3)
Cryptococcus gattii	*Cryptococcus gattii*	VGI (AFLP4)
		VGII (AFLP6)
		VGIII (AFLP5)
		VGIV (AFLP7)
		VGIV/VGIIIc (AFLP10)
	Cryptococcus gattii DB hybrid	AFLP 8
	Cryptococcus gattii AB hybrid	AFLP 9
	Cryptococcus gattii AB hybrid	AFLP 11

VNB)], *C. deneoformans* [*C. neoformans* var. *neoformans* (serotype D or VNIV)], 5 *C. gattii* cryptic species with serotypes B/C or VGI–IV [*C. gattii* (VGI), *C. bacillisporus* (VGIII), *C. deuterogattii* (VGII), *C. tetragattii* (VGIV), and *C. decagattii* (VGIV/VGIIIc)], as well as the following hybrids, *C. neoformans* × *C. deneoformans* hybrid, *C. deneoformans* × *C. gattii* hybrid, *C. neoformans* × *C. gattii* hybrid, and *C. deneoformans* × *C. deuterogattii* hybrid. However, a panel of experts representing numerous laboratories around the world agreed that in the absence of biological differences among clades, a new nomenclature system for *Cryptococcus* was preferable to naming each clade as a separate species. Accordingly, "*C. neoformans* species complex" and "*C. gattii* species complex" were both proposed (Table 12.1), as the etiological agents of cryptococcosis (Kwon-Chung et al. 2017). Furthermore, a new hypothesis also suggested continental drift as a possible speciation driver within the *Cryptococcus* species complexes, adding greatly to the ongoing discussion of speciation between these complexes (Casadevall et al. 2017). On the basis of the geographic distribution of these two species complexes and the coincidence of the evolutionary divergence and Pangea breakup times, it was proposed that a spatial separation caused by continental drift resulted in the emergence of the *C. gattii* and *C. neoformans* species complexes from a Pangean ancestor (Casadevall et al. 2017).

12.3 Ecology and Environmental Niche

C. neoformans was first cultivated by Sanfelice from fermenting peach juice in 1894, and it was initially named *Saccharomyces neoformans*. In 1901, Jean-Paul Vuillemin renamed the organism *C. neoformans* because it did not produce ascospores, a defining characteristic of the *Saccharomyces*

(Barnett 2010). Emmons later (Emmons 1951, 1955) isolated the fungus from soil and pigeon excreta. The species *C. neoformans* has also been recovered from the guano of other avian species, including chickens, parrots, sparrows, skylarks, starlings, canaries, and turtledoves (Muchmore et al. 1963; Littman and Walter 1968; Bauwens et al. 1986, DeVroey and Swinne 1986, Hedayati et al. 2011), as well as from the nares and upper respiratory tract of cats, dogs, koalas, and ferrets with or without cryptococcosis (Malik et al. 1995, 1997, 2002; Sorrell et al. 1996; Lester et al. 2004; Morera et al. 2014; Danesi et al. 2014b). Soil has been suggested to represent a more inhospitable environment for *C. neoformans* than avian excrement (Ruiz et al. 1981). Moreover, *C. neoformans* is occasionally isolated from various nonavian sources (Pal and Baxter 1985; Swinne et al. 1986) and decaying wood debris within living tree hollows; its association with tree hollows is, however, weaker than that of *C. gattii* (Klein 1901; Nawange et al. 2006; Grover et al. 2007).

The first isolation of *C. gattii* (VGI) from the environment was reported from the wood, bark, leaves, and plant debris of *Eucalyptus camaldulensis* trees in Australia (Ellis and Pfeiffer 1990). Thereafter, *C. gattii* has been isolated from *Eucalyptus* trees in tropical and subtropical regions of Brazil (Montenegro and Paula 2000), other countries in South America (Ellis and Pfeiffer 1990; Bava and Negroni 1992; Castanon-Olivares et al. 2000), Mexico (Castanon-Olivares Castanon-Olivares et al. 2000), North America (Kwon-Chung and Bennett 1984; Clancy et al. 1990; Levitz 1991), the region around the Mediterranean Basin (Mseddi et al. 2011; Elhariri et al. 2016; Cogliati et al. 2016), Malaysia (Keah et al. 1994), Australia (Pfeiffer and Ellis 1992), India (Padhye et al. 1993; Chakrabarti et al. 1997), and China (Li et al. 1993). Of note, the two known host trees of *C. gattii*, *E. camaldulensis* and *Eucalyptus tereticornis*, do not occur naturally in the "Top End" of the Northern Territory of Australia. Therefore, the isolation of a separate molecular type of *C. gattii* as "VGII" from different trees in this region strongly suggested the existence of an alternative environmental niche for *C. gattii* (Chen et al. 1997a). *C. gattii* has been isolated from more than 50 tree species, mostly angiosperms (such as *Ceratonia*, *Eucalyptus* spp., *Ficus* spp., *Mangifera indica*, *Olea* spp., *Pithecellobium dulce*, *Quercus garryana*, *Syzygium cumini*, *Tamarindus indica*, and *Terminalia* spp.), and less frequently from the gymnosperms (such as *Abies* spp., *Arbutus menziesii* var. *menziesii*, *Cedrus* spp., *Cupressus lusitanica*, *Cupressus sempervirens*, *Picea* spp., *Pinus* spp., *Pseudotsuga menziesii* var. *menziesii*, and *Thuja plicata*) in North America, South America, Africa, India, and the Mediterranean Basin (Randhawa et al. 2003; Kidd et al. 2004; Grover et al. 2007; Springer and Chaturvedi 2010; Cogliati 2013, 2016). Gymnosperms and angiosperms have in common the ability to develop decayed hollows but differ in biochemical composition, available nutrients, water content, microbial communities, and fungal associations (Lazera et al. 1998; Springer and Chaturvedi 2010).

C. gattii has also been recovered from several wild animals, including African grey parrots (Sorrell et al. 1996) and koalas (Ellis and Pfeiffer 1990; Krockenberger et al. 2003), as well as from bat guano (Lazera et al. 1993), ostrich feathers, and camel hair (Montagna et al. 1996). In Australia, the persistence and amplification of *C. gattii* in the environment have been attributed to koalas feeding on *Eucalyptus* trees (Ellis and Pfeiffer 1990; Krockenberger et al. 2003), in concert with the effects of their abrading wooden surfaces and providing nutrients to fungi from their urine and/or feces.

12.4 Epidemiology

In humans, *C. neoformans* causes cryptococcal meningitis worldwide and is primarily associated with desiccated bird droppings in environments protected from the sun (Casadevall and Perfect 1998). The species *C. neoformans* and *C. gattii* cause systemic infections in both immunocompromised and immunocompetent individuals (Casadevall and Pirofski 1999). Globally, more than one million people get infected with cryptococcosis worldwide every year (Holmes et al. 2003; Cogliati 2013). The disease is more common in sub-Saharan Africa and tropical countries (such as Brazil, Thailand, Malaysia, and Papua New Guinea)(Seaton et al. 1996; Laurenson et al. 1997). Among HIV-negative individuals in the USA, the average annual incidence has remained almost constant at about one case per 100,000 individuals (Pappas 2013). The incidence of cryptococcosis has increased to about 5% in Western countries, while in sub-Saharan Africa, its incidence is as high as 30% in individuals with malignancies or those receiving corticosteroids or immunosuppressive therapy (Idnurm et al. 2005). Before the advent of molecular identification techniques, it was believed that *C. neoformans* infections occurred mainly in immunocompromised patients. However, it has since been observed that the epidemiology of *C. neoformans* is far more complex. Moreover, in contrast to *C. neoformans* var. *grubii*, *C. neoformans* var. *neoformans* appears to be more prevalent in Western and Mediterranean Europe (Viviani et al. 2006; Desnos-Ollivier et al. 2010; Guinea et al. 2010).

12.4.1 Animal Cryptococcosis in Australia and Asia

Veterinary studies from Papua New Guinea, Australia, New Zealand, and the Hawaiian Islands have reported the isolation of *C. neoformans* and *C. gattii* from cats, dogs, horses, koalas, ferrets, sheep, goats, kiwis, cows, African grey parrots, cockatoos, other parrots, spinner dolphins, and echidnas (Riley et al. 1992; Connolly et al. 1999; Malik et al. 2002, 2003; McGill et al. 2009; Rotstein et al. 2010). However, in Asia, *C. neoformans*, but not *C. gattii*, has been recovered from cats, dogs, bandicoot rats, and house mice (Singh et al. 2007; Okabayashi et al. 2009; Singh et al. 2017).

12.4.2 Animal Cryptococcosis in Europe

The species *C. neoformans* has been isolated from cats, dogs, magpies, and striped grass mice (Bauwens et al. 2004; Lagrou et al. 2005; Belluco et al. 2008; Morera et al. 2014; Danesi et al. 2014a), while *C. gattii* has been isolated from pet ferrets and goats in Spain, with the latter occurring in association with stands of *E. camaldulensis* trees (Frasés et al. 2009; Morera et al. 2011).

12.4.3 Animal Cryptococcosis in the Americas

The species *C. neoformans* has been recovered from insects, bulls, ferrets, and sheep (Lemos et al. 2007; Riet-Correa et al. 2011; de Jesus et al. 2012), while infection due to *C. gattii* has been reported in cheetahs, goats, and psittacine birds (Raso et al. 2004; Illnait-Zaragozí et al. 2010). Three cases of cryptococcal infection caused by *C. neoformans* have been reported in North American ferrets; furthermore, dogs, cats, horses, ferrets, marine mammals, and birds living on or near Vancouver Island (British Columbia, Canada) have been reported to be infected with *C. gattii* (Hanley et al. 2006; MacDougall et al. 2007; Bartlett et al. 2012).

12.5 Virulence Factors in *C. neoformans*/*C. gattii* Species Complexes

Virulence factors are defined as features that allow an organism to survive and cause disease within a susceptible host (Kozel 1995). There is much evidence supporting the hypothesis that cryptococcal virulence originated because of environmental selection pressure (Casadevall et al. 2003). First, numerous environmental isolates of *C. neoformans* are virulent in laboratory mice and rats, indicating that these virulence factors developed without the prerequisite for prior interaction with host animals. Second, a broad range of animals is susceptible to these organisms, and the hosts are not required for replication or viability of the pathogen. Third, several virulence factors appear to have "dual use" capacities that confer survival advantages in both animal hosts and the environment.

The pathogenesis of *Cryptococcus* is determined by three broad factors: the status of host defenses, the virulence of the strain, and the inoculum size (Mitchell and Perfect 1995). A summary of well-defined virulence factors of pathogenic members of *C. neoformans* and *C. gattii* species complexes is presented in Table 12.2.

The polysaccharide capsule plays an important role as a virulent factor in animal models of cryptococcosis, as well as in clinical settings (Schelenz et al. 1995). The main component of the polysaccharide capsule is glucuronoxylomannan. Capsule-free or poorly capsulated isolates have been found to elicit a strong immune response and less severe disease in humans (Levinson et al. 1974; Farmer and Komorowski 1973). However, the degree of virulence has been shown to be unrelated to the degree of encapsulation (Kwon-Chung and Bennett 1992). Four capsule's genes, namely, *CAP64*, *CAP60*, *CAP59*, and *CAP10*, are required for virulence in a murine model (Chang and Kwon-Chung 1998). The capsule appears to inhibit the ingestion of yeast cells by phagocytes in the absence of opsonins (Vecchiarelli 2000). The capsule also plays an important role in eliciting damage to the host (Steenbergen and Casadevall 2003). The correlation between the amount of polysaccharide in the

Table 12.2 Well-defined virulence factors of pathogenic members of *C. neoformans*/*C. gattii* species complexes

Virulence factor	Role in pathogenesis
Capsule	Antiphagocytic, immunomodulatory, intracellular aggression
Laccase	Interference with oxidative burst
Melanin	Resistance to oxidative killing, antiphagocytic, immunomodulator, resistance to microbicidal peptides, antifungal drug resistance
Phospholipase	Intracellular growth
Proteases	Tissue damage
Urease	Intracellular growth
Phenotypic switching	Immune evasion
Mating type	Virulence factor regulation
Calcineurin and cAMP signaling	Virulence factor regulation
Superoxide dismutase	Intracellular growth
Polyploid titan cells	Produce resistant aneuploids

serum and severity of cryptococcosis provides circumstantial evidence for relationships among polysaccharide release, host damage, and disease progression (Vecchiarelli 2000).

An encapsulated melanin-forming isolate, which failed to grow at 37 °C, was found to be avirulent (Kwon-Chung et al. 2014). The generation time is known to contribute to virulence: isolates with a longer generation time were unable to cause fatal infection in mice even when high inoculum (10^7 cells/mouse) was used (Kwon-Chung et al. 1982). The species *C. neoformans* is inhibited or killed at 41 °C; this temperature restriction might represent an important determinant of its pathogenicity (Mitchell and Perfect 1995) and limit disseminated infection of most birds, which have a high core body temperature, limiting infection to cool portions of the upper respiratory tract. Several virulence genes that are upregulated and downregulated at the site of infection are important in the pathobiology of *C. neoformans*.

Melanin formation, which is used for accurate and rapid identification of this pathogen, has been identified as a second key virulence factor for *C. neoformans* (Rinaldi et al. 1986). In addition, melanin functions as an antioxidant and might protect *C. neoformans* from oxidative host defenses (Polak 1990; Jacobson and Tinnell 1993; Jacobson et al. 1994). Melanized cells of *C. neoformans* are less susceptible to antifungals (van Duin et al. 2002). The products of the laccase genes LAC1 and LAC2 are recognized as key enzymes of melanin biosynthesis (Torres-Guererro and Edman 1994; Zhu and Williamson 2004; Missall et al. 2005). Kronstad et al. (2011) identified 33 novel genes for melanization in *Cryptococcus*.

The ability to grow at 37 °C and above is an essential contributing factor to infection in warm-blooded animals and humans. *Cryptococcus* possesses the CNA1 gene, encoding the calcineurin A catalytic subunit, which confers the ability to survive at 37 °C (Odom et al. 1997). Numerous cryptococcal genes are known to be upregulated at 37 °C, including MGA2, although these genes are not essential for growth at 37 °C (Ma and May 2009). Proteinases and signal transduction pathways are also important for virulence. Both clinical and environmental isolates of *C. neoformans* possess protease activity, which has been shown to degrade host proteins including collagen, elastin, fibrinogen, immunoglobulin, and complement factors, and cause tissue damage, thereby providing nutrients to the pathogen and protecting it from the host (Chen et al. 1996). Replication of *C. neoformans* within macrophages has also been reported to be accompanied by the production of proteinases and phospholipases, which cause the degradation of host cell membranes (Chen et al. 1997b; Ghannoum 2000; Tucker and Casadevall 2002). Two other degradative enzymes—urease and superoxide dismutase—are also known to contribute to the virulence of this organism (Buchanan and Murphy 1998; Ma and May 2009).

Phenotypic switching has been shown to cause changes in virulence. Switching from smooth colonies to mucoid colonies has been observed in *C. gattii* during pulmonary infection (Jain et al. 2006). Similarly, the mucoid colony phenotype of *C. neoformans* triggers a macrophage- and neutrophil-dominated immune response, whereas the smooth-colony phenotype initiates a lymphocyte-dominated immune response (Pietrella et al. 2005).

C. neoformans and *C. gattii* are haploid yeasts that predominantly reproduce asexually (by budding). However, they also possess a bipolar mating system, with mating types MATa and MATα (Kwon-Chung 1975, 1976). Mating involves fusion between cells of opposite mating type (a and α), resulting in the conversion from a haploid budding yeast form to a dikaryotic mycelial form that produces basidia and basidiospores, which may serve as infectious propagules. Most of the clinical and environmental isolates are predominantly mating type α, which has been linked to virulence (Metin et al. 2010). The discovery that monokaryotic fruiting under laboratory conditions represents a novel type of sexual reproduction involving only one of the two mating types, most commonly MATα, demonstrated that same-sex or unisexual reproduction could profoundly influence the organism's population structure. Haploid fruiting usually occurs in response to nitrogen starvation and/or desiccation. Recent population genetic studies provide robust evidence that both a–α opposite sex mating and α–α unisexual mating occur in nature in both *C. neoformans* and *C. gattii*, with the potential to influence the evolutionary trajectory and the production of infectious spores (Lin and Heitman 2006; Lin et al. 2008). The MATα mating type is generally more virulent and invariably more prevalent than the MATa type (Kwon-Chung and Bennett 1978). Same-sex mating between two α cells forms stable *a/a* diploids and α haploid progeny and is considered to confer a survival advantage, particularly under harsh conditions (Lin and Heitman 2006).

Three signal transduction pathways that regulate virulence and morphogenesis have been described in *C. neoformans*. These include the cAMP-PKA pathway—a pheromone-regulated MAP kinase pathway involved in mating—and a calcineurin-regulated pathway (Lengeler et al. 2002). These pathways act specifically on a variety of virulence attributes: the cAMP-PKA pathway regulates capsule production, melanin formation, mating, and virulence (Lengeler et al. 2002), whereas the calcineurin-regulated pathway is essential for growth at 37 °C, mating, morphogenesis, and virulence (Steenbergen and Casadevall 2003). Moreover, the integrated signaling and regulatory pathways are especially important in the control of virulence in *Cryptococcus* yeasts (Ma and May 2009).

Mitochondrial tubular morphology has also been shown to protect *Cryptococcus* cells from cell death. The *C. gattii* strains, which can promote mitochondrial fusion to form long tubular mitochondria, are able to more efficiently repair mtDNA damage caused by the oxidative species and hypoxic conditions present within the macrophage phagosome (Ma et al. 2009; Springer et al. 2014). Mitochondrial tubularization is generally thought to result from mitochondrial fusion, a phenomenon that allows mitochondria within a cell to cooperate with each other, and protects cells from the detrimental effect of mtDNA mutations by allowing functional complementation of mtDNA gene products (Chen et al. 2003).

Recent studies also showed that *C. neoformans* produces large polyploid "titan cells" in response to the stress of the host environment (Okagaki et al. 2010). Typical cryptococcal cells are 5–7 μm in diameter and have a haploid (1C) genome, whereas titan cells can be five to ten times larger than normal cells and are predominantly tetraploid (4C) or octoploid (8C). Titan cell formation plays a key role in disease progression and enhances the virulence of *C. neoformans* to establish the initial pulmonary infection (Crabtree et al. 2012). Titan cells are shown to be more resistant to stress and antifungals, such as fluconazole, and produce populations of more-resistant aneuploids (Gerstein et al. 2015). Because titan cells cannot be phagocytosed, they may also prevent the phagocytosis of "normal" *C. neoformans* cells in their vicinity, especially at early stages of infection (Crabtree et al. 2012). Titan cells are also resistant to oxidative and nitrosative stresses; they hamper infection clearance and might play a key role in latent infection (Zaragoza et al. 2010).

12.6 Host Range

The species *C. neoformans* has been reported in a large variety of birds and mammals, including companion animals (pets) and domestic and free-range animals, worldwide. The species *C. gattii* has also been isolated from various animal species, including cats, dogs, marine mammals, ferrets, and camelids, in regions affected by the outbreak that started in Vancouver Island and subsequently spread to the Pacific Northwest of the USA (Stephen et al. 2002). Species other than *C. neoformans* and *C. gattii* have been reported sporadically. *Cryptococcus albidus* was found in horse with a genital infection, another horse with keratitis, and a dog, cat, and California sea lion with fatal disseminated infection (Mcleland et al. 2012). *Cryptococcus magnus* has been isolated from a cat with otitis externa and from a cat with recurrent painful mass lesion of the limbs (Poth et al. 2010). The species *C. laurentii* was detected in a dog with panniculitis and osteomyelitis (Refai et al. 2014).

12.6.1 Cryptococcosis in Invertebrates

Several nonmammalian models have been reported to allow the analysis of cryptococcal pathogenesis through survival and cryptococcal growth assays in invertebrates (Tenor et al. 2015), including the amoeba *Acanthamoeba castellanii* (Steenbergen et al. 2001; Malliaris et al. 2004), the nematode *Caenorhabditis elegans* (Mylonakis et al. 2002), the wax moth *Galleria mellonella* (Mylonakis et al. 2005), and the fruit fly *Drosophila melanogaster* (Apidianakis et al. 2004). These invertebrate models allow rapid analysis of cryptococcal virulence determinants and host genetics to various degrees (Desalermos et al. 2015).

12.6.2 Cryptococcosis in Cold-Blooded Vertebrates

12.6.2.1 Fishes

Zebra fish (*Danio rerio*) has been used as a visually and genetically accessible vertebrate model system to investigate cryptococcal pathogenesis and host–fungus interactions. Live imaging of the cranial blood vessels of infected larvae showed that zebra fish are extremely susceptible to mortality following infection and *C. neoformans* is able to penetrate the zebra fish brain following intravenous infection (Tenor et al. 2015). A zebra fish model of cryptococcal infection revealed significant roles for macrophages, endothelial cells, and neutrophils in the establishment and control of hematogenous dissemination of *C. neoformans* (Davis et al. 2016).

12.6.2.2 Amphibians

A case of pulmonary cryptococcosis caused by *C. neoformans* has been reported from Portugal in a free-living adult female toad (*Bufo bufo*). The animal was apparently healthy and was killed by a vehicle (Seixas et al. 2008). The causative agent was confirmed by immunohistochemistry, using the monoclonal antibody anti-*C. neoformans*, and by a PCR-based method using *C. neoformans*-specific primers.

12.6.2.3 Reptiles

Compared with mammals, cryptococcal infections are rarely seen in reptiles. Cryptococcosis has been described in few cases of reptiles, including lizards and snakes (Spickler 2013). An eastern water skink (*Eulamprus quoyii*) was found to be infected with *C. neoformans*, resulting in a subcutaneous lesion (Hough 1998). Systemic cryptococcosis has also been reported in a common anaconda (*Eunectes murinus*), with *C. neoformans* as the etiologic agent (McNamara et al. 1994).

12.6.3 Diseases in Warm-Blooded Vertebrates

12.6.3.1 Birds

Excrement of avian species, especially from Columbiformes birds, provides the required nutrients for *Cryptococcus* to proliferate (Emmons 1955). The species *C. neoformans* temporarily colonizes the intestinal tract of some avian species following ingestion and has been detected in the guano or cloacae of swans, raptors, rheas, chickens, and starlings, as well as some Psittaciformes and Passeriformes birds (Sorrell et al. 1996; Cafarchia et al. 2006; Tsiodras et al. 2000). The cloacae of four migratory birds of the species *Anas crecca*, *Anas platyrhynchos*, and *Fulica atra* were also found to be positive for *C. neoformans* (Amirrajab et al. 2016). *C. neoformans* has also been isolated from four common kestrels (*Falco tinnunculus*) (Cafarchia et al. 2006).

Despite the very high degree of exposure of pigeons to massive quantities of *C. neoformans*, cryptococcosis seems to be very rare in poultry and pigeons. *Cryptococcus* species accounted for localized and disseminated infections of the upper respiratory tract of immunocompetent captive parrots of widely differing ages living in Australia, which resulted in signs of mycotic rhinitis or the involvement of structures contiguous with the nasal cavity, such as the beak, sinuses, choana, retrobulbar space, and palate (Malik et al. 2003). Cryptococcal rhinitis caused by *C. gattii* was also identified in the African grey parrot (Refai et al. 2014). Several cases of cryptococcosis have been reported in captive North Island brown kiwis (*Apteryx australis mantelli*) in Australia and in New Zealand (Hill et al. 1995; Malik et al. 2003); these birds are unusual in that they have a lower core body temperature than other avian species. Psittacines are occasionally infected by *C. neoformans* var. *grubii* or *C. gattii* (Cafarchia et al. 2006; Refai et al. 2014); e.g., *C. gattii* infection has been described in the pink-fronted cockatoo (Lester et al. 2004; Spickler 2013).

Clinical and laboratory findings have demonstrated minimally invasive subcutaneous disease in an Australian racing pigeon caused by *C. neoformans* var. *grubii* (Malik et al. 2003). In contrast, pigeons and other birds studied principally in America and Europe display a different pattern of disease, more suggestive of opportunistic infection of immunodeficient hosts.

12.6.3.2 Marsupials

Overall, cryptococcosis is the second most common infectious disease of koalas (*Phascolarctos cinereus*) after chlamydiosis. However, it appears more frequently than chlamydiosis in captive populations owing to border security and quarantine measures. *Cryptococcus* spp. can be cultured from the mucosal surface of the nasal cavity or the skin of most healthy koalas, i.e., animals without any evidence of tissue invasion or disease. Some cases progress to limited tissue invasion (as reflected by a positive antigen titer) but with no outward evidence of disease (Krockenberger et al. 2002). The species *C. gattii* is responsible for cryptococcosis in captive and wild koalas in Australia (Krockenberger et al. 2003). No age or sex predisposition has been observed, although animals are not colonized or infected until they leave the pouch. The respiratory tract is the primary target of disease. Although the lower respiratory tract is most commonly affected (60% of cases), 30% of cases had upper respiratory tract lesions, and 14% had both. Dissemination to the central nervous system (CNS) was common. Other tissues showing cryptococcal invasion included the lymph nodes, gastrointestinal tract, kidneys, spleen, and skin. Meningoencephalitis due to *C. gattii* had also been observed in a 10-year-old captive male koala (Spencer et al. 1993).

12.6.3.3 Dogs

In dogs, cryptococcosis typically begins in the nasal cavity (Fig. 12.2), overtly or covertly, gradually affecting the CNS, ocular structures (especially the retinae and optic nerves), lymph nodes, digestive system, bones, and other organs (Duncan et al. 2006). Dogs may present with signs of disease in the upper and/or lower respiratory tract and with subcutaneous granulomata and lesions reflecting disseminated disease (Fig. 12.3). CNS and ocular involvement is most common, however, and is associated with high mortality rate (Castella et al. 2008; de Abreu et al. 2017).

Cryptococcal infection of the upper respiratory tract was reported in 57 dogs residing in Western Australia (McGill et al. 2009), and the predominant etiological agent was found to be the VGII genotype of *C. gattii*, whereas molecular genotyping suggested that *C. gattii* infections in domestic animals on the eastern seaboard of Australia are less common than *C. neoformans* infections and mainly due to the VGI genotype. Lester et al. (2004) also detected cryptococcosis in 15 dogs from British Columbia, Canada, which were initially from Vancouver Island. Several additional

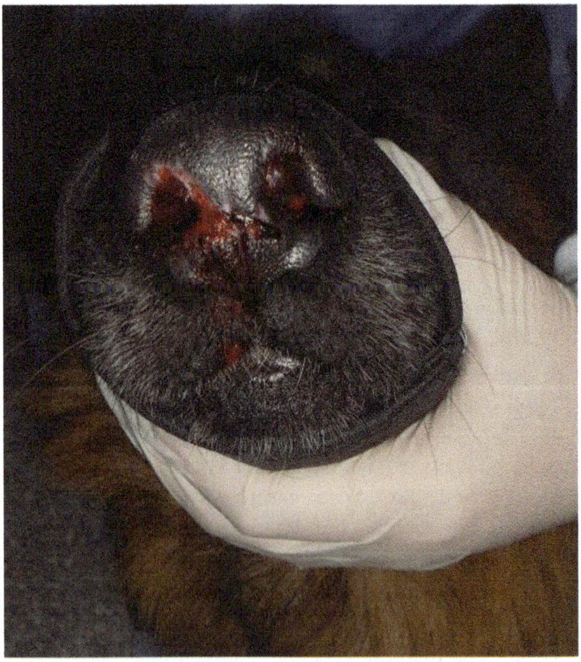

Fig. 12.2 Photograph of a young adult German shepherd dog with localized cryptococcosis of the nasal cavity with involvement of the nasal planum. This case was successfully treated with intralesional amphotericin B and oral fluconazole. Photo courtesy of Steve Metcalfe

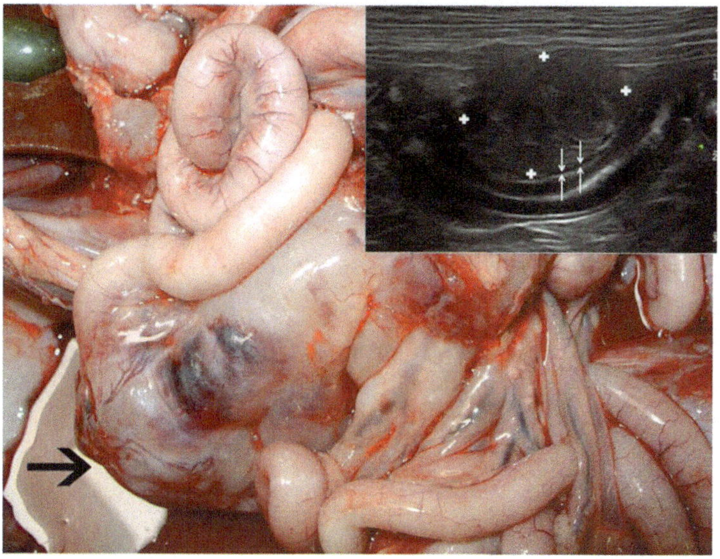

Fig. 12.3 Intestinal cryptococcomas (arrow) in a dog from Perth, Western Australia, due to *C. gattii* VGII infection. An ultrasound image showing the lesion is shown as an insert on the top right of the photograph. Images courtesy of Anna Tebb

cases of cryptococcal rhinitis, meningitis, disseminated cryptococcosis, and the CNS disease have been identified in dogs (Bowles and Fry 2009; Bryan et al. 2011; Trivedi et al. 2011).

Disseminated cryptococcosis is reported to be more common in dogs than in cats (Spickler 2013). However, subclinical infection and asymptomatic colonization of the sinonasal cavity by *C. gattii* have also been observed in dogs exposed to environments with high loads of *C. gattii* (Duncan et al. 2005). Dogs do not show any sex-related predisposition for cryptococcosis. American Cocker Spaniels, Labrador Retrievers, Doberman Pinschers, and Great Danes were overrepresented compared with other dog breeds, perhaps for genetic reasons or because of their tendency to be exposed to the great outdoors.

12.6.3.4 Cats

Feline cryptococcosis can be either focal or disseminated, and it mostly begins as mycotic rhinitis following asymptomatic colonization of the nasal cavity (Malik et al. 2003). Cryptococcosis represents the most common systemic mycosis of cats (Sorrell 2001; Duncan et al. 2006; McGill et al. 2009). Cats with cryptococcosis are most often brought to a veterinarian because of single or multiple, slowly enlarging cutaneous masses or ulcerative lesions (frequently located on the head in locations adjacent or contiguous with the sinonasal cavity or its draining lymph nodes) (Fig. 12.4). Upper respiratory signs observed include stertor, persistent sneezing, and/or mucopurulent nasal discharge. After long-standing disease, or when the

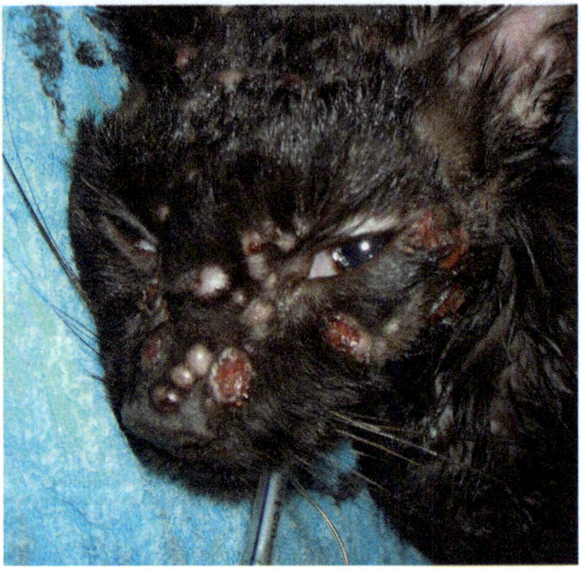

Fig. 12.4 Photograph of a cat with disseminated cryptococcosis. The presence of numerous cutaneous masses of different sizes is usually referable to hematogenous dissemination in a cat with immunodeficiency, e.g., due to long-standing feline immunodeficiency virus (FIV) infection. Note that some of the masses have ulcerated. Similar lesions were present over the entire integument of this patient

infection commences in the back of the nasal cavity, neurological signs can develop. In some cats with nasal cavity involvement, fleshy masses protrude from the nostril, or gross deformation of the nasal planum or bridge of the nose is apparent. Cats with cutaneous lesions or nasal cavity involvement often feel otherwise well but can be lethargic and lack appetite. Fever is uncommon and typically low-grade. Lower respiratory tract signs such as tachypnea are unusual but can occur as a result of cryptococcomas, pleural effusion, or severe mediastinal lymphadenomegaly. Neurological signs vary depending on the location of lesions in and around the CNS but include obtundation, abnormal behavior, ataxia, apparent blindness (typically peripheral, with widely dilated pupils that fail to respond to light), placing deficits, tremors, circling, head tilt, nystagmus, spinal pain, and/or focal or generalized seizures. Optic neuritis and chorioretinitis develop in as many as one-third of affected cats and almost always indicate CNS involvement (Malik et al. 2006b). Other reported signs include peripheral lymphadenomegaly (unassociated with skin lesions and often asymmetrical), lameness due to osteomyelitis or cryptococcal arthritis, and swollen digits. Rarely, the involvement of the abdominal lymph nodes, kidneys, spleen, liver, thyroid, and salivary glands occurs in cats (Sykes and Malik 2014).

No significant age- or sex-related propensity of the disease has been reported (Kerl 2003), although males might be affected slightly more often than females (Gerds-Grogan and Dayrell-Hart 1997). Among cats, Ragdoll and Birman breeds were found to be overrepresented (Refai et al. 2014), and this is most likely attributable to a breed-related defect in innate or cell-mediated immunity.

12.6.3.5 Other Carnivores

C. neoformans has been recovered from the brain of wild foxes (*Vulpes vulpes*) (Staib et al. 1985; Weller et al. 1985). Two cases of pulmonary cryptococcosis and extensive meningoencephalomyelitis have been documented in captive cheetahs (*Acinonyx jubatus*), the causal organism being *C. gattii* (Bolton et al. 1999). Pulmonary cryptococcosis and cryptococcal meningoencephalomyelitis have also been described in a king cheetah (*A. jubatus*) (Berry et al. 1997). Millward and Williams (2005) also reported a case of *C. neoformans* infection in a wild free-ranging cheetah. It is well known that cheetahs have suffered an evolutionary genetic bottleneck, and the marked overrepresentation of cases compared to other large felids (lions, tigers, leopards) is akin to the overrepresentation of specific breeds of cat referred to earlier.

Cryptococcosis has also been diagnosed in seven domestic ferrets (*Mustela putorius furo*) (five from Australia and two from western Vancouver Island) displaying a wide range of clinical signs (Malik et al. 2002). Cryptococcal rhinitis; lymphadenopathy and respiratory signs; localized masses on the nose, spine, or digits; gastrointestinal and neurological signs; pneumonia; meningitis; and disseminated disease were documented in the affected ferrets. *C. gattii* VGII molecular type accounted for the Australian ferrets. The first case of ferret cryptococcosis in the USA was documented in 2006; cytological examination and culture of an enlarged mandibular lymph node revealed *C. neoformans* var. *grubii* (serotype A) (Hanley et al. 2006). In Spain, a domestic ferret presenting with lymphadenopathy and acute bilateral blindness was found to be infected with *C. gattii* (Morera et al. 2011).

12.6.3.6 Rodents

Clinical presentations, similar to those in other species, have been found in various rodents. In 1922, Sangiorgi reported for the first time the presence of *Cryptococcus* spp. in a rat (*Rattus norvegicus*). Subsequently, *Cryptococcus* spp. have been isolated from both mice and rats (Emmons et al. 1947). The isolation of *C. neoformans* from the striped grass mouse (*Lemniscomys barbarus*) has also been reported (Bauwens et al. 2004); subsequently, *C. neoformans* var. *grubii* was recovered from apparently healthy greater bandicoot rats (*Bandicota indica*) and house mice (*Mus musculus castaneus*) (Singh et al. 2007, 2017). The yeast was also isolated from fecal samples of striped grass mice, bandicoot rats, and house mice. Of note, experimental rodent models such as mouse (*Mus musculus*) (Fig. 12.5), rat (*Rattus rattus*), and guinea pig (*Cavia porcellus*) have been used extensively in cryptococcal pathogenesis research and mammalian host responses to *Cryptococcus* (Sabiiti et al. 2012).

12.6.3.7 Ruminants

Cases of meningoencephalitis, mastitis, and pulmonary disease have been reported in goats (Stilwell and Pissarra 2014), while in sheep, cryptococcal infection has been shown to manifest as mastitis or rhinitis (Lemos et al. 2007). Only mastitis has been documented in water buffaloes (Spickler 2013). Clinical manifestations of cryptococcosis in camelids include the involvement of the lower respiratory tract and CNS

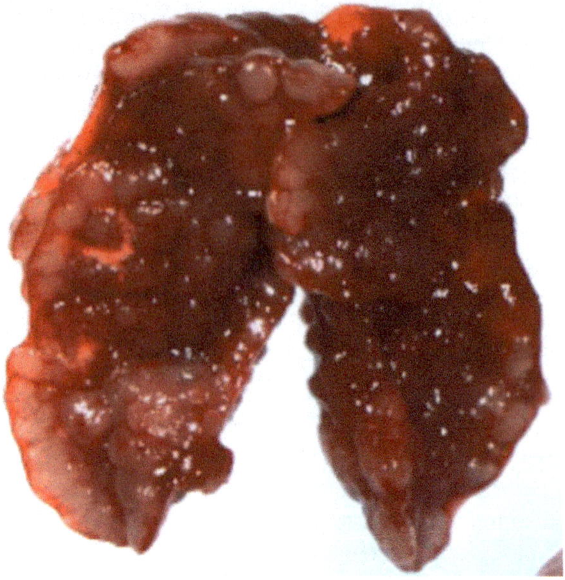

Fig. 12.5 Photograph of multiple pulmonary cryptococcosis in mouse experimentally inoculated with *C. gattii* via intranasal instillation

disease, as well as widely disseminated infections (Bildfell et al. 2002; Goodchild et al. 1996). In 1953, an outbreak of bovine cryptococcosis was reported (Simos et al. 1953). Pulmonary infections, CNS involvement, and mastitis have been observed in cattle infected with *Cryptococcus* spp. Cryptococcosis has additionally been recorded in an elk (*Cervus canadensis*), in which the diencephalon, mesencephalon, and brain stem were found to be affected (Refai et al. 2014). Cryptococcosis caused by *C. gattii* was reported in a white-tailed deer (*Odocoileus virginianus*) with nasopulmonary, lymph node, and CNS involvement (Overy et al. 2016). *C. gattii* has also been isolated from a black male alpaca in South American (Goodchild et al. 1996).

12.6.3.8 Equids

Riley et al. (1992) described the clinical and laboratory findings of cryptococcosis in seven horses from Australia (Riley et al. 1992). Most horses in this report had been exposed to areas where *Eucalyptus camaldulensis*, or the closely related *Eucalyptus rudis*, were growing. The species *C. gattii* was recovered from two of the horses. In pulmonary cryptococcosis, the lesions were either diffuse and multiple, with bilateral lung involvement, or localized mainly to the dorsocaudal region of one lung. The cases of widely disseminated cryptococcosis were associated with hematogenous spread of the fungus most likely after primary gastrointestinal infection and dissemination from abdominal lymph nodes. The localized form of the disease was associated with the inhalation of *C. gattii* VGIIb (Riley et al. 1992).

Studies on animals in Vancouver Island revealed *C. gattii* in the nasal passage of horses (Duncan et al. 2011). Although upper respiratory tract infection is the common manifestation of cryptococcosis in horses in many regions, other clinical signs, such as meningitis, pneumonia, osteomyelitis, endometritis, and mycotic placentitis, have also been documented (Refai et al. 2014). Another case of cryptococcosis in the form of cutaneous lesions was observed in a donkey (Refai et al. 2014).

12.6.3.9 Marine Mammals

Cryptococcosis has also been identified in an increasing number of marine mammal species on southwestern Vancouver Island, including Pacific white-sided dolphins (*Lagenorhynchus obliquidens*) and both harbor and Dall's porpoises (*Phocoena phocoena* and *Phocoenoides dalli*) (Bartlett et al. 2012; Stephen et al. 2002). *C. gattii* VGIIa has been documented in a harbor seal (*Phoca vitulina*) pup and in an unrelated adult (Rosenberg et al. 2016). Both animals showed generalized weakness, dehydration, respiratory compromise, minimally responsive mentation, and suboptimal body condition. Necropsy and histopathology findings showed generalized lymphadenopathy, bronchopneumonia, and meningoencephalitis with intralesional yeast and fungemia. Clinical cases of cryptococcosis have also been reported in a 7-year-old Atlantic bottlenose dolphin (*Tursiops truncatus*), a female Pacific white-sided dolphin (*L. obliquidens*), and a striped dolphin (*Stenella coeruleoalba*) kept in captivity (Migaki and Jones 1983; Nick Gales et al. 1985; Miller et al. 2002; Kidd et al. 2004). Rosenberg et al. (2016) recovered *C. gattii* VGIIa in a harbor seal juvenile pup and an adult. *C. albidus*, a traditionally nonpathogenic species, was recovered from a juvenile California sea lion (*Zalophus californianus*) (McLeland et al. 2012).

12.6.3.10 Nonhuman Primates and Other Mammals

Pulmonary cryptococcosis has been observed in a rhesus monkey (*Macaca mulatta*) (Pal et al. 1984). Multiple cryptococcosis were identified in the brain and heart of a 23-year-old captive-bred red-tailed guenon (*Cercopithecus ascanius*) (Helke et al. 2006). Cryptococcosis involving the respiratory tract and CNS was diagnosed in an Allen's swamp monkey (*Allenopithecus nigroviridis*), purple-faced leaf monkey (*Presbytis senex*), lion-tailed macaque (*Macaca silenus*), proboscis monkey (*Nasalis larvatus*) (Griner 1983; Barrie and Stadler 1995), tree shrews (*Tupaia tana* and *Tupaia minor*), and elephant shrews (*Macroscelides proboscideus*) (Tell et al. 1997).

12.7 Public Health Considerations and Potential Risk of Transmission to Humans

12.7.1 *Cryptococcus* Isolation from Synanthropic Rodents

Rodents play a pivotal role in the epidemiology of several fungal diseases. Although they usually inhabit wild environments, murid omnivores are well known for their association with humans (Guimarães et al. 2014). Rodents enrich the soil via their excrement, thereby supporting the growth of geophilic fungi. Almost a century ago,

Sangiorgi described the presence of *Cryptococcus* in the large mononuclear cells of the liver and spleen of a brown rat (*Rattus norvegicus*) (Sangiorgi 1922). During an investigation of histoplasmosis, Emmons et al. (1947) isolated *Cryptococcus* from mice and rats (Emmons et al. 1947). After a long gap, a wild black rat (*R. rattus*) trapped in Papua New Guinea was noted to have chronic, cystic pulmonary cryptococcosis (Scrimgeour and Purohit 1984). Naturally acquired cryptococcosis was again reported but, this time, in the greater bandicoot rat (*B. indica*) (Singh et al. 2007). Although pathological lesions were observed only in the liver and lungs, other organs such as the kidneys, spleen, and brain were found to be positive for *C. neoformans* var. *grubii*. Singh et al. (2007) additionally isolated *C. neoformans* var. *grubii* from bandicoot burrows and the surrounding bamboo debris. These findings suggest that *B. indica* acts as a sentinel species, potentially amplifying the pathogen in the environment, and it could serve as a possible vector of human cryptococcosis.

Recently, a cluster of cases of cryptococcosis was observed in a synanthropic Southeastern Asian murid (*Mus musculus castaneus*) (Singh et al. 2017). Unlike bandicoot rats, no lesions were recorded in any organ of these animals. Interestingly, *C. neoformans* var. *grubii* was recovered from cultures of tissue homogenates of the brain, lungs, liver, and kidneys. The habitat soil and fresh feces of the animals were also positive for the fungus. It is fascinating to note that, despite the presence of *Cryptococcus* in the central vein, neither the liver nor any other organs exhibited pathological signs. Apparently, the pathogen passes through the animal host without affecting it, and all isolates recovered from *M. musculus* were only weakly pathogenic to experimental mice. These findings define the status of *M. musculus* as a "passenger host" for *C. neoformans* var. *grubii* in a more appropriate manner. It is noteworthy that across the range of species studies, *Cryptococcus* yeasts have been isolated from apparently healthy rodents in most cases.

Of note, rodents—considered household pests and nuisance animals—may act as a continuous source of infection for humans and their pets, and in this context, it is important to think about cats as the major predator of rodents in many such situations. While rodents, especially rats and mice, have expanded their geographic range dramatically and significantly extended the territory of the pathogens they harbor (Aplin et al. 2011), they might play a role in preventing human infections by acting as sentinels for the presence of *Cryptococcus* in the environment (Morera et al. 2014). On the basis of the degree of interaction between the host and harbored pathogens, rodents might act as natural reservoirs, alternate hosts, sentinel animals, carriers, or even passenger hosts. Pulmonary exposure of experimental rats and mice with *C. neoformans* remains subclinical and does not usually lead to overt disease despite widespread dissemination, while *C. gattii* strains produce progressive and ultimately fatal pneumonia with late dissemination to the meninges as the terminal event (Coelho et al. 2014; Zaragoza et al. 2007; Goldman et al. 1994, 2000; Krockenberger et al. 2010).

12.7.2 Zoonotic Potential

Both *C. neoformans* and *C. gattii* cause infection in humans and animals, and both are thought to be acquired from the environment rather than from infected hosts (Spickler 2013). The feces of pet birds have been implicated as a possible source of *C. neoformans* infection in immunocompromised individuals. Case clusters of cryptococcosis tend to occur worldwide in individuals exposed to large quantities of bird guano (Muchmore et al. 1963; Procknow et al. 1965). There is no evidence, however, of an increased incidence of active cryptococcosis among these groups (Levitz 1991). Naturally acquired cryptococcosis occurs in both animals (wild and domestic) and humans, although no mammal-to-mammal transmission has been documented. A bird (magpie)-to-human transmission (Lagrou et al. 2005) and two possible cases of person-to-person transmission have been reported (Levitz 1991). The incidence of cryptococcal infection cannot be correlated with race or age. However, in certain situations, asymptomatic animals (koalas) have been found to contribute to the amplification of cryptococci in certain man-made environments and possibly in natural habitat as well (Malik et al. 1997; Krockenberger et al. 2002; Singh et al. 2007, 2017; Morera et al. 2014).

12.8 Diagnosis, Species Identification, and Genotyping

Preliminary diagnosis of cryptococcosis is made by direct microscopic examination of India ink preparations of cerebrospinal fluid samples or rapid staining (DiffQuik, Giemsa, etc.) of smears made from aspirates and nasal exudates (Kwon-Chung and Bennett 1992) (Fig. 12.6). In fresh material, India ink preparations often display

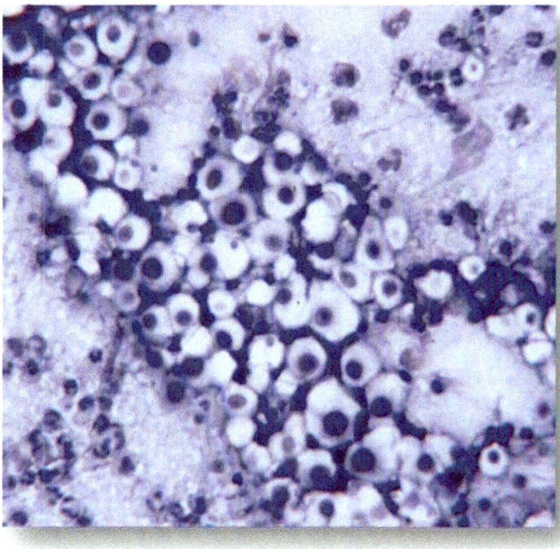

Fig. 12.6 Photomicrograph of a DiffQuik®-stained smear from nasal exudate from a cat with sinonasal cryptococcosis. The negative-staining capsule is characteristic, as is the rare narrow-necked budding

spherical vacuoles in Brownian motion (Mitchell and Perfect 1995) (Fig. 12.7). The capsules are generally spherical and may appear thick and vary in diameter.

If mycologically suitable and if suitable substrate is present in the media, culturing is the most suitable method for demonstrating the presence of pathogenic fungi (Vanessa et al. 2005). *Cryptococcus* may be cultured on routine mycological culture medium (Sabouraud glucose agar), on which it produces white, variably mucoid colonies when inoculated at 37 °C. Unlike other saprophytic species, *C. neoformans* produces laccase (phenol oxidase), which leads to the formation of melanin. Colonies turn brown to black and may thus be readily be identified (either *C. gattii* or *C. neoformans*); therefore, any medium, e.g., caffeic acid, Staib's medium, or tobacco agar medium, containing a substrate for phenol oxidase may serve as the primary isolation medium for *C. neoformans* and *C. gattii.* This is especially useful when sampling non-sterile sites (e.g., the nasal cavity) or the environment, where other fungi are often present. Both species may be recognized by color differentiation when grown on Christensen urea agar and L-canavanine-glycine-bromothymol blue (CGB) agar. Despite the incidence of false-positive and false-negative results, CGB medium is most widely used for differentiating between *C. gattii* and *C. neoformans* (McTaggart et al. 2011; Klein et al. 2009).

Histopathology, which may be used to investigate the presence of the fungus in tissue sections via corresponding tissue reactions, provides useful insights (Chander 2009). Cryptococcal cells may be easily detected in tissue sections of the lungs, brain, and liver using specific stains such as Gomori's methenamine silver nitrate, periodic acid–Schiff and Mayer's mucicarmine (Fig. 12.8). However, in

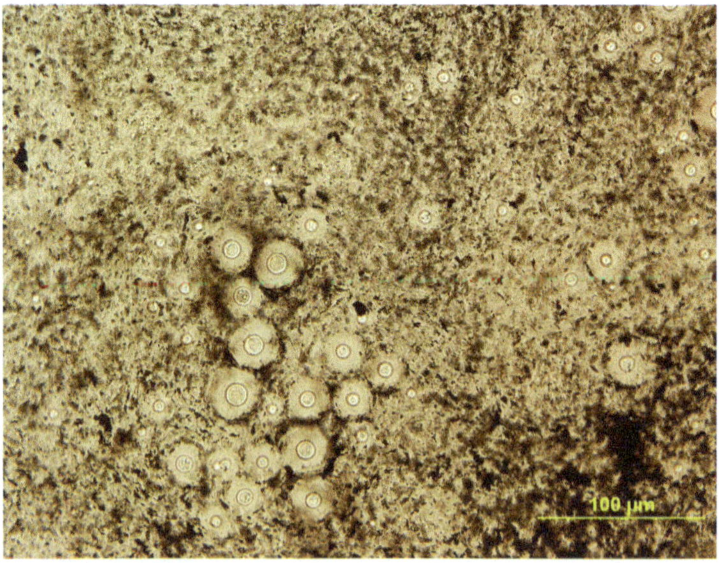

Fig. 12.7 Indian ink preparation of cerebrospinal fluid (CSF) from a dog with cryptococcal meningoencephalitis. Note the negatively stained capsules thrown into relief by the dark carbon material in the Indian ink stain

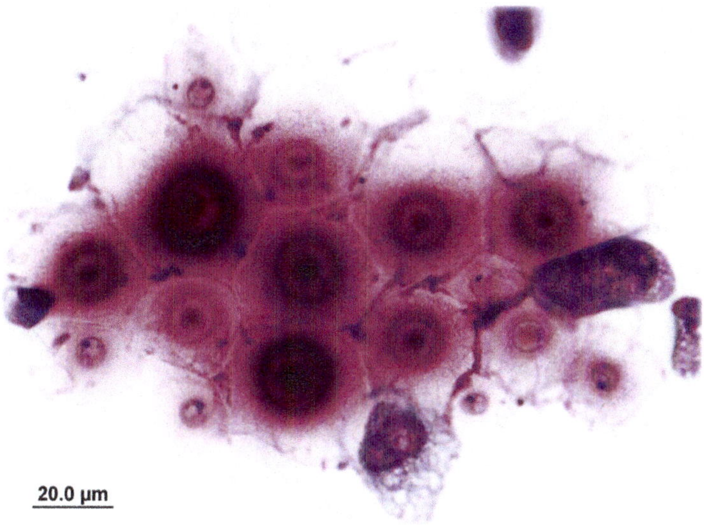

Fig. 12.8 Photomicrograph of histology from a dog with widely disseminated cryptococcosis. The mucicarmine stains the capsule bright pink. Image courtesy of Alan Kessel

hematoxylin–eosin-stained sections, tissue reactions are poorly demonstrated, and diagnosis therefore depends on the observation of cryptococcal capsules within the tissues.

Some physiological features of *C. neoformans*, such as urease production (in the presence of urea), have been used as the basis of differentiations between isolates (Singh et al. 2013). The presence of this enzyme can be determined and semiquantitated by the rapid urease test and using Christensen's agar medium. Similarly, in yeast carbon base medium, *C. gattii* assimilates D-proline, whereas *C. neoformans* usually fails to do so (Dufait et al. 1987). Wickerham and Burton (1948) have reemphasized the role of carbon assimilation tests in the taxonomy of *Cryptococcus*. In most clinical laboratories worldwide, this method has been replaced by several commercial kits (Kwon-Chung and Bennett 1992). The physiological tests commonly used for identification are fermentation of seven to eight carbohydrates, growth on various carbon and nitrogen sources, determination of growth requirements, growth on media with a high content of sugar, and testing for urea hydrolysis (Rij 1984). There is no correlation between the growth of yeast on a particular carbohydrate and fermentation of the same carbohydrate by that yeast. Yeasts vary in their ability to ferment sugars, which may be measured by the production of CO_2. *C. neoformans* and *C. gattii* assimilate sucrose and maltose but are usually lactose negative (Kwon-Chung and Bennett 1992).

Carbon assimilation tests determine the ability of yeast to grow aerobically on a particular carbon compound (Rij 1984). These tests may be performed either on solid or liquid media (Wickerham 1951). Yeast nitrogen base medium is most commonly used in carbon assimilation tests. Similarly, yeasts are additionally capable of utilizing a wide variety of nitrogen sources. Various nitrogen compounds, such as nitrate, nitrite, ethylamine hydrochloride imidazole, ammonium sulfate, glucosamine, creatine, and creatinine, are also tested by carbon assimilation tests.

The rapid detection of the presence of yeast in clinical specimens has also been performed by several molecular methods. Numerous molecular approaches have shown great potential in the field of mycology diagnostics during the last decade. Molecular methods based on the analysis of yeast nucleic acids were first used for the investigation of total nuclear DNA and the determination of the existence of co-specific relationships between morphologically and physiologically different strains (Alcoba-Flórez et al. 2007). Historically, different isolates of the same species were distinguished by biotyping or serotyping. Currently, various molecular techniques have superseded older methodologies, including PCR-fingerprinting, restriction fragment length polymorphism (RFLP) analysis, amplified fragment length polymorphism (AFLP) analysis, multilocus sequence typing (MLST), and microsatellite length polymorphism (MLP) using short tandem repeat (STR) markers. These methods enable the researcher to determine whether isolates belong to a single strain (biotype). In addition, next-generation sequencing of the whole fungal genome is becoming the most appropriate method for fungi with no MLP or MLST schemes available (Loftus et al. 2005; Alanio et al. 2017).

Karyotyping was one of the first molecular methods employed for the identification of *Cryptococcus* (Perfect et al. 1989). *C. neoformans* and *C. gattii* have been distinguished in terms of number and lengths of chromosomes by contour-clamped homogenous field gel electrophoresis and orthogonal-field-alternation gel electrophoresis (Polacheck and Lebens 1989; Boekhout et al. 1997; Hagen 2011).

Highly sensitive and rapid PCR amplification possess advantages over various molecular genotyping methods. Several PCR protocols have been used for the detection and identification of fungal species (Chang et al. 2001). However, methods of DNA extraction and purification have not been standardized to date (Alcoba-Flórez et al. 2007). Various PCR-based approaches, such as PCR with specific primers at the genus or species level, nested PCR, real-time PCR, multiplex PCR, PCR and posterior restriction analysis of amplicons, and PCR followed by sequencing, have been applied (Ahmad et al. 2002; Alcoba-Flórez et al. 2005, 2007). RFLP fingerprinting utilizes the rDNA locus to fingerprint the members of genus *Cryptococcus*, although *C. neoformans* could not be differentiated from *C. gattii* following the digestion of PCR products with four restriction enzymes (Vilgalys and Hester 1990; Hagen 2011). Researchers tested several hybridization probes such as the M13 phage core sequence $(GACA)_4$ and $(GTG)_5$, which are based on hypervariable, repetitive, and highly conserved DNA sequences/regions (Meyer et al. 1993; Meyer and Mitchel 1995). Hypervariable regions differing in sequences and/or length permit the identification of *Cryptococcus* at the species level. Overall, four genotypes were designated: VNI and VNII for *C. neoformans* var. *grubii* (serotype A), VNIII for hybrid (serotype AD), and VNIV for *C. neoformans* var. *neoformans* (serotype D) using random amplification of polymorphic DNA (RAPD) (Meyer et al. 1993). Subsequently, four genotypes were also introduced for *C. gattii*, namely, VGI, VGII, VGIII, and VGIV (Sorrell et al. 1996). The species

C. neoformans was subsequently investigated by the analysis of the intergenic spacer (IGS) region (rDNA sequence) and internal transcribed spacer (ITS), a phylogenetically important region of DNA (Diaz and Fell 2000; Diaz et al. 2005; Imai et al. 2000). IGS revealed significant differences between and within the varieties of *C. neoformans* and further indicated a similar grouping of *C. neoformans* isolates (Katsu et al. 2004; Hagen 2011). Furthermore, researchers combined the sequences of IGS, ITS, and topoisomerase and applied MLST (Sugita et al. 2001). Subsequently, a consensus MLST scheme was proposed (Meyer et al. 2009), comprising of seven unlinked nuclear loci CAP_{59}, GPD_1, IGS_1, LAC_1, PLB_1, SOD_1, and URA_5, with optional discriminatory power. Recently, multiple multilocus microsatellite typing (MLMT) schemes have been developed for the two sibling species of *Cryptococcus*; these are more reliable and cost-effective than MLST (Karaoglu et al. 2008; Illnait-Zaragozi et al. 2010).

A relatively novel method using the matrix-assisted laser desorption/ionization–time-of-flight (MALDI-TOF) mass spectrometry (MS)-based strategy has shown promising results to discriminate major molecular types within the *C. neoformans* and *C. gattii* species complexes obtained from human (Firacative et al. 2012) and veterinary sources (Danesi et al. 2014b). Currently, two MALDI-TOF MS systems are approved by the US Food and Drug Administration for the identification of yeasts: the Bruker Biotyper (Bruker Daltonics, Bremen, Germany) and the Vitek MS (bioMérieux, Marcy l'Etoile, France) systems. The Bruker instrument provides the MALDI Biotyper (software and database), and bioMérieux includes the Vitek MS and the SARAMIS (AnagnosTec, Germany) databases, referred to as the Vitek MS IVD system. Biotyper software generates log score values ranging from 0 to 3, with a value of ≥ 2 indicating species-level identification; scores between 1.7 and 1.99, genus-level identification; and scores below 1.7 unreliable identification. Meanwhile, using Vitek MS software, a confidence score value of $\geq 60\%$ is recommended for species-level identification (Posteraro et al. 2012). The species coverage of MALDI-TOF MS reference databases represents the Achilles' heel of each MALDI-TOF MS system. It is therefore recommended for users to create and supplement their own libraries of reference mass spectra by including locally important strains or species/strains not represented or not sufficiently represented in the commercial libraries (Posteraro et al. 2013).

Diagnosis of animal cryptococcosis can also be achieved by using serological assays that detect the capsular glucuronoxylomannan polysaccharide of *Cryptococcus* species in body fluids. Serum is the specimen of choice for antigen testing, although cerebrospinal fluid and urine can also be used (Boulware et al. 2014; Jarvis et al. 2011). Multiple assays have been developed and approved by the FDA for the detection of the capsular antigen of *Cryptococcus* species, including a *Cryptococcus* antigen (CrAg) latex agglutination system (CALAS) (Meridian Bioscience, Cincinnati, OH, USA) and a CrAg lateral flow assay (LFA;IMMY, Norman, OK, USA) (Temstet et al. 1992; Binnicker et al. 2012). In veterinary medicine, the CALAS assays have been most widely used to date. Antigen titers are frequently high to extremely high (>1:65,536) in cats, but even a titer of 1:2 can indicate cryptococcal infection (Sykes and Malik 2014). Titers typically decline with successful treatment, although the drop in titer generally lags behind the clinical improvement. False-negative results in cats appear to be very rare (sensitivity >95%) but can occur in cases of very localized infections; therefore, a negative titer does not completely rule out cryptococcosis (Belluco et al. 2008). The causes of false-positive test results in these situations remain unclear. Before treatment is initiated, the diagnosis of cryptococcosis should be confirmed using other diagnostic tests (cytology and culture) in cats with weakly positive antigen titers. The CALAS test is expensive and requires extra technological input; thus, the method is not practical in many developing countries, where cryptococcosis is endemic or enzootic. In contrast, the novel point-of-care CrAg LFA offers several advantages over the CALAS, including enhanced sensitivity for the detection of CrAg and a rapid turnaround time of approximately 15 min. In addition, the assay does not require pronase pretreatment of serum samples (Jarvis et al. 2011; McMullan et al. 2012).

12.9 Treatment

In animals, cryptococcosis may be treated with various antifungal drugs including amphotericin B, 5-fluorocytosine (also known as 5-flucytosine), fluconazole, itraconazole, ketoconazole, and terbinafine (Spickler 2013). The use of amphotericin B represented the first effective therapy for the treatment of cryptococcal infections. Flucytosine in combination with amphotericin B or an azole is recommended for the treatment of cryptococcosis in cats. However, the treatment of dogs with flucytosine is not recommended owing to the risk of potential severe toxic reactions such as toxic epidermal necrolysis (Malik et al. 1995; Panciera and Bevier 1987). Flucytosine should never be used alone because of the rapid development of resistance, and its use may be prohibitively expensive. In numerous areas, fluconazole is widely used in dogs and cats with cryptococcosis (Malik et al. 1992; O'Brien et al. 2006). As the original patent has expired, there are many excellent generic formulations available at quite a reasonable cost in most countries. In dogs, both fluconazole and itraconazole appear to be effective in the treatment of cryptococcosis, although they are less reliable than amphotericin B and best utilized for consolidation therapy (O'Brien et al. 2006). Ketoconazole is licensed for use in dogs only in France. In dogs and cats, ketoconazole has also been used successfully to treat systemic cryptococcosis (O'Brien et al. 2006), and it seems useful in some of the VGII cases encountered in Vancouver Island and surrounding areas.

In vitro studies have also shown that voriconazole, posaconazole, and isavuconazole are more active against *C. gattii* VGII and VGII than fluconazole (Yamazumi et al. 2000; Illnait-Zaragozi et al. 2008; Thompson et al. 2009). Importantly, fluconazole resistance has been reported at a much higher frequency and may have originated from the colonization of patients with AIDS undergoing long-term azole maintenance therapy (Pfaller et al. 2011). It is worth emphasizing that among *C. neoformans* and *C. gattii* isolates, the MIC and epidemiological cutoff values for fluconazole against *C. gattii* VGII and VGIII are substantially higher than in VGI and *C. neoformans* var. *grubii* isolates (Espinel-Ingroff et al. 2012). Alterations in

ERG11, which encodes a lanosterol 14α-demethylase, have been reported (Sionov et al. 2012).

In addition, surgery can be performed to reduce the size of mass lesions (Spickler 2013). Supportive therapy may be required to treat increased intracranial pressure during meningitis using judicious doses of corticosteroids for the first few days of therapy in patients with CNS disease.

12.10 Prevention

Animals should be restricted from contact with possible sources of *C. neoformans* infection such as areas that contain abundant pigeon droppings. Furthermore, the accumulation of bird droppings (especially from pigeons) should be avoided. Prevention of *C. gattii* infections is challenging because it is difficult to restrict contact with contaminated soil and tree bark. Suggested disinfectants for housing facilities include accelerated hydrogen peroxide, potassium peroxymonosulfate, 1.2% sodium hypochlorite, 0.5% chlorhexidine, and phenolic disinfectants, although the latter must be washed off as it irritates the skin and mucosa of dogs, cats, and koalas. *C. neoformans* is also killed by exposure to moist heat at 121 °C for a minimum of 20 min or dry heat at 165–170 °C for 2 h.

12.11 Conclusion

Pathogenic members of both *Cryptococcus* species complexes are capable of causing different clinical diseases in a wide range of living organisms worldwide, including domestic and experimental animals, as well as wildlife. Only experimental infection has been documented in invertebrates and fishes.

The *C. neoformans* species complex has a global distribution but is particularly prevalent in southeastern and Western Australia, British Columbia in Canada, and the West Coast of the USA. The *C. gattii* species complex appears to be present in hotspots around the world; recent outbreaks have been reported in the Pacific Northwest USA and British Columbia. The complete prevention of exposure might be impossible to achieve, although in some circumstances, it may be possible to decrease the level of exposure from certain environmental sources. Species other than *C. neoformans* and *C. gattii* have only rarely been reported to affect animals.

Despite similarities, the clinical manifestations of cryptococcosis in animals differ from those in humans. Animals and humans might be infected from the same environmental source. To date, mammal-to-mammal transmission has not been documented. Household rodents may also serve as natural reservoirs and source of infection for humans and their pets.

Conflict of Interests The work of Seyedmojtaba Seyedmousavi was supported by the Intramural Research Program of the NIH, NIAID.

References

Ahmad S, Khan Z, Mustafa AS, Khan ZU (2002) Seminested PCR for diagnosis of candidemia: comparison with culture, antigen detection, and biochemical methods for species identification. J Clin Microbiol 40(7):2483–2489

Ajello L (1958) Occurrence of *Cryptococcus neoformans* in soils. Am J Epidemiol 67(1):72–77

Alanio A, Desnos-Ollivier M, Garcia-Hermoso D, Bretagne S (2017) Investigating clinical issues by genotyping of medically important fungi: why and how? Clin Microbiol Rev 30(3):671–707

Alcoba-Flórez J, Méndez-Álvarez S, Cano J, Guarro J, Pérez-Roth E, del Pilar Arévalo M (2005) Phenotypic and molecular characterization of *Candida nivariensis* sp. nov., a possible new opportunistic fungus. J Clin Microbiol 43(8):4107–4111

Alcoba-Flórez J, del Pilar Arévalo-Morales M, Pérez-Roth E, Laich F, Rivero-Pérez B, Méndez-Álvarez S (2007) Yeast molecular identification and typing. Communicating current research and educational topics and trends in applied microbiology. Formatex Research Center, Extremadura, pp 535–546

Amirrajab N, Haghani I, Rasuli M, Shokohi T (2016) Migratory birds as a potential reservoirs of *Cryptococcus neoformans*. Int J Environ Res 10(3):459–464

Apidianakis Y, Rahme LG, Heitman J, Ausubel FM, Calderwood SB, Mylonakis E (2004) Challenge of Drosophila melanogaster with *Cryptococcus neoformans* and role of the innate immune response. Eukaryot Cell 3:413–419

Aplin KP, Suzuki H, Chinen AA, Chesser RT, Ten Have J, Donnellan SC, Austin J, Frost A, Gonzalez JP, Herbreteau V, Catzeflis F, Soubrier J, Fang YP, Robins J, Matisoo-Smith E, Bastos AD, Maryanto I, Sinaga MH, Denys C, Van Den Bussche RA, Conroy C, Rowe K, Cooper A (2011) Multiple geographic origins of commensalism and complex dispersal history of Black Rats. PLoS One 6:e26357

Barclay WP, Delahunta A (1979) Cryptococcal meningitis in a horse. J Am Vet Med Assoc 174(11):1236

Barnett JA (2010) A history of research on yeasts 14: medical yeasts part 2, *Cryptococcus neoformans*. Yeast 27:875–904

Barrie MT, Stadler CK (1995) Successful treatment of *Cryptococcus neoformans* infection in an Allen's swamp monkey (*Allenopithecus nigroviridis*) using fluconazole and flucytosine. J Zoo Wildl Med 26:109–114

Bartlett KH, Cheng PY, Duncan C, Galanis E, Hoang L, Kidd S, Lee MK, Lester S, MacDougall L, Mak S, Morshed M, Taylor M, Kronstad J (2012) A decade of experience: *Cryptococcus gattii* in British Columbia. Mycopathologia 173(5–6):311–319

Bauwens L, Swinne D, Vroey C, Meurichy WD (1986) Isolation of *Cryptococcus neoformans* var. *neoformans* in the Aviaries of the Antwerp Zoological Gardens/Isolation von *Cryptococcus neoformans* var. *neoformans* im Vogelhaus des Antwerpener Zoos. Mycoses 29(7):291–294

Bauwens L, Vercammen F, Wuytack C, Van Looveren K, Swinne D (2004) Isolation of *Cryptococcus neoformans* in Antwerp Zoo's nocturnal house. Mycoses 47(7):292–296

Bava AJ, Negroni R (1992) The epidemiological characteristics of 105 cases of cryptococcosis diagnosed in the Republic of Argentina between 1981–1990. Rev Inst Med Trop Sao Paulo 34:335–340

Belluco S, Thibaud JL, Guillot J, Krockenberger MB, Wyers M, Blot S, Colle MA (2008) Spinal cryptococcoma in an immunocompetent cat. J Comp Pathol 139:246–251

Berry WL, Jardine JE, Espie IW (1997) Pulmonary cryptococcoma and cryptococcal meningoencephalomyelitis in a king cheetah (Acinonyx jubatus). J Zoo Wildl Med 28:485–490

Bildfell RJ, Long P, Sonn R (2002) Cryptococcosis in a llama (Lama glama). J Vet Diagn Investig 14:337–339

Binnicker MJ, Jespersen DJ, Bestrom JE, Rollins LO (2012) Comparison of four assays for the detection of cryptococcal antigen. Clin Vaccine Immunol 19(12):1988–1990

Boekhout T, Van Belkum A, Leenders AC, Verbrugh HA, Mukamurangwa P, Swinne D, Scheffers WA (1997) Molecular typing of *Cryptococcus neoformans*: taxonomic and epidemiological aspects. Int J Syst Bacteriol 47(2):432–442

Boekhout T, Theelen B, Diaz M, Fell JW, Hop WC, Abeln EC, Dromer F, Meyer W (2001) Hybrid genotypes in the pathogenic yeast *Cryptococcus neoformans*. Microbiology 147:891–907

Bolton LA, Lobetti RG, Evezard DN, Picard JA, Nesbit JW, Van Heerden J, Burroughs REJ (1999) Cryptococcosis in captive cheetah (*Acinonyx jubatus*): two cases: case report. J S Afr Vet Assoc 70(1):35–39. https://doi.org/10.4102/jsava.v70i1.748

Boulware DR, Rolfes MA, Rajasingham R, Vonohenberg M, Qin Z, Taseera K, Schutz C, Kwizera R, Butler EK, Meintjes G, Muzoora C, Bischof JC, Meya DB (2014) Multisite validation of cryptococcal antigen lateral flow assay and quantification by laser thermal contrast. Emerg Infect Dis 20:45–53

Bowles DB, Fry DR (2009) Nasal cryptococcosis in two dogs in New Zealand. N Z Vet J 57(1): 53–57

Bryan HM, Darimont CT, Paquet PC, Ellis JA, Goji N, Maëlle Gouix M, Smits JE (2011) Exposure to infectious agents in dogs in remote coastal British Columbia: possible sentinels of diseases in wildlife and humans. Can J Vet Res 75(1):11–17

Buchanan KL, Murphy JW (1998) What makes *Cryptococcus neoformans* a pathogen? Emerg Infect Dis 4(1):71–83

Byrnes EJ, Li WJ, Lewit Y, Ma HS, Voelz K, Ren P, Carter DA, Chaturvedi V, Bildfell RJ, May RC, Heitman J (2010) Emergence and pathogenicity of highly virulent *Cryptococcus gattii* genotypes in the Northwest United States. PLoS Pathog 6(4):e1000850

Cafarchia C, Romito D, Iatta R, Camarda A, Montagna MT, Otranto D (2006) Role of birds of prey as carriers and spreaders of *Cryptococcus neoformans* and other zoonotic yeasts. Med Mycol 44 (6):485–492

Casadevall A, Perfect JR (1998) *Cryptococcus neoformans*, vol 595. ASM Press, Washington, DC

Casadevall A, Pirofski LA (1999) Host-pathogen interactions: redefining the basic concepts of virulence and pathogenicity. Infect Immun 67(8):3703–3713

Casadevall A, Steenbergen JN, Nosanchuk JD (2003) 'Ready made' virulence and 'dual use' virulence factors in pathogenic environmental fungi—the *Cryptococcus neoformans* paradigm. Curr Opin Microbiol 6(4):332–337

Casadevall A, Freij JB, Hann-Soden C, Taylor J (2017) Continental drift and speciation of the Cryptococcus neoformans and *Cryptococcus gattii* species complexes. mSphere 2: e00103–e00117

Castanon-Olivares LR, Arreguin-Espinosa R, Ruiz-Palacios Y, Santos G, Lopez-Martinez R (2000) Frequency of *Cryptococcus* species and varieties in Mexico and their comparison with some Latin American countries. Rev Latinoam Microbiol 42:35–40

Castella G, Abarca ML, Cabanes FJ (2008) Cryptococcosis and pets. Rev Iberoam Micol 25(1): S19–S24

Chakrabarti A, Jatana M, Kumar P, Chatha L, Kaushal A, Padhye AA (1997) Isolation of *Cryptococcus neoformans* var. *gattii* from Eucalyptus camaldulensis in India. J Clin Microbiol 35:3340–3342

Chander J (2009) A text book of medical mycology. Mehta Publishers, New Delhi, pp 266–290

Chang YC, Kwon-Chung KJ (1998) Isolation of the third capsule associated gene, CAP60, required for virulence in *Cryptococcus neoformans*. Infect Immun 66:2230–2236

Chang HC, Lea SN, Huang AH, Wu TL, Chang TC (2001) Rapid identification of yeasts in positive blood cultures by a multiplex PCR method. J Clin Microbiol 39(10):3466–3471

Chapman HM, Robinson WF, Bolton JR, Robertson JP (1990) *Cryptococcus neoformans* infection in goats. Aust Vet J 67(7):263–265

Chen H, Detmer SA, Ewald AJ, Griffin EE, Fraser SE, Chan DC (2003) Mitofusins Mfn1 and Mfn2 coordinately regulate mitochondrial fusion and are essential for embryonic development. J Cell Biol 160(2):189–200

Chen LC, Blank ES, Casadevall A (1996) Extracellular proteinase activity of *Cryptococcus neoformans*. Clin Diagn Lab Immunol 3:570–574

Chen SC, Currie BJ, Campbell HM, Fisher DA, Pfeiffer TJ, Ellis DH, Sorrell TC (1997a) *Cryptococcus neoformans* var. *gattii* infection in northern Australia: existence of an environmental source other than known host eucalypts. Trans R Soc Trop Med Hyg 91(5):547–550

Chen SCA, Muller M, Jin Zhong Z, Wright LC, Sorrell TC (1997b) Phospholipase activity in *Cryptococcus neoformans*: a new virulence factor? J Infect Dis 175:414–420

Clancy MN, Fleischmann J, Howard DH, Kwon-Chung KJ, Shimizo RY (1990) Isolation of *Cryptococcus neoformans gattii* from a patient with AIDS in Southern California. J Infect Dis 161:809

Coelho C, Bocca AL, Casadevall A (2014) The intracellular life of *Cryptococcus neoformans*. Annu Rev Pathol 9:219–238

Cogliati M (2013) Global molecular epidemiology of *Cryptococcus neoformans* and *Cryptococcus gattii*: an atlas of the molecular types. Scientifica 2013:675213

Cogliati M, D'Amicis R, Zani A, Montagna MT, Caggiano C, De Giglio O et al (2016) Environmental distribution of *Cryptococcus neoformans* and *C. gattii* around the Mediterranean basin. FEMS Yeast Res 16(4):fow045

Connolly JH, Krockenberger MB, Malik R, Canfield PJ, Wigney DI, Muir DB (1999) Asymptomatic carriage of Cryptococcus neoformans in the nasal cavity of the koala (*Phascolarctos cinereus*). Med Mycol 37:331–338

Crabtree JN, Okagaki LH, Wiesner DL, Strain AK, Nielsen JN, Nielsen K (2012) Titan cell production enhances the virulence of *Cryptococcus neoformans*. Infect Immun 80:3776–3785

Danesi P, Furnari C, Granato A, Schivo A, Otranto D, Capelli G, Cafarchia C (2014a) Molecular identity and prevalence of *Cryptococcus* spp. nasal carriage in asymptomatic feral cats in Italy. Med Mycol 52(7):667–673

Danesi P, Drigo I, Iatta R, Firacative C, Capelli G, Cafarchia C, Meyer W (2014b) MALDI-TOF MS for the identification of veterinary non-*C. neoformans–C. gattii Cryptococcus* spp. isolates from Italy. Med Mycol 52(6):659–666

Davis JM, Huang M, Botts MR, Hull CM, Huttenlocher AA (2016) Zebrafish model of cryptococcal infection reveals roles for macrophages, endothelial cells, and neutrophils in the establishment and control of sustained fungemia. Infect Immun 84(10):3047–3062

de Abreu DPB, Machado CH, Makita MT, Botelho CFM, Oliveira FG, da Veiga CCP, Martins MDA, Baroni FA (2017) Intestinal lesion in a dog due to *Cryptococcus gattii* type VGII and review of published cases of canine gastrointestinal cryptococcosis. Mycopathologia 182(5–6): 597–602

de Jesus MS, Rodrigues WC, Barbosa G et al (2012) *Cryptococcus neoformans* carried by Odontomachus bauri ants. Mem Inst Oswaldo Cruz 107(4):466–469

Desalermos A, Tan X, Rajamuthiah R, Arvanitis M, Wang Y, Li D, Kourkoumpetis TK, Fuchs BB, Mylonakis E (2015) A multi-host approach for the systematic analysis of virulence factors in *Cryptococcus neoformans*. J Infect Dis 211:298–305

Desnos-Ollivier M, Patel S, Spaulding AR, Charlier C, Garcia-Hermoso D, Nielsen K, Dromer F (2010) Mixed infections and *in vivo* evolution in the human fungal pathogen *Cryptococcus neoformans*. MBio 1(1):e00091–e00010

DeVroey C, Swinne D (1986) Isolements de *Cryptococcus neoformans* a l'oc-casion de concours de chant de canaries. Bull Soc Fr Mycol Med 15:353–356

Diaz MR, Fell JW (2000) Molecular analyses of the IGS & ITS regions of rDNA of the psychrophilic yeasts in the genus Mrakia. Antonie Van Leeuwenhoek 77(1):7–12

Diaz MR, Boekhout T, Kiesling T, Fell JW (2005) Comparative analysis of the intergenic spacer regions and population structure of the species complex of the pathogenic yeast *Cryptococcus neoformans*. FEMS Yeast Res 5:1129–1140

Dufait R, Velho R, Vroey C (1987) Rapid identification of the two varieties of *Cryptococcus neoformans* by D-Proline assimilation/Eine Schnellmethode zur Identifizierung der beiden Varietäten von *Cryptococcus neoformans* mit Hilfe der D-Prolin-assimilation. Mycoses 30(10):483–483

Duncan C, Stephen C, Lester S, Bartlett KH (2005) Sub-clinical infection and asymptomatic carriage of *Cryptococcus gattii* in dogs and cats during an outbreak of cryptococcosis. Med Mycol 43(6):511–516

Duncan C, Schwantje H, Stephen C, Campbell J, Bartlett K (2006) *Cryptococcus gattii* in wildlife of Vancouver Island, British Columbia, Canada. J Wildl Dis 42(1):175–178

Duncan C, Bartlett KH, Lester S, Bobsien B, Campbell J, Stephen C, Raverty S (2011) Surveillance for *Cryptococcus gattii* in horses of Vancouver Island, British Columbia, Canada. Med Mycol 49(7):734–738

Elhariri M, Hamza D, Elhelw R, Refai M (2016) Eucalyptus tree: a potential source of *Cryptococcus neoformans* in Egyptian environment. Int J Microbiol 2016:4080725

Ellis DH, Pfeiffer TJ (1990) Natural habitat of *Cryptococcus neoformans* var. *gattii*. J Clin Microbiol 28(7):1642–1644

Emmons CW (1951) Isolation of *Cryptococcus neoformans* from soil. J Bacteriol 62(6):685

Emmons CW (1952) *Cryptococcus neoformans* strains from a severe outbreak of bovine mastitis. Mycopathol Mycol Appl 6(3):231–234

Emmons CW (1955) Saprophytic sources of *Cryptococcus neoformans* associated with the pigeon (Columba livia). Am J Epidemiol 62(3):227–232

Emmons CW, Bell JA, Olson BJ (1947) Naturally occurring histoplasmosis in Mus-musculus and Rattus-norvegicus. Pub Health Rep 62(46):1642–1646

Espinel-Ingroff A, Chowdhary A, Cuenca-Estrella M et al (2012) *Cryptococcus neoformans–Cryptococcus gattii* species complex: an international study of wild-type susceptibility endpoint distributions and epidemiological cutoff values for amphotericin B and flucytosine. Antimicrob Agents Chemother 56:3107–3113

Farmer SG, Komorowski RA (1973) Histologic response to capsule-deficient *Cryptococcus neoformans*. Arch Pathol 96(6):383–387

Firacative C, Trilles L, Meyer W (2012) MALDI-TOF MS enables the rapid identification of the major molecular types within the *Cryptococcus neoformans/C. gattii* species complex. PLoS One 7(5):e37566

Frasés S, Ferrer C, Sánchez M, Colom-Valiente MF (2009) Molecular epidemiology of isolates of the *Cryptococcus neoformans* species complex from Spain. Rev Iberoam Micol 26(2):112–117

Gerds-Grogan S, Dayrell-Hart B (1997) Feline cryptococcosis: a retrospective evaluation. J Am Anim Hosp Assoc 33(2):118–122

Gerstein AC, Fu MS, Mukaremera L, Li Z, Ormerod KL, Fraser JA, Berman J, Nielsen K (2015) Polyploid titan cells produce haploid and aneuploidy progeny to promote stress adaptation. mBio 6:e01340–e01315

Ghannoum MA (2000) Potential role of phospholipases in virulence and fungal pathogenesis. Clin Microbiol Rev 13:122–143

Goldman D, Lee SC, Casadevall A (1994) Pathogenesis of pulmonary *Cryptococcus neoformans* infection in the rat. Infect Immun 62(11):4755–4761

Goldman DL, Lee SC, Mednick AJ, Montella L, Casadevall A (2000) Persistent *Cryptococcus neoformans* pulmonary infection in the rat is associated with intracellular parasitism, decreased inducible nitric oxide synthase expression, and altered antibody responsiveness to cryptococcal polysaccharide. Infect Immun 68(2):832–838

Goodchild LM, Dart AJ, Collins MB, Dart CM, Hodgson JL, Hodgson DR (1996) Cryptococcal meningitis in an alpaca. Aust Vet J 74:428–430

Griner LA (1983) Pathology of zoo animals. Zoological Society of San Diego, San Diego

Grover N, Nawange SR, Naidu J et al (2007) Ecological niche of *Cryptococcus neoformans* var. *grubii* and *Cryptococcus gattii* in decaying wood of trunk hollows of living trees in Jabalpur City of Central India. Mycopathologia 164:159–170

Guimarães AO, Valença FM, Sousa JBS, Souza SA, Madi RR, de Melo CM (2014) Arasitic and fungal infections in synanthropic rodents in an area of urban expansion, Aracaju, Sergipe State, Brazil. Acta Sci Biol Sci 36:113–120

Guinea J, Hagen F, Peláez T, Boekhout T, Tahoune H, Torres-Narbona M, Bouza E (2010) Antifungal susceptibility, serotyping, and genotyping of clinical *Cryptococcus neoformans* isolates collected during 18 years in a single institution in Madrid, Spain. Med Mycol 48(7): 942–948

Hagen F (2011) *Cryptococcus gattii* and *Cryptococcus neoformans*-cosmopolitans on the move. Ph.D. Thesis, Utrecht University, Utrecht

Hagen F, Khayhan K, Theelen B, Kolecka A, Polacheck I, Sionov E, Falk R, Parnmen S, Lumbsch HT, Boekhout T (2015) Recognition of seven species in the *Cryptococcus gattii/Cryptococcus neoformans* species complex. Fungal Genet Biol 78:16–48

Hanley CS, MacWilliams P, Giles S, Paré J (2006) Diagnosis and successful treatment of *Cryptococcus neoformans* variety grubii in a domestic ferret. Can Vet J 47(10):1015–1017

Harris J, Lockhart S, Chiller T (2012) *Cryptococcus gattii*: where do we go from here? Med Mycol 50:113–129

Headley SA, Mota FC, Lindsay S, de Oliveira LM, Medeiros AA, Pretto-Giordano LG, Saut JP, Krockenberger M (2016) *Cryptococcus neoformans* var. *grubii*-induced arthritis with encephalitic dissemination in a dog and review of published literature. Mycopathologia 181(7–8): 595–601

Hedayati MT, Mayahi S, Fakhar M, Shokohi T, Majidi M (2011) *Cryptococcus neoformans* isolation from swallow (*Hirundo rustica*) excreta in Iran. Rev Inst Med Trop Sao Paulo 53(3):125–127

Helke KL, Denver MC, Bronson E, Mankowski JL (2006) Disseminate cryptococcosis in a guenon (Cercopithecus ascanius). Vet Pathol 43:75–78

Hill FI, Woodgyer AJ, Lintott MA (1995) Cryptococcosis in a North Island brown kiwi (Apteryx australis mantelli) in New Zealand. J Med Vet Mycol 33:305–309

Holmes CB, Losina E, Walensky RP, Yazdanpanah Y, Freedberg KA (2003) Review of human immunodeficiency virus type 1-related opportunistic infections in sub-Saharan Africa. Clin Infect Dis 36(5):652–662

Hough I (1998) Cryptococcosis in an eastern water skink. Aust Vet J 76:471–472

Idnurm A, Bahn YS, Nielsen K, Lin X, Fraser JA, Heitman J (2005) Deciphering the model pathogenic fungus *Cryptococcus neoformans*. Nat Rev Microbiol 3(10):753–764

Illnait-Zaragozi MT, Martínez GF, Curfs-Breuker I, Fernàndez CM, Boekhout T, Meis JF (2008) *In vitro* activity of the new azole isavuconazole (BAL4815) compared with six other antifungal agents against 162 *Cryptococcus neoformans* isolates from Cuba. Antimicrob Agents Chemother 52:1580–1582

Illnait-Zaragozi MT, Martínez-Machín GF, Fernández-Andreu CM, Boekhout T, Meis JF, Klaassen CHW (2010) Microsatellite typing of clinical and environmental *Cryptococcus neoformans* var. *grubii* isolates from Cuba shows multiple genetic lineages. PLoS One 5(2):e9124

Imai T, Watanabe K, Tamura M et al (2000) Geographic grouping of *Cryptococcus neoformans* var. *gattii* by random amplified polymorphic DNA fingerprinting patterns and ITS sequence divergence. Clin Lab 46:345–354

Jacobson ES, Tinnell SB (1993) Antioxidant function of fungal melanin. J Bacteriol 175(21): 7102–7104

Jacobson ES, Jenkins ND, Todd JM (1994) Relationship between superoxide dismutase and melanin in a pathogenic fungus. Infect Immun 62(9):4085–4086

Jain N, Li L, McFadden DC, Banarjee U, Wang X, Cook E, Fries BC (2006) Phenotypic switching in a *Cryptococcus neoformans* variety gattii strain is associated with changes in virulence and promotes dissemination to the central nervous system. Infect Immun 74:896–903

Jarvis JN, Percival A, Bauman S et al (2011) Evaluation of a novel point-of-care cryptococcal antigen test on serum, plasma, and urine from patients with HIV-associated cryptococcal meningitis. Clin Infect Dis 53:1019–1023

Karaoglu H, Lee CMY, Carter D, Meyer W (2008) Development of polymorphic microsatellite markers for *Cryptococcus neoformans*. Mol Ecol Resour 8(5):1136–1138
Katsu M, Kidd S, Ando A, Moretti-Branchini ML, Mikami Y, Nishimura K, Meyer W (2004) The internal transcribed spacers and 58 S rRNA gene show extensive diversity among isolates of the *Cryptococcus neoformans* species complex. FEMS Yeast Res 4(4–5):377–388
Keah KC, Parameswari S, Cheong YM (1994) Serotypes of clinical isolates of *Cryptococcus neoformans* in Malaysia. Trop Biomed 11:205–207
Kerl ME (2003) Update on canine and feline fungal diseases. Vet Clin N Am Small Anim Pract 33(4):721–747
Kidd SE, Hagen F, Tscharke RL, Huynh M, Bartlett KH, Fyfe M, MacDougall L, Boekhout T, Kwon-Chung KJ, Meyer W (2004) A rare genotype of *Cryptococcus gattii* caused the cryptococcosis outbreak on Vancouver Island (British Columbia, Canada). Proc Natl Acad Sci USA 101:17258–17263
Klein E (1901) Pathogenic microbes in milk. J Hyg (Lond) 1:78–95
Klein KR, Hall L, Deml SM, Rysavy JM, Wohlfiel SL, Wengenack NL (2009) Identification of *Cryptococcus gattii* by use of L-canavanine glycine bromothymol blue medium and DNA sequencing. J Clin Microbiol 47(11):3669–3672
Kozel TR (1995) Virulence factors of *Cryptococcus neoformans*. Trends Microbiol 3(8):295–299
Krockenberger M, Canfield PJ, Malik R (2002) *Cryptococcus neoformans* in the koala (*Phascolarctos cinereus*): colonization by *C. n.* var. *gattii* and investigation of environmental sources. Med Mycol 40:263–272
Krockenberger MB, Canfield PJ, Malik R (2003) *Cryptococcus neoformans* var. *gattii* in the koala (*Phascolarctos cinereus*): a review of 43 cases of cryptococcosis. Med Mycol 41:225–234
Krockenberger MB, Malik R, Ngamskulrungroj P, Trilles L, Escandon P, Dowd S, Allen C, Himmelreich U, Canfield PJ, Sorrell TC, Meyer W (2010) Pathogenesis of pulmonary *Cryptococcus gattii* infection: a rat model. Mycopathologia 170(5):315–330
Kronstad JW, Attarian R, Cadieux B, Choi J, D'Souza CA, Griffiths EJ, Geddes JM, Hu G, Jung WH, Kretschmer M, Saikia S, Wang J (2011) Expanding fungal pathogenesis: *Cryptococcus* breaks out of the opportunistic box. Nat Rev Microbiol 9(3):193–203
Kwon-Chung KJ (1975) A new genus, Filobasidiella, the perfect state of *Cryptococcus neoformans*. Mycologia 67:1197–1200
Kwon-Chung KJ (1976) Morphogenesis of *Filobasidiella neoformans*, the sexual state of *Cryptococcus neoformans*. Mycologia 68:821–833
Kwon-Chung KJ, Bennett JE (1978) Distribution of alpha and alpha mating types of *Cryptococcus neoformans* among natural and clinical isolates. Am J Epidemiol 108:337–340
Kwon-Chung KJ, Bennett JE (1984) Epidemiologic differences between the two varieties of *Cryptococcus neoformans*. Am J Epidemiol 120:123–130
Kwon-Chung KJ, Bennett JE (1992) Medical Mycology. Lea & Febiger, Philadelphia
Kwon-Chung KJ, Varma A (2006) Do major species concepts support one, two or more species within *Cryptococcus* neoformans? FEMS Yeast Res 6:574–587
Kwon-Chung KJ, Bennett JE, Rhodes JC (1982) Taxonomic studies on *Filobasidiella* species and their anamorphs. Antonie Van Leeuwenhoek 48(1):25–38
Kwon-Chung KJ, Fraser JA, Doering TL et al (2014) *Cryptococcus neoformans* and *Cryptococcus gattii*, the etiologic agents of cryptococcosis. Cold Spring Harb Perspect Med 4(7):a019760
Kwon-Chung KJ, Bennett JE, Wickes BL, Meyer W, Cuomo CA, Wollenburg KR, Bicanic TA, Castaneda E, Chang YC, Chen J, Cogliati M, Dromer F, Ellis D, Filler SG, Fisher MC, Harrison TS, Holland SM, Kohno S, Kronstad JW, Lazera M, Levitz SM, Lionakis MS, May RC, Ngamskulrongroj P, Pappas PG, Perfect JR, Rickerts V, Sorrell TC, Walsh TJ, Williamson PR, Xu J, Zelazny AM, Casadevall A (2017) The case for adopting the "species complex" nomenclature for the etiologic agents of cryptococcosis. mSphere 2(1):e00357
Lagrou K, Van Eldere J, Keuleers S, Hagen F, Merckx R, Verhaegen J, Peetermans WE, Boekhout T (2005) Zoonotic transmission of *Cryptococcus neoformans* from a magpie to an immunocompetent patient. J Intern Med 257(4):385–388

Laurenson IF, Lalloo DG, Naraqi S et al (1997) *Cryptococcus neoformans* in Papua New Guinea: a common pathogen but an elusive source. J Med Vet Mycol 35:437–440

Lazera MS, Wanke B, Nishikawa MM (1993) Isolation of both varieties of *Cryptococcus neoformans* from saprophytic sourcesin the city of Rio de Janeiro, Brazil. J Med Vet Mycol 31: 449–454

Lazera MS, Cavalcanti MA, Trilles L, Nishikawa MM, Wanke B (1998) *Cryptococcus neoformans* var. *gattii* – evidence for a natural habitat related to decaying wood in a pottery tree hollow. Med Mycol 36:119–122

Lemos LS, Siqueira de O dos Santos A, Vieira-da-Motta O, Texeira GN, de Carvalho ECQ (2007) Pulmonary cryptococcosis in slaughtered sheep: anatomopathology and culture. Vet Microbiol 125(3–4):350–354

Lengeler KB, Fox DS, Fraser JA, Allen A, Forrester K, Dietrich FS, Heitman J (2002) Mating-type locus of *Cryptococcus neoformans*: a step in the evolution of sex chromosomes. Eukaryot Cell 1:704–718

Lester SJ, Kowalewich NJ, Bartlett KH, Krockenberger MB, Fairfax TM, Malik R (2004) Clinicopathologic features of an unusual outbreak of cryptococcosis in dogs, cats, ferrets and a bird: 38 cases (January 2003 to July 2003). J Am Vet Med Assoc 225:1716–1722

Levinson DJ, Silcox DC, Rippon JW, Thomsen S (1974) Septic arthritis due to nonencapsulated *Cryptococcus neoformans* with coexisting sarcoidosis. Arthritis Rheum 17(6):1037–1047

Levitz SM (1991) The ecology of *Cryptococcus neoformans* and the epidemiology of cryptococcosis. Rev Infect Dis 13(6):1163–1169

Lewington JH (1982) Isolation of *Cryptococcus neoformans* from a ferret. Aust Vet J 58(3): 124–124

Li A, Nishimura K, Taguchi H, Tanaka R, Wu S, Miyaji M (1993) The isolation of *Cryptococcus neoformans* from pigeon droppings and serotyping of naturally and clinically sourced isolates in China. Mycopathologia 124:1–5

Lin X, Heitman J (2006) The biology of the *Cryptococcus neoformans* species complex. Annu Rev Microbiol 60:69–105

Lin X, Nielsen K, Patel S, Heitman J (2008) Impact of mating type, serotype, and ploidy on virulence of *Cryptococcus neoformans*. Infect Immun 76:2923–2938

Littman ML, Walter JE (1968) Cryptococcosis: current status. Am J Med 45(6):922–932

Litvintseva AP, Thakur R, Reller LB, Mitchell TG (2005) Prevalence of clinical isolates of *Cryptococcus gattii* serotype C among patients with AIDS in sub-Saharan Africa. J Infect Dis 192(5):888–892

Loftus BJ, Fung E, Roncaglia P (2005) The genome of the basidiomycetous yeast and human pathogen *Cryptococcus neoformans*. Science 307:1321–1324

Ma H, May RC (2009) Virulence in *Cryptococcus* species. Adv Appl Microbiol 67:131–190

Ma H, Hagen F, Stekel DJ et al (2009) The fatal fungal outbreak on Vancouver Island is characterized by enhanced intracellular parasitism driven by mitochondrial regulation. Proc Natl Acad Sci U S A 106(31):12980–12985

MacDougall L, Kidd SE, Galanis E et al (2007) Spread of *Cryptococcus gattii* in British Columbia, Canada, and detection in the Pacific Northwest, USA. Emerg Infect Dis 13(1):42–50

Malik R, Wigney DI, Muir DB, Gregory DJ, Love DN (1992) Cryptococcosis in cats – clinical and mycological assessment of 29 cases and evaluation of treatment using orally-administered fluconazole. J Med Vet Mycol 30(2):133–144

Malik R, Dill-Macky E, Martin P, Wigney DI, Muir DB, Love DN (1995) Cryptococcosis in dogs: a retrospective study of 20 consecutive cases. Med Mycol 33(5):291–297

Malik R, Wigney DI, Muir DB, Love DN (1997) Asymptomatic carriage of *Cryptococcus neoformans* in the nasal cavity of dogs and cats. J Med Vet Mycol 35(1):27–31

Malik R, Alderton B, Finlaison D et al (2002) Cryptococcosis in ferrets: a diverse spectrum of clinical disease. Aust Vet J 80(12):749–755

Malik R, Krockenberger MB, Cross G, Doneley R, Madill DN, Black D, Mcwhirter P, Rozenwax A, Rose K, Alley M, Forshaw D, Russel-Brown I, Johnstone AC, Martin P, O'Brien CR, Love DN (2003) Avian cryptococcosis. Med Mycol 41:115–124

Malik R, Krockenberger M, O'Brien CR et al (2006a) Cryptococcosis. In: Greene CE (ed) Infectious diseases of the dog and cat, 3rd edn. Elsevier, St. Louis, pp 584–598

Malik R, Norris J, White J, Jantulik B (2006b) Wound cat. J Feline Med Surg 8:135–140

Malliaris SD, Steenbergen JN, Casadevall A (2004) *Cryptococcus neoformans* var. *gattii* can exploit Acanthamoeba castellanii for growth. Med Mycol 42:149–158

McGill S, Malik R, Saul N, Beetson S, Secombe C, Robertson I, Irwin P (2009) Cryptococcosis in domestic animals in Western Australia: a retrospective study from 1995–2006. Med Mycol 47: 625–639

McLeland S, Duncan C, Spraker T, Wheeler E, Lockhart SR, Gulland F (2012) *Cryptococcus albidus* infection in a California sea lion (Zalophus californianus). J Wildl Dis 48:1030–1034

McMullan BJ, Halliday C, Sorrell TC et al (2012) Clinical utility of the cryptococcal antigen lateral flow assay in a diagnostic mycology laboratory. PLoS One 7(11):e49541

McNamara TS, Cook RA, Behler JL, Ajello L, Padhye AA (1994) Cryptococcosis in a common Anaconda (*Eunectes murinus*). J Zoo Wildl Med 25:128–132

McTaggart L, Richardson SE, Seah C, Hoang L, Fothergill A, Zhang SX (2011) Rapid identification of *Cryptococcus neoformans* var. grubii, *C neoformans* var. *neoformans*, and *C. gattii* by use of rapid biochemical tests, differential media, and DNA sequencing. J Clin Microbiol 49(7): 2522–2527

Metin B, Findley K, Heitman J (2010) The mating type locus (MAT) and sexual reproduction of *Cryptococcus heveanensis*: insights into the evolution of sex and sex-determining chromosomal regions in fungi. PLoS Genet 6(5):e1000961

Meyer W, Mitchel TG (1995) Polymerase chain reaction fingerprinting in fungi using single primers specific to minisatellites and simple repetitive DNA sequences: strain variation in *Cryptococcus neoformans*. Electrophoresis 16(1):1648–1656

Meyer W, Mitchell TG, Freedman EZ, Vilgalys R (1993) Hybridization probes for conventional DNA fingerprinting used as single primers in the polymerase chain reaction to distinguish strains of *Cryptococcus neoformans*. J Clin Microbiol 31(9):2274–2280

Meyer W, Aanensen DM, Boekhout T, Cogliati M, Diaz MR, Esposto MC, Fisher M, Gilgado F, Hagen F, Kaocharoen S, Litvintseva AP, Mitchell TG, Simwami SP, Trilles L, Viviani MA, Kwon-Chung J (2009) Consensus multi-locus sequence typing scheme for *Cryptococcus neoformans* and *Cryptococcus gattii*. Med Mycol 47(6):561–570

Migaki G, Jones SR (1983) Mycotic diseases in marine animals. In: Howard EB (ed) Pathobiology of marine mammal diseases, vol II. CRC Press, Boca Raton, pp 1–127

Miller WG, Padhye AA, van Bonn W, Jensen E, Brandt ME, Ridgway SH (2002) Cryptococcosis in a bottlenose dolphin (Tursiops truncatus) caused by *Cryptococcus neoformans* var. gattii. J Clin Microbiol 40(2):721–724

Millward IR, Williams MC (2005) *Cryptococcus neoformans* granuloma in the lung and spinal cord of a free-ranging cheetah (Acinonyx jubatus). A clinical report and literature review. J S Afr Vet Assoc 76(4):228–232

Missall TA, Moran JM, Corbett JA, Lodge JK (2005) Distinct stress responses of two functional laccases in *Cryptococcus neoformans* are revealed in the absence of the thiol-specific antioxidant Tsa1. Eukaryot Cell 4:202–208

Mitchell TG, Perfect JR (1995) Cryptococcosis in the era of AIDS--100 years after the discovery of *Cryptococcus neoformans*. Clin Microbiol Rev 8(4):515–548

Montagna MT, Mele MS, De Donno A, Marcuccio C, Pulito A (1996) Criptococcosi e AIDS. Nota I. Indagini sulla diffusione di *Cryptococcus neoformans* nelle citta di Bari e Lecce. Riv Ital Igiene 56:69–77

Montenegro H, Paula CR (2000) Environmental isolation of *Cryptococcus neoformans* var. *gattii* and C. neoformans var. neoformans in the city of São Paulo, Brazil. Med Mycol 38:385–390

Morera N, Juan-Sallés C, Torres JM, Andreu M, Sánchez M, Zamora MÁ, Colom MF (2011) *Cryptococcus gattii* infection in a Spanish pet ferret (Mustela putorius furo) and asymptomatic carriage in ferrets and humans from its environment. Med Mycol 49(7):779–784
Morera N, Hagen F, Juan-Sallés C, Artigas C, Patricio R, Serra JI, Colom MF (2014) Ferrets as sentinels of the presence of pathogenic *Cryptococcus* species in the Mediterranean environment. Mycopathologia 178(1–2):145–151
Mseddi F, Sellami A, Jarboui MA, Sellami H, Makni F, Ayadi A (2011) First environmental isolations of *Cryptococcus neoformans* and *Cryptococcus gattii* in Tunisia and review of published studies on environmental isolations in Africa. Mycopathologia 171(5):355–360
Muchmore HG, Rhoades ER, Nix GE, Felton FG, Carpenter RE (1963) Occurrence of Cryptococcus neoformans in the environment of three geographically associated cases of cryptococcal meningitis. N Engl J Med 268(20):1112–1114
Mylonakis E, Ausubel FM, Perfect JR, Heitman J, Calderwood SB (2002) Killing of Caenorhabditis elegans by *Cryptococcus neoformans* as a model of yeast pathogenesis. Proc Natl Acad Sci U S A 99:15675–15680
Mylonakis E, Moreno R, El Khoury JB, Idnurm A, Heitman J, Calderwood SB, Ausubel FM, Diener A (2005) Galleria mellonella as a model system to study *Cryptococcus neoformans* pathogenesis. Infect Immun 73:3842–3850
Nawange SR, Shakya K, Naidu J, Singh SM, Jharia N, Garg S (2006) Decayed wood inside trunk hollows of living trees of T. indica, S. cumini and M. indica as natural habitat of *Cryptococcus neoformans* and their serotypes in Jabalpur city of Central India. J Mycol Med 16:63–71
Nick Gales N, Wallace G, Dickson J (1985) Pulmonary Cryptococcosis in a striped dolphin *(Stenella coeruleoalba)*. J Wildl Dis 21(4):443–446
O'Brien CR, Krockenberger MB, Martin P, Wigney DI, Malik R (2006) Long-term outcome of therapy for 59 cats and 11 dogs with cryptococcosis. Aust Vet J 84(11):384–392
Odom A, Muir S, Lim E, Toffaletti DL, Perfect J, Heitman J (1997) Calcineurin is required for virulence of *Cryptococcus neoformans*. EMBO J 16:2576–2589
Okabayashi K, Imaji M, Osumi T et al (2009) Antifungal activity of itraconazole and voriconazole against clinical isolates obtained from animals with mycoses. Jpn J Med Mycol 50(2):91–94
Okagaki LH, Strain AK, Nielsen JN, Charlier C, Baltes NJ, Chrétien F, Heitman J, Dromer F, Nielsen K (2010) Cryptococcal cell morphology affects host cell interactions and pathogenicity. PLoS Pathog 6:e1000953
Overy DP, McBurney S, Muckle A, Lund L, Lewis PJ, Strang R (2016) *Cryptococcus gattii* VGIIb-like variant in white-tailed deer, Nova Scotia, Canada. Emerg Infect Dis 22(6):1131–1133
Padhye AA, Chakrabarti A, Chander J, Kaufman L (1993) *Cryptococcus neoformans* var. *gattii* in India. J Med Vet Mycol 31:165–168
Pal M, Baxter M (1985) Isolation of *Cryptococcus neoformans* using a simplified sunflower seed medium. Proc NZ Microbiol Soc 29:155–158
Pal M, Dube GD, Mehrotra BS (1984) Pulmonary cryptococcosis in a rhesus monkey (Macaca mulatta). Mycoses 27(6):309–312
Panciera DL, Bevier D (1987) Management of cryptococcosis and toxic epidermal necrolysis in a dog. J Am Vet Med Assoc 191(9):1125–1127
Pappas PG (2013) Cryptococcal infections in non-HIV-infected patients. Trans Am Clin Climatol Assoc 124:61–79
Passoni LFC (1999) Wood, animals and human beings as reservoirs for human *Cryptococcus neoformans* infection. Rev Iberoam Micol 16(6):77–81
Pennisi MG, Hartmann K, Lloret A, Ferrer L, Addie D, Belák S, Boucraut-Baralon C, Egberink H, Frymus T, Gruffydd-Jones T, Hosie MJ, Lutz H, Marsilio F, Möstl K, Radford AD, Thiry E, Truyen U, Horzinek MC (2013) Cryptococcosis in cats: ABCD guidelines on prevention and management. J Feline Med Surg 15(7):611–618
Perfect JR, Magee BB, Magee MT (1989) Separation of chromosomes of *Cryptococcus neoformans* by pulsed field gel electrophoresis. Infect Immun 57:2624–2627

Pfaller MA, Castanheira M, Diekema DJ, Messer SA, Jones RN (2011) Wild-type MIC distributions and epidemiologic cutoff values for fluconazole, posaconazole, and voriconazole when testing *Cryptococcus neoformans* as determined by the CLSI broth microdilution method. Diagn Microbiol Infect Dis 71:252–259

Pfeiffer TJ, Ellis DH (1992) Environmental isolation of *Cryptococcus neoformans* var. *gattii* from Eucalyptus tereticornis. J Med Vet Mycol 30:407–408

Pietrella D, Corbucci C, Perito S, Bistoni G, Vecchiarelli A (2005) Mannoproteins from *Cryptococcus neoformans* promote dendritic cell maturation and activation. Infect Immun 73:820–827

Polacheck I, Lebens GA (1989) Electrophoretic karyotype of the pathogenic yeast *Cryptococcus neoformans*. J Gen Microbiol 135(1):65–71

Polak A (1990) Melanin as a virulence factor in pathogenic fungi. Mycoses 33(5):215–224

Posteraro B, Vella A, Cogliati M et al (2012) Matrix-assisted laser desorption ionization–time of flight mass spectrometry-based method for discrimination between molecular types of *Cryptococcus neoformans* and *Cryptococcus gattii*. J Clin Microbiol 50(7):2472–2476

Posteraro B, De Carolis E, Vella A, Sanguinetti M (2013) MALDI-TOF mass spectrometry in the clinical mycology laboratory: identification of fungi and beyond. Expert Rev Proteomics 10: 151–164

Poth T, Seibold M, Werckenthin C, Hermanns W (2010) First report of a *Cryptococcus magnus* infection in a cat. Med Mycol 48(7):1000–1004

Procknow JJ, Benfield JR, Rippon JW, Diener CF, Archer FL (1965) Cryptococcal hepatitis presenting as a surgical emergency: first isolation of *Cryptococcus neoformans* from point source in Chicago. JAMA 191(4):269–274

Randhawa HS, Kowshik T, Khan ZU (2003) Decayed wood of Syzygium cumini and Ficus religiosa living trees in Delhi/New Delhi metropolitan area as natural habitat of *Cryptococcus neoformans*. Med Mycol 41(3):199–209

Raso TF, Werther K, Miranda ET, Mendes-Giannini MJS (2004) Cryptococcosis outbreak in psittacine birds in Brazil. Med Mycol 42(4):355–362

Refai M, El-Hariri M, Alarousy R (2014) Monograph on *Cryptococcus* and Cryptococcosis in man, animals and birds: a guide for postgraduate students in developing countries. https://www.researchgate.net/publication/266402090

Riet-Correa F, Krockenberger M, Dantas AFM, Oliveira DM (2011) Bovine cryptococcal meningoencephalitis. J Vet Diagn Investig 23(5):1056–1060

Rij KV (1984) The yeasts a taxonomic study, 3rd edn. Elsevier, Amsterdam

Riley CB, Bolton JR, Mills JN, Thomas JB (1992) Cryptococcosis in seven horses. Aust Vet J 69:135–139

Rinaldi MG, Drutz DJ, Howell A, Sande MA, Wofsy CB, Hadley WK (1986) Serotypes of *Cryptococcus neoformans* in patients with AIDS. J Infect Dis 153(3):642

Rosenberg JF, Haulena M, Hoang LM, Morshed M, Zabek E, Raverty SA (2016) *Cryptococcus gattii* type VGIIa infection in harbor seals (Phoca vitulina) in British Columbia, Canada. J Wildl Dis 52(3):677–681

Rotstein DS, West K, Levine G et al (2010) *Cryptococcus gattii* VGI in a spinner dolphin (Stenella longirostris) from Hawaii. J Zoo Wildl Med 41(1):181–183

Ruiz A, Fromtling RA, Bulmer GS (1981) Distribution of *Cryptococcus neoformans* in a natural site. Infect Immun 31(2):560–563

Sabiiti W, May RC, Pursall ER (2012) Experimental models of cryptococcosis. Int J Microbiol 2012:626745

Sangiorgi G (1922) Blastomicosi spontanea nei muridi. Pathology 14:493–495

Schelenz S, Malhotra R, Sim RB, Holmskov U, Bancroft GJ (1995) Binding of host collectins to the pathogenic yeast *Cryptococcus neoformans*: human surfactant protein D acts as an agglutinin for acapsular yeast cells. Infect Immun 63:3360–3366

Scrimgeour EM, Purohit RG (1984) Chronic pulmonary cryptococcosis in a *Rattus rattus* from Rabaul, Papua New Guinea. Trans R Soc Trop Med Hyg 78(6):827–828

Seaton RA, Hamilton AJ, Hay RJ, Warrell DA (1996) Exposure to *Cryptococcus neoformans* var. *gattii* – a seroepidemiological study. Trans R Soc Trop Med Hyg 90:508–512

Seixas F, Martins ML, Pinto LM et al (2008) A case of pulmonary cryptococcosis in a free-living toad *(Bufo bufo)*. J Wildl Dis 44:460–463
Simos J, Nichols RE, Morse EW (1953) An outbreak of bovine cryptococcosis. J Am Vet Med Assoc 122:31–35
Singh A, Panting RJ, Varma A, Saijo T, Waldron KJ, Jong A, Ngamskulrungroj P, Chang YC, Rutherford JC, Kwon-Chung KJ (2013) Factors required for activation of urease as a virulence determinant in *Cryptococcus neoformans*. mBio 4(3):e00220–e00213
Singh K, Rani J, Neelabh Rai GK, Singh M (2017) The Southeastern Asian house mouse (*Mus musculus castaneus* Linn.) as a new passenger host for *Cryptococcus neoformans* var. *grubii* molecular type VNI. Med Mycol 55:1–8
Singh SM, Naidu J, Sharma A, Nawange SR, Singh K (2007) First case of cryptococcosis in a new species of bandicoot (Bandicota indica) caused by *Cryptococcus neoformans* var. *grubii*. Med Mycol 45(1):89–93
Sionov E, Chang YC, Garraffo HM, Dolan MA, Ghannoum MA, KwonChung KJ (2012) Identification of a *Cryptococcus neoformans* cytochrome P450 lanosterol 14-demethylase (Erg11) residue critical for differential susceptibility between fluconazole/voriconazole and itraconazole/posaconazole. Antimicrob Agents Chemother 56:1162–1169
Sorrell TC (2001) *Cryptococcus neoformans* variety *gattii*. Med Mycol 39:155–168
Sorrell TC, Brownlee AG, Ruma P, Malik R, Pfeiffer TJ, Ellis DH (1996) Natural environmental sources of Cryptococcus neoformans var. gattii. J Clin Microbiol 34(5):1261–1263
Spencer A, Ley C, Canfield P, Martin P, Perry R (1993) Meningoencephalitis in a koala *(Phascolarctos cinereus)* due to *Cryptococcus neoformans* var. *gattii* infection. J Zoo Wildl Med 24:519–522
Spickler AR (2013) Cryptococcosis. http://www.cfsph.iastate.edu/DiseaseInfo/factsheets.php
Springer DJ, Chaturvedi V (2010) Projecting global occurrence of *Cryptococcus gattii*. Emerg Infect Dis 16(1):14–20
Springer DJ, Billmyre RB, Filler EE et al (2014) *Cryptococcus gattii* VGIII isolates causing infections in HIV/AIDS patients in Southern California: identification of the local environmental source as arboreal. PLoS Pathog 10(8):e1004285
Staib F, Weller W, Brem S, Schindlmayr R, Schmittdiel E (1985) A *Cryptococcus neoformans* strain from the brain of a wildlife fox *(Vulpes vulpes)* suspected of rabies: mycological observations and comments. Zentralbl Bakteriol Mikrobiol Hyg A 260:566–571
Steenbergen JN, Casadevall A (2003) The origin and maintenance of virulence for the human pathogenic fungus *Cryptococcus neoformans*. Microbes Infect 5(7):667–675
Steenbergen JN, Shuman HA, Casadevall A (2001) *Cryptococcus neoformans* interactions with amoebae suggest an explanation for its virulence and intracellular pathogenic strategy in macrophages. Proc Natl Acad Sci U S A 98(26):15245–15250
Stephen C, Lester S, Black W, Fyfe M, Raverty S (2002) British Columbia: multispecies outbreak of cryptococcosis on Southern Vancouver Island, British Columbia. Can Vet J 43(10):792–794
Stilwell G, Pissarra H (2014) Cryptococcal meningitis in a goat – a case report. BMC Vet Res 10:84
Sugita T, Suto H, Unno T, Tsuboi R, Ogawa H, Shinoda T, Nishikawa A (2001) Molecular analysis of Malassezia microflora on the skin of atopic dermatitis patients and healthy subjects. J Clin Microbiol 39(10):3486–3490
Sutton RH (1981) Cryptococcosis in dogs: a report on 6 cases. Aust Vet J 57(12):558–564
Swinne D, Kayembe K, Niyimi M (1986) Isolation of saprophytic *Cryptococcus neoformans* var. *neoformans* in Kinshasa, Zaire. Ann Soc Belg Med Trop 66:57–61
Sykes JE, Malik R (2014) Cryptococcosis. In: Sykes JE (ed) Canine and feline infectious diseases, 1st edn. Elsevier, St Louis, pp 599–612
Tell LA, Nichols DK, Fleming WP, Bush M (1997) Cryptococcosis in tree shrews (*Tupaia tana* and *Tupaia minor*) and elephant shrews (Macroscelides proboscides). J Zoo Wildl Med 28(2): 175–181

Temstet A, Roux P, Poirot JL et al (1992) Evaluation of a monoclonal antibody-based latex agglutination test for diagnosis of cryptococcosis: comparison with two tests using polyclonal antibodies. J Clin Microbiol 30:2544–2550

Tenor JL, Oehlers SH, Yang JL, Tobin DM, Perfect JR (2015) Live imaging of host-parasite interactions in a Zebrafish infection model reveals *Cryptococcal determinants* of virulence and central nervous system invasion. mBio 6(5):e01425–e01415

Thompson GR, Wiederhold NP, Fothergill AW, Vallor AC, Wickes BL, Patterson TF (2009) Anti-fungal susceptibilities among different serotypes of *Cryptococcus gattii* and *Cryptococcus neoformans*. Antimicrob Agents Chemother 53:309–311

Torres-Guererro H, Edman JC (1994) Melanin-deficient mutants of *Cryptococcus neoformans*. J Med Vet Mycol 32:303–313

Trivedi SR, Sykes JE, Cannon MS, Wisner ER, Meyer W, Sturges BK, Dickinson PJ, Johnson LR (2011) Clinical features and epidemiology of cryptococcosis in cats and dogs in California: 93 cases (1988–2010). J Am Vet Med Assoc 239:357–369

Tsiodras S, Samonis G, Keating MJ, Kontoyiannis DP (2000) Infection and immunity in chronic lymphocytic leukemia. Mayo Clin Proc 75:1039–1054

Tucker SC, Casadevall A (2002) Replication of *Cryptococcus neoformans* in macrophages is accompanied by phagosomal permeabilization and accumulation of vesicles containing poly-saccharide in the cytoplasm. Proc Natl Acad Sci U S A 99:3165–3170

van Duin D, Casadevall A, Nosanchuk JD (2002) Melanization of *Cryptococcus neoformans* and Histoplasma capsulatum reduces their susceptibilities to amphotericin B and caspofungin. Antimicrob Agents Chemother 46:3394–3400

Vanessa N, Pérez C, Mata-Essayag S, Colella MT, Roselló A, de Capriles CH, Landaeta ME, Olaizola C, Magaldi S (2005) Identificación de Aislados de *Cryptococcus neoformans* Usando Agar Staib sin Creatinina. Kasmera 33(2):102–108

Vecchiarelli A (2000) Immunoregulation by capsular components of *Cryptococcus neoformans*. Med Mycol 38(6):407–417

Vilgalys R, Hester M (1990) Rapid genetic identification and mapping of enzymatically amplified ribosomal DNA from several *Cryptococcus* species. J Bacteriol 172(8):4238–4246

Viviani MA, Cogliati M, Esposto MC, Lemmer K, Tintelnot K, Valiente MFC, Velho R (2006) Molecular analysis of 311 *Cryptococcus neoformans isolates* from a 30-month ECMM survey of cryptococcosis in Europe. FEMS Yeast Res 6(4):614–619

Weller W, Brem S, Schindlmayr R, Schmittdiel E (1985) Cryptococcosis in a red fox (*Vulpes vulpes*). Berl Munch Tierarztl Wochenschr 98(1):14–15. [Article in German]

Wickerham LJ (1951) Taxonomy of yeasts. Tech Bull No 1029. US Dept Agric, Washington DC, pp 1–55

Wickerham LJ, Burton KA (1948) Carbon assimilation tests for the classification of yeasts. J Bacteriol 56(3):363

Yamazumi T, Pfaller MA, Messer SA, Houston A, Hollis RJ, Jones RN (2000) *In vitro* activities of ravuconazole (BMS-207147) against 541 clinical isolates of *Cryptococcus neoformans*. Antimicrob Agents Chemother 44:2883–2886

Zaragoza O, Alvarez M, Telzak A, Rivera J, Casadevall A (2007) The relative susceptibility of mouse strains to pulmonary *Cryptococcus neoformans* infection is associated with pleiotropic differences in the immune response. Infect Immun 75(6):2729–2739

Zaragoza O, García-Rodas R, Nosanchuk JD, Cuenca-Estrella M, Rodríguez-Tudela JL, Casadevall A (2010) Fungal cell gigantism during mammalian infection. PLoS Pathog 6:e1000945

Zhu X, Williamson PR (2004) Role of laccase in the biology and virulence of *Cryptococcus neoformans*. FEMS Yeast Res 5:1–10

White-Nose Syndrome in Hibernating Bats 13

Gudrun Wibbelt

Abstract

White-nose syndrome (WNS) is a fungal disease exclusively found in hibernating bats. The causative agent, *Pseudogymnoascus* (*Geomyces*) *destructans* (*Pd*), is a psychrophilic fungus, which doesn't tolerate temperatures above 24 °C. Since the emergence of this formerly unknown pathogen in 2006 in Northeastern America, it has killed millions of bats. The infection is limited to the skin with a rather characteristic growth appearance of white powdery patches around the muzzle of the bats, but white fungal aerial hyphae are also found on ears and particularly wing membranes. Confirmation of the disease requires histopathological investigations with proof of hallmark lesions, i.e. cup-like erosions containing fungal hyphae and invasion of the dermal connective tissue. Evidence of typical asymmetrically curved conidia or nucleic acids of *Pd* is considered only suspicious for WNS. Subsequent search for the fungus in Europe and Asia found the fungus to be enzootic and causing similar pathological lesions, but it is not associated with fatalities. Molecular investigations revealed a single clonal genotype for North America, while European isolates diversified. However, the most common haplotype in Europe is shared with the North American strain providing strong support for the hypothesis of an introduction of *Pd* from Europe into a naïve bat population in North America. Despite 10 years of intensive research, the chain of causation in the pathogenesis of WNS leading to the death of the animals is still not satisfactorily answered.

G. Wibbelt (✉)
Department of Wildlife Diseases, Leibniz Institute for Zoo and Wildlife Research, Berlin, Germany
e-mail: wibbelt@izw-berlin.de

S. Seyedmousavi et al. (eds.), *Emerging and Epizootic Fungal Infections in Animals*,
https://doi.org/10.1007/978-3-319-72093-7_13

13.1 Brief History of an Emerging Disease

When "white-nose syndrome" (WNS) emerged, nobody anticipated that it would be the start of one of the most devastating mass mortalities in mammals ever recorded. In 2006 a group of hibernating little brown bats (*Myotis lucifugus*) was photographed in Howes Cave, New York State, USA, depicting the animals with distinct white powdery patches around their muzzles and on their wing membranes, name-giving for WNS. During the following winters of 2006–2008, it was noticed that bats displayed aberrant hibernation behaviour in the vicinity of several cavernous hibernacula like day-flight or aggregation of bats close to the cavern entries. In mid-March 2007, a tenfold increase of midday flight compared to the previous 25-year record high was reported (Blehert et al. 2009). But the most dramatic observation was a decline of up to 75% of hibernating bats in this region at that time. A fevered search for the causative agent followed, and several laboratories tried to identify a possible microbial agent or toxic component. Several bat carcasses underwent pathological investigations including necropsy and histopathology. One of the obstacles during the search was that during shipment from the field to the investigation centres, the white powdery substance seen on the affected bats rendered invisible (Meteyer et al. 2009), and it took many attempts until finally a fungal organism was consistently cultured and identified. In late 2008, Blehert et al. described a novel fungus closely related to the psychrophilic and keratinophilic species *Geomyces pannorum*. But because of distinct, characteristically formed conidia as well as differences in the ITS genes, the isolated fungus was identified as a new species: *Geomyces destructans* (Blehert et al. 2009; Gargas et al. 2009). Meanwhile no other microbial agent or toxic component was consistently found in all investigated bats, making *G. destructans* the likely cause of disease. A few years later, Koch's postulates were fulfilled in experimental infection trials proving the fungus as the causative agent (Lorch et al. 2011). Soon it became evident that the fatalities associated with the fungus were rapidly spreading in a radial fashion away from the index cave to further wintering sites. Such pattern is well known as a characteristic feature of pathogens introduced into a naïve population. Today, 10 years after the first detection, more than six million hibernating bats succumbed to the infection (US Fish and Wildlife Service March 2016a) grimly making earlier modelling studies, that predicted regional extinction of the once abundant little brown bat within 16 years, become true (Frick et al. 2010). Seven North American bat species are affected and hibernacula positive for *G. destructans* are currently found in 26 US states and 5 Canadian provinces (Hayman et al. 2016), and it is likely to proceed further (Fig. 13.1). In 2013, improved phylogenetic techniques indicated that the former classification of the fungus needed to be reclassified as *Pseudogymnoascus destructans* (Minnis and Lindner 2013).

While these dramatic events enfolded in North America, Europe feared similar consequences if the fungus should enter European hibernacula. Hence, it was rather surprising when it became evident that *P. destructans* is enzootic in Europe and hibernating bats in various European countries carry the fungus (Fig. 13.2) (Puechmaille et al. 2011; Zukal et al. 2016). However, in Europe, *P. destructans*

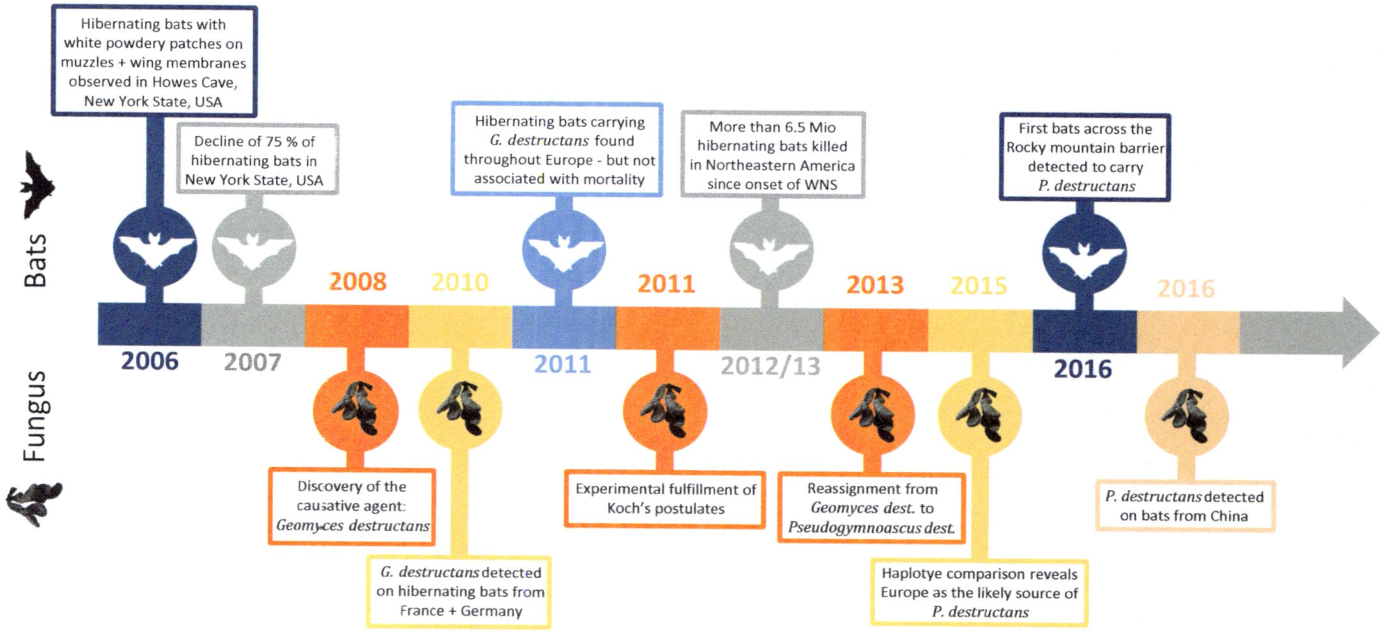

Fig. 13.1 Timeline of *Pseudogymnoascus destructans* detection and subsequent developments

Fig. 13.2 An European greater mouse-eared bat (*Myotis myotis*) carrying white fungal patches of *Pseudogymnoascus destructans* around its nose, the ears and forearms (Courtesy of C. Jungmann)

infection is not associated with mass mortalities (Wibbelt 2015). Recently *P. destructans* was also discovered on four different hibernating bat species in Northeastern China, similar to the situation in Europe without causing fatalities, but rather indicating host resistance to the pathogen (Hoyt et al. 2016). Meanwhile, also in North America, little brown bats were found which recovered from the disease, and during summer bats do clear the infection entirely from their body (Meteyer et al. 2011).

So far hibernating bats are the only host species for which pathogenicity of *P. destructans* was shown. The slow growth and psychrophilic nature of the fungus make these animals an ideal substrate as the bats' body temperature decreases to almost ambient temperature of the cavernous wintering sites and during their torpor phases of 12–14 days each, the animals remain motionless and leave the fungus undisturbed. During their active life, their body temperature rises like in all warm-blooded animals creating an inhospitable environment for the fungus. With the differences between North America and Europe/Asia, the use of the term WNS remains slightly confusing. At the time when the name developed, it referred to a true syndrome: distinct clinical signs leading to (mass) mortalities with unknown aetiology. Although today the causative agent and infection patterns are identified, we are still faced with two different scenarios: hibernating bats infected with *P. destructans* and ongoing mass mortalities in North America and hibernating bats infected with *P. destructans* not associated with fatalities in Europe and Palearctic Asia. The indiscriminate use of the term WNS for either situation as well as for bats in North American recovering from WNS might blur future epidemiologic approaches, where a distinction between the two groups infected and dying versus infected and surviving might be difficult to reconstruct. A sharp distinction between histopathologically proven *P. destructans* infection associated with various clinical signs, including mortality equalling WNS (in North America), and confirmed *P. destructans* infection of otherwise unremarkable (European) bats equalling *P. destructans* disease would help to keep a clear line between these important epidemiologic features.

13.2 The Causative Agent

13.2.1 Classification

To date the WNS fungus is classified as *Pseudogymnoascus destructans*. The genus *Pseudogymnoascus* is a member of the family *Pseudeurotiaceae* within the class *Leotiomycetes* of the phylum *Ascomycota*.

First classifications after the fungus' detection based on rRNA investigations compared the internal transcribed spacer (ITS) and small subunit (SSU) gene sequences and placed the new fungus in the genus *Geomyces* (Blehert et al. 2009; Gargas et al. 2009; Chaturvedi et al. 2010), but parsimonious trees for ITS and SSU already indicated a close relationship with *Pseudogymnoascus*. Lorch et al. (2010) designed primers which selectively amplified DNA targeting a conserved 1506 bp intron and ITS sequences. Soon investigations on soil samples from North American bat hibernacula not only revealed a huge diversity of so far unknown *Geomyces* species but also showed cross-reactivity with primers thought to be selective for *G. destructans* (Lindner et al. 2011). Faced by the large number of different *Geomyces* species and allies in cave soil and a missing taxonomic lead, Minnis and Lindner (2013) employed not only ITS but also nuclear large subunit (LSU) rDNA, DNA replication licensing factor MCM7, RNA polymerase II second largest subunit (RPB2) and translation elongation factor EF-1α (TEF1) to investigate the taxonomic relationship of the various fungal species detected in American hibernacula. The result of this work was a shift of the genera *Geomyces*, *Gymnostellatospora* and *Pseudogymnoascus* from the historically assigned family *Myxotrichaceae* to the family *Pseudeurotiaceae* (Minnis and Lindner 2013). At the same time, it was shown that *G. destructans* was indeed member of the genus *Pseudogymnoascus* and not *Geomyces* and was subsequently renamed in *Pseudogymnoascus destructans* (Minnis and Lindner 2013).

13.2.2 Culture Conditions

Pseudogymnoascus destructans is a true psychrophile, slowly growing between 0° and 24 °C. The temperature optimum lies between 12° and 15 °C with some minor strain-dependent differences, and no growth occurs above 24 °C (Gargas et al. 2009; Verant et al. 2012). There are no specific demands for culture media, but Sabouraud glucose agar containing chloramphenicol and gentamicin seems to be preferred by most studies. However *P. destructans* grows equally well on corn meal agar, malt extract agar, creatine sucrose agar, yeast and malt extract agar, potato dextrose agar and minimal media agar (Gargas et al. 2009; Chaturvedi et al. 2010; Martínková et al. 2010; Puechmaille et al. 2010; Wibbelt et al. 2010; Verant et al. 2012; Lorch et al. 2013a; Khankhet et al. 2014).

The type description by Gargas et al. (2009) notes colony diameters after 16 days on corn meal agar with 1 mm at 3.5 °C, 5 mm at 7 °C, 8 mm at 14 °C and no growth at 24 °C. Fungal growth starts with white, smooth-domed colonies, which will

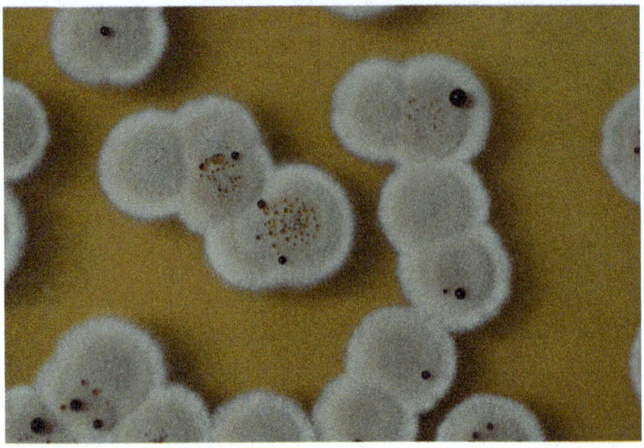

Fig. 13.3 Culture of *Pseudogymnoascus destructans* on Sabouraud glucose agar with excretion droplets

eventually turn into greyish-olive radially from the centre to the outer rim (Gargas et al. 2009; Puechmaille et al. 2010; Verant et al. 2012). However, depending on the isolate, culture medium and temperature, phenotypic differences may occur in growth velocity, exudate production, secretion (Fig. 13.3) and diffusion of soluble pigments and colony reverse colour, the latter ranging from white, grey, grey-green, brown to black (Gargas et al. 2009; Chaturvedi et al. 2010; Verant et al. 2012; Khankhet et al. 2014). *Pseudogymnoascus destructans* is considered a biosafety level 2 pathogen, and culture and isolation should be performed under the relevant requirements.

13.2.3 Morphology and Reproduction

Gargas et al. (2009) described the fungal morphology for the *P. destructans* type strain: asymmetrically curved conidia are borne singly at the tips, on the sides or in short chains on verticillately branched conidiophores (Fig. 13.4). Within chains of conidia that undergo rhexolytic dehiscence sometimes, intercalary conidia (arthroconidia) occur with conspicuous separating cells. Conidiophores are erect, hyaline, smooth and thin-walled and narrow (1.5–2 μm wide by 35–90 μm or more in length) and commonly bear verticils of 2–4 branches borne at an acute angle to the stipe. The size of the conidia is 5–12 × 2.0–3.5 μm, tapering basally to 1.5–2.0 μm and apically to 0.5–1.5 μm. They truncate with prominent scars at one or both ends and are smooth and lightly pigmented. Their shape is predominantly curved, sometimes oval, obovoid or cymbiform; at maturity they are moderately thick-walled and readily seceding.

However, when *P. destructans* is grown at temperatures above 12 °C, stress-related changes appear, e.g. hyphae become notably thickened and diffusely septate with differential sequestration of cellular material among segments. Also the conidia change their shape to mostly pyriform or globoid. Increasing temperature seems to result in pronounced deformity of the fungal structures (Verant et al. 2012).

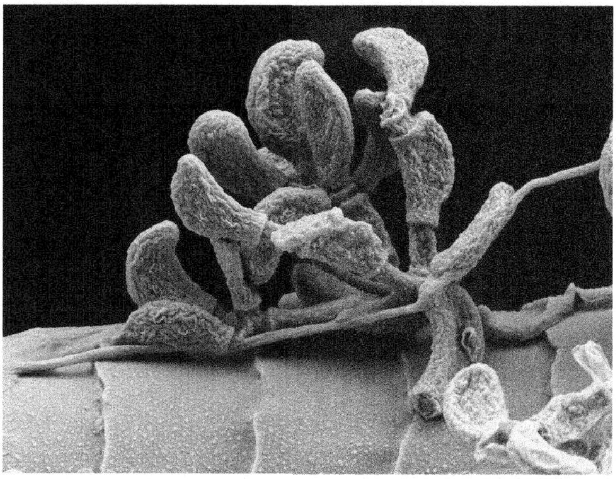

Fig. 13.4 Scanning electron microscopy image of *Pseudogymnoascus destructans* colonising the hair of a bat

The mating system of *P. destructans* still remains cryptic, but first molecular investigations of the *P. destructans* reference strain's (20631-21) genome are suggestive of a heterothallic mating system, because this strain misses the mating type locus MAT1-2-1 high mobility group (HMG) box protein, which in homothallic fungi would be contained in the genome of each isolate. Since *P. destructans* populations in North America are of single clonal origin (Rajkumar et al. 2011) comprising only a single mating type, isolates of both mating types MAT1-1 and MAT1-2 are coexisting in European hibernacula, indicating that there is the potential for mating populations (Palmer et al. 2014). Examination of mRNA from isolates of these two mating types grown in mixed culture revealed that cocultivation of MAT1-1 and MAT1-2 strains results in a weak induction of the MAT1-1-3 HMG domain-containing gene. All these data indicate a heterothallic sexual reproduction pathway (Palmer et al. 2014). Currently in North America, *P. destructans* spreads by asexual reproduction, as a true mating partner is missing. The introduction of additional isolates could result in sexual reproduction, and possible resulting recombination would allow further adaptation to environment and host species (Palmer et al. 2014).

13.2.4 Biochemical Characteristics

Several investigations on the biochemical properties of *P. destructans* were performed to elucidate potential pathogenicity factors of the fungus. Assays with substrates containing urea agar revealed the presence of urease, while albumin, casein and gelatin showed that the fungus secretes proteinases (Chaturvedi et al. 2010).

Currently the most important enzyme identified in *P. destructans* is a serine protease called "*P. destructans* serine protease 1" (PdSP1) (Pannkuk et al. 2015) or "Destructin-1" (O'Donoghue et al. 2015). This subtilisin-like peptidase and its two isoforms PdSP2 and PdSP3 have major proteolytic activities produced by

P. destructans, at least under culture conditions. The ability of the fungus to produce proteolytic and hydrolysing enzymes was further tested by API-ZYM tests, and it was found that *P. destructans* isolates also produce acid phosphatase, alkaline phosphatase, N-acetyl-β-glucosaminidase, β-glucosidase, esterase, esterase lipase, lipase, leucine arylamidase, naphthol-AS-B1-phophohydrolase and valine arylamidase. Enzyme assays on 6% gelatin implicated that lipase and esterase activity are more rapid than proteinase activity, while gelatin degradation was accompanied by alkalinisation (Raudabaugh and Miller 2013).

Substrate suitability assays demonstrated that *P. destructans* is capable of growth and sporulation on dead fish, insect and mushroom tissues, as well as on media ranging from pH 5 to 11. It also tolerates media supplemented with 2000 mg/L of calcium and 700 mg/L of sulphur compounds like thiosulphate, L-cysteine or sulphite; thiosulphate is reduced via the generation of hydrogen sulphide gas. The fungus exhibits Class 2 nitrogen utilisation with growth-dependent interactions among different pH and nitrogen sources (Raudabaugh and Miller 2013). Investigations on the influence of the matric potential of substrates used polyethylene glycol for adjustment. It was shown by delayed germination and growth that *P. destructans* was intolerant to polyethylene glycol-induced matric potential at 22.5 MPa and at 25 MPa no germination was at all visible. However, when surface tension of 25 MPa polyethylene glycol medium was decreased, germination and growth of *P. destructans* were permitted, suggesting a link between substrate suitability and aqueous surface tension altering substances (Raudabaugh and Miller 2013). *Pseudogymnoascus destructans* is not capable of producing any extracellular biosurfactants or surface tension-reducing compounds, but assays suggest that substrates containing such agents would be beneficial for the growth of the WNS fungus (Raudabaugh and Miller 2013). In a cave environment, surface tension-reducing substrates would contain free fatty acids or lipids that could be hydrolysed to fatty acids by the fungus. Lipids of bat skin and glandular secretions contain free fatty acids (Pannkuk et al. 2013, 2014) making the surface of bat skin an excellent habitat for *P. destructans* (Raudabaugh and Miller 2013). Members of *Pseudoeurotiaceae* grow saprophytically on woody tissue and rotting vegetation, and β-glucosidase produced by *Pd* makes it highly likely that it can degrade cellulosic substrates as long as sufficient moisture will reduce the water potential of debris for *P. destructans* requirements (Raudabaugh and Miller 2013).

Investigations on the triacylglyceride composition and fatty acyl saturation profiles of *P. destructans* gave insight into its metabolic capabilities. The fungus produces higher proportions of unsaturated 18 °C fatty acids and triacylglycerides than *G. pannorum*. At 5 °C it also produces up to a twofold increase in 18:3 fatty acids in comparison to higher culture temperatures, while the triacylglyceride proportion at upper as well as lower temperature growth limits was greater than 50% of total dried mycelia mass (Pannkuk et al. 2014). It is thought that the fungus uses the alteration of acyl lipid unsaturation as a strategy to adapt to cold temperatures as particularly triacylglycerides serve as the main energy reserve in fungi. Moreover, they prevent free fatty acids to reach toxic levels and stimulate reproduction (Brennan and Lösel 1978).

13.3 The White-Nose Syndrome

13.3.1 Clinical Signs

Due to its psychrophilic temperature requirements, *P. destructans* can only infect hibernating bats, while in their active life in summer, bats can discard the infection. In winter during the hibernation period, bats observed at sites affected by WNS display a number of clinical signs either concurrent or singly, but these concern mostly aberrant hibernation behaviour:

(a) White powdery fungus, especially on the bats' nose but also on the wings, ears or tail.
(b) Bats flying outside during the day in temperatures at or below zero.
(c) Shifts of large numbers of bats to locations near the hibernacula entrances or unusually cold areas.
(d) Dead or dying bats on the ground or on buildings, trees or other structures. On occasion, large numbers of fatalities occur either inside the hibernacula, in the immediate vicinity of the entrance.
(e) Bats show a general unresponsiveness to human disturbance.
(f) Bats arouse with markedly increased frequency from torpor (survival strategy where metabolism is reduced to a minimum) (Reeder et al. 2012; Warnecke et al. 2012).
(g) WNS-affected bats groom themselves in higher rates, with particular focus on their wing membranes (Brownlee-Bouboulis and Reeder 2013).

13.3.2 Histopathological Changes

The infection of hibernating bats with *P. destructans* is restricted to the glabrous skin of muzzle, ears and wing membranes, but in contrast to most dermatophytes, *P. destructans* invades the epidermis deep into the underlying connective tissue (Meteyer et al. 2009). Hallmarks of the disease are so-called cupping erosions in the epidermis containing dense pockets of fungal hyphae (Fig. 13.5) and/or ulcerations caused by fungal hyphae penetrating into the dermis, sometimes spanning the whole thickness of the wing membrane (Meteyer et al. 2009). At the muzzle fungal hyphae usually fill hair follicles and invade sebaceous and apocrine glands while also reaching into the adjacent connective tissue. Hibernating North American bats have no inflammatory response regardless of the severity of the lesions. This is attributed to the hibernation status where all body functions are reduced to a minimum including the immune system (Bouma et al. 2010). However, North American bats with severe wing damage, which were caught shortly after hibernation, as well as European bats in late hibernation revealed influx of neutrophilic granulocytes and macrophages as well as intradermal abscesses (Meteyer et al. 2009; Wibbelt et al. 2013; Zukal et al. 2014). Whether the differences in inflammatory

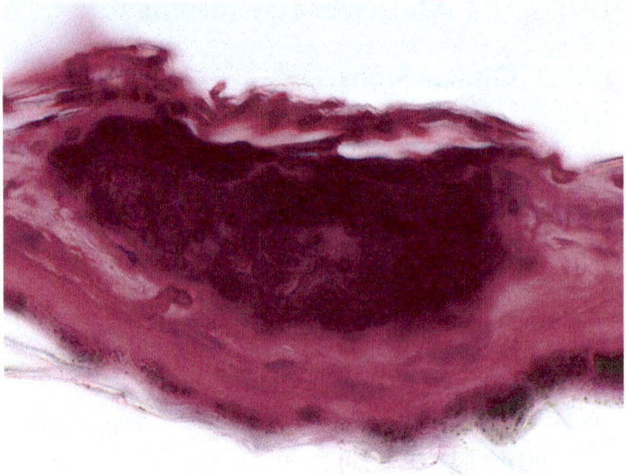

Fig. 13.5 Light microscopy image of epidermal "cupping erosion" caused by hyphae of *Pseudogymnoascus destructans* (PAS stain)

response between hibernating North American and European bats are merely due to the time of sampling with regard to hibernation length or a true species difference is currently unclear.

13.3.3 Pathogenesis

Despite the successful proof of *P. destructans* as the causative agent for WNS lesions in infection experiments (Lorch et al. 2011; Warnecke et al. 2013), the exact pathogenesis of the deadly outcome of this fungal infection on North American bats still remains unclear.

First investigations of deceased bats reported emaciated body conditions in little brown bats (*Myotis lucifugus*) (Meteyer et al. 2009; Courtin et al. 2010). An extensive study on free-ranging little brown bats used temperature-sensitive data logger fixed to the back of individual bats to investigate the frequency of arousal from torpor bouts, as during arousal bats will resume euthermic body temperature for a short period of time. Evidently, bats affected by *P. destructans* or bats that died due to the infection aroused significantly more often than unaffected bats and had shortened torpor bouts (Reeder et al. 2012). The same results were drawn from experimental infections, where two groups of little brown bats infected with a North American and a European fungal strain, respectively, presented a markedly increased arousal pattern (Warnecke et al. 2012). As arousals involve a high energy demand, it is thought that WNS-affected bats most likely burn their fuel storage prematurely before the end of winter and subsequently die of starvation. Additionally many WNS-positive bats are severely dehydrated. Current hypotheses assume that either due to increased arousals more time is spent in an euthermic, water-consuming state or that severe epidermal damage by *P. destructans* disrupts the wing membrane integrity. As the wing membranes play an important part in homeostasis and water balance, this could result in increased evaporative water loss (Cryan et al. 2010). The small body size of little brown bats (5–14 g body weight) is one of the many obstacles to answer these questions, for example, through the analysis of blood parameters. A number of investigations attempted interpretation of blood values from *P. destructans*-infected free-ranging as well as from captive little brown bats (Warnecke et al. 2013; Cryan et al. 2013; Verant et al. 2014), but results partly contradicted each other and did not satisfactorily explain the mechanisms behind the cause of death in WNS-affected bats.

European bats sometimes display large areas of wing membrane covered with numerous fungal colonies (Puechmaille et al. 2011; Wibbelt et al. 2013; Zukal et al. 2016) and still seem in good body condition. Bats from China had vastly reduced fungal loads compared to North American species and were similarly thriving as their European conspecifics. These circumstances could indicate a possible host resistance of these bat species being exposed to *P. destructans* for very long time (Hoyt et al. 2016). A thorough study of the transcriptome of WNS-affected and WNS-unaffected North American bats found numerous upregulated genes responsible for inflammatory responses in affected bats, but nevertheless local inflammatory response recruitment of leukocytes to the infection sites is missing (Field et al. 2015). Additionally, the transcriptome of the fungus was retrieved, detecting several putative virulence factors indicating host-pathogen interactions. Still the retrieved data cannot fully explain the complex mechanisms behind the pathogenesis.

An interesting approach is a new modelling study, which takes the environmental requirements of *P. destructans*, i.e. growth temperature and relative humidity, as well as hibernation requirements of different bat species into account. The results are surprisingly similar to observations from field studies and experimental infections and suggest that environmental conditions and basic host traits alone may explain much of the variability in disease outcomes among species of bats infected by *P. destructans* in North America and Europe. Moreover, the key features of fungal ecology are inextricably linked to bat mortality from WNS (Hayman et al. 2016).

13.3.4 Sampling and Diagnosis

13.3.4.1 Time of Sampling

P. destructans is a very slow-growing fungus, and colonies on the skin of hibernating bats will most likely be visible only when several weeks of the hibernation period have already passed. Sampling efforts should be attempted about 3 months after bats entered hibernation. The later towards the end of hibernation, the higher the chance that aerial mycelia have accumulated in visible patches. However, if bats start to arise from hibernation, they can quickly groom off all visible traces (Puechmaille et al. 2011).

13.3.4.2 Touch Imprints

Loops of adhesive tape can be used for quick analysis of suspicious fungal infection. Lowering the sticky side of a tape loop towards the surface of a fungal patch on

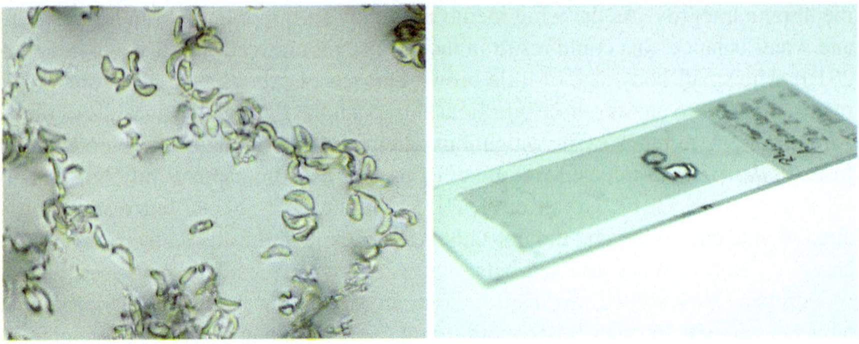

Fig. 13.6 Light microscopy image of unstained touch imprint from *Pseudogymnoascus destructans* (left); remaining adhesive tape of touch imprint after being dissected for culture or PCR use (right)

hibernating bats allows enough material to adhere (Meteyer et al. 2009; Wibbelt et al. 2010). Immediate transfer onto a glass slide and examination by light microscopy with narrowed condenser aperture and 100 times magnification make *P. destructans*' characteristic conidia easily visible (Fig. 13.6). After thorough cleansing of the tape's outer surface with 96% ethanol, cut-out areas laden with conidia can be used for isolation of pure cultures and PCR likewise (Wibbelt et al. 2010).

13.3.4.3 Swabs

Cotton-, polyester- or nylon-flocked swabs are equally used for sampling fungal growth from hibernating bats for culture isolation and PCR. Although immediate visual examination is not possible, swabs have been employed, for example, for quantifying the fungal load on wing membranes (Janicki et al. 2015; Langwig et al. 2015a, b; Zukal et al. 2016) via quantitative PCR (Shuey et al. 2014).

13.4 The Identification of the Causative Agent

Despite advanced methods to specifically identify nucleic acids of *P. destructans*, histopathological investigations remain the gold standard for confirming the disease. The hallmark lesions, i.e. cup-like erosions, have to be documented for diagnosing the disease. The colonisation of only the superficial epidermis without evidence of invasion into deeper tissues (Fig. 13.7) is not accounted for confirming the diagnosis of the disease. Evidence of *P. destructans* either by swab, touch imprint, culture and/or specific PCR is considered only suspicious of the disease, if confirmation of hallmark lesions by histopathological examination was not pursued.

To detect and differentiate *P. destructans* from closely related fungi in environmental samples, a quantitative dual-probe real-time PCR assay was established based on a single-nucleotide polymorphism (SNP) of the ITS1 region of the genome which was specific to *P. destructans* (Shuey et al. 2014). As environmental samples

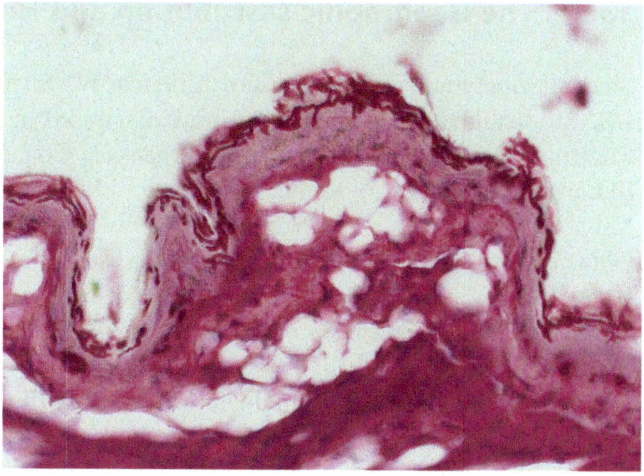

Fig. 13.7 Light microscopy image of superficial colonisation by *Pseudogymnoascus destructans* (PAS stain)—these lesions are not diagnostic for WNS

may contain various PCR inhibitors, further investigations were performed to optimise the methodologies. The comparison of different commercial test kits and the addition of a combination of chemical, enzymatic (lyticase and proteinase K) and mechanical procedures for the extraction and purification of nucleic acid from *P. destructans* in different environmental sample types have led to improved and efficient methods for high-throughput and quantification analyses (Verant et al. 2016).

Conventional PCR tests targeting the internal transcribed spacer (ITS) region of the rRNA gene complex can be used, but they make subsequent sequencing necessary as otherwise cross-reaction with the plethora of *P. destructans* allies cannot be discriminated (Lorch et al. 2010). A highly sensitive and consistent real-time TaqMan PCR test was developed that detects as little as 3.3 fg of genomic DNA from *P. destructans* per 25 mL reaction equalling approximately 0.1 genome equivalents. It is designed to target a portion of the multicopy intergenic spacer region of the rRNA gene complex (Muller et al. 2013). This method was employed in a large study to evaluate the fungal load of individual bats by rubbing a swab five times across muzzle and forearm. However, these results require careful interpretation, since Field et al. (2015) compared fungal loads with the amount of cupping erosions detected by histopathology. They found high measures of fungal loads and relatively low numbers of cupping erosions and vice versa, indicating that fungal load and actual tissue damage do not necessarily correlate well.

Long-wave ultraviolet (UV) light with a wavelength of 366–385 nm elicits a distinct orange-yellow fluorescence in bat wing membranes that corresponds directly with the fungal cupping erosions in histological sections (Turner et al. 2014) and can be used to visualise epidermal clusters of (invading) fungal hyphae. A non-invasive approach to estimate the severity of *P. destructans* infection on live bats is based on correlating the fungal load of individual bats assessed via quantitative PCR (Shuey et al. 2014) and the number of foci on wing membranes fluorescing in UV light (Zukal et al. 2016).

13.5 The Geographic Distribution and Host Range

Since its emergence in Northeastern America, WNS spreads from New York State, USA, to neighbouring states and is now diagnosed in the following US states and Canadian (C) provinces: Alabama, Arkansas, Connecticut, Delaware, Georgia, Illinois, Indiana, Iowa, Kentucky, Maine, Maryland, Massachusetts, Michigan, Minnesota, Missouri, New Hampshire, New Jersey, New York, North Carolina, Ohio, Pennsylvania, South Carolina, Tennessee, Vermont, Virginia, Washington State, West Virginia, Wisconsin, New Brunswick (C), Nova Scotia (C), Prince Edward Island (C), Ontario (C) and Quebec (C) (US Fish and Wildlife Service March 2016a).

Seven bat species have been diagnosed with WNS in North America: big brown bat (*Eptesicus fuscus*), grey bat (*Myotis grisescens*), eastern small-footed bat (*Myotis leibii*), little brown bat (*Myotis lucifugus*), northern long-eared bat (*Myotis septentrionalis*), Indiana bat (*Myotis sodalis*) and tricolored bat (*Perimyotis subflavus*); *Myotis lucifugus* being one of the most affected species. Further five American bat species have been detected with *P. destructans* but without confirming the disease: Rafinesque's big-eared bat (*Corynorhinus rafinesquii*), Virginia big-eared bat (*Corynorhinus townsendii virginianus*), eastern red bat (*Lasiurus borealis*), silver-haired bat (*Lasionycteris noctivagans*) and southeastern bat (*Myotis austroriparius*) (US Fish and Wildlife Service March 2016a).

In Europe *P. destructans* has been found on hibernating bats in almost every country from the most western border to the West Siberian plain of Palearctic Asia, namely, Austria, Belgium, Croatia, Czech Republic, Estonia, France, Germany, Great Britain, Hungary, Latvia, Luxemburg, Poland, Portugal, Russia, Slovakia, Switzerland, the Netherlands and Ukraine. Photographs of bats suspicious for *P. destructans* infection were provided from Denmark, Romania and Turkey (Puechmaille et al. 2011; Šimonovičová et al. 2011; Bürger et al. 2013; das Neves Paiva-Cardoso et al. 2014; Barlow et al. 2015; Leopardi et al. 2015; Zukal et al. 2016).

Seventeen bat species were found to carry the fungus in Europe: western barbastelle (*Barbastella barbastellus*), northern bat (*Eptesicus nilssonii*), Schreiber's bent-winged bat (*Miniopterus schreibersii*), Alcathoe myotis (*Myotis alcathoe*), Bechstein's bat (*Myotis bechsteinii*), lesser mouse-eared bat (*Myotis blythii*), Brandt's bat (*Myotis brandtii*), pond bat (*Myotis dasycneme*), Daubenton's bat (*Myotis daubentonii*), Geoffroy's bat (*Myotis emarginatus*), greater mouse-eared bat (*Myotis myotis*), whiskered bat (*Myotis mystacinus*), Natterer's bat (*Myotis nattereri*), brown long-eared bat (*Plecotus auritus*), grey long-eared bat (*Plecotus austriacus*), Mediterranean horseshoe bat (*Rhinolophus euryale*) and lesser horseshoe bat (*Rhinolophus hipposideros*) (Puechmaille et al. 2011; Bandouchova et al. 2015; Zukal et al. 2016); *Myotis myotis* is most often detected to carry the fungus. The most recent discovery of *P. destructans* in Northeastern China on individuals of Alashanian pipistrelle (*Hypsugo alaschanicus*), greater tube-nosed bat (*Murina leucogaster*), eastern water bat (*Myotis petax*) and greater horseshoe bat

(*Rhinolophus ferrumequinum*) (Hoyt et al. 2016) extends the known geographic distribution of the fungus by far.

Phylogenetic investigations of the novel pathogen soon revealed that *P. destructans* isolates found in North America originate from a single clonal genotype (Rajkumar et al. 2011; Ren et al. 2012) supporting the idea of a recent entry of this pathogen into a new environment. Investigations of European isolates conversely found clear diversification of the different isolates indicating that *P. destructans* is enzootic in Europe for a long time (Leopardi et al. 2015). In the latter study, eight genomic loci were sequenced, and seven of these were found to be polymorphic among the European isolates, which is in sharp contrast to the reported absence of variation in North American isolates. Via combination of these gene loci for each of the investigated isolates, eight *P. destructans* haplotypes were discovered across Europe. Moreover, the most common of these haplotypes was shared with all North American isolates, and phylogenetic analyses of this haplotype revealed that samples from France, Germany and Belgium were the most basal (Leopardi et al. 2015). These results make the early hypothesis of an introduction of the fungus from Europe into North America indeed very likely.

13.6 The Transmission

It is thought that the introduction of *P. destructans* into North America possibly occurred accidently by gear or clothing contaminated with conidia brought from Europe to America. Although the relevance of anthropogenic fungal spread is unclear, the fast spread of the fungus over thousands of kilometres from the epicentre leaves an open question on whether, for example, contaminated caving gear might have contributed (Turner et al. 2011). But the bats themselves can undoubtedly serve as vectors for distributing the fungus. They transfer the fungus not only via bat-to-bat contact (Lorch et al. 2011) but also by contaminating formerly *P. destructans*-free hibernacula, where viable fungus can be found in environmental samples even after bats were absent from a hibernaculum for 1–2 years (Lorch et al. 2013b). By entering these contaminated caves, unaffected bats can get infected through contact with the fungal spores. During the swarming period in autumn, bats meet in large congregations in front of potential hibernacula for mating, and while they move in and out of the caves, they are likely to sweep up conidia and carry them to the next cave (Turner et al. 2011). As fungal spores are easily transferred from one location to another, strict decontamination measures have been implied for any entry in North American cavernous hibernacula (US Fish and Wildlife Service April 2016b).

13.7 Outlook

The unceasing spread of WNS in North America and continuous loss of further animals adding to the millions of dead bats will not only result in a dramatic loss of biodiversity but also in unforeseen ecological and economical consequences. Bats are one of the most significant nocturnal pest controls for agricultural- and forestry-related pest insects (Boyles et al. 2011). Moreover by foraging on crop pests, they suppress pest-associated fungal growth and mycotoxin in corn multiplying their economic value and service to mankind with an estimated money equivalent of more than one billion USD globally (Maine and Boyles 2015).

The most urgent quest would be the mitigation of the infection. Several attempts were undertaken to find a solution for either treating the bats or decontaminating the cavernous hibernacula, but as wildlife settings are in sharp contrast to livestock husbandry, the management of an infectious disease caused by a spore-forming pathogen seems an unfathomable mission. The highly complex environment of cavernous ecosystems is particularly fragile, and the application of, e.g. disinfectants could lead to unpredictable negative consequences (Wibbelt 2015).

Still, many open questions remain regarding the exact pathological mechanisms from the initial infection and development of disease to finally the death of the chiropteran host, like why is the fungal infection limited to glabrous skin? Is the sheer extent of the lesions responsible for the mortality? And if so, is the affected area or the depth of invasion the crucial factor? If all bat carcasses of a hibernaculum struck by WNS would be histopathologically investigated—would they all show the highest possible severity score in wing membrane lesions? Would all animals be emaciated? And if not, what other factors would dominate the fatal outcome of the disease?

Numerous pieces of this big puzzle have been found and placed: some had to be replaced or relocated, others had to be entirely removed, and while the image's framework is clearly laid out, there still are many missing pieces waiting to be found.

References

Bandouchova H, Bartonicka T, Berkova H et al (2015) *Pseudogymnoascus destructans*: evidence of virulent skin invasion for bats under natural conditions, Europe. Transbound Emerg Dis 62 (1):1–5. https://doi.org/10.1111/tbed.12282

Barlow AM, Worledge L, Miller H et al (2015) First confirmation of *Pseudogymnoascus destructans* in British bats and hibernacula. Vet Rec 177(3):73. https://doi.org/10.1136/vr.102923

Blehert DS, Hicks AC, Behr M et al (2009) Bat white-nose syndrome: an emerging fungal pathogen? Science 323(5911):227. https://doi.org/10.1126/science.1163874

Bouma HR, Carey HV, Kroese FG (2010) Hibernation: the immune system at rest? J Leukoc Biol 88(4):619–624. https://doi.org/10.1189/jlb.0310174

Boyles JG, Cryan PM, McCracken GF, Kunz TH (2011) Conservation. Economic importance of bats in agriculture. Science 332(6025):41–42. https://doi.org/10.1126/science.1201366

Brennan PJ, Lösel DM (1978) Role of lipid in fungal-host relationships. In: Rose AH, Morris JG (eds) Advances in microbial physiology, vol 17. Academic Press, New York, pp 140–180

Brownlee-Bouboulis SA, Reeder DM (2013) White-nose syndrome-affected little brown myotis (*Myotis lucifugus*) increase grooming and other active behaviors during arousals from hibernation. J Wildl Dis 49(4):850–859. https://doi.org/10.7589/2012-10-242

Bürger K, Gebhardt O, Wohlfahrt S et al (2013) First confirmed records of *Geomyces destructans* (Blehert & Gargas 2009) in Austria. Ber Nat-Med Ver Innsbruck 98:127–135

Chaturvedi V, Springer DJ, Behr MJ et al (2010) Morphological and molecular characterizations of psychrophilic fungus *Geomyces destructans* from New York bats with White Nose Syndrome (WNS). PLoS One 5(5):e10783. https://doi.org/10.1371/journal.pone.0010783
Courtin F, Stone WB, Risatti G et al (2010) Pathologic findings and liver elements in hibernating bats with white-nose syndrome. Vet Pathol 47(2):214–219. https://doi.org/10.1177/0300985809358614
Cryan PM, Meteyer CU, Boyles JG, Blehert DS (2010) Wing pathology of white-nose syndrome in bats suggests life-threatening disruption of physiology. BMC Biol 8:135. https://doi.org/10.1186/1741-7007-8-135
Cryan PM, Meteyer CU, Boyles JG, Blehert DS (2013) White-nose syndrome in bats: illuminating the darkness. BMC Biol 11:47. https://doi.org/10.1186/1741-7007-11-47
das Neves Paiva-Cardoso M, Morinha F, Barros P et al (2014) First isolation of *Pseudogymnoascus destructans* in bats from Portugal. Eur J Wildl Res 60:645–649
Field KA, Johnson JS, Lilley TM et al (2015) The white-nose syndrome transcriptome: activation of anti-fungal host responses in wing tissue of hibernating little brown myotis. PLoS Pathog 11(10):e1005168. https://doi.org/10.1371/journal.ppat.1005168
Frick WF, Pollock JF, Hicks AC et al (2010) An emerging disease causes regional population collapse of a common North American bat species. Science 329(5992):679–682. https://doi.org/10.1126/science.1188594
Gargas A, Trest MT, Christensen M (2009) *Geomyces destructans* sp. nov. associated with bat white-nose syndrome. Mycotaxon 108:147–154
Hayman DT, Pulliam JR, Marshall JC et al (2016) Environment, host, and fungal traits predict continental-scale white-nose syndrome in bats. Sci Adv 2(1):e1500831. https://doi.org/10.1126/sciadv.1500831
Hoyt JR, Sun K, Parise KL (2016) Widespread bat white-nose syndrome fungus, Northeastern China. Emerg Infect Dis 22:140–142
Janicki AF, Frick WF, Kilpatrick AM et al (2015) Efficacy of visual surveys for white-nose syndrome at bat hibernacula. PLoS One 10(7):e0133390. https://doi.org/10.1371/journal.pone.0133390
Khankhet J, Vanderwolf KJ, McAlpine DF et al (2014) Clonal expansion of the *Pseudogymnoascus destructans* genotype in North America is accompanied by significant variation in phenotypic expression. PLoS One 9(8):e104684. https://doi.org/10.1371/journal.pone.0104684
Langwig KE, Frick WF, Reynolds R et al (2015a) Host and pathogen ecology drive the seasonal dynamics of a fungal disease, white-nose syndrome. Proc Biol Sci 282(1799):20142335. https://doi.org/10.1098/rspb.2014.2335
Langwig KE, Hoyt JR, Parise KL et al (2015b) Invasion dynamics of white-nose syndrome fungus, midwestern United States, 2012–2014. Emerg Infect Dis 21(6):1023–1026. https://doi.org/10.3201/eid2106.150123
Leopardi S, Blake D, Puechmaille SJ (2015) White-Nose Syndrome fungus introduced from Europe to North America. Curr Biol 25(6):R217–R219. https://doi.org/10.1016/j.cub.2015.01.047
Lindner DL, Gargas A, Lorch JM et al (2011) DNA-based detection of the fungal pathogen *Geomyces destructans* in soils from bat hibernacula. Mycologia 103(2):241–246. https://doi.org/10.3852/10-262
Lorch JM, Gargas A, Meteyer CU et al (2010) Rapid polymerase chain reaction diagnosis of white-nose syndrome in bats. J Vet Diagn Investig 22(2):224–230
Lorch JM, Meteyer CU, Behr MJ et al (2011) Experimental infection of bats with *Geomyces destructans* causes white-nose syndrome. Nature 480(7377):376–378. https://doi.org/10.1038/nature10590
Lorch JM, Lindner DL, Gargas A (2013a) A culture-based survey of fungi in soil from bat hibernacula in the eastern United States and its implications for detection of *Geomyces*

destructans, the causal agent of bat white-nose syndrome. Mycologia 105(2):237–252. https://doi.org/10.3852/12-207

Lorch JM, Muller LK, Russell RE et al (2013b) Distribution and environmental persistence of the causative agent of white-nose syndrome, *Geomyces destructans*, in bat hibernacula of the eastern United States. Appl Environ Microbiol 79(4):1293–1301. https://doi.org/10.1128/AEM.02939-12

Maine JJ, Boyles JG (2015) Bats initiate vital agroecological interactions in corn. Proc Natl Acad Sci U S A 112:12438–12443

Martínková N, Bačkor P, Bartonička T et al (2010) Increasing incidence of *Geomyces destructans* fungus in bats from the Czech Republic and Slovakia. PLoS One 5(11):e13853. https://doi.org/10.1371/journal.pone.0013853

Meteyer CU, Buckles EL, Blehert DS et al (2009) Histopathologic criteria to confirm white-nose syndrome in bats. J Vet Diagn Investig 21(4):411–414

Meteyer CU, Valent M, Kashmer J et al (2011) Recovery of little brown bats (*Myotis lucifugus*) from natural infection with *Geomyces destructans*, white-nose syndrome. J Wildl Dis 47 (3):618–626

Minnis AM, Lindner DL (2013) Phylogenetic evaluation of *Geomyces* and allies reveals no close relatives of *Pseudogymnoascus destructans*, comb. nov., in bat hibernacula of eastern North America. Fungal Biol 117(9):638–649. https://doi.org/10.1016/j.funbio.2013.07.001

Muller LK, Lorch JM, Lindner DL et al (2013) Bat white-nose syndrome: a real-time TaqMan polymerase chain reaction test targeting the intergenic spacer region of *Geomyces destructans*. Mycologia 105(2):253–259. https://doi.org/10.3852/12-242

O'Donoghue AJ, Knudsen GM, Beekman C et al (2015) Destructin-1 is a collagen-degrading endopeptidase secreted by *Pseudogymnoascus destructans*, the causative agent of white-nose syndrome. Proc Natl Acad Sci U S A 112(24):7478–7483. https://doi.org/10.1073/pnas.1507082112

Palmer JM, Kubatova A, Novakova A et al (2014) Molecular characterization of a heterothallic mating system in *Pseudogymnoascus destructans*, the fungus causing white-nose syndrome of bats. G3 (Bethesda) 4(9):1755–1763. https://doi.org/10.1534/g3.114.012641

Pannkuk EL, Gilmore DF, Fuller NW et al (2013) Sebaceous lipid profiling of bat integumentary tissues: quantitative analysis of free fatty acids, monoacylglycerides, squalene, and sterols. Chem Biodivers 10(12):2122–1232. https://doi.org/10.1002/cbdv.201300319

Pannkuk EL, Fuller NW, Moore PR et al (2014) Fatty acid methyl ester profiles of bat wing surface lipids. Lipids 49(11):1143–1150. https://doi.org/10.1007/s11745-014-3951-2

Pannkuk EL, Risch TS, Savary BJ (2015) Isolation and identification of an extracellular subtilisin-like serine protease secreted by the bat pathogen *Pseudogymnoascus destructans*. PLoS One 10 (3):e0120508. https://doi.org/10.1371/journal.pone.0120508

Puechmaille SJ, Verdeyroux P, Fuller H et al (2010) White-nose syndrome fungus (*Geomyces destructans*) in bat, France. Emerg Infect Dis 16(2):290–293. https://doi.org/10.3201/eid1602.091391

Puechmaille SJ, Wibbelt G, Korn V et al (2011) Pan-European distribution of white-nose syndrome fungus (*Geomyces destructans*) not associated with mass mortality. PLoS One 6(4):e19167. https://doi.org/10.1371/journal.pone.0019167

Rajkumar SS, Li X, Rudd RJ et al (2011) Clonal genotype of *Geomyces destructans* among bats with White Nose Syndrome, New York, USA. Emerg Infect Dis 17(7):1273–1276. https://doi.org/10.3201/eid1707.102056

Raudabaugh DB, Miller AN (2013) Nutritional capability of and substrate suitability for *Pseudogymnoascus destructans*, the causal agent of bat white-nose syndrome. PLoS One 8 (10):e78300. https://doi.org/10.1371/journal.pone.0078300

Reeder DM, Frank CL, Turner GG et al (2012) Frequent arousal from hibernation linked to severity of infection and mortality in bats with white-nose syndrome. PLoS One 7(6):e38920. https://doi.org/10.1371/journal.pone.0038920

Ren P, Haman KH, Last LA et al (2012) Clonal spread of *Geomyces destructans* among bats, midwestern and southern United States. Emerg Infect Dis 18(5):883–885. https://doi.org/10.3201/eid1805.111711

Shuey MM, Drees KP, Lindner DL et al (2014) Highly sensitive quantitative PCR for the detection and differentiation of *Pseudogymnoascus destructans* and other *Pseudogymnoascus* species. Appl Environ Microbiol 80(5):1726–1731. https://doi.org/10.1128/AEM.02897-13

Šimonovičová A, Pangallo D, Chovanová K, Lehotská B (2011) *Geomyces destructans* associated with bat disease WNS detected in Slovakia. Biologia 66(3):562–564. https://doi.org/10.2478/s11756-011-0041-2

Turner GG, Reeder DM, Coleman JTH (2011) A five-year assessment of mortality and geographic spread of white-nose syndrome in North American bats and a look to the future. Bat Res News 52(2):13–27

Turner GG, Meteyer CU, Barton H et al (2014) Nonlethal screening of bat-wing skin with the use of ultraviolet fluorescence to detect lesions indicative of white-nose syndrome. J Wildl Dis 50(3):566–573. https://doi.org/10.7589/2014-03-058

US Fish and Wildlife Service (2016a) White-nose syndrome fact sheet March 2016. https://www.whitenosesyndrome.org/sites/default/files/resource/white-nose_fact_sheet_5-2016_2.pdf

US Fish and Wildlife Service (2016b) White-nose syndrome decontamination protocol April 2016. https://www.whitenosesyndrome.org/sites/default/files/files/national_wns_decon_protocol_04.12.2016.pdf

Verant ML, Boyles JG, Waldrep W Jr et al (2012) Temperature-dependent growth of *Geomyces destructans*, the fungus that causes bat white-nose syndrome. PLoS One 7(9):e46280. https://doi.org/10.1371/journal.pone.0046280

Verant ML, Meteyer CU, Speakman JR et al (2014) White-nose syndrome initiates a cascade of physiologic disturbances in the hibernating bat host. BMC Physiol 14:10. https://doi.org/10.1186/s12899-014-0010-4

Verant ML, Bohuski EA, Lorch JM, Blehert DS (2016) Optimized methods for total nucleic acid extraction and quantification of the bat white-nose syndrome fungus, *Pseudogymnoascus destructans*, from swab and environmental samples. J Vet Diagn Investig 28(2):110–118. https://doi.org/10.1177/1040638715626963

Warnecke L, Turner JM, Bollinger TK et al (2012) Inoculation of bats with European *Geomyces destructans* supports the novel pathogen hypothesis for the origin of white-nose syndrome. Proc Natl Acad Sci U S A 109(18):6999–7003. https://doi.org/10.1073/pnas.1200374109

Warnecke L, Turner JM, Bollinger TK et al (2013) Pathophysiology of white-nose syndrome in bats: a mechanistic model linking wing damage to mortality. Biol Lett 9(4):20130177. https://doi.org/10.1098/rsbl.2013.0177

Wibbelt G (2015) Out of the dark abyss: white-nose syndrome in bats. Vet Rec 177(3):70–72. https://doi.org/10.1136/vr.h3782

Wibbelt G, Kurth A, Hellmann D et al (2010) White-nose syndrome fungus (*Geomyces destructans*) in bats, Europe. Emerg Infect Dis 16(8):1237–1243. https://doi.org/10.3201/eid1608.100002

Wibbelt G, Puechmaille SJ, Ohlendorf B et al (2013) Skin lesions in European hibernating bats associated with *Geomyces destructans*, the etiologic agent of white-nose syndrome. PLoS One 8(9):e74105. https://doi.org/10.1371/journal.pone.0074105

Zukal J, Bandouchova H, Bartonicka T et al (2014) White-nose syndrome fungus: a generalist pathogen of hibernating bats. PLoS One 9(5):e97224. https://doi.org/10.1371/journal.pone.0097224

Zukal J, Bandouchova H, Brichta J et al (2016) White-nose syndrome without borders: *Pseudogymnoascus destructans* infection tolerated in Europe and Palearctic Asia but not in North America. Sci Rep 6:19829. https://doi.org/10.1038/srep19829

Chytridiomycosis

14

An Martel, Frank Pasmans, Matthew C. Fisher, Laura F. Grogan, Lee F. Skerratt, and Lee Berger

Abstract

The amphibian fungal disease chytridiomycosis is considered one of the greatest threats to biodiversity. This lethal skin disease is caused by chytridiomycete fungi belonging to the genus *Batrachochytrium*. Although sudden amphibian population declines had occurred since the 1970s in the Americas and Australia, mass mortalities were not observed until the 1990s. The fungus *Batrachochytrium dendrobatidis* (*Bd*) was identified as the cause of these declines. It is estimated that *Bd* has caused the rapid decline or extinction of at least 200 amphibian species, which is probably an underestimation due to the cryptic behaviour of many amphibians such as many salamanders and also the lack of monitoring. A second chytrid species, *B. salamandrivorans* (*Bsal*), has recently emerged and caused mass mortality in salamanders in Belgium, the Netherlands and Germany, affecting most salamander and newt taxa in the amphibian community and is considered a major threat to the western Palearctic amphibian biodiversity. In this

A. Martel (✉) · F. Pasmans
Faculty of Veterinary Medicine, Department of Pathology, Bacteriology and Avian Diseases, Ghent University, Ghent, Belgium
e-mail: an.martel@ugent.be; frank.pasmans@ugent.be

M. C. Fisher
Department of Infectious Disease Epidemiology, School of Public Health, Imperial College London, London, UK
e-mail: matthew.fisher@imperial.ac.uk

L. F. Grogan
Griffith University, Environmental Futures Research Institute, School of Environment, Brisbane, QLD, Australia
e-mail: l.grogan@griffith.edu.au

L. F. Skerratt · L. Berger
One Health Research Group, College of Public Health, Medical and Veterinary Sciences, James Cook University, Townsville, QLD, Australia
e-mail: lee.skerratt@jcu.edu.au; lee.berger@jcu.edu.au

S. Seyedmousavi et al. (eds.), *Emerging and Epizootic Fungal Infections in Animals*,
https://doi.org/10.1007/978-3-319-72093-7_14

chapter we review the epidemiology, host pathogen interactions and mitigation strategies of both chytrid pathogens.

14.1 Chytrids

The causative agents of the disease chytridiomycosis belong to a "lower" fungal phylum: the *Chytridiomycota* (Longcore et al. 1999). *Bd* and *Bsal* belong to the order of the *Rhizophydiales* and are, together with the enigmatic fish pathogen *Ichthyochytrium vulgare* (Plehn 1920), to date the only known members of this phylum, which are adapted to vertebrate hosts.

Chytridiomycota are unusual among the fungi in that they produce zoospores. Asexual reproduction in *Bd* and *Bsal* occurs through the release of motile flagellated spores (zoospores) from the reproductive body or thallus in which they are produced (zoosporangium) (Fig. 14.1) (Longcore et al. 1999). Neither *Bd* nor *Bsal* have yet been observed in culture to reproduce sexually. However, as several genetic studies have identified *Bd* isolates with a hybrid genotype, sexual recombination and hybridization must have been important mechanisms in its evolutionary history (Farrer et al. 2011; Schloegel et al. 2012; Rosenblum et al. 2013).

In vitro, both pathogens grow well in tryptone broth and have a similar life cycle (Van Rooij et al. 2015). The zoospore first encysts by developing a cell wall and absorbing its flagellum and forms a germling with fine threadlike rhizoids. The germling matures into a zoosporangium in which the cytoplasm cleaves mitotically to form new zoospores. Discharge papillae or tubes are formed during the growth of the sporangium. At maturity, the plug blocking the papillae dissolves, and the zoospores are released into the environment to continue their life cycle (Longcore et al. 1999; Berger et al. 2005; Van Rooij et al. 2015). *Bd* zoosporangia are, in contrast with *Bsal* thalli, predominantly monocentric (a thallus containing a single

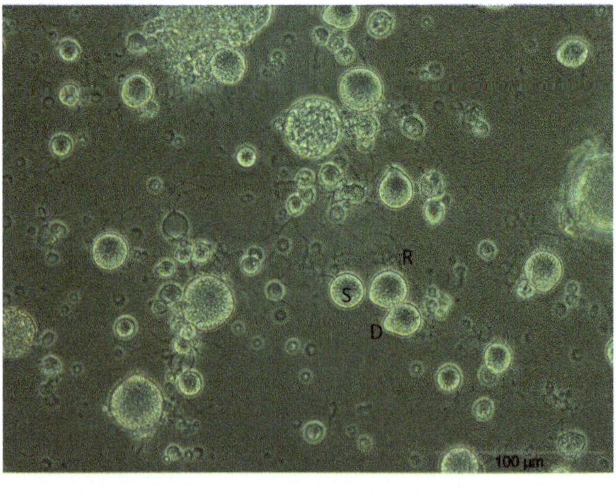

Fig. 14.1 In vitro culture of *Bsal* in TGhL broth at 15 °C. Monocentric thalli predominate, with the rare presence of colonial thalli. Sporangia (S) with rhizoids (R) develop discharge tubes (D) to release zoospores (Scale bar, 100 µm) (© Pasmans and Martel 2015. All Rights Reserved)

sporangium) rather than colonial (a thallus containing more than one sporangium). Despite the difference in thermal preference (*Bd* 22 °C, *Bsal* 15 °C), the life cycle of both chytrids in culture is completed within 5 days (Berger et al. 2005; Martel et al. 2013). The life cycle of *Bd* in amphibian skin is largely similar to that observed in culture (Van Rooij et al. 2015).

14.2 Global Epidemiology of *Batrachochytrium dendrobatidis* Emergence

Following the discovery that *Bd* was a key driver of declines in amphibian species in Australia and the Americas (Berger et al. 1998; Longcore et al. 1999), attention has been focused on determining where outbreaks of chytridiomycosis have occurred and whether there were any spatiotemporal patterns that indicate the original sources of infection as well as pathways of spread. A key technological advance has been the development of a *Bd*-specific quantitative PCR reaction that has been widely used to screen amphibian populations for infection (Boyle et al. 2004). A global mapping project (http://www.bd-maps.net) (Olson et al. 2013) showed *Bd* occurred in 56 of 82 (68%) countries and, in 516 of 1240 (42%) species tested, determined using a dataset of more than 36,000 individuals (Bd-maps.net data snapshot August 2012). Across the world, broadscale distribution patterns are evident, with *Bd* widely detected in the Americas and detected only patchily in Australia, Africa, Asia, and Europe.

Combining data on amphibian population health alongside molecular qPCR surveys has uncovered regional epidemiological trends. In eastern Australia, prospective and retrospective sampling of amphibian populations showed that populations were *Bd*-negative prior to 1978 followed by an expansion north and south from an inferred centre in Southern Queensland, reaching its northern limits in the mid-1990s. Western Australia was *Bd*-negative until mid-1985 where upon the arrival and spread of disease was documented (Berger et al. 2009). Mesoamerica has witnessed a very rapid wave-like front of expansion from a putative origin in Mexico in the 1970s (James et al. 2006), through Guatemala to Monteverde, Costa Rica, in the 1980s, and then southwards at estimated rates of up to 50 km y^{-1}, until the infection jumped the Panama Canal. This epidemic front of chytridiomycosis along the North-South transect of Central America was predictable to the extent that researchers were able to anticipate the arrival of *Bd* in uninfected regions, such as El Copé in Panama in 2004, and to then document the subsequent collapse of the amphibian community (Lips et al. 2008). North America saw steep declines across populations of the Sierra mountain yellow-legged frogs (*Rana muscosa* and *R. sierrae*), most notably in the Sequoia-Kings Canyon National Park, where regional populations were sequentially extirpated as *Bd* invaded lakes in 2004 (Vredenburg et al. 2010). However, historical declines across other regions of the Sierras suggest that the infection had been present since at least the 1970s, and elsewhere in the United States, retrospective surveys of museum collections have demonstrated a widespread prevalence since as early as 1888 in Illinois (Talley et al. 2015).

Although the pattern of declines in Ecuador, Venezuela, Bolivia and Peru are consistent with chytridiomycosis, most of South America lacks convincing evidence for outbreaks of chytridiomycosis, and retrospective analyses of museum specimens have shown *Bd* to be enzootic in the Brazilian Atlantic Forests for over a century (Rodriguez et al. 2014). Surveillance in Europe showed that outbreaks occurred around the 1990s in Spain (Bosch et al. 2001; Walker et al. 2010); however, it is suspected that asymptomatic infections were widespread prior to the 1990s as outbreaks of the disease have only been witnessed at high altitudes (Walker et al. 2010). In Southeast Asia, *Bd* has a low prevalence, patchy distribution and outbreaks of chytridiomycosis, or cryptic population declines have not been recorded (Swei et al. 2011).

Given these patterns of declines, is it possible to determine whether there is a single original 'source' of the panzootic of chytridiomycosis or are we rather witnessing multiple regional and perhaps unlinked foci? Answers to this question have been sought by attempting to identify geographic regions where *Bd* has had a long and stable association with host species (which is possibly indicative of long-term co-evolutionary associations), as well as through phylogenomic analyses. Histological studies from Southern African museum collections (Weldon et al. 2004; Soto-Azat et al. 2010) identified Africa as a potential source of the panzootic, leading to the '*Bd* Out of Africa' hypothesis being coined to suggest that *Bd* was spread around the world via the extensive trade in the African clawed frog *Xenopus laevis* from the 1930s onwards. However, the widespread occurrence of century-old infections from similar retrospective studies means that Brazil, the United States and Asia are now also included as possible origins (Goka et al. 2009; Rodriguez et al. 2014; Talley et al. 2015; Fong et al. 2015).

A paradigm shift in our understanding of *Bd*'s epidemiology occurred with the onset of high-throughput whole-genome sequencing (WGS). The phylogenetic resolution afforded by being able to align and detect single-nucleotide polymorphisms (SNPs) across the ~24 Mb genome of *Bd* has uncovered over 500,000 polymorphic sites. An initial phylogenetic analysis of 20 global isolates showed that *Bd* is composed of at least 3 deeply diverged lineages (Farrer et al. 2011). Of these, one was found to have a global distribution and showed low levels of genetic polymorphism as well as a lack of genetic structure among continents. These data are consistent with a rapid global emergence that was dated to sometime in the twentieth century, and accordingly the lineage was named the '*Bd* Global Panzootic Lineage' (*Bd* GPL). Alongside *Bd* GPL, two other regionally endemic lineages were identified: one was found in Switzerland (*Bd* CH), while the other was found widely occurring in South Africa (*Bd* CAPE). Subsequent genome sequencing of *Bd* isolates from other global regions identified a fourth endemic lineage that infects amphibians in the Brazilian Atlantic forest (*Bd* Brazil) (Rodriguez et al. 2014). Further genome sequencing from a broader spatial collection of isolates will undoubtedly uncover further endemic lineages, consolidating the finding that *Bd* is an amphibian pathogen with a broad global distribution and ancient associations with its amphibian hosts.

These phylogenomic analyses have shown that the worldwide emergence of chytridiomycosis is mostly likely explained by the rapid transmission of *Bd* GPL across a global scale as this lineage has now been found to co-occur in all continents alongside endemic lineages. In support of this hypothesis is that all mass mortality and extinction events thus far attributed to chytridiomycosis are associated with the presence of *Bd* GPL. In vivo laboratory assessments of *Bd*'s virulence have shown that the lineage is hypervirulent when compared against other lineages, a feature that confers "superbug" status on *Bd* GPL (Farrer et al. 2011). However, the origin of *Bd* GPL remains currently unknown, as are details on the historical timing of this lineage's emergence across different continents. Answers to these questions await the accumulation of genome sequences from a more representative set of spatial regions and a more nuanced understanding of the molecular clock rates that govern *Bd*'s evolution (Rosenblum et al. 2013; Farrer et al. 2013).

What is clear, however, is that the global trade in amphibians is a potent force in spreading *Bd* into naïve populations and species. This statement is especially true for the so-called 'Typhoid Mary' species such as *X. laevis* and the North American bullfrog *Lithobates catesbeianus*; these species carry *Bd* infections but rarely become diseased (Fisher et al. 2009). They are also widely traded and are often highly invasive when introduced by accident or deliberately into new environments (Lips et al. 2008). Therefore, these two species constitute ideal vectors for introducing *Bd* into uninfected regions of the globe (Garner et al. 2006) and are likely a major source of new *Bd* infections when released into naïve environments. Molecular epidemiology from whole-genome sequencing is now being used to identify the sources of regional outbreaks. The best understood example of this is the introduction of *Bd* onto the Balearic island of Mallorca where genome sequencing showed that *Bd* CAPE was found to be infecting two populations of Mallorcan midwife toads *Alytes muletensis*. Retrospective analyses identified a spillover event within a zoo captive assurance population of *A. muletensis* that had been subsequently released onto the island in order to supplement the wild populations. In this example, the co-housing of Southern African *Xenopus* species alongside *A. muletensis* led to cross-transmission and the vectoring of *Bd* CAPE onto the island (Walker et al. 2008; Doddington et al. 2013).

14.3 Global Epidemiology of *Batrachochytrium salamandrivorans* Emergence

The outbreak of chytridiomycosis caused by *Bsal* in the European low countries shares many features with the emergence of *Bd*. While population genomic analyses have not yet been published, the rapid development of a qPCR-based molecular diagnostic has enabled the use of global surveillance to map the distribution of the pathogen which has indicated likely sources of European infection (Blooi et al. 2013). In total, over 5000 amphibians were screened by qPCR from across four continents (Martel et al. 2014). These data showed that detections of *Bsal* were limited to East Asia, namely, Thailand, Japan and Vietnam. The data also point to a likely Asian origin of *Bsal*, and assessments of host associations have strengthened this hypothesis by demonstrating that, while Palearctic caudates from Europe and North America suffer clinical disease following infection by *Bsal*, Asian species tolerate or clear their infections. Molecular surveys of museum collections have shown that *Bsal* DNA can be detected in museum specimens of Asian *Cynops ensicauda* that were collected over 150 years ago and molecular dating shows that *Bsal* separated from *Bd* over 30 million years ago. Taken together, these data strongly indicate an Asian evolutionary centre of origin for *Bsal* infections that have spilled over to infect European salamander and newt species.

As with *Bd*, there has been much focus on how the transcontinental vectoring of *Bsal* may have occurred. It is clear now that the international trade in Asian caudate species is enormous, with, for instance, over 2.3 million individuals of *Cynops orientalis* imported into the United States from 2001 to 2009. Further, the infection is associated with outbreaks of disease in captive European caudates, which constitute a continuous threat to vectoring the disease to wild amphibians (Cunningham et al. 2015; Sabino-Pinto et al. 2015). Therefore, the international movement of traded species of amphibians is a key mechanism contributing to the spread of *Bsal* and its invasion of naïve disease-free ecosystems.

14.4 Pathogenesis

Whereas the pathogenesis of *Bd* infections is relatively well known, *Bsal* has only recently been discovered, and its pathogenesis is incompletely understood. This chapter will focus on the pathogenesis of *Bd*, while highlighting any known differences to *Bsal*.

14.4.1 Skin Colonization

The different steps in the infection process comprise attraction of the free swimming zoospore to a suitable host with subsequent attachment to and invasion into the host skin, leading to impairment of the skin function. *Bd* is attracted to keratin and to carbohydrate components of the mucus and epidermis (Meyer et al. 2007; Moss et al. 2008; Van Rooij et al. 2015). If the zoospores resist the defence mechanisms of the mucus (see below), adhesion to the skin surface occurs within 2–4 h after exposure (Van Rooij et al. 2012), after which the zoospores mature into thick-walled cysts. Cysts are anchored to the skin surface by fine fibrillar projections. Several adhesion proteins such as vinculin, fibronectin and fasciclin are expected to contribute to this process (Rosenblum et al. 2008, 2012). Besides, *Bd* is equipped with a chitin binding module (CBM18) that is hypothesized to facilitate survival on its amphibian host by limiting access for foreign chitinases by binding to the chitin of their proper cell wall. In addition, CBM18 would also allow attachment of the pathogen to non-host chitinous structures (e.g. insect or crustacean exoskeletons) allowing vectored disease spread (Abramyan and Stajich 2012; McMahon et al. 2013; Liu and Stajich 2015; Van Rooij et al. 2015).

Bd further develops endobiotically (with sporangia located inside the host cell). Colonization is established via a tubular extension or germ tube arising from the zoospore cyst that penetrates the host cell membrane and enables transfer of genetic material into the host cell. The distal end of the germ tube swells and gives rise to a new intracellular chytrid thallus. The pathogen then uses the same tactics to dig its way to deeper skin layers: older "mother" thalli develop rhizoid-like structures spreading to deeper skin layers, forming a swelling inside the host cell to finally give rise to a new "daughter" thallus (Berger et al. 2005; Van Rooij et al. 2012; Greenspan et al. 2012; Van Rooij et al. 2015). This intracellular proliferation occurs within the cells of the stratum corneum and the stratum granulosum. Immature sporangia are carried from the deeper skin layers to the skin surface by differentiating epidermal cells. At the time sporangia have developed discharge tubes and contain mature zoospores, they occur in the stratum corneum, where the zoospores are released in the environment (Longcore et al. 1999; Berger et al. 2005; Van Rooij et al. 2015).

Conversely, in explanted skin of the infection tolerant *X. laevis*, the pathogen develops merely epibiotically (with sporangia developing upon the skin) (Van Rooij et al. 2015). Here, the affected epidermal cells seem to be solely used as nutrient source for the growing sporangium upon the epidermis (Van Rooij et al. 2012). Due to the lack of conclusive histological evidence, it is not clear how infections manifest in this species under natural conditions (Van Rooij et al. 2015). As this "saprobic" type of development has only been observed in vitro, further in vivo evidence is needed. Skin invasion by *Bsal* was also shown to occur within 24 h (Martel et al. 2014), although the mechanism of *Bsal* invasion is not known.

Bd selectively colonizes keratinized, stratified epidermis. In anuran larvae, colonization is limited to the keratinized mouthparts, i.e. tooth rows and jaw sheaths, but absent from the body, limbs, tail, mouth and gills. Studies in *Mixophyes fasciolatus* and *Osteopilus septentrionalis* tadpoles show that during metamorphosis, colonization of the skin by *Bd* progresses following the distribution of keratin (Marantelli et al. 2004; McMahon and Rohr 2015). In contrast with anuran larvae, no colonization of salamander larvae by *Bd* has been demonstrated so far (Van Rooij et al. 2015).

Susceptibility to *Bsal* infection was equally shown to be life stage dependent. Whereas *Bsal* infects all fire salamander (*Salamandra salamandra*) life stages post metamorphosis, larvae seem to be refractory to infection (Van Rooij et al. 2015).

14.4.2 Pathophysiology

Severe chytridiomycosis causes functional disruption of the epidermal barrier. *Bd* possesses a large number of proteolytic enzymes, including serine-type proteases and fungalysin metallopeptidases, which can cause damage to skin integrity by disturbing the host's intracellular junctions (Joneson et al. 2011; Brutyn et al.

2012 Rosenblum et al. 2013). Furthermore, infection due to *Bd* triggers a decreased expression of host genes encoding essential skin integrity components such as keratin, collagen, elastin and fibrinogen (Rosenblum et al. 2012).

Physical disruption of the epidermis directly affects the osmoregulatory function of the skin: it impairs the electrolyte transport across the skin, accompanied by a reduction in transepithelial resistance and leakage of ions, giving rise to ion imbalances, and a reduced ability of amphibians to osmoregulate or rehydrate. Blood samples from amphibians with clinical chytridiomycosis show significantly reduced plasma sodium, potassium and chloride ion concentrations and reduced overall blood plasma osmolality. The low plasma potassium concentrations (or hypokalaemia) are linked to abnormal cardiac electrical activity and cardiac arrest and are thought to be the proximate cause of death in diseased amphibians (Campbell et al. 2012, Van Rooij et al. 2015).

Besides the detrimental effect on the skin, *Bd* also actively suppresses host responses (Rosenblum et al. 2009; Ellison et al. 2014; Ribas et al. 2009; Woodhams et al. 2012a, b; Young et al. 2014). *Bd* culture supernatant inhibits lymphocyte proliferation and induces apoptosis (Fites et al. 2013, 2014). Fungal proteases, such as a subtilisin-like serine protease, inhibit antimicrobial peptide (AMP) release from dermal granular glands and/or selectively degrade AMPs (Woodhams et al. 2012a, b; Thekkiniath et al. 2013).

14.5 Clinical Signs and Lesions

Clinical signs of chytridiomycosis due to *Bd* in juvenile and adult anurans may include erythema of ventral surfaces, abnormal posture such as splayed limbs, depression, slow righting reflex, abnormal skin shedding and tetanic spasms upon handling (Berger et al. 2009). Signs generally occur in the terminal stages of disease and correlate with heavy infections, severe skin pathology and loss of plasma electrolytes due to disturbance of epidermal ion transport (Voyles et al. 2009). Therefore most of the course of infection remains subclinical, and dead frogs are often in good body condition. With *Bsal*, there is also lethargy and excessive skin shedding, but widespread multifocal ulceration is a marked and typical sign (Martel et al. 2013; Van Rooij et al. 2015).

Both *Batrachochytrium* species occur only within epidermal cells, but sporangia of *Bd* infect the superficial epidermal layers (including the stratum granulosum), whereas *Bsal* grows in all layers (Berger et al. 2005; Martel et al. 2013). *Bd* generally has a predilection for ventral skin areas and feet, but *Bsal* occurs over all the skin including the dorsum. Histopathology of *Bd* infection includes hyperkeratosis, disordered epidermal cell layers, spongiosis, erosions and occasional ulcerations. Hyperplasia may occur resulting in diffuse or focal thickening, whereas other regions of epidermis may be thin and eroded. Bacteria often colonize the layers of sloughing keratin and grow within "empty" sporangia. A mild inflammatory response may occur as a slight increase in mononuclear cells in the dermis and epidermis but is often negligible (Berger et al. 2009). The deeper infection by *Bsal* leads to more erosive and ulcerative changes without hyperkeratosis or hyperplasia (Van Rooij et al. 2015).

Electron microscopy revealed a zone of apparently condensed host cytoplasm, up to 2.5 μm thick, around some *Bd* sporangia that appeared fibrillar and excluded organelles (Berger et al. 2005). It is possible this host reaction provides protection of sporangia from antimicrobials or immune response. Keratinization appears to occur prematurely in infected cells, compared with uninfected cells in the same epidermal layer (Berger et al. 2005).

Specific internal lesions are not observed in frogs with severe chytridiomycosis, besides some terminal changes. In an Australian survey, concurrent diseases were diagnosed in 12% of frogs with severe chytridiomycosis, but most frogs had no other specific lesions (Berger et al. 2004).

14.6 Susceptibility to Chytridiomycosis

Considerable variation in susceptibility to chytridiomycosis has been observed between hosts at the individual, population and species levels and is attributable to both extrinsic and intrinsic host factors including immunological defences (Blaustein et al. 2005; Tobler and Schmidt 2010; Pilliod et al. 2010; Searle et al. 2011).

14.6.1 Extrinsic Host Factors

At the broad scale, population declines due to *Bd* appear strongly linked to species' distribution and habitat use. Declines have been particularly severe among tropical high-altitude, stream-associated anurans, although in many cases, susceptibility and population declines are likely associated with environmental suitability for the fungus rather than innate variations between species (Lips et al. 2006; Fisher et al. 2009). Even among syntopic species, however, habitat use may differ which may explain some degree of variation in perceived susceptibility to chytridiomycosis (Rowley and Alford 2007).

In Central and South America, the *Atelopus* genus of neotropical toads appears to have been most severely affected (La Marca et al. 2005). In Australia, diverse species of Myobatrachid and Hylid frogs have become extinct or are critically endangered (Skerratt et al. 2016). Globally, *Bd*-associated declines and extinctions have been recorded in numerous other amphibian taxa, particularly other anurans (Fisher et al. 2009). Many salamanders may also have been significantly affected (Rovito et al. 2009; Cheng et al. 2011). There are several host species that are predominantly resistant to or tolerant of infection and may serve as reservoirs or carriers of infection. These species may have been responsible for the widespread dissemination of *Bd* and include, for example, *Pseudacris regilla*, *L. catesbeianus*, *L. pipiens*, *X. laevis*, *Eleutherodactylus coqui*, *Litoria lesueurii* complex, *L. ewingii* and *Crinia signifera* (Retallick et al. 2004; Weldon et al. 2004; Beard and O'Neill 2005; Ricardo 2006; Schloegel et al. 2010; Reeder et al. 2012).

Behaviour that increases exposure to the pathogen may constitute an important extrinsic host factor contributing to variation in susceptibility. This may be through types and degrees of host contact with conspecifics, contaminated water and environmental substrates containing *Bd* (Rowley and Alford 2007). Importantly, however, other general movement patterns, including thermoregulatory behaviours such as basking, and the use of retreat sites may also play a role by altering existing infection intensities or pathogen growth rates due to the optimal thermal range of the fungus (Rowley et al. 2007; Richards-Zawacki 2010; Puschendorf et al. 2011; Daskin et al. 2011). Behaviour may differ particularly between life stages due to aquatic and terrestrial or arboreal habitats as well as between male and female adults due to their differing seasonal utilisation of the environment (Duellman and Trueb 1994). Other factors may include the density of individuals within the population (higher densities may increase environmental zoospore load; Hudson and Dobson 1998), the age structure of the population (tadpoles may act as reservoir hosts, or dispersal patterns and habitat use differ between stages; Rachowicz and Vredenburg 2004) and the presence of and interactions with any syntopic species and vectors (Reeder et al. 2012; Rivas 1964).

14.6.2 Intrinsic Host Factors

Intrinsic host factors associated with signalment (life stage, age and body size) have been found to differentiate susceptibility to chytridiomycosis between individuals in laboratory experiments and field observations. Infections are limited to the keratinized mouthparts of larval anurans and are typically not fatal during this stage (Berger et al. 1998; Pessier et al. 1999; Fellers et al. 2001; Rachowicz and Vredenburg 2004), although infection prevalence and intensity has been found to increase with larval development (Smith et al. 2007). Tadpole mouthpart abnormalities may affect feeding efficiency leading to the smaller body size seen in experimentally infected animals (Parris and Beaudoin 2004). Despite interspecific variation, metamorphosis from larval to adult form and the immediate post-metamorphic phase appear to be the most susceptible periods (Berger et al. 1998), which may be associated with immune restructuring. The immune system of tadpoles, while competent, is functionally less well developed compared with that of adult anurans and undergoes substantial remodelling accompanied by immunosuppression during metamorphosis (Rollins-Smith et al. 2011). There is little evidence from experimental infections to suggest an intrinsic difference between sexes in susceptibility to chytridiomycosis (Grogan 2014), suggesting that any observed variation from field studies may be more likely associated with extrinsic factors such as differences in behaviour, and hence fungal exposure or infection development (Johnson and Hoverman 2014). Other putative intrinsic determinants of susceptibility may include concurrent infection, nutritional level and the presence of stressors (Murphy et al. 2011; Kinderman et al. 2012; Young, unpublished).

Controlling for the factors described above, experimental evidence suggests the importance of both genetic and phenotypic immunologic mechanisms in differentiating susceptibility between individuals, clutches, populations and species (Rosenblum et al. 2012; Grogan 2014). These variations in response may be associated with inherent or evolved differences in innate immunity (Savage and Zamudio 2011; Bataille et al. 2015), previous exposure history and the development of adaptive immunity (Cashins et al. 2013; McMahon et al. 2014) or phenotypic differences in pathogenesis and differing functional expression of immune responses. While predisposing immunosuppression is not necessary for epidemics to occur, apparent immunosuppression has been observed in *Bd*-infected individuals (detected via skin histopathology, and corroborated via gene expression and in vitro immune studies (Berger et al. 2005; Ribas et al. 2009; Rollins-Smith et al. 2011; Rosenblum et al. 2012). Thus it appears that *Bd* may have low inherent antigenicity perhaps due to intracellular localization, may suppress adaptive immune responses in susceptible hosts and may elicit immunopathologic responses in late infection (Ellison et al. 2015).

14.6.2.1 Innate Immune Response

Many components of the innate immune response to *Bd* infection differ between experimental groups and may be associated with differential host susceptibility. The secreted mucus and epidermal layer of amphibian skin provide a constitutive physical and chemical defence barrier against pathogen invasion and may include variable expression of lysozyme (Rollins-Smith 2009), mucosal antibodies, inducible AMPs (such as defensins, cathelicidins and histatins; Ramsey et al. 2010; Pask et al. 2013) as well as commensal symbiotic bacterial communities and their antimicrobial metabolites (Lauer et al. 2007; Harris et al. 2009a, b).

14.6.2.2 Antimicrobial Peptides

AMPs are small (12–50 amino acid residues), cationic and hydrophobic peptides produced in the dermal granular glands. Most of our current knowledge concerning amphibian AMPs stems from studies on Anura (Van Rooij et al. 2015). To date, approximately 40 anuran AMPs inhibiting *Bd* have been characterized. Both purified and natural mixtures of these AMPs effectively inhibit in vitro growth of both *Bd* zoospores and sporangia (Woodhams et al. 2007; Rollins-Smith 2009; Ramsey et al. 2010). However, it is not clear to which extent these peptides provide protection against chytridiomycosis in vivo (Van Rooij et al. 2015). Species with peptides active in vitro such as the mountain yellow-legged frog (*Rana muscosa*) may still be very susceptible to *Bd* (Rollins-Smith et al. 2006). Moreover, the efficacy of skin peptide defences may vary at species and population level (Tennessen et al. 2009).

14.6.2.3 Bacteria and Their Antifungal Metabolites

Several secondary metabolites secreted by symbiotic bacteria present on amphibian skin have been shown to inhibit *Bd* growth in vitro and in vivo (Van Rooij et al. 2015). In the bacteria *Janthinobacterium lividum*, *Lysobacter gummosus* and *Pseudomonas fluorescens*, these metabolites are 2,4-diacetylphloroglucinol (2,4-DAPG), indol-3-carboxaldehyde (I3C) and violacein (Brucker et al. 2008; Myers et al. 2012; Lam et al. 2010; Harris et al. 2009a, b). In addition, the metabolites 2,4-DAPG and I3C seem to exert a repellent action on *Bd* zoospores (Lam et al. 2011). Besides, coculture of skin bacterial isolates can lead to secretion of new, more potent metabolites than when grown in monoculture. As such, the inhibitory metabolite tryptophol was found to emerge from coculturing an unknown *Bacillus* skin bacterium and *Chitinophaga arvensicola* (Loudon et al. 2014). Myers et al. (2012) showed that these metabolites work synergistically with AMPs to inhibit growth of *Bd*, at lower minimal inhibitory concentrations (MIC) than necessary for inhibition by either metabolites or AMPs. As for AMPs, variation in infection susceptibility among populations is thought to result in part from differences in skin bacterial communities (Van Rooij et al. 2015). By comparing bacterial communities on the skin of a declining *R. muscosa* population and a population coexisting with *Bd*, researchers found a significantly higher number of individuals with culturable bacterial species displaying antifungal properties in coexisting populations than in those at decline. Alteration of this microbial community composition, for example, by environmental factors, can considerably increase susceptibility to disease (Lam et al. 2010).

14.6.2.4 Other Inducible Pathways of the Innate Immune System

Zoospores that survive to invade beyond the skin mucus layer induce host innate immune responses via antigens that are either secreted or expressed on the pathogen cell surface or released through lysosomal degradation within host cells. These antigens often contain epitopes of widely recognized pathogen-associated molecular patterns (PAMPs) that are common to many different microorganisms. These PAMPs bind to host pattern recognition receptors (PRRs; including toll-like receptors, mannose receptors, scavenger receptors, glucan receptors, C-type lectin receptors and NOD-like receptors, among others), activating the inflammatory and complement cascades and stimulating the release of cytokines (including interleukins, tumour necrosis factors, interferons, chemokines, and stress proteins among others). Gene expression studies in *Bd*-infected frogs have demonstrated mixed results on the expression of various PRRs and cytokines, and more research is needed to tease apart these interactions, their roles in host defence and their association with infection susceptibility (Ribas et al. 2009; Rosenblum et al. 2012; Ellison et al. 2014). Leukocyte recruitment and infiltration to the site of infection is typically lower than expected in infected susceptible amphibians (examined via haematology and histopathology; Woodhams et al. 2007; Davis et al. 2010; Peterson et al. 2013; Young et al. 2014).

14.6.2.5 Adaptive Immune Response

There is currently little evidence to suggest the effective activation of the adaptive immune response to chytridiomycosis in terms of a systemic or localized lymphocyte response (Pessier et al. 1999; Berger et al. 2005; Peterson et al. 2013; Young et al. 2014; Nichols et al. 2001) that may differentiate susceptible from resistant individuals. Mapping of transcriptomic changes in immunologically important tissues (skin, liver, spleen, small intestine) from frogs with chytridiomycosis demonstrated the absence of a robust adaptive immune response at various time points post exposure in various species (e.g. lymphocyte, immunoglobulin, MHC and classical complement pathway genes; Ribas et al. 2009; Rosenblum et al. 2009; Ellison et al. 2014). Attempts to immunize frogs by subcutaneous or intraperitoneal injection of formalin (*R. muscosa*; Stice and Briggs 2010) or heat-killed *Bd* (*Bufo boreas*; Rollins-Smith 2009) failed to elicit a protective immune response. Only in *X. laevis*, *Bd*-specific IgM, IgX (mammalian IgA-like) and IgY (mammalian IgG-like) antibodies were found in skin mucus after injection with heat-killed zoospores (Ramsey et al. 2010). Although repeated exposure to *Bd* did not result in increased resistance in all experiments (Cashins et al. 2013), repeated cycles of exposure to *Bd* with subsequent temperature treatment of infection resulted in a marginally higher survival rate, reduction in the infection load (Ramsey et al. 2010; McMahon et al. 2014), which coincided with increased lymphocyte proliferation and abundance in the spleen. As such, some species of frogs are capable of acquiring at least some degree of immunity against *Bd* (McMahon et al. 2014), although in general, adaptive immune responses are suppressed by *Bd*.

14.6.3 Host Defences Against *Bsal*

Little is known about amphibian defences against *Bsal*. Unlike for anurans, information about the AMP arsenal in skin secretions of urodelans is scant (Van Rooij et al. 2015). To date only a single antimicrobial peptide (the defensin CFBD) has been described from *Cynops fudingensis* (Fuding fire belly newt) (Meng et al. 2013). Although the antimicrobial action of CFBD against both *Batrachochytrium* species has not (yet) been evaluated, it is suggested that AMPs may be involved in the pronounced anti-*Bd* activity of salamander skin secretions (Sheafor et al. 2008; Pasmans et al. 2013). Infection trials demonstrated that host responses vary dramatically, not only within but also between urodelan species (Martel et al. 2014). Interestingly, in some species that are likely *Bsal* reservoirs, a proportion of the exposed individuals do not develop lethal skin disease but are capable of self-cure after initial infection, in some cases with complete fungal elimination. This process can take several months and is most likely due to either increased innate defence mechanisms or the buildup of protective acquired immunity. Other urodelan species, however, which were shown to invariably develop lethal disease, do not seem to increase resistance against *Bsal*, even after five cycles of exposure and thermal treatment (unpublished results).

14.7 Diagnosis

Laboratory tests are needed to diagnose chytridiomycosis, as tolerant species are infected without being sick, and because in susceptible hosts clinical signs of chytridiomycosis only occur in late terminal stages and are often non-specific.

With heavy infections, changes that are suggestive of chytridiomycosis may include erythema of ventral surfaces, abnormal posture such as splayed limbs, slow righting reflex and abnormal skin shedding (Fig. 14.2a) (Berger et al. 2009). With *Bsal*, susceptible salamanders usually develop obvious ulcers (Fig. 14.2b) (Martel et al. 2014). These signs are typical in sick amphibians, and two major diseases that can present similarly are ranaviral disease and bacterial septicemia "redleg." In tadpoles with *Bd*, infections cause mouthpart abnormalities including loss of dark tooth rows which can lead to emaciation (Rachowicz and Vredenburg 2004).

Details of diagnostic methods for *Bd* are available in Berger et al. (2009) and the online OIE Manual of Diagnostic Tests for Aquatic Animals 2017 (Chapter 2.1.1), and for *Bsal* see Martel et al. (2013). A nationwide survey protocol for detecting *Bd* is also available (Skerratt et al. 2008).

Quantitative polymerase chain reaction (qPCR) is the gold standard for testing. As sampling for PCR involves taking skin swabs, which can be stored at room temperature, this method is non-invasive and is convenient for researchers, veterinarians and conservation managers testing wild or captive animals. For *Bd*, qPCR was shown to be much more sensitive (72.9%) than histology (26.5%) although was less specific (94.2% versus 99.5%) (Skerratt et al. 2011a, b). Quantitative PCR permits relative quantification of conserved *Bd* 18S and 28S ribosomal DNA from skin swabs down to a resolution of one genomic zoospore equivalent from as soon as 7 days post infection (Annis et al. 2004; Boyle et al. 2004; Kriger et al. 2006; Hyatt et al. 2007). Standard PCR, however, is also accurate and may be cheaper (Garland et al. 2011). A qPCR to detect *Bsal* is based on the 5.8S rRNA gene (Martel et al. 2013). Besides, a duplex qPCR detecting *Bd* and *Bsal* is available (Blooi et al. 2013, 2016).

Microscopy was the original diagnostic method for chytridiomycosis and is highly accurate in sick animals which have heavy burdens. Microscopy includes histology, wet preparations and immunostaining, and requires pieces of whole or shed skin. Also histological examination of all organs by a pathologist is important if ill or dead frogs are found, as part of general disease surveillance (Duffus 2009). For

Fig. 14.2 (**a**) *Bd*-infected frogs (*Phyllobates bicolor*) that develop chytridiomycosis show abnormal posture such as splayed limbs and abnormal skin shedding, with remnants of shed skin. (**b**) After infection with *Bsal*, susceptible salamanders (*Salamandra salamandra*) usually develop obvious skin ulcers, here coinciding with haemorrhages (© Pasmans and Martel 2015. All Rights Reserved)

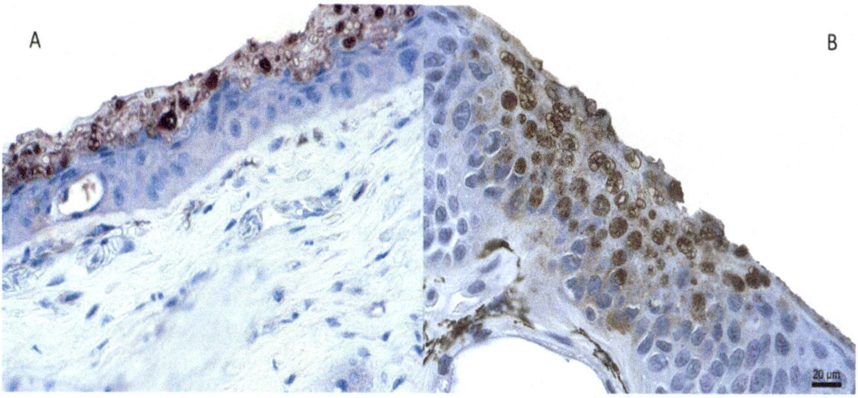

Fig. 14.3 Immunohistochemical staining of the skin of a *Bd*-infected frog, which died due to chytridiomycosis (*Alytes obstetricans*) (reprinted from Berger et al. 2002) (**a**) and a *Bsal*-infected fire salamander (**b**). *Bd* causes focal epidermal hyperplasia and hyperkeratosis (with abundant thalli present in the keratin), whereas *Bsal* causes focal necrosis with subsequent erosive and ulcerative lesions. For both chytrids, intracellular thalli (stained dark brown) abound in the lesions (© Pasmans and Martel 2015. All Rights Reserved)

histology, skin from body, digits, or tadpole mouthparts can be examined with haematoxylin and eosin (H&E) staining (Berger et al. 2000, 2009). Sporangia occur in clusters within the epidermis and appear spherical or oval (5–13 µm) with a smooth eosinophilic wall. Discharge papillae are occasionally seen and project towards the skin surface. Focal hyperkeratosis and erosions are common in *Bd*-infected areas, whereas *Bsal* causes necrosis with subsequent erosive and ulcerative lesions (Fig. 14.3). To confirm suspect cases, sporangia can be highlighted using other stains, such as periodic acid-Schiff (PAS) or silver, or an immunoperoxidase stain using polyclonal antibodies with high affinity for *Bd* and *Bsal* (Berger et al. 2002; Van Ells et al. 2003; Martel et al. 2013). Wet preparations are quick to prepare using shedding stratum corneum, whole skin or excised tadpole mouthparts that are spread on a slide and cover-slipped (Berger et al. 2009). Diagnosis requires some practice, but the observation of internal septa within sporangia increases confidence in the diagnosis. As *Batrachochytrium* species are slow-growing compared with microorganism contaminants, culture is difficult and is not used for diagnosis (Berger et al. 2009).

14.8 Treatment

14.8.1 Treatment with Antimicrobial Compounds

The most important class of antifungal drugs used to treat chytridiomycosis in amphibians is that of the imidazole, triazole and thiazole or "azole" group. The most widely used azole antifungal in treating chytridiomycosis is itraconazole. In general, most itraconazole treatment protocols are based on a study performed by Nichols et al. (2000) describing successful treatment of chytridiomycosis by bathing *Bd*-infected amphibians in a 0.01% solution (100 mg/L) of itraconazole diluted in 0.6% saline, daily for 5 min during 11 days (Nichols et al. 2000; Forzan et al. 2008; Une et al. 2008; Pessier and Mendelson 2010; Georoff et al. 2013). Studies describing successful treatment of chytridiomycosis with minor adaptations to this protocol (other concentration, other diluting agent and longer exposure time or treatment period) also exist (Garner et al. 2009; Tamukai et al. 2011; Jones et al. 2012; Georoff et al. 2013). However, treatment failure and adverse side effects due to itraconazole toxicity at this concentration (and even lower concentrations) have also been reported for some amphibian species (Brannelly et al. 2012; Woodhams et al. 2012a, b). Variable outcome of itraconazole treatment in clearing *Bd* infections might be explained not only by the used concentration but also by the frequency of exposure to itraconazole (Woodward et al. 2014). Other antifungals belonging to the azole group used to treat chytridiomycosis are miconazole and voriconazole. Nichols et al. (2000) demonstrated that treatment by using miconazole baths at a concentration of 100 μg/ml, once daily for 5 min during 8 days was effective in clearing *Bd* infections. Voriconazole has been shown to have potent *Bd* inhibitory effects both in vitro and in vivo (Woodward et al. 2014, Martel et al. 2011), with successful clearance of *Bd* in experimentally and naturally infected amphibians using a treatment protocol composed of topically spraying voriconazole once daily during 7 days at a concentration of 1.25 μg/ml (Martel et al. 2011).

Topical treatment of *Bsal*-infected animals with a combination of polymyxin E (2000 IU/ml) and voriconazole (12.5 μg/ml) at an ambient temperature of 20 °C results in clearance of *Bsal* infections (Blooi et al. 2015a, b). For all treatments, post-treatment assessment for clearance of infection is necessary, and treatments may need to be repeated.

14.8.2 Physical Therapy

Short exposure to relatively high ambient temperatures (37 °C less than 16 h (Woodhams et al. 2003) and 30 °C, 10-day exposure (Chatfield and Richards-Zawacki 2011) and longer exposure to lower ambient temperatures (27 °C, clearance after 98 days; Berger et al. 2004) have been able to clear *Bd* infections from adult amphibians. Exposure to 26 °C during 5 days was able to clear *Bd* infections in midwife toad (*Alytes obstetricans*) larvae (Geiger et al. 2011). Ten-day exposure at 25 °C is able to clear salamanders from *Bsal* infections (Blooi et al. 2015a, b). The main disadvantages linked to temperature treatment of chytrid infections is that elevated temperature might not be endured by all amphibian species and that thermal shock might occur (especially when taking into account that the treatment is applied on sick individuals). Furthermore, strain dependent thermal preferences of *Bd* or *Bsal* may compromise thermal treatment protocols.

14.9 Mitigation

Mitigation of chytridiomycosis has two broad aims: (1) reducing spread and (2) reducing impacts. By the time *Bd* was recognized widely as the cause of global declines, it had already spread to most areas that contain suitable habitats. The faster recognition of *Bsal*, however, has meant that there is potential for biosecurity measures to keep countries disease-free, such as the United States (Martel et al. 2014). However, for both fungi, there is a risk that spread of different strains could lead to dangerous recombinations; hence, reducing movement across already-infected areas is recommended. Long-distance spread is most likely to have occurred due to movement of infected amphibians, particularly through the pet trade but also via accidental movement in produce and in the frog meat industry (although the latter is likely more important for viruses such as ranaviruses, since most frog products are frozen; Kolby et al. 2014). The listing of chytridiomycosis as an internationally notifiable disease by the OIE represents the first disease listed that is solely a biodiversity concern, with the aim to improve trade safety. However, although rigorous quarantine and surveillance protocols are often in place for livestock diseases, improved standards are needed for wildlife (Grogan et al. 2014). Fortunately, the US Fish and Wildlife Service announced restrictions from January 2016 on the importation and interstate movement of salamanders in the United States. In the United States, planning and surveillance for early detection of *Bsal* incursions and emergency responses is underway (Grant et al. 2016). Ideally, however, all amphibian trade should be restricted. A good example is Australia, where exotic frogs are rarely allowed to be imported and are restricted to biosecured facilities such as zoos or research institutes.

Within a region, risk of anthropogenic spread of chytridiomycosis to naive populations, and between infected regions, may be mitigated through containment. This is a priority for isolated populations such as those on islands or in habitats where natural spread is unlikely to or could not occur (Berger and Skerratt 2012). In moist wilderness areas with abundant wildlife, attempts to stop natural spread appear unlikely to succeed. Control of anthropogenic spread involves restricting access to sites and the use of stringent hygiene protocols on equipment (Murray et al. 2013; Phillott et al. 2010). However, the efficacy of reducing the risk of spread by focusing on humans has not been assessed. Educating the community about basic disease management and the risks of transport of potentially infected amphibians and water is important.

Much research has focused on reducing impacts of *Bd*, which also has high relevance to *Bsal* where studies have recently commenced. Although small-scale eradication of *Bd* has been achieved in a specific type of isolated or ephemeral habitat (Bosch et al. 2015), this approach is not broadly applicable. Hence in areas where chytridiomycosis has established, the emphasis is on ensuring the persistence of amphibian populations and species (Skerratt et al. 2016). Extinction has been prevented via establishing amphibian ex situ captive assurance colonies, but methods to ensure self-sustaining wild populations are obviously the goal (McFadden et al. 2013; Scheele et al. 2014; Skerratt et al. 2016).

Currently, reintroductions have had low success, with the continued presence of the pathogen in the environment leading to eventual mortality of reintroduced individuals. Research and trials are currently underway on potential management strategies to improve survival rates in the wild; however these are largely in the experimental phase.

As infected frog populations can thrive in naturally suboptimal habitats for the fungus, which may include warmer, drier or more saline regions (Heard et al. 2015; Puschendorf et al. 2011; Stockwell et al. 2015a, b), eradication of disease is not necessary for successful mitigation. This also shows the importance of assessing suitability for *Batrachochytrium* species when choosing habitats to preserve for amphibians. This may lead to conservation of areas that may not have previously been considered prime habitat (Skerratt et al. 2016).

Another angle involves altering the environmental suitability for chytridiomycosis. Physical modifications of the environment might be used on a local scale for critically threatened amphibians in situ and may render the habitat less suitable for *Bd*. These include drying, drainage or alteration of waterflow, provision of shallow warm-water areas, reduction in canopy cover to increase temperature or the addition of basking sites or artificial heat (Scheele et al. 2014). A number of chemical treatments have been proposed or trialled in the field, including the application of salt (Stockwell et al. 2015a, b) and agricultural fungicides (Johnson et al. 2003; Woodhams et al. 2011). Reducing transmission might also be achieved by removal of reservoir hosts or by making habitat less favourable for reservoir species (Scheele et al. 2014; Skerratt et al. 2016).

Other ideas for fighting the pathogen directly include manipulating microbial competition via bioaugmentation of the host or environment with probiotic bacteria that express antifungal metabolites (Becker et al. 2009; Muletz et al. 2012; Vredenburg et al. 2011), augmenting the numbers of *Bd* predators such as zooplankton (Schmeller et al. 2014) and pathogens such as mycoviruses (Skerratt et al. 2016) and the identification or engineering and release of nonvirulent strains of *Bd* (Woodhams et al. 2011).

Management strategies aimed at improving host immunity have long-term potential. Manipulation of the host adaptive and innate immune response (via vaccination and assisted selection for disease resistance) is a proven strategy in humans and domestic animals, with potential to reduce the impact of chytridiomycosis in the field. Although adaptive immunity is not heritable and hence immunization may be perceived as a short-term approach, it could assist in providing a population size buffer for the natural evolution of innate immunity, although the artificial maintenance of susceptible genotypes may counter this beneficial effect (Harding et al. 2005). Unfortunately, reinfection trials in the few species examined to date have not demonstrated strong acquired immunity against *Bd* (Cashins et al. 2013; McMahon et al. 2014).

The innate immune system is generally considered responsible for the evolution of inter-generational immunity, and disease resistance or tolerance may be upregulated within a population via assisted selection for less susceptible individuals (Venesky et al. 2012). Comparative techniques (e.g. marker-assisted selection and estimated breeding values) have been widely and successfully used for breeding of disease resistance in plant and domestic animal agriculture (Heringstad et al. 2007; Miedaner and Korzun 2012). This may present a sustainable approach for repatriating the numerous amphibian species that are now extinct in the wild and only persist in ex situ captive programs. Two main approaches that might be feasible in practice for promoting disease resistance include (1) direct selection via exposure of post-metamorphic individuals to *Bd*, then breeding from those with lower susceptibility, and (2) identifying molecular markers of resistance to advance selection to earlier life stages, removing the need to regularly expose individuals to infection. Some progress has been made towards this latter goal via studies of the major histocompatibility complex (MHC; Bataille et al. 2015; Savage and Zamudio 2011). The increasing longevity of some recovering wild populations suggests evolution of resistance may occur naturally (Newell et al. 2013), although in other species individual annual survival rates remain very low despite a long history of infection (i.e. 15–20 years) (Phillott et al. 2013; Brannelly et al. 2015).

Scheele et al. (2014) present immediately applicable suggestions for bolstering overall population size and recruitment to counteract disease-associated mortality. This strategy is based on the observation of species with high mortality rates persisting via high recruitment (Phillott et al. 2013, Scheele et al. 2015). Interventions may involve removal of other threatening processes from small and declining populations such as improving habitat or excluding competitors and introduced pests (Scheele et al. 2014). Increase of population size can be achieved through reintroductions and minimizing the effect of early predation by head-starting larval stages through metamorphosis in captivity (Hunter et al. 1999; Scheele et al. 2014). An alternative to establishing long-term captive assurance colonies is to rear wild-caught eggs or tadpoles in captivity to ensure higher survival rates before release. Direct translocation to 'disease refugia' sites could be an efficient way to create sustainable populations (Puschendorf et al. 2011; Skerratt et al. 2016).

As disease ecology varies greatly between species and habitats, management will be context-specific; hence research aimed at understanding each situation is needed to devise effective local strategies. A proactive approach is crucial, as many of the most endangered species occur in already protected areas and will not survive without intervention (Skerratt et al. 2016).

Acknowledgements LB was supported by the Australian Research Council (grant FT100100375).

References

Abramyan J, Stajich JE (2012) Species-specific chitin binding module 18 expansion in the amphibian pathogen *Batrachochytrium dendrobatidis*. MBio 3:e00150–e00112

Annis SL, Dastoor FP, Ziel H et al (2004) A DNA-based assay identifies *Batrachochytrium dendrobatidis* in amphibians. J Wildl Dis 40:420–428

Bataille A, Cashins SD, Grogan L et al (2015) Susceptibility of amphibians to chytridiomycosis is associated with MHC class II conformation. Proc Biol Sci 22:282

Beard KH, O'Neill EM (2005) Infection of an invasive frog *Eleutherodactylus coqui* by the chytrid fungus *Batrachochytrium dendrobatidis* in Hawaii. Biol Conserv 126:591–595

Becker MH, Brucker RM, Schwantes CR et al (2009) The bacterially produced metabolite violacein is associated with survival of amphibians infected with a lethal fungus. Appl Environ Microbiol 75:6635–6638

Berger L, Skerratt L (2012) Disease strategy chytridiomycosis (infection with *Batrachochytrium dendrobatidis*) Version 1, 2012. Department of Sustainability, Environment, Water, Populations and Communities, Public Affairs, Commonwealth of Australia, Canberra. Available at: http://www.environment.gov.au/system/files/resources/387d3e66-3cdc-4676-8fed-759328277da4/files/chytrid-fungus-manual.pdf

Berger L, Speare R, Daszak P et al (1998) Chytridiomycosis causes amphibian mortality associated with population declines in the rain forests of Australia and Central America. Proc Natl Acad Sci U S A 95:9031–9036

Berger L, Speare R, Kent A (2000) Diagnosis of chytridiomycosis in amphibians by histological examination. Zoos Print J 15:184–190

Berger L, Hyatt AD, Olsen V et al (2002) Production of polyclonal antibodies to *Batrachochytrium dendrobatidis* and their use in an immunoperoxidase test for chytridiomycosis in amphibians. Dis Aquat Org 48:213–220

Berger L, Speare R, Hines HB et al (2004) Effect of season and temperature on mortality in amphibians due to chytridiomycosis. Aust Vet J 82:434–439

Berger L, Hyatt AD, Speare R et al (2005) Life cycle stages of the amphibian chytrid *Batrachochytrium dendrobatidis*. Dis Aquat Org 68:51–63

Berger L, Longcore J, Speare R, Hyatt A, Skerratt LF (2009) Fungal diseases in amphibians. In: Heatwole H, Wilkinson JW (eds) Amphibian biology, volume 8 amphibian decline: disease, parasites, maladies, and pollution. Surrey Beatty and Sons, Baulkham Hills, NSW, pp 2986–3052

Blaustein AR, Romansic JM, Scheessele EA et al (2005) Interspecific variation in susceptibility of frog tadpoles to the pathogenic fungus *Batrachochytrium dendrobatidis*. Conserv Biol 19: 1460–1146

Blooi M, Pasmans F, Longcore JE et al (2013) Duplex real-time PCR for rapid simultaneous detection of *Batrachochytrium dendrobatidis* and *Batrachochytrium salamandrivorans* in amphibian samples. J Clin Microbiol 51:4173–4177

Blooi M, Martel A, Haesebrouck F et al (2015a) Treatment of urodelans based on temperature dependent infection dynamics of *Batrachochytrium salamandrivorans*. Sci Rep 5:8037

Blooi M, Pasmans F, Rouffaer L et al (2015b) Succesful treatment of Batrachochytrium salamandrivorans infections in salamanders requires synergy between voriconazole, polymyxin E and temperature. Sci Rep 5:11788

Blooi M, Pasmans F, Longcore JE et al (2016) Duplex real-time PCR for rapid simultaneous detection of *Batrachochytrium dendrobatidis* and *Batrachochytrium salamandrivorans* in amphibian samples. J Clin Microbiol 54:246–246

Bosch J, Martínez-Solano I, García-París M (2001) Evidence of a chytrid fungus infection involved in the decline of the common midwife toad (*Alytes obstetricans*) in protected areas of central Spain. Biol Conserv 97:331–337

Bosch J, Sanchez-Tome E, Fernandez-Loras A et al (2015) Successful elimination of a lethal wildlife infectious disease in nature. Biol Lett 11(11):20150874

Boyle DG, Boyle DB, Olsen V et al (2004) Rapid quantitative detection of chytridiomycosis (*Batrachochytrium dendrobatidis*) in amphibian samples using real-time Taqman PCR assay. Dis Aquat Org 60:141–148

Brannelly LA, Richards-Zawacki CL et al (2012) Clinical trials with itraconazole as a treatment for chytrid fungal infections in amphibians. Dis Aquat Org 101:95–104

Brannelly LA, Hunter DA, Skerratt LF et al (2015) Chytrid infection and post-release fitness in the reintroduction of an endangered alpine tree frog. Anim Conserv 19(2):153–162. https://doi.org/10.1111/acv.12230

Brucker RM, Harris RN, Schwantes CR et al (2008) Amphibian chemical defense: antifungal metabolites of the microsymbiont *Janthinobacterium lividum* on the salamander *Plethodon cinereus*. J Chem Ecol 34:1422–1429

Brutyn M, D'Herde K, Dhaenens M et al (2012) *Batrachochytrium dendrobatidis* zoospore secretions rapidly disturb intercellular junctions in frog skin. Fungal Genet Biol 49:830–837

Campbell CR, Voyles J, Cook DI et al (2012) Frog skin epithelium: electrolyte transport and chytridiomycosis. Int J Biochem Cell Biol 44:431–434

Cashins SD, Grogan LF, McFadden M et al (2013) Prior infection does not improve survival against the amphibian disease Chytridiomycosis. PLoS One 8:e56747

Chatfield MWH, Richards-Zawacki CL (2011) Elevated temperature as a treatment for *Batrachochytrium dendrobatidis* infection in captive frogs. Dis Aquat Org 94:235–238

Cheng TL, Rovito SM, Wake DB et al (2011) Coincident mass extirpation of neotropical amphibians with the emergence of the infectious fungal pathogen *Batrachochytrium dendrobatidis*. Proc Natl Acad Sci U S A 108:9502–9507

Cunningham AA, Beckmann K, Perkins M et al (2015) Surveillance emerging disease in UK amphibians. Vet Rec 176:468–468

Daskin JH, Alford RA, Puschendorf R (2011) Short-term exposure to warm microhabitats could explain amphibian persistence with *Batrachochytrium dendrobatidis*. PLoS One 6:e26215

Davis AK, Keel MK, Ferreira A et al (2010) Effects of chytridiomycosis on circulating white blood cell distributions of bullfrog larvae (*Rana catesbeiana*). Comp Clin Pathol 19:49–55

Doddington BJ, Bosch J, Oliver JA et al (2013) Context dependent amphibian host population response to an invading pathogen. Ecology 94:1795–1804

Duellman WE, Trueb L (1994) Biology of amphibians. The Johns Hopkins University Press, Baltimore

Duffus ALJ (2009) Chytrid blinders: what other disease risks to amphibians are we missing? EcoHealth 6:335–339

Ellison AR, Savage AE, DiRenzo GV et al (2014) Fighting a losing battle: vigorous immune response countered by pathogen suppression of host defenses in the chytridiomycosis-susceptible frog *Atelopus zeteki*. G3 (Bethesda) 4:1275–1289

Ellison AR et al (2015) More than skin deep: functional genomic basis for resistance to amphibian chytridiomycosis. Genome Biol Evol 7:286–298. https://doi.org/10.1093/gbe/evu285

Farrer RA, Weinert LA, Bielby J et al (2011) Multiple emergences of genetically diverse amphibian-infecting chytrids include a globalized hypervirulent recombinant lineage. Proc Natl Acad Sci U S A 108:18732–18736

Farrer RA, Henk DA, Garner TWJ et al (2013) Chromosomal copy number variation, selection and uneven rates of recombination reveal cryptic genome diversity linked to pathogenicity. PLoS Genet 9(8):e1003703

Fellers GM, Green DE, Longcore JE (2001) Oral chytridiomycosis in the mountain yellow-legged frog (*Rana muscosa*). Copeia 2001:945–953

Fisher MC, Garner TWJ, Walker SF (2009) Global emergence of *Batrachochytrium dendrobatidis* and amphibian chytridiomycosis in space, time, and host. Annu Rev Microbiol 63:291–310

Fites JS, Ramsey JP, Holden WM et al (2013) The invasive chytrid fungus of amphibians paralyzes lymphocyte responses. Science 342:366–369

Fites JS, Reinert LK, Chappell TM et al (2014) Inhibition of local immune responses by the frog-killing fungus *Batrachochytrium dendrobatidis*. Infect Immun 82:4698–4706

Fong JJ, Cheng TL, Bataille A et al (2015) Early 1900s detection of *Batrachochytrium dendrobatidis* in Korean amphibians. PLoS One 10(3):e0115656

Forzan MJ, Gunn H, Scott P (2008) Chytridiomycosis in an aquarium collection of frogs: diagnosis, treatment, and control. J Zoo Wildl Med 39:406–411

Garland S, Wood J, Skerratt LF (2011) Comparison of sensitivity between real-time detection of a TaqMan assay for Batrachochytrium dendrobatidis and conventional detection. Dis Aquat Organ 94:101–105

Garner TW, Perkins MW, Govindarajulu P et al (2006) The emerging amphibian pathogen *Batrachochytrium dendrobatidis* globally infects introduced populations of the north American bullfrog, *Rana catesbeiana*. Biol Lett 2:455–459

Garner TW, Garcia G, Carroll B et al (2009) Using itraconazole to clear *Batrachochytrium dendrobatidis* infection, and subsequent depigmentation of *Alytes muletensis* tadpoles. Dis Aquat Organ 83:257–260

Geiger CC, Kupfer E, Schar S et al (2011) Elevated temperature clears chytrid fungus infections from tadpoles of the midwife toad, *Alytes obstetricans*. Amphibia-Reptilia 32:276–280

Georoff TA, Moore RP, Rodriguez C et al (2013) Efficacy of treatment and long-term follow-up of *Batrachochytrium dendrobatidis* PCR-positive anurans following itraconazole bath treatment. J Zoo Wildl Med 44:395–403

Goka K, Yokoyama J, Une Y et al (2009) Amphibian chytridiomycosis in Japan: distribution, haplotypes and possible route of entry into Japan. Mol Ecol 18(23):4757–4774

Grant EHG, Muths E, Katz RA, et al (2016) Salamander chytrid fungus (*Batrachochytrium salamandrivorans*) in the United States—Developing research, monitoring, and management strategies. USGS Report https://doi.org/10.3133/ofr20151233

Greenspan SE, Longcore JE, Calhoun AJ (2012) Host invasion by *Batrachochytrium dendrobatidis*: fungal and epidermal ultrastructure in model anurans. Dis Aquat Org 100:201–210

Grogan LF (2014) Understanding host and environmental factors in the immunology and epidemiology of chytridiomycosis in anuran populations in Australia. PhD thesis, James Cook University

Grogan LF, Berger L, Rose K et al (2014) Surveillance for emerging biodiversity diseases of wildlife. PLoS Pathog 10:1–4

Harding KC, Hansen BJL, Goodman SJ (2005) Acquired immunity and stochasticity in epidemic intervals impede the evolution of host disease resistance. Am Nat 166:722–730

Harris RN, Lauer A, Simon MA et al (2009a) Addition of antifungal skin bacteria to salamanders ameliorates the effects of chytridiomycosis. Dis Aquat Org 83:11–16

Harris RN, Brucker RM, Walke JB et al (2009b) Skin microbes on frogs prevent morbidity and mortality caused by a lethal skin fungus. ISME J 3:818–824

Heard GW, Thomas CD, Hodgson JA et al (2015) Refugia and connectivity sustain amphibian metapopulations afflicted by disease. Ecol Lett 18:853–863

Heringstad B, Klemetsdal G, Steine T (2007) Selection responses for disease resistance in two selection experiments with Norwegian red cows. J Dairy Sci 90:2419–2426

Hudson PJ, Dobson AP (1998) In: Grenfell BT, Dobson AP (eds) Ecology of infectious diseases in natural populations. Cambridge University Press, Cambridge

Hunter D, Osborne W, Marantelli G et al (1999) Implementation of a population augmentation project for remnant populations of the southern corroboree frog (Pseudophryne corroboree). In: Campbell A (ed) Declines and disappearances of Australian frogs. Environment Australia, Canberra

Hyatt AD, Boyle DG, Olsen V et al (2007) Diagnostic assays and sampling protocols for the detection of *Batrachochytrium dendrobatidis*. Dis Aquat Org 73:175–192

James TY, Kauf F, Schoch CL et al (2006) Reconstructing the early evolution of fungi using a six-gene phylogeny. Nature 443:818–822

Johnson ML, Berger L, Philips L et al (2003) Fungicidal effects of chemical disinfectans, UV light, desiccation and heat on the amphibian chytrid *Batrachochytrium dendrobatidis*. Dis Aquat Org 57:255–260

Johnson PTJ, Hoverman JT (2014) Heterogeneous hosts: how variation in host size, behaviour and immunity affects parasite aggregation. J Anim Ecol 83:1103–1112

Jones ME, Paddock D, Bender L et al (2012) Treatment of chytridiomycosis with reduced-dose itraconazole. Dis Aquat Org 99:243–249

Joneson S, Stajich JE, Shiu SH et al (2011) Genomic transition to pathogenicity in Chytrid fungi. PLoS Pathog 7:e1002338

Kindermann C, Narayan EJ, Hero JM (2012) Urinary corticosterone metabolites and chytridiomycosis disease prevalence in a free-living population of male Stony Creek frogs (*Litoria wilcoxii*). Comp Biochem Physiol A 162:171–176

Kolby JE, Smith KM, Berger L et al (2014) First evidence of amphibian chytrid fungus (*Batrachochytrium dendrobatidis*) and ranavirus in Hong Kong amphibian trade. PLoS One 9:e90750

Kriger KM, Hines HB, Hyatt AD et al (2006) Techniques for detecting chytridiomycosis in wild frogs: comparing histology with real-time Taqman PCR. Dis Aquat Org 71:141–148

La Marca E, Lips KR, Lötters S et al (2005) Catastrophic population declines and extinctions in Neotropical harlequin frogs (Bufonidae: *Atelopus*). Biotropica 37:190–201

Lam BA, Walke JB, Vredenburg VT et al (2010) Proportion of individuals with anti-*Batrachochytrium dendrobatidis* skin bacteria is associated with population persistence in the frog *Rana muscosa*. Biol Conserv 143:529–531

Lam BA, Walton DB, Harris RN (2011) Motile zoospores of *Batrachochytrium dendrobatidis* move away from antifungal metabolites produced by amphibian skin bacteria. EcoHealth 8: 36–45

Lauer A et al (2007) Common cutaneous bacteria from the eastern red-backed salamander can inhibit pathogenic fungi. Copeia 3:630–640

Lips KR, Brem F, Brenes R et al (2006) Emerging infectious disease and the loss of biodiversity in a Neotropical amphibian community. Proc Natl Acad Sci U S A 103:3165–3170

Lips KR, Diffendorfer J, Mendelson JR et al (2008) Riding the wave: reconciling the roles of disease and climate change in amphibian declines. PLoS Biol 6:e72

Liu P, Stajich JE (2015) Characterization of the carbohydrate binding module 18 gene family in the amphibian pathogen *Batrachochytrium dendrobatidis*. Fungal Genet Biol 77:31–39

Longcore J, Pessier A, Nichols D (1999) *Batrachochytrium dendrobatidis* gen et sp nov, a chytrid pathogenic to amphibians. Mycologia 91:219–227

Loudon AH, Holland JA, Umile TP et al (2014) Interactions between amphibians' symbiotic bacteria cause the production of emergent anti-fungal metabolites. Front Microbiol 5:441

Marantelli G, Berger L, Speare R et al (2004) Distribution of the amphibian chytrid *Batrachochytrium dendrobatidis* and keratin during tadpole development. Pac Conserv Biol 10: 173–179

Martel A, Van Rooij P, Vercauteren G et al (2011) Developing a safe antifungal treatment protocol to eliminate *Batrachochytrium dendrobatidis* from amphibians. Med Mycol 49:143–149

Martel A, Spitzen-van der Sluijs A, Blooi M et al (2013) *Batrachochytrium salamandrivorans* sp. nov. causes lethal chytridiomycosis in amphibians. Proc Natl Acad Sci U S A 110: 15325–15329

Martel A, Blooi M, Adriaensen C et al (2014) Recent introduction of a chytrid fungus endangers Western Palearctic salamanders. Science 346:630–631

McFadden M, Hobbs R, Marantelli G et al (2013) Captive management and breeding of the critically endangered southern corroboree frog (*Pseudophryne corroboree*) (Moore 1953) at Taronga and Melbourne zoos. Amphib Reptile Conserv 5:70–87

McMahon TA, Rohr JR (2015) Transition of chytrid dungus infection from mouthparts to hind limbs during amphibian metamorphosis. EcoHealth 12:88–193

McMahon TA, Brannelly LA, Chatfield MWH et al (2013) Chytrid fungus *Batrachochytrium dendrobatidis* has nonamphibian hosts and releases chemicals that cause pathology in absence of infection. Proc Natl Acad Sci U S A 110:210–215

McMahon TA, Sears BF, Venesky MD et al (2014) Amphibians acquire resistance to live and dead fungus overcoming fungal immunosuppression. Nature 511:224–227

Meng P, Yang S, Shen C et al (2013) The first salamander defensing antimicrobial peptide. PLoS One 8:e83044

Meyer W, Seegers U, Schnapper A et al (2007) Possible antimicrobial defense by free sugars on the epidermal surface of aquatic vertebrates. Aquat Biol 1:167–175

Miedaner T, Korzun V (2012) Marker-assisted selection for disease resistance in wheat and barley breeding. Phytopathology 102:560–566

Moss AS, Reddy NS, Dortaj IM et al (2008) Chemotaxis of the amphibian pathogen *Batrachochytrium dendrobatidis* and its response to a variety of attractants. Mycologia 100:1–5

Muletz CR, Myers JM, Domangue RJ et al (2012) Soil bioaugmentation with amphibian cutaneous bacteria protects amphibian hosts from infection by *Batrachochytrium dendrobatidis*. Biol Conserv 152:119–126

Murphy PJ, St-Hilaire S, Corn PS (2011) Temperature, hydric environment, and prior pathogen exposure alter the experimental severity of chytridiomycosis in boreal toads. Dis Aquat Org 95: 31–42

Murray KA, Skerratt LF, Garland S et al (2013) Whether the weather drives patterns of endemic amphibian chytridiomycosis: a pathogen proliferation approach. PLoS One 8(4):e61061

Myers JM, Ramsey JP, Blackman AL et al (2012) Synergistic inhibition of the lethal fungal pathogen *Batrachochytrium dendrobatidis*: the combined effect of symbiotic bacterial metabolites and antimicrobial peptides of the frog *Rana muscosa*. J Chem Ecol 38:958–965

Newell DA, Goldingay RL, Brooks LO (2013) Population recovery following decline in an endangered stream-breeding frog (*Mixophyes fleayi*) from subtropical Australia. PLoS One 8: e58559

Nichols DK, Lamirande EW, Pessier AP et al (2000) Experimental transmission and treatment of cutaneous chytridiomycosis in poison dart frogs (*Dendrobates auratus* and *Dendrobates tinctorius*). In: Proceedings of the Joint Conference of American Association of Zoo Veterinarians and International Association for Aquatic Animal Medicine, pp 42–44

Nichols DK, Lamirande EW, Pessier AP et al (2001) Experimental transmission of cutaneous chytridiomycosis in dendrobatid frogs. J Wildl Dis 37:1–11

Olson DH, Aanensen DM, Ronnenberg KL et al (2013) Mapping the global emergence of *Batrachochytrium dendrobatidis*, the amphibian chytrid fungus. PLoS One 8:e56802

Parris MJ, Beaudoin JG (2004) Chytridiomycosis impacts predator-prey interactions in larval amphibian communities. Oecologia 140:626–632

Pask JD, Cary TL, Rollins-Smith LA (2013) Skin peptides protect juvenile leopard frogs (*Rana pipiens*) against chytridiomycosis. J Exp Biol 216:2908–2916

Pasmans F, Van Rooij P, Blooi M et al (2013) Resistance to chytridiomycosis in European plethodontid salamanders of the genus *Speleomantes*. PLoS One 8:e63639

Pessier AP, Mendelson JR (2010) A manual for control of infectious diseases in amphibian survival assurance colonies and reintroduction programs. IUCN/SSC Conservation Breeding Specialist Group

Pessier AP, Nichols DK, Longcore JE et al (1999) Cutaneous chytridiomycosis in poison dart frogs (*Dendrobates spp.*) and White's tree frogs (*Litoria caerulea*). J Vet Diagn Investig 11:194–199

Peterson JD, Steffen JE, Reinert LK et al (2013) Host stress response is important for the pathogenesis of the deadly amphibian disease, chytridiomycosis, in *Litoria caerulea*. PLoS One 8: e62146

Phillott AD, Speare R, Hines HB et al (2010) Minimising exposure of amphibians to pathogens during field studies. Dis Aquat Org 92:175–185

Phillott AD, Grogan LF, Cashins SD et al (2013) Chytridiomycosis and seasonal mortality of tropical stream-associated frogs 15 years after introduction of *Batrachochytrium dendrobatidis*. Conserv Biol 27:1058–1068

Pilliod DS, Muths E, Scherer RD et al (2010) Effects of amphibian chytrid fungus on individual survival probability in wild boreal toads. Conserv Biol 24:1259–1267

Plehn M (1920) Neue Parasiten in Haut and Kiemen von Fischen. *Ichthyochytrium* und *Mucophilus* Zentralblatt für Bakteriologie und Parasitenkunde Abteilung, vol 1, pp 275–281

Puschendorf R, Hoskin CJ, Cashins SD et al (2011) Environmental refuge from disease-driven amphibian extinction. Conserv Biol 25:956–964

Rachowicz LJ, Vredenburg VT (2004) Transmission of *Batrachochytrium dendrobatidis* within and between amphibian life stages. Dis Aquat Org 61:75–83

Ramsey JP, Reinert LK, Harper LK et al (2010) Immune defenses against *Batrachochytrium dendrobatidis*, a fungus linked to global amphibian declines, in the south African clawed frog, *Xenopus laevis*. Infect Immun 78:3981–3992

Reeder NMM, Pessier AP, Vredenburg VT (2012) A reservoir species for the emerging amphibian pathogen *Batrachochytrium dendrobatidis* thrives in a landscape decimated by disease. PLoS One 7:e33567

Retallick RWR, McCallum H, Speare R (2004) Endemic infection of the amphibian chytrid fungus in a frog community post-decline. PLoS Biol 2:1965–1971

Ribas L, Li MS, Doddington BJ, Robert J et al (2009) Expression profiling the temperature-dependent amphibian response to infection by *Batrachochytrium dendrobatidis*. PLoS One 4: e8408

Ricardo H (2006) Distribution and ecology of chytrid in Tasmania. Honours thesis, University of Tasmania

Richards-Zawacki CL (2010) Thermoregulatory behaviour affects prevalence of chytrid fungal infection in a wild population of Panamanian golden frogs. Proc R Soc Lond B Biol Sci 277: 519–528

Rivas LR (1964) A reinterpretation of the concepts "sympatric" and "allopatric" with proposal of the additional terms "syntopic" and "allotopic". Syst Zool 13:42–43

Rodriguez D, Becker CG, Pupin NC et al (2014) Long-term endemism of two highly divergent lineages of the amphibian-killing fungus in the Atlantic Forest of Brazil. Mol Ecol 23:774–787

Rollins-Smith LA (2009) The role of amphibian antimicrobial peptides in protection of amphibians from pathogens linked to global amphibian declines. Biochim Biophys Acta 1788:1593–1599

Rollins-Smith LA, Woodhams DC, Reinert LK et al (2006) Antimicrobial peptide defenses of the mountain yellow-legged frog (*Rana muscosa*). Dev Comp Immunol 30:831–842

Rollins-Smith LA, Ramsey JP, Pask JD et al (2011) Amphibian immune defenses against chytridiomycosis: impacts of changing environments. Integr Comp Biol 51:552–562

Rosenblum EB, Stajich JE, Maddox N et al (2008) Global gene expression profiles for life stages of the deadly amphibian pathogen *Batrachochytrium dendrobatidis*. Proc Natl Acad Sci U S A 105:17034–17039

Rosenblum EB, Poorten TJ, Settles M et al (2009) Genome-wide transcriptional response of *Silurana (Xenopus) tropicalis* to infection with the deadly chytrid fungus. PLoS One 4:e6494

Rosenblum EB, Poorten TJ, Settles M et al (2012) Only skin deep: shared genetic response to the deadly chytrid fungus in susceptible frog species. Mol Ecol 21:3110–3120

Rosenblum EB, James TY, Zamudio KR et al (2013) Complex history of the amphibian-killing chytrid fungus revealed with genome resequencing data. Proc Natl Acad Sci U S A 110: 9385–9390

Rovito SM, Parra-Olea G, Vasquez-Almazan CR et al (2009) Dramatic declines in neotropical salamander populations are an important part of the global amphibian crisis. Proc Natl Acad Sci U S A 106:3231–3236

Rowley JJL, Alford RA (2007) Behaviour of Australian rain forest stream frogs may affect the transmission of chytridiomycosis. Dis Aquat Org 77:1–9

Rowley JJL, Skerratt LF, Alford RA et al (2007) Retreat sites of rain forest stream frogs are not a reservoir for *Batrachochytrium dendrobatidis* in northern Queensland, Australia. Dis Aquat Organ 74:7–12

Sabino-Pinto JS, Bletz M, Hendrix R et al (2015) First detection of the emerging fungal pathogen in *Batrachochytrium salamandrivorans* in Germany. Amphibia-Reptilia 36(4):411–416. https://doi.org/10.1163/15685381-00003008

Savage AE, Zamudio KR (2011) MHC genotypes associate with resistance to a frog-killing fungus. Proc Natl Acad Sci U S A 108:16705–16710

Scheele BC, Guarino F, Osbourne W et al (2014) Decline and re-expansion of an amphibian with high prevalence of chytrid fungus. Biol Conserv 170:86–91

Scheele BC, Hunter DA, Skerratt LF et al (2015) Low impact of chytridiomycosis on frog recruitment enables persistence in refuges despite high adult mortality. Biol Conserv 182:36–43

Schloegel LM, Ferreira CM, James TY et al (2010) The north American bullfrog as a reservoir for the spread of *Batrachochytrium dendrobatidis* in Brazil. Anim Conserv 13:53–61

Schloegel LM, Toledo LF, Longcore JE et al (2012) Novel, panzootic and hybrid genotypes of amphibian chytridiomycosis associated with the bullfrog trade. Mol Ecol 21:5162–5177

Schmeller DS, Blooi M, Martel A et al (2014) Microscopic aquatic predatorsstrongly affect infection dynamics of a globally emerged pathogen. Curr Biol 24:176–180

Searle CL, Gervasi SS, Hua J et al (2011) Differential host susceptibility to *Batrachochytrium dendrobatidis*, an emerging amphibian pathogen. Conserv Biol 25:965–974

Sheafor B, Davidson EW, Parr L et al (2008) Antimicrobial peptide defenses in the salamander *Ambystoma tigrinum*, against emerging amphibian pathogens. J Wildl Dis 44:226–236

Skerratt L, Speare R, Berger L (2011a) Mitigating the impact of diseases affecting biodiversity—retrospective on the outbreak investigation for chytridiomycosis. Ecohealth 7:S26

Skerratt LF, Berger L, Hines HB et al (2008) Survey protocol for detecting chytridiomycosis in all Australian frog populations. Dis Aquat Org 80:85–94

Skerratt LF, Mendez D, McDonald KR et al (2011b) Validation of diagnostic tests in wildlife: the case of chytridiomycosis in wild amphibians. J Herpetol 45:444–450

Skerratt LF, Berger L, Clemann N et al (2016) Priorities for management of chytridiomycosis in Australia: saving frogs from extinction. Wildlife Res 43(2):105–120

Smith KG, Weldon C, Conradie W et al (2007) Relationships among size, development, and *Batrachochytrium dendrobatidis* infection in African tadpoles. Dis Aquat Organ 74:159–164

Soto-Azat C, Clarke BT, Poynton JC (2010) Widespread historical presence of *Batrachochytrium dendrobatidis* in African pipid frogs. Divers Distrib 16:126–131

Stice MJ, Briggs CJ (2010) Immunization is ineffective at preventing infection and mortality due to the amphibian chytrid fungus *Batrachochytrium dendrobatidis*. J Wildl Dis 46:70–77

Stockwell MP, Clulow J, Mahony MJ (2015a) Evidence of a salt refuge: chytrid infection loads are suppressed in hosts exposed to salt. Oecologia 177:901–910

Stockwell MP, Storrie LJ, Pollard CJ et al (2015b) Effects of pond salinization on survival rate of amphibian hosts infected with the chytrid fungus. Conserv Biol 29:391–399

Swei A, Rowley JJL, Rodder D et al (2011) Is chytridiomycosis an emerging infectious disease in Asia? PLoS One 6(8):e23179

Talley BL, Muletz CR, Vredenburg VT et al (2015) A century of *Batrachochytrium dendrobatidis* in Illinois amphibians (1888–1989). Biol Conserv 182:254–261

Tamukai K, Une Y, Tominaga A et al (2011) Treatment of spontaneous chytridiomycosis in captive amphibians using itraconazole. J Vet Med Sci 73:155–159

Tennessen JA, Woodhams DC, Chaurand P et al (2009) Variations in the expresses antimicrobial peptide repertoire of northern leopard frog (*Rana pipiens*) populations suggest intraspecies differences in resistance to pathogens. Dev Comp Immunol 33:1247–1257

Thekkiniath JC, Zabet-Moghaddam M, San Francisco SK et al (2013) A novel subtilisin-like serine protease of *Batrachochytrium dendrobatidis* is induced by thyroid hormone and degrades antimicrobial peptides. Fungal Biol 117:451–461

Tobler U, Schmidt BR (2010) Within- and among-population variation in chytridiomycosis-induced mortality in the toad *Alytes obstetricans*. PLoS One 5:e10927

Une Y, Kadekaru S, Tamukai K et al (2008) First report of spontaneous chytridiomycosis in frogs in Asia. Dis Aquat Organ 82:157–160

Van Ells T, Stanton J, Strieby A et al (2003) Use of immunohistochemistry to diagnose chytridiomycosis in dyeing poison dart frogs (Dendrobates tinctorius). J Wildl Dis 39:742–745

Van Rooij P, Martel A, D'Herde K et al (2012) Germ tube mediated invasion of *Batrachochytrium dendrobatidis* in amphibian skin is host dependent. PLoS One 7:1–8

Van Rooij P, Martel A, Haesebrouck F et al (2015) Amphibian chytridiomycosis: a review with focus on fungus-host interactions. Vet Res 46:137

Venesky MD, Mendelson JR, Sears BF et al (2012) Selecting for tolerance against pathogens and herbivores to enhance success of reintroduction and translocation. Conserv Biol 26:586–592

Voyles J, Young S, Berger L, Campbell C, Voyles WF, Dinudom A, Cook D, Webb R, Alford RA, Skerratt LF, Speare R (2009) Pathogenesis of Chytridiomycosis, a Cause of Catastrophic Amphibian Declines. Science 326: 582–585.

Vredenburg VT, Knapp RA, Tunstall TS et al (2010) Dynamics of an emerging disease drive large-scale amphibian population extinctions. Proc Nat Acad Sci U S A 107:9689–9694

Vredenburg VT, Briggs CJ, Harris RN (2011) Host-pathogen dynamics of amphibian chytridiomycosis: the role of the skin microbiome in health and disease. In: Fungal diseases: an emerging threat to human, animal and plant health: workshop summary. National Academy Press, Washington

Walker S, Bosch J, James TY et al (2008) Invasive pathogens threaten species recovery programs. Curr Biol 18:853–R854

Walker SF, Bosch J, Gomez V et al (2010) Factors driving pathogenicity vs. prevalence of amphibian panzootic chytridiomycosis in Iberia. Ecol Lett 13(3):372–382

Weldon C, du Preez LH, Hyatt AD et al (2004) Origin of the amphibian chytrid fungus. Emerg Infect Dis 10:2100–2105

Woodhams DC, Alford RA, Marantelli G (2003) Emerging disease of amphibians cured by elevated body temperature. Dis Aquat Org 55:65–67

Woodhams DC, Ardipradja K, Alford RA et al (2007) Resistance to chytridiomycosis varies among amphibian species and is correlated with skin peptide defenses. Anim Conserv 10:409–417

Woodhams DC, Bosch J, Briggs CJ et al (2011) Mitigating amphibian disease: strategies to maintain wild populations and control chytridiomycosis. Front Zool 8(1):8

Woodhams DC, Bell SC, Kenyon N et al (2012a) Immune evasion or avoidance: fungal skin infection linked to reduced defence peptides in Australian green-eyed treefrogs, *Litoria serrata*. Fungal Biol 116:1203–1211

Woodhams DC, Geiger CC, Reinert LK et al (2012b) Treatment of amphibians infected with chytrid fungus: learning from failed trials with itraconazole, antimicrobial peptides, bacteria, and heat therapy. Dis Aquat Organ 98:11–25

Woodward A, Berger L, Skerratt LF (2014) *In vitro* sensitivity of the amphibian pathogen *Batrachochytrium dendrobatidis* to antifungal therapeutics. Res Vet Sci 97:364–366

Young S, Whitehorn P, Berger L et al (2014) Defects in host immune function in tree frogs with chronic chytridiomycosis. PLoS One 9:e107284

Feline Aspergillosis

15

Vanessa R. Barrs

Abstract

Fungal rhinosinusitis (FRS), the most common form of aspergillosis in immunocompetent cats, comprises two anatomic forms: sinonasal aspergillosis (SNA) and sino-orbital aspergillosis (SOA). SNA is confined to the sinonasal cavity, while SOA also involves paranasal structures including the orbit. Although *Aspergillus fumigatus* sensu stricto is the most common cause of SNA, cryptic species in sections *Fumigati* and *Nigri* can also cause disease. SOA is an increasingly recognised invasive mycosis that is most frequently caused by the recently discovered opportunist *A. felis* and by *A. udagawae*, two cryptic species in section *Fumigati.* Serological detection of *Aspergillus*-specific IgG has high diagnostic utility for feline FRS, but galactomannan assays are only positive in a quarter of cases. Fungal pathogens causing FRS can be cultured from sinonasal fungal plaques or tissue biopsies for identification. Comparative sequence analysis of the rDNA internal transcribed spacer region and partial calmodulin gene enables differentiation of cryptic species from *A. fumigatus s. str.* Accurate species identification is required to guide therapy since minimum inhibitory concentrations of most antifungal drugs are higher for cryptic species than for *A. fumigatus s. str.* and azole cross-resistance can occur. Current treatment recommendations for SNA include endosurgical debridement of sinonasal plaques and topical intranasal azole infusion. The prognosis for SOA is poor overall. Individual cases have been successfully treated with itraconazole or posaconazole monotherapy or combined with amphotericin B and/or terbinafine.

V. R. Barrs (✉)
Sydney School of Veterinary Science, Faculty of Science, The University of Sydney, Camperdown, NSW, Australia

Marie Bashir Institute of Infectious Diseases and Biosecurity, The University of Sydney, Camperdown, NSW, Australia
e-mail: vanessa.barrs@sydney.edu.au

S. Seyedmousavi et al. (eds.), *Emerging and Epizootic Fungal Infections in Animals*,
https://doi.org/10.1007/978-3-319-72093-7_15

15.1 Introduction

Aspergillus species are filamentous fungi that are commonly found in soil, where they thrive as saprophytes and have the potential to opportunistically infect living hosts including plants, insects, birds and mammals. Aspergillosis is an umbrella term covering a wide range of diseases from localised conditions to fatal disseminated infections of humans and various animals (Seyedmousavi et al. 2015). *Aspergillus fumigatus* is the most prevalent environmental *Aspergillus* species and the most common cause of aspergillosis in humans. However, cryptic species, specifically close molecular siblings of *A. fumigatus*, are being increasingly identified to cause disease in humans and animals and are frequently identified in respiratory aspergillosis of cats. In this chapter, we describe the fungal pathogens responsible for feline respiratory aspergillosis and review the epidemiology, pathology, host-pathogen interactions, clinical presentation, diagnosis, treatment strategies and prognosis of these infections.

15.2 Classification of Feline Aspergillosis

Similar to disease in humans, aspergillosis in cats can be classified by anatomic location, invasiveness, duration of infection, host immune status, pathology and pathogenesis. The most common site of disease is the respiratory tract, reflecting the primary inhalational route of infection. Respiratory involvement is usually confined to the upper respiratory tract as chronic fungal rhinosinusitis (FRS). However, invasive pulmonary aspergillosis (IPA) can occur as a focal infection (Pakes et al. 1967; Hazell et al. 2011) or as part of disseminated invasive aspergillosis (DIA) (Fox et al. 1978; Ossent 1987; Burk et al. 1990). Focal invasive infections of the gut or urinary bladder have also been described (Stokes 1973; Adamama-Moraitou et al. 2001).

Two forms of FRS occur in cats: sinonasal aspergillosis (SNA) and sino-orbital aspergillosis (SOA). Together, these conditions are referred to as feline upper respiratory tract aspergillosis (FURTA). SNA, which is usually non-invasive, is confined to the nasal cavity and paranasal sinuses. SOA is an invasive mycosis involving the sinonasal cavity, orbit and other paranasal tissues. It is the more common form of FRS, accounting for two thirds of cases of FURTA (Wilkinson et al. 1982; Hamilton et al. 2000; Malik et al. 2004; Barachetti et al. 2009; Giordano et al. 2010; Halenda and Reed 1997; Kano et al. 2008, 2013, 2015; McLellan et al. 2006; Smith and Hoffman 2010; Barrs et al. 2007, 2012, 2013, 2014, 2015; Karnik et al. 2009; Quimby et al. 2010).

15.3 The Causative Agents of Feline Aspergillosis

In contrast to dogs with SNA, in which *A. fumigatus* sensu stricto causes over 95% of infections (Talbot et al. 2014), a more diverse range of *Aspergillus* species has been identified in FURTA (Table 15.1). Overall, the *Aspergillus viridinutans* species

Table 15.1 Causative agents in feline upper respiratory tract aspergillosis based on molecular identification

Section	Species	Form of disease
Fumigati	*A. fumigatus*	SNA
	A. lentulus	SNA
	A. fischeri	SOA
	A. thermomutatus	SNA, SOA
	A. felis	SNA, SOA
	A. parafelis	SOA
	A. udagawae	SOA
	A. wyomingensis	SOA
Nigri	*A. niger*	SNA
Flavi	*A. flavus*	SNA

SNA sinonasal aspergillosis, *SOA* sino-orbital aspergillosis

complex and the *A. fumigatus* species complex in section *Fumigati* are most frequently implicated in FURTA (Fig. 15.1).

Investigations into the causal agents of SOA in cats resulted in the recent discovery of a new species *A. felis* (Barrs et al. 2013), which is the most common cause of SOA followed by *A. udagawae* (Kano et al. 2008, 2013; Barrs et al. 2013, 2014, 2015). Two other molecular siblings of *A. viridinutans*, *A. parafelis* and *A. wyomingensis*, have been isolated from cats with SOA (Barrs et al. 2014; Talbot et al. 2017). The species *A. viridinutans* was reported to cause SOA in a cat from Japan (Kano et al. 2013); however the *A. viridinutans* sequences deposited on GenBank as HE578084 and AB24899 for which the isolate had greatest ß-tubulin sequence homology with have been subsequently reclassified as *A. felis* on further phylogenetic analysis (Novakova et al. 2014). Other section *Fumigati* species isolated from cats with SOA include *A. thermomutatus* (Barrs et al. 2012) and *A. fischeri* (Kano et al. 2015).

Aspergillus fumigatus s. str. and *A. niger* and relatives (section *Nigri*) are the most common isolates in SNA (Furrow and Groman 2009; Barrs et al. 2012, 2014, 2015; Whitney et al. 2005). Other section *Fumigati* species occasionally isolated from cats with SNA include *A. lentulus*, *A. felis* and *A. thermomutatus* (Barrs et al. 2013, 2014). *Aspergillus flavus* (section *Flavi*) was isolated from one cat with SNA in association with a plant foreign body (Barrs et al. 2015).

There are over 60 species of *Aspergillus* in section *Fumigati* with many new species additions in the last decade. That number is likely to continue to increase as molecular methods facilitate more accurate species delimitation, and since many more fungal species are thought to remain undiscovered. Cryptic species within section *Fumigati* cannot be reliably distinguished from *A. fumigatus s. str.* using phenotypic methods. However, accurate identification has important therapeutic implications since antifungal resistance profiles of cryptic species differ to that of *A. fumigatus s. str.*, with some isolates having inherently higher MICs of azoles and other antifungal drugs. Molecular methods such as comparative sequence analysis and more recently MALDI-TOF are increasingly used for identification of human

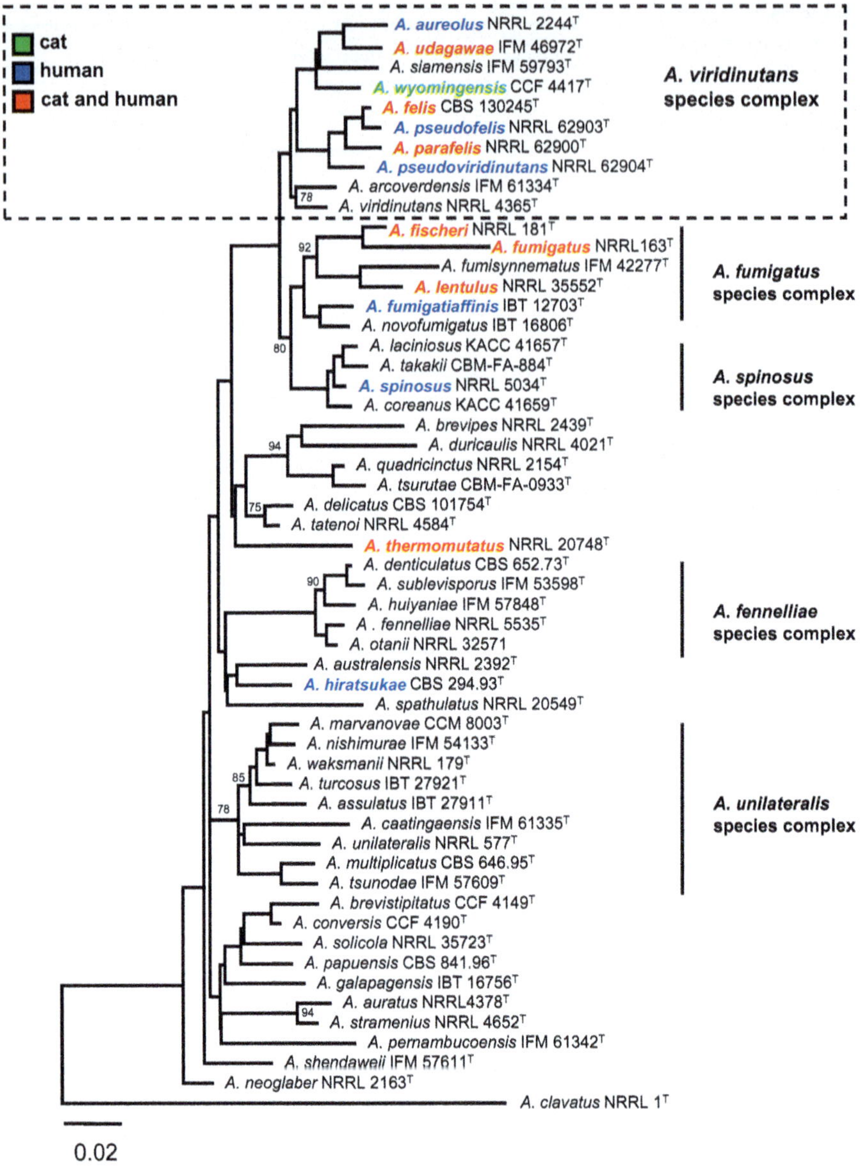

Fig. 15.1 Best-scoring maximum likelihood tree (kindly provided by Vit Hubka, Charles University, Prague) based on combined dataset of calmodulin and β-tubulin sequences showing the relationship of *Aspergillus viridinutans* complex to other *Aspergillus* species from section *Fumigati*. To filter both gaps and variable regions in the alignment, GBLOCKS v0.91b (Talavera and Castresana 2007) was used with less stringent selection allowing smaller final blocks and gap positions within the final blocks. The tree was constructed with the IQ-TREE version 1.4.0 (Nguyen et al. 2015). Dataset contained 58 taxa and a total of 908 characters of which 450 were variable and 292 parsimony-informative. Optimal partitioning scheme and substitution models were selected using PartitionFinder v1.1.0 (Lanfear et al. 2012) with setting allowing introns, exons and codon

clinical *Aspergillus* isolates. Uptake of these methods in veterinary medicine has been slower. Misidentification of heterothallic cryptic species causing feline URTA including *A. felis*, *A. udagawae* and *A. lentulus* is likely when only morphological typing methods are used (Barrs et al. 2012, 2013; Balajee et al. 2006). For example, in one case of SOA where *A. fumigatus* was reportedly identified using micro- and macromorphology, misidentification of a cryptic species was likely based on high MICs of amphotericin and itraconazole for this isolate (McLellan et al. 2006). Compared to *A. fumigatus s. str.*, the MICs of amphotericin B for cryptic species including *A. lentulus* and *A. udagawae* are high (Balajee et al. 2006; Alcazar-Fuoli et al. 2008).

For molecular identification of section *Fumigati* species using comparative sequence analysis, a universal DNA barcode such as the nuclear ribosomal internal transcribed spacer (ITS) region is unreliable and should be combined with a secondary identification marker such as calmodulin, ß-tubulin or the RNA polymerase II second largest subunit (*RPB2*) (Samson et al. 2014). Of these three, calmodulin is best for identification of clinical isolates since amplification of *RPB2* is more difficult and amplification of ß-tubulin can be complicated by the presence of paralogous genes and variable numbers of introns (Hubka and Kolarik 2012). For molecular identification of isolates from clinical samples, DNA can be extracted from fungal culture material for PCR and sequencing. Alternatively, fungal DNA can be extracted directly from fresh or frozen clinical specimens or from formalin-fixed paraffin-embedded tissues (FFPET). However, DNA fragments from FFPET may be poor in quality and short, limiting the gene targets for amplification. The entire ribosomal DNA ITS1-5.8S-ITS2 region of approximately 600 nucleotides can easily be amplified from DNA extracted from fungal culture material, using primers ITS1 and ITS4 (Barrs et al. 2013). For FFPET, amplification of the shorter ITS1 (primers ITS1 and ITS2) region (~290 nt) or ITS2 (primers ITS3 and ITS4) region (~330 bp) is often possible (Barrs et al. 2012; Meason-Smith et al. 2017). In one study using amplification of the ITS2 region from FFPET for a range of fungal organisms, identification of fungi to genus level was possible in 65% of samples, and in 96% of these, the identified fungus was the same as that identified morphologically on histological examination (Meason-Smith et al. 2017).

While all members of section *Fumigati* reproduce asexually by mitosis (anamorph), many are also capable of sexual reproduction by meiosis (teleomorph) under certain optimal growth conditions. Anamorphs are mould-like and are comprised of filamentous hyphae bearing mitotic spores. Teleomorphs comprise

Fig. 15.1 (continued) positions to be independent datasets. Support values at branches were obtained from 1000 bootstrap replicates. Only branches with bootstrap support ≥70% are shown, and branches with support ≥95% are double thick; ex-type strains are designated by a superscript T. The tree is rooted with *Aspergillus clavatus* NRRL 1. Pathogenic species in cats that cause fungal rhinosinusitis (sinonasal and/or sino-orbital aspergillosis) are shaded green or red. Species shaded red are also pathogenic in humans. Species shaded in blue have been documented to act as pathogens in humans but not as yet in cats

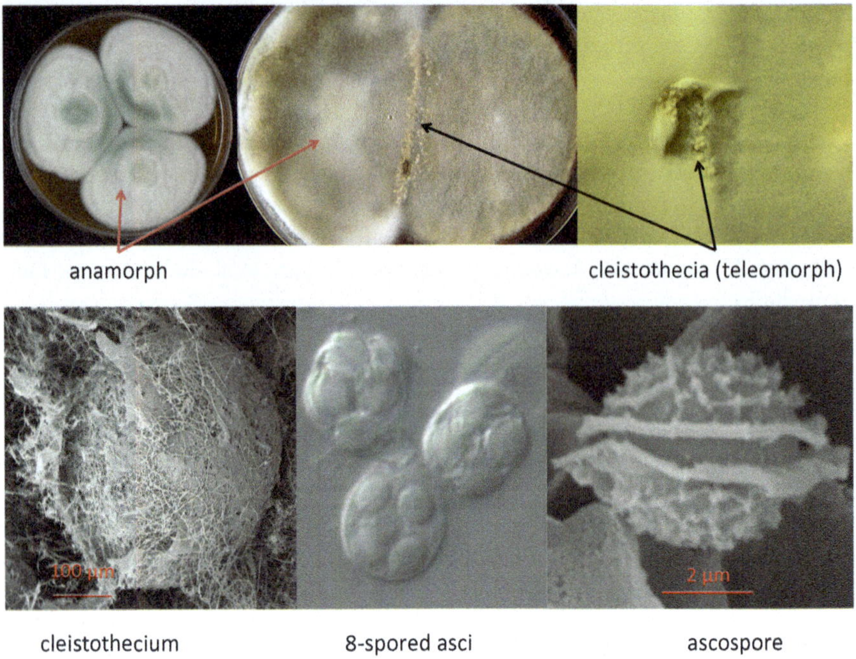

Fig. 15.2 Holomorph of an *A. viridinutans complex* species, *A. felis* comprising anamorph (asexual) and teleomorph (sexual) phases. For heterothallic fungi the teleomorph forms at the junction of two colonies of complementary mating type. Ascospores develop within sacs (asci) that are contained within the mycelial walls of globose fruiting bodies (ascomata or cleistothecia)

fruiting bodies (cleistothecia) consisting of spheres of matted fungal hyphae that contain sacs (asci) within which meiotic spores (ascospores) develop. The anamorph and teleomorph phases of *A. felis* are depicted in Fig. 15.2.

Fungal taxonomy has recently become simplified through reforms of the *International Code of Nomenclature for Algae, Fungi and Plants* to adopt a "one-fungus, one-name" principle, in July 2011 (Miller et al. 2011). Before this, a dual nomenclature system was in place, in which the anamorph and teleomorph of the same fungus were assigned to a different genus. For example, in section *Fumigati*, the anamorph was in genus *Aspergillus*, and the teleomorph was in genus *Neosartorya*. When referring to the organism as a whole (the holomorph), taxonomic precedence was given to the teleomorphic name, which was convenient for rapidly distinguishing ascospore-producing fungi, e.g. *Neosartorya pseudofischeri* (Pitt and Samson 2007). The advantage of this system was its ability to distinguish ascospore-producing fungi. Over time, the system became confusing because of asynchronous discovery of the teleomorph of some species compared to the anamorph. For example, although the teleomorph *Neosartorya fumigata* was discovered in 2009, the holomorph continued to be referred to by its anamorph name, *Aspergillus fumigatus* (O'Gorman et al. 2009). Now, as per the Amsterdam declaration on fungal nomenclature, the presence of teleomorphs is indicated by an informal cross-reference name, e.g. *Aspergillus fumigatus* (neosartorya-morph) (Hawksworth et al. 2011).

15.4 Epidemiology

The first case of FURTA was described in Australia in 1982 more than a century after SNA was described in dogs (Wilkinson et al. 1982). About half of all cases reported are from Australia (Wilkinson et al. 1982; Malik et al. 2004; Katz et al. 2005; Barrs et al. 2012, 2013), and a third are from the USA (Hamilton et al. 2000; Tomsa et al. 2003; Whitney et al. 2005; McLellan et al. 2006; Furrow and Groman 2009; Smith and Hoffman 2010; Karnik et al. 2009). Reports of disease in the UK, mainland Europe (Goodall et al. 1984; Barachetti et al. 2009; Tomsa et al. 2003; Declercq et al. 2012) and Japan (Kano et al. 2008, 2013, 2015) reflect the global environmental presence of saprotrophic *Aspergillus* species responsible for these infections.

Cats of any age can be affected. The median age at diagnosis of reported cases was 6.5 years (Wilkinson et al. 1982; Goodall et al. 1984; Hamilton et al. 2000; Tomsa et al. 2003; Malik et al. 2004; Davies and Troy 1996; Karnik et al. 2009; Furrow and Groman 2009; Barachetti et al. 2009; Giordano et al. 2010; Smith and Hoffman 2010; Barrs et al. 2012; Declercq et al. 2012; Kano et al. 2013). A marked predisposition for FURTA in pure-bred brachycephalic cats of Persian lineage, including Persian, Himalayan, exotic shorthair, British shorthair and Scottish shorthair, has been demonstrated (Barrs et al. 2015). Cats of these breeds account for over a third of all FURTA cases. Another striking feature of FURTA is its propensity to occur in systemically immunocompetent cats, in contrast to DIA in this species, which is invariably associated with immune compromise from concurrent disease, e.g. feline panleukopenia virus infection, or prolonged corticosteroid therapy (Ossent 1987). Only two cases of FURTA have been associated with retroviral infection: feline leukaemia virus (Goodall et al. 1984) or feline immunodeficiency virus (Barrs et al. 2015). It is becoming increasingly apparent that diabetes mellitus, a recognised risk factor for aspergillosis in humans, is also a risk factor for FURTA with several cases now reported in cats with diabetes mellitus (Malik et al. 2004; Furrow and Groman 2009; Kano et al. 2015). FURTA occurs occasionally in association with facial trauma, nasal neoplasia and inhaled plant material, as has been reported in canine SNA (Sharp et al. 1991; Peeters and Clercx 2007; Day 2009; Barrs and Talbot 2014).

15.5 Pathogenesis and Pathology

FURTA can be non-invasive or deeply invasive. SNA is usually non-invasive, while SOA is locally invasive and spreads across tissue planes. SOA is the result of extension of a primary sinonasal infection to involve paranasal structures including but not limited to the orbit. The orbital bone in cats is very thin and bone lysis subsequent to infection allows direct communication with the nasal cavity or frontal sinuses.

In non-invasive forms of SNA, fungal hyphae do not penetrate the mucosal epithelium. Histological changes include severe inflammatory rhinitis with lymphoplasmacytic or mixed-cell inflammatory cell infiltrates, epithelial ulceration and necrosis, which can be extensive and mats or plaques of fungal hyphae adjacent to mucosal epithelium and/or within nasal exudates (Whitney et al. 2005; Tomsa et al. 2003; Furrow and Groman 2009; Barrs et al. 2012).

Cats with SOA have granulomatous invasive mycotic rhinosinusitis with submucosal hyphal invasion (Wilkinson et al. 1982; Barachetti et al. 2009; Barrs et al. 2012) and extension to involve orbital and other paranasal tissues. Orbital fungal granulomas have a central area of coagulative necrosis in which parallel-walled, dichotomously branching, septate fungal hyphae are confined and stain positively with fungal stains, e.g. periodic acid-Schiff (PAS), Grocott's methenamine silver (GMS) or Gridley's stain (Fig. 15.3) (Hamilton et al. 2000; Wilkinson et al. 1982; McLellan et al. 2006; Barachetti et al. 2009; Smith and Hoffman 2010; Barrs et al. 2012). There are surrounding zones of inflammatory cells including predominantly neutrophilic or eosinophilic infiltrates, activated and epithelioid macrophages and peripheral fibroblasts, lymphocytes and plasma cells. Inflammatory lesions can efface adjacent skeletal muscle and bone (Barrs et al. 2012). The globe is resistant to infiltration by fungal hyphae but may be affected by keratitis and anterior uveitis, and there may be invasion of adjacent structures including the nictitating membrane, eyelid, optic nerve and optic chiasm (McLellan et al. 2006; Giordano et al. 2010; Barachetti et al. 2009; Hamilton et al. 2000; Barrs et al. 2012). Brain involvement has been demonstrated on MRI in cats with advanced disease and seizures, but histological evaluation was not performed (Giordano et al. 2010; Smith and Hoffman 2010). Haematogenous dissemination is possible but has only been seen in one case by the author, a young Ragdoll cat that presented with DIA due to *A. felis* infection with sino-orbital, cervical lymph node and pulmonary involvement.

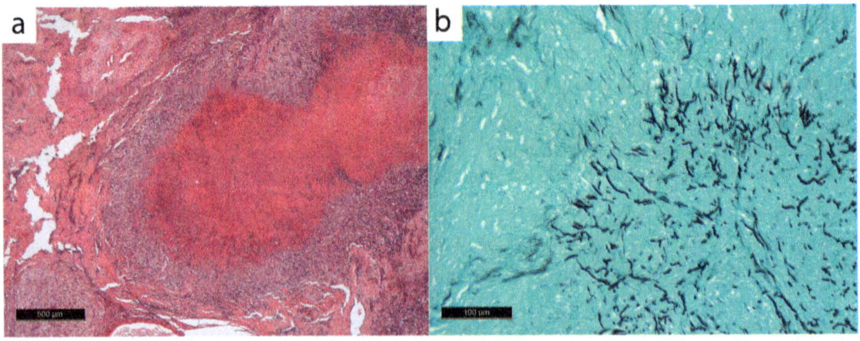

Fig. 15.3 Histopathology of an orbital fungal granuloma showing a central area of coagulative necrosis in which parallel-walled dichotomously branching, septate fungal hyphae are confined (**a**) and stain positively with Grocott's methenamine silver (**b**)

Aspergillus conidia that escape mucociliary clearance after inhalation are phagocytosed by macrophages and dendritic cells that express pattern recognition receptors (PRRs). PRRs recognise specific fungal epitopes or pathogen-associated molecular patterns (PAMPS) and activate Th_1-, Th_2- and Th_{17}-related intracellular molecular pathways triggering the production of proinflammatory and anti-inflammatory cytokines, chemokines and their receptors and proangiogenic factors (Segal 2009). Single-nucleotide polymorphisms (SNPs) in PRRs in humans can increase susceptibility to IA, including SNPs in Toll-like receptors (TLRs) 1, 3, 4 and 6; in C-type lectin receptors (CLRs), Dectin-1 and DC-SIGN; in the secreted PPR mannan-binding lectin (MBL); and in PTX3 (Gresnigt et al. 2012; Romani 2011; Lupianez et al. 2016). In addition, SNPs in cytokines, chemokines and immune receptors increase susceptibility to IA (Lupianez et al. 2016). Whether similar genetic mutations could be associated with increased susceptibility to FURTA in brachycephalic cats of Persian lineage has not been investigated. Interestingly, Persian cats are also at increased risk for invasive dermatophyte infections (pseudomycetomas) (Miller 2010). Expression of TLR 2, 4 and 9 mRNA was upregulated in the nasal mucosa of dogs with SNA, prompting a mutation analysis of these genes. However, SNPs correlating with an increased susceptibility to canine SNA were not identified (Mercier et al. 2012, 2014).

Conformational abnormalities of the upper respiratory system may contribute to the increased risk of URTA in brachycephalic breeds of cats. Congenital shortening of the facial and neurocranial bones, together with dorsal rotation of the jaw, results in deformation and displacement of ventral nasal and ethmoid turbinates as well as impaired nasolacrimal drainage (Schlueter et al. 2009). Decreased sinus aeration and respiratory secretion drainage secondary to infection, polyps and allergic rhinosinusitis are a risk factor for invasive FRS in humans (Siddiqui et al. 2004). Similarly, impaired drainage of nasal secretions has been proposed as a risk factor for fungal colonisation in brachycephalic cats (Tomsa et al. 2003). By contrast, in dogs with SNA, brachycephalic dogs are under-represented, and disease is most frequent in dolichocephalic and mesaticephalic breeds (Peeters and Clercx 2007). Additional risk factors such as previous viral respiratory infection or recurrent antimicrobial therapy may increase the risk of sinonasal fungal colonisation in brachycephalic cats (Tomsa et al. 2003; Barrs et al. 2012; Goodall et al. 1984).

A role for species-specific fungal virulence factors in the development of invasive FURTA is suspected since different fungal species are implicated in SNA and SOA in cats (Barrs et al. 2013). This has been demonstrated to some extent in vitro. *Galleria mellonella* larvae and two murine hosts (BALB/c mice immunosuppressed with hydrocortisone and mice with chronic granulomatous disease (GCD)) were inoculated with conidial suspensions of *A. fumigatus* and *A. felis* (Sugui et al. 2014). The *A. felis* type strain (isolated from a cat with SOA) showed significantly higher virulence in *G. mellonella* larvae than *A. fumigatus*. The species *A. felis* was slightly more virulent in immunosuppressed BALB/c mice than *A. fumigatus* but was less virulent in mice with GCD. *Aspergillus fumigatus* caused 100% mortality in GCD mice in 11 days, whereas *A. felis* caused no fatalities, demonstrating decreased virulence in hosts deficient in production of reactive oxygen species.

Secondary fungal metabolites or extrolites are numerous, often bioactive and species-specific and may contribute to infection and invasion. For example, gliotoxin inhibits immune responses, phagocytosis and angiogenesis and is produced by *A. fumigatus* and *A. thermomutatus*. Fumagillin also suppresses immune responses, neutrophil function and angiogenesis and is synthesised by *A. felis*, *A. fumigatus* and *A. udagawae*. Sulochrin, an eosinophil activation inhibitor, is synthesised by these same three species, as well as by *A. lentulus*. Helvolic acid can adversely affect respiratory epithelium and macrophage metabolism and is synthesised by many of the section *Fumigati* species that cause FURTA (Frisvad and Larsen 2016). The role of many extrolites in fungal pathogenicity remains to be elucidated.

15.6 Clinical Presentations

Cats with SNA are usually presented for sneezing and unilateral or bilateral nasal discharge. Intermittent epistaxis occurs in a third of cases and fever is variably present. In severe or very chronic infections, erosion of nasal or frontal bones results in a discharging sinus or facial deformity.

Most cats with SOA are presented for clinical signs associated with a retrobulbar fungal granuloma, unilateral exophthalmos with dorsolateral deviation of the globe, a prolapsed inflamed third eyelid and exposure keratitis (Fig. 15.4) (Wilkinson et al. 1982; Hamilton et al. 2000; McLellan et al. 2006; Barrs et al. 2012; Declercq et al. 2012; Kano et al. 2008, 2013; Barachetti et al. 2009; Giordano et al. 2010; Smith and Hoffman 2010). Resistance to retropulsion of the globe and normal intraocular pressure readings on tonometry enables differentiation of exophthalmos from buphthalmos (abnormal enlargement of the globe) (Smith and Hoffman 2010). Anterior and/or posterior uveitis may be present. Vision loss occurs late in disease due to severe ulcerative keratitis, as well as infiltration of the optic nerve and/or optic

Fig. 15.4 Brachycephalic pure-bred Ragdoll cat with sino-orbital aspergillosis. A retrobulbar fungal granuloma in the right orbit has caused exophthalmos, resulting in prolapse of the third eyelid and exposure keratitis with central corneal ulceration

chiasm (Hamilton et al. 2000; Tomsa et al. 2003; Barachetti et al. 2009; Barrs et al. 2012). In severe chronic infections, exophthalmos may become bilateral (Barachetti et al. 2009; Barrs et al. 2012; Wilkinson et al. 1982). As the ventral floor of the orbit is not encased by bone, the orbital granuloma often invades the oral cavity where it first becomes visible as an ulcerated area of mucosa or a submucosal mass in the pterygopalatine fossa adjacent the last molar tooth. Involvement of maxillary subcutaneous tissues causes facial swelling, although this may be subtle. Nasal signs are absent in 40% of SOA cases at presentation; however the medical history usually reveals sneezing or nasal discharge in the preceding 6 months. Pain on opening the mouth and neurological signs are uncommon at initial presentation. Neurological signs develop in chronic end-stage infection and can include seizures, nystagmus, circling, facial muscle fasciculation and hyperesthesia (Barrs et al. 2012; Giordano et al. 2010; Smith and Hoffman 2010). Similar to SNA, the presence of fever is variable.

15.7 Diagnosis

Brachycephalic conformation should increase suspicion for aspergillosis, although these cats are also over-represented for viral respiratory infections. Where epistaxis is present, neoplasia, mycotic rhinitis or severe chronic rhinosinusitis is more likely. Other fungal, bacterial or neoplastic processes extending from the nasal cavity to the orbit can have a similar presentation to SOA. Haematological and serum biochemical changes are usually mild and non-specific. Mild peripheral eosinophilia is present occasionally, and mild to marked hyperglobulinaemia is relatively common in cats with SOA (Hamilton et al. 2000; Tomsa et al. 2003; Furrow and Groman 2009; McLellan et al. 2006; Smith and Hoffman 2010; Barachetti et al. 2009; Barrs et al. 2012; Giordano et al. 2010; Kano et al. 2013). Serological tests are used widely for non-invasive diagnosis of aspergillosis in humans and animals (Seyedmousavi et al. 2015). However, the diagnostic utility of serological tests to detect the fungal cell-wall antigen galactomannan (GM) or *Aspergillus*-specific antibodies is heavily influenced by host immunocompetence. A commercial ELISA to detect GM in serum (Platelia *Aspergillus* EIA, Bio-Rad) has a sensitivity of up to 90% in immunocompromised patients, including neutropenic human patients with pulmonary aspergillosis and dogs with DIA (Pfeiffer et al. 2006; Garcia et al. 2012). However, in non-neutropenic patients with pulmonary aspergillosis, assay sensitivity is less than 30% (Hachem et al. 2009; Kitasato et al. 2009; Pfeiffer et al. 2006). In dogs with SNA, a patient cohort that is systemically immunocompetent, GM also performs poorly (24% sensitivity) (Billen et al. 2009). In immunocompetent patients, GM is cleared by neutrophils, which possess mannose-binding receptors, or by antibody complexing (Mennink-Kersten et al. 2004; Herbrecht et al. 2002). Also, lack of tissue invasion in non-invasive mycoses such as SNA likely contributes to lack of detectable circulating GM. Interestingly, in cats with URTA, the sensitivity of GM detection is low in both non-invasive (SNA) and invasive (SOA) forms of disease with an overall sensitivity of only 23%. This result likely reflects the general systemic immunocompetence of affected cats (Whitney et al. 2013). Except in the setting of ruling out URTA in cats with respiratory disease (90% specificity), serum GM is not useful as a routine diagnostic test for feline URTA.

Detection of *Aspergillus*-specific IgG by ELISA has high sensitivity in immunocompetent patients with aspergillosis, including dogs with SNA (88%) and humans with chronic pulmonary aspergillosis (90%). Specificity is also high (>90%), since optimal cut-off values can generally differentiate between exposure and infection (Billen et al. 2009; Page et al. 2015). Similar results were found using an indirect ELISA and a commercial antigen preparation derived from mycelial elements of *A. fumigatus*, *A. flavus* and *A. niger* to diagnose URTA in cats (sensitivity 95%, specificity 93%). Most cats in the study had infections with cryptic species (*A. felis*, *A. thermomutatus*, *A. lentulus* or *A. udagawae*), demonstrating antibody cross-reactivity with the antigen preparation (Barrs et al. 2015). A more recent study showed that a high proportion of cats with URTA also have detectable *Aspergillus*-specific IgA, but in the cohort studied, paired measurement of serum *Aspergillus*-specific IgA and IgG was of no benefit for diagnosis of URTA over IgG alone (Taylor et al. 2016). Human patients with chronic pulmonary aspergillosis that test negative for IgG may test positive of IgA, since *Aspergillus*-specific IgA can bind different fungal antigens than IgG (Page et al. 2015). Advanced imaging (computed tomography or magnetic resonance imaging) is recommended for all cases of suspected feline URTA and is the most useful test to differentiate SNA from SOA. Orbital involvement may not be apparent at presentation in early disease, but its detection necessitates systemic antifungal therapy as part of the management plan. For cats with only sinonasal cavity involvement, assessment of cribriform plate integrity is required before treatment using intranasal azole preparations.

Computed tomographic features of feline SNA are similar to canine SNA and include severe cavitated turbinate lysis (Fig. 15.5). Nasal cavity involvement is frequently bilateral and asymmetric, while frontal sinus involvement is usually unilateral. Other common findings in SNA include punctate areas of orbital bone lysis, which may be bilateral, and reactive bony change (sclerosis) of the nasal and frontal bones (Barrs et al. 2014). Cats with SOA have similar signs of sinonasal cavity involvement and sphenoid sinus involvement is frequent. Orbital masses cause dorsolateral displacement of the globe and extend laterally into paranasal maxillary soft tissues and ventrally into the pterygopalatine fossa of the oral cavity. Masses show heterogeneous contrast enhancement, including central coalescing hypoattenuating areas and peripheral rim enhancement (Fig. 15.5). Mass lesions within the nasal cavity, nasopharynx or paranasal sinuses are also more commonly observed in cryptic species infections than in infections caused by *A. fumigatus*, which typically cause cavitated turbinate lysis (Barrs and Talbot 2014). In SOA lysis and reactive change of the paranasal bones are also common. Changes seen in URTA are not pathognomonic for the disease and may be similar to other mycotic, chronic inflammatory or neoplastic diseases. MRI is the imaging modality of choice for patients with SOA presenting with neurological signs. On MRI, feline SOA orbital masses are T2-hyperintense.

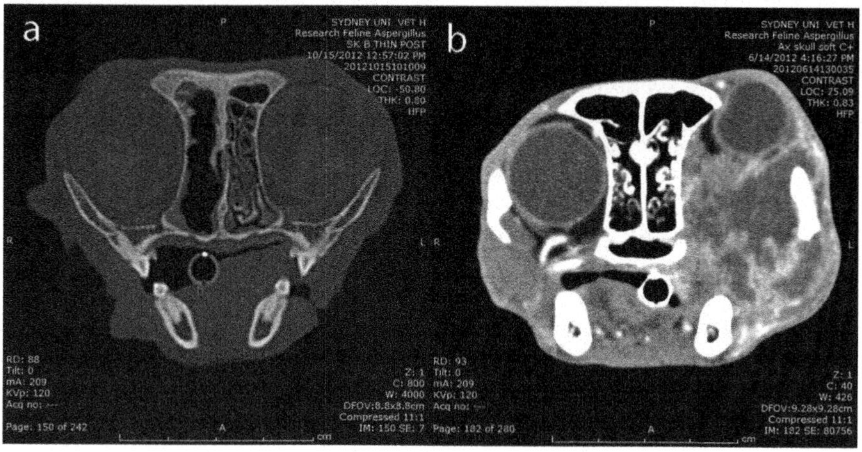

Fig. 15.5 (**a**) Transverse skull CT image of a cat with sinonasal aspergillosis due to *A. fumigatus* infection. There is severe unilateral right-sided cavitated nasal turbinate lysis. (**b**) Transverse post-contrast soft-tissue images of the head in a cat with sino-orbital aspergillosis from infection with *A. felis*. A left orbital fungal granuloma has caused compression and dorsal displacement of the globe, with extension into the oral cavity, and adjacent maxillary soft tissues. The orbital mass shows heterogeneous contrast enhancement, with central coalescing hypoattenuating foci and peripheral rim enhancement

Endoscopic assessment of the sinonasal cavity using rigid rhinoscopy and nasopharyngoscopy is indicated for both diagnosis of SNA and therapeutic debridement of sinonasal fungal plaques. Biopsy specimens can be obtained for fungal culture and cytological/histological identification of hyphae. Biopsy specimens should be stored frozen for PCR if URTA is suspected but fungal culture is negative. Sinuscopy is indicated for diagnosis when CT findings indicate sinus involvement and fungal plaques are not visualised on rhinoscopy. While anatomical landmarks have been defined for sinus trephination in cats (Winstanley 1974), CT findings can provide more precise information.

For cats with SOA, retrobulbar masses can be biopsied via the oral cavity where there is pterygopalatine invasion (Barrs et al. 2012). Nasal, nasopharyngeal and paranasal lesions can also be biopsied, as indicated by CT findings (Smith and Hoffman 2010).

The fungal pathogens that cause feline URTA can usually be isolated from tissue biopsies or sinonasal fungal plaques on commercial culture media such as Sabouraud's dextrose agar or malt extract agar. Twenty-two of 23 cases of feline URTA were culture positive in one study (Barrs et al. 2012). Cryptic species that cause URTA including *A. felis*, *A. udagawae* and *A. lentulus* are usually unable to grow at 50 °C, in contrast to *A. fumigatus s. str.* These species also tend to have a slow-sporulating phenotype giving colonies a whitish appearance before sporulation in contrast to the grey-green rapidly sporulating colonies of *A. fumigatus s. str.*

Aspergillus viridinutans complex species have angled or "nodding" conidial heads. However, the prevalence of this morphological feature varies between species, and other non-related section *Fumigati* species including *A. brevipes*, *A. unilateralis* and *A. duricaulis* also share this feature (Novakova et al. 2014). Thus, molecular confirmation of species is essential for accurate identification of fungal isolates from cats with URTA, with the exception of *A. fumigatus*, where consistent morphological features and growth at 50 °C may be acceptable.

15.8 Treatment and Prognosis

Due to lack of prospective treatment trials, evidence-based treatment protocols are not currently available for SNA. Prognosis appears favourable based on the small numbers of cases for which treatment outcomes have been reported. Signs resolved in 11 of 15 treated cases with follow-up available (Goodall et al. 1984; Tomsa et al. 2003; Whitney et al. 2005; Furrow and Groman 2009; Barrs et al. 2012; Kano et al. 2015). Successful treatment regimens included systemic antifungal therapy in five cases (itraconazole or posaconazole monotherapy or combined with amphotericin B), systemic triazole therapy (itraconazole or posaconazole) combined with topical intranasal clotrimazole or enilconazole infusion in two cases and topical intranasal clotrimazole infusion alone in two cases. Similar to canine SNA (Zonderland et al. 2002), debridement of fungal lesions in the nasal cavity was an integral component of therapy for most cases of feline SNA (Barrs et al. 2012; Tomsa et al. 2003; Goodall et al. 1984; Furrow and Groman 2009). In humans with non-invasive fungal rhinosinusitis due to *A. fumigatus* infection, sinus fungal plaques can be extensive and form tangled masses of hyphae termed “fungal balls” (Montone et al. 2012). Aggressive endosurgical debridement is usually curative and post-operative or perioperative antifungal treatment is not warranted. In contrast to SNA of dogs and cats, sinonasal fungal balls in humans are not usually associated with nasal bone lysis on CT. However, other similarities of this disease between humans and animals highlight the importance of endoscopic debridement of all visible fungal elements in the therapeutic approach to non-invasive fungal rhinosinusitis (Dufour et al. 2006). A therapeutic strategy for feline SNA, based on previous reports, treatment of canine SNA and consideration of whether infection is invasive or non-invasive, is presented in Table 15.2. Techniques for intranasal clotrimazole infusion in cats are similar to those used for treatment of canine SNA with slight modifications (Furrow and Groman 2009; Tomsa et al. 2003; Peeters and Clercx 2007). Polypropylene glycol must not be used as the vehicle for clotrimazole as it can cause severe mucosal oedema and ulceration (Barr et al. 2010). Polyethylene glycol is a safe vehicle for clotrimazole infusion. As for canine SNA, it may be necessary to repeat endoscopic fungal plaque debridement and intranasal clotrimazole infusion on one or more occasions to effect a cure, and nasal discharge may persist where turbinate destruction is severe (Barrs et al. 2012).

Optimal evidence-based treatment protocols for feline SOA are yet to be identified. Prognosis is poor even with aggressive treatment including orbital exenteration combined with systemic antifungal therapy. In the largest case series of 12 cats with SOA for which treatment outcomes could be assessed, all cats were treated with

Table 15.2 Therapeutic approach for treatment of feline sinonasal aspergillosis

• Determine the identity of the fungal isolate and its antifungal susceptibility
• Assess whether infection is invasive or non-invasive based on histopathology and CT findings
• Determine the integrity of the cribriform plate on CT
• Debride all visible fungal plaques/lesions from the nasal cavity and frontal sinuses using endoscopic techniques and saline irrigation
• For non-invasive infections instil an intranasal infusion of 1% clotrimazole in polyethylene glycol (1 h soak under general anaesthesia)
• For invasive infections and/or where *A. felis* is identified, give additional systemic antifungal therapy (see SOA treatment)

systemic triazole therapy, and five also had orbital exenteration. Treatment was successful in only one case, which did not have exenteration. Relapse of infection occurred 19 months after treatment was stopped, and infection eventually resolved after treatment with caspofungin and posaconazole (Barrs et al. 2012). Of six other cases that responded to systemic antifungal therapy, three of these also had surgical debridement of orbital granulomas, and orbital tissues of one were lavaged at surgery with 1% voriconazole (Smith and Hoffman 2010; Hamilton et al. 2000; McLellan et al. 2006). In only one of these cases was resolution of infection confirmed by follow-up CT (McLellan et al. 2006), and another cat was euthanased 4 months after exenteration with likely progressive disease (Hamilton et al. 2000). Most cats with SOA that responded to systemic antifungal therapy were treated with posaconazole or itraconazole monotherapy or combined with terbinafine and/or amphotericin B for treatment intervals of 6 months or more (Barrs et al. 2012; Kano et al. 2013; McLellan et al. 2006; Hamilton et al. 2000). The importance of antifungal susceptibility testing in guiding therapy is illustrated by one case that failed sequential treatment with itraconazole and amphotericin B but was cured with posaconazole. The section *Fumigati* isolate from this cat, which was not identified molecularly, had high MICs of amphotericin and itraconazole and a low MIC of posaconazole (McLellan et al. 2006). High MICs of itraconazole and cross-resistance with voriconazole are not uncommon among *A. felis* and other cryptic species isolates that cause feline URTA. A recent investigation, in which the results of a reference broth microdilution antifungal susceptibility testing method (EUCAST) and a commonly used commercial broth microdilution method (Sensititre YeastOne, Trek Diagnostic System Ltd., East Grinstead, UK) were compared for 90 environmental and clinical isolates from the *Aspergillus viridinutans* complex, showed that MICs of itraconazole and voriconazole were high. Also, there was poor correlation between the two methods for itraconazole, with the commercial method frequently failing to detect high MICs of itraconazole (Lysokva et al. 2017). *A. felis* is usually susceptible to amphotericin B (Barrs et al. 2013). Posaconazole is well tolerated after oral administration, and MICs for *A. felis* are usually lower than voriconazole MICs (Barrs et al. 2013; Lysokva et al. 2017). Voriconazole administration has been associated with severe adverse effects in cats including paraplegia, blindness and anorexia (Quimby et al. 2010; Smith and Hoffman 2010; Barrs et al. 2012). In cats voriconazole has a long oral half-life (>43 h) and non-linear pharmacokinetics are suspected to be the cause of toxicosis at higher drug doses due to saturation of metabolising enzymes and decreased drug clearance (Vishkautsan et al. 2016). Caspofungin was well tolerated and efficacious in one cat with SOA that failed treatment with AMB and posaconazole (Barrs et al. 2012) although in another case treatment with micafungin was unsuccessful (Kano et al. 2008).

Serum GM is used to monitor fungal antigen load in humans with aspergillosis, and although it is only elevated in around 30% of cats with SOA, in cats that test positive, this application could be useful to help determine the therapeutic endpoint.

15.9 Conclusion

Feline upper respiratory aspergillosis represents a naturally occurring model of focal respiratory aspergillosis in an immunocompetent host, with both non-invasive and invasive forms of disease occurring. The pathogenic potential of fungi within *Aspergillus* section *Fumigati* has been realised for most species in the *Aspergillus viridinutans* complex, and these species have a propensity to cause chronic invasive disease in cats.

References

Adamama-Moraitou KK, Paitaki CG, Rallis TS, Tontis D (2001) *Aspergillus* species cystitis in a cat. J Feline Med Surg 3:31–34

Alcazar-Fuoli L, Mellado E, Aslastruey-Izquierdo A, Cuenca-Estrella M, Rodriguez-Tudela JL (2008) *Aspergillus* section *Fumigati*: antifungal susceptibility patterns and sequence-based identification. Antimicrob Agents Chemother 52(4):1244–1251

Balajee SA, Nickle D, Varga J, Marr KA (2006) Molecular studies reveal frequent misidentification of *Aspergillus fumigatus* by morphotyping. Eukaryot Cell 5(10):1705–1712

Barachetti L, Mortellaro CM, Di Giancamillo M, Giudice C et al (2009) Bilateral orbital and nasal aspergillosis in a cat. Vet Ophthalmol 12(3):176–182. https://doi.org/10.1111/j.1463-5224.2009.00695.x

Barr SC, Rishniw M, Lynch M (2010) Questions contents of clotrimazole solution. J Am Vet Med Assoc 236(2):163–164

Barrs VR, Talbot JJ (2014) Feline aspergillosis. Vet Clin North Am Small Anim Pract 44:51–73

Barrs VR, Beatty JA, Lingard AE, Malik R et al (2007) Feline sino-orbital aspergillosis: an emerging clinical syndrome. Aust Vet J 85(3):N23–N23

Barrs VR, Halliday C, Martin P, Wilson B et al (2012) Sinonasal and sino-orbital aspergillosis in 23 cats: aetiology, clinicopathological features and treatment outcomes. Vet J 191(1):58–64. https://doi.org/10.1016/j.tvjl.2011.02.009

Barrs VR, van Doorn T, Houbraken J, Kidd SE et al (2013) *Aspergillus felis* sp. nov., an emerging agent of invasive aspergillosis in humans, cats and dogs. PLoS One 8(6):e64871. https://doi.org/10.1371/journal.pone.0064871

Barrs VR, Dhand NK, Talbot JJ, Bell E, Abraham LA, Chapman P, Bennett S, Makara M (2014) Computed tomographic features of feline sino-nasal and sino-orbital aspergillosis. Vet J 201:215–222

Barrs VR, Ujvari B, Dhand NK, Peters IR et al (2015) Detection of *Aspergillus*-specific antibodies by agar gel double immunodiffusion and IgG ELISA in feline upper respiratory tract aspergillosis. Vet J 203(3):285–289. https://doi.org/10.1016/j.tvjl.2014.12.020

Billen F, Peeters D, Peters IR, Helps CR et al (2009) Comparison of the value of measurement of serum galactomannan and *Aspergillus*-specific antibodies in the diagnosis of canine sino-nasal aspergillosis. Vet Microbiol 133(4):358–365. https://doi.org/10.1016/j.vetmic.2008.07.018

Burk RL, Joseph R, Baer K (1990) Systemic aspergillosis in a cat. Vet Radiol 31(1):26–28

Davies C, Troy GC (1996) Deep mycotic infections in cats. J Am Anim Hosp Assoc 32(5):380–391

Day MJ (2009) Canine sino-nasal aspergillosis: parallels with human disease. Med Mycol 47:S315–S323. https://doi.org/10.1080/13693780802056038

Declercq J, Declercq L, Fincioen S (2012) Unilateral sino-orbital and subcutaneous aspergillosis in a cat. Vlaams Diergeneeskundig Tijdschrift 81(6):357–362

Dufour X, Kauffmann-Lacroix C, Ferrie JC, Goujon JM et al (2006) Paranasal sinus fungus ball: epidemiology, clinical features and diagnosis. A retrospective analysis of 173 cases from a single medical center in France, 1989–2002. Med Mycol 44(1):61–67. https://doi.org/10.1080/13693780500235728

Fox JG, Murphy JC, Shalev M (1978) Systemic fungal infections in cats. J Am Vet Med Assoc 173(9):1191–1195

Frisvad JC, Larsen TO (2016) Extrolites of *Aspergillus fumigatus* and other pathogenic species in *Aspergillus* section *Fumigati*. Front Microbiol 6:1485. https://doi.org/10.3389/fmicb.2015.01485

Furrow E, Groman RP (2009) Intranasal infusion of clotrimazole for the treatment of nasal aspergillosis in two cats. J Am Vet Med Assoc 235(10):1188–1193

Garcia RS, Wheat LJ, Cook AK, Kirsch EJ, Sykes JE (2012) Sensitivity and specificity of a blood and urine galactomannan antigen assay for diagnosis of systemic aspergillosis in dogs. J Vet Intern Med 26(4):911–919. https://doi.org/10.1111/j.1939-1676.2012.00935.x

Giordano C, Gianella P, Bo S, Vercelli A et al (2010) Invasive mould infections of the naso-orbital region of cats: a case involving *Aspergillus fumigatus* and an aetiological review. J Feline Med Surg 12(9):714–723. https://doi.org/10.1016/j.jfms.2010.07.015

Goodall SA, Lane JG, Warnock DW (1984) The diagnosis and treatment of a case of nasal aspergillosis in a cat. J Small Anim Pract 25(10):627–633

Gresnigt MS, Netea MG, van de Veerdonk FL (2012) Pattern recognition receptors and their role in invasive aspergillosis. Ann N Y Acad Sci 1273:60–67. https://doi.org/10.1111/j.1749-6632.2012.06759.x

Hachem RY, Kontoyiannais DP, Chemaly RF, Jiang Y et al (2009) Utility of galactomannan enzyme immunoassay and (1,3) B-D-glucan in diagnosis of invasive fungal infections: low sensitivity for *Aspergillus fumigatus* infection in hematologic malignancy patients. J Clin Microbiol 47(1):129–133

Halenda RM, Reed AL (1997) Ultrasound computed tomography diagnosis – fungal, sinusitis and retrobulbar myofascitis in a cat. Vet Radiol Ultrasound 38(3):208–210. https://doi.org/10.1111/j.1740-8261.1997.tb00842.x

Hamilton HL, Whitley RD, McLaughlin SA (2000) Exophthalmos secondary to aspergillosis in a cat. J Am Anim Hosp Assoc 36(4):343–347

Hawksworth DL, Crous PW, Redhead SA, Reynolds DR et al (2011) The Amsterdam declaration on fungal nomenclature. Mycotaxon 116:491–500. https://doi.org/10.5248/116.491

Hazell KLA, Swift IM, Sullivan N (2011) Successful treatment of pulmonary aspergillosis in a cat. Aust Vet J 89(3):101–104

Herbrecht R, Letscher-Bru V, Oprea C, Lioure B et al (2002) *Aspergillus* galactomannan detection in the diagnosis of invasive aspergillosis in cancer patients. J Clin Oncol 20(7):1898–1906. https://doi.org/10.1200/jco.2002.07.004

Hubka V, Kolarik M (2012) Beta-tubulin paralogue tubC is frequently misidentified as the benA gene in *Aspergillus* section *Nigri* taxonomy: primer specificity testing ad taxonomic consequences. Persoonia 29:1–10. https://doi.org/10.3767/003158512x658123

Kano R, Itamoto K, Okuda M, Inokuma H et al (2008) Isolation of *Aspergillus udagawae* from a fatal case of feline orbital aspergillosis. Mycoses 51(4):360–361. https://doi.org/10.1111/j.1439-0507.2008.01493.x

Kano R, Shibahashi A, Fujino Y, Sakai H et al (2013) Two cases of feline orbital aspergillosis due to *A. udagawae* and *A. viridinutans*. J Vet Med Sci 75:7–10

Kano R, Takahashi T, Hayakawa T, Yamaya Y et al (2015) The first case of feline sinonasal aspergillosis due to *Aspergillus fischeri* in Japan. J Vet Med Sci 77(9):1183–1185. https://doi.org/10.1292/jvms.14-0454

Karnik K, Reichle JK, Fischetti AJ, Goggin JM (2009) Computed tomographic findings of fungal rhinitis and sinusitis in cats. Vet Radiol Ultrasound 50(1):65–68. https://doi.org/10.1111/j.1740-8261.2008.01491.x

Katz ME, Dougall AM, Weeks K, Cheetham BF (2005) Multiple genetically distinct groups revealed among clinical isolates identified as atypical *Aspergillus fumigatus*. J Clin Microbiol 43(2):551–555. https://doi.org/10.1128/jcm.43.2.551-555.2005

Kitasato Y, Tao Y, Hoshino T, Tachibana K et al (2009) Comparison of *Aspergillus* galactomannan antigen testing with a new cut-off index and *Aspergillus* precipitating antibody testing for the diagnosis of chronic pulmonary aspergillosis. Respirology 14(5):701–708. https://doi.org/10.1111/j.1440-1843.2009.01548.x

Lanfear R, Calcott B, Ho SYW, Guindon S (2012) PartitionFinder: combined selection of partitioning schemes and substitution models for phylogenetic analyses. Mol Biol Evol 29:1695–1701

Lupianez CB, Canet LM, Carvalho A, Alcazar-Fuoli L et al (2016) Polymorphisms in host immunity-modulating genes and risk of invasive aspergillosis: results from the AspBIOmics consortium. Infect Immun 84(3):643–657. https://doi.org/10.1128/iai.01359-15

Lysokva P, Hubka V, Svobodova L et al (2017) In vitro activities of seven antifungal agents against opportunistic pathogens from the *Aspergillus viridinutans* complex, a cryptic species of A. fumigatus. Antimicrob Agents Chemother. In Review Jan 2018

Malik R, Vogelnest L, O'Brien CR, White J et al (2004) Infections and some other conditions affecting the skin and subcutis of the naso-ocular region of cats – clinical experience 1987–2003. J Feline Med Surg 6(6):383–390. https://doi.org/10.1016/j.jfms.2004.02.001

McLellan GJ, Aquino SM, Mason DR, Kinyon JM, Myers RK (2006) Use of posaconazole in the management of invasive orbital aspergillosis in a cat. J Am Anim Hosp Assoc 42(4):302–307

Meason-Smith C, Edwards EE, Older CE et al (2017) Panfungal polymerase chain reaction for identificatio nof fungal pathogens in formalin-fixed animal tissues. Vet Pathol 54(4):640–648. https://doi.org/10.1177/030098517698207

Mennink-Kersten M, Donnelly JP, Verweij PE (2004) Detection of circulating galactomannan for the diagnosis and management of invasive aspergillosis. Lancet Infect Dis 4(6):349–357. https://doi.org/10.1016/s1473-3099(04)01045-x

Mercier E, Peters IR, Day MJ, Clercx C, Peeters D (2012) Toll- and NOD-like receptor mRNA expression in canine sino-nasal aspergillosis and idiopathic lymphoplasmacytic rhinitis. Vet Immunol Immunopathol 145(3–4):618–624. https://doi.org/10.1016/j.vetimm.2012.01.009

Mercier E, Peters IR, Farnir F, Lavoue R et al (2014) Assessment of toll-like receptor 2, 4 and 9 SNP genotypes in canine sino-nasal aspergillosis. BMC Vet Res 10:187. https://doi.org/10.1186/s12917-014-0187-6

Miller JS, Funk VA, Wagner WL, Barrie F et al (2011) Outcomes of the 2011 botanical nomenclature section at the XVIII international botanical congress. Phytokeys 5:1–3

Miller RI (2010) Nodular granulomatous fungal skin diseases of cats in the United Kingdom: a retrospective review. Vet Dermatol 21(2):130–135. https://doi.org/10.1111/j.1365-3164.2009.00801.x

Montone KA, Livolsi VA, Feldman MD, Palmer J et al (2012) Fungal rhinosinusitis: a retrospective microbiologic and pathologic review of 400 patients at a single university medical center. Int J Otolaryngol 2012:1–9. https://doi.org/10.1155/2012/684835

Nguyen LT, Schmidt HA, von Haeseler A, Minh BQ (2015) IQ-TREE: a fast and effective stochastic algorithm for estimating maximum-likelihood phylogenies. Mol Biol Evol 32:268–274

Novakova A, Hubka V, Dudova Z, Matsuzawa T et al (2014) New species in *Aspergillus* section *Fumigati* from reclamation sites in Wyoming (USA) and revision of *A. viridinutans* complex. Fungal Divers 64(1):253–274. https://doi.org/10.1007/s13225-013-0262-5

O'Gorman CM, Fuller HT, Dyer PS (2009) Discovery of a sexual cycle in the opportunistic fungal pathogen *Aspergillus fumigatus*. Nature 457:471–474

Ossent P (1987) Systemic aspergillosis and mucormycosis in 23 cats. Vet Rec 120(14):330–333

Page ID, Richardson M, Denning DW (2015) Antibody testing in aspergillosis—quo vadis? Med Mycol 53(5):417–439

Pakes SP, New AE, Benbrook SC (1967) Pulmonary aspergillosis in a cat. J Am Vet Med Assoc 151(7):950–953

Peeters D, Clercx C (2007) Update on canine sinonasal aspergillosis. Vet Clin North Am Small Anim Pract 37(5):901–916

Pfeiffer CD, Fine JP, Safdar N (2006) Diagnosis of invasive aspergillosis using a galactomannan assay: a meta-analysis. Clin Infect Dis 42(10):1417–1427. https://doi.org/10.1086/503427

Pitt JI, Samson RA (2007) Nomenclatural considerations in naming species of *Aspergillus* and its teleomorphs. Stud Mycol 59:67–70. https://doi.org/10.3114/sim.2007.59.08

Quimby JM, Hoffman SB, Duke J, Lappin MR (2010) Adverse neurologic events associated with voriconazole use in 3 cats. J Vet Intern Med 24(3):647–649. https://doi.org/10.1111/j.1939-1676.2010.0504.x

Romani L (2011) Immunity to fungal infections. Nat Rev Immunol 11(4):275–288. https://doi.org/10.1038/nri2939

Samson RA, Visagie CM, Houbraken J, Hong SB et al (2014) Phylogeny, identification and nomenclature of the genus *Aspergillus*. Stud Mycol 78:141–173. https://doi.org/10.1016/j.simyco.2014.07.004

Schlueter C, Budras KD, Ludewig E, Mayrhofer E et al (2009) Brachycephalic feline noses CT and anatomical study of the relationship between head conformation and the nasolacrimal drainage system. J Feline Med Surg 11(11):891–900. https://doi.org/10.1016/j.jfms.2009.09.010

Segal BH (2009) Medical progress aspergillosis. N Engl J Med 360(18):1870–1884. https://doi.org/10.1056/NEJMra0808853

Seyedmousavi S, Guillot J, Arné P, de Hoog GS et al (2015) *Aspergillus* and aspergilloses in wild and domestic animals: a global health concern with parallels to human disease. Med Mycol 53:765–797

Sharp NJH, Harvey CE, Sullivan M (1991) Canine nasal aspergillosis and penicilliosis. Compend Contin Educ Pract Vet 13(1):41–46

Siddiqui AA, Shah AA, Bashir SH (2004) Craniocerebral aspergillosis of sinonasal origin in immunocompetent patients: clinical spectrum and outcome in 25 cases. J Neurosurg 55(3):602–611. https://doi.org/10.1227/01.neu.0000134597.94269.48

Smith LN, Hoffman SB (2010) A case series of unilateral orbital aspergillosis in three cats and treatment with voriconazole. Vet Ophthalmol 13(3):190–203

Stokes R (1973) Letter: intestinal mycosis in a cat. Aust Vet J 49(10):499–500

Sugui JA, Peterson SW, Flgat A, Hansen B et al (2014) Genetic relatedness versus biological compatibility between *Aspergillus fumigatus* and related species. J Clin Microbiol 52(10):3707–3721. https://doi.org/10.1128/JCM.01704-14

Talavera G, Castresana J (2007) Improvement of phylogenies after removing divergent and ambiguously aligned blocks from protein sequence alignments. Syst Biol 56:564–577

Talbot JJ, Johnson LR, Martin P, Beatty JA et al (2014) What causes sino-nasal aspergillosis in dogs? A molecular approach to species identification. Vet J 200:17–21

Talbot JJ, Houbraken J, Frisvad JC, Samson RA et al (2017) Discovery of *Aspergillus frankstonensis* sp. nov. during environmental sampling for animal and human fungal pathogens. Public Libr Sci 12(8):e0181660. https://doi.org/10.1371/journal.pone.0181660

Taylor A, Peters I, Dhand NK, Whitney J et al (2016) Evaluation of serum *Aspergillus*-specific immunoglobulin A by indirect ELISA for diagnosis of feline upper respiratory tract aspergillosis. J Vet Intern Med 30:1708–1714. https://doi.org/10.1111/jvim.14567

Tomsa K, Glaus TA, Zimmer C, Greene CE (2003) Fungal rhinitis and sinusitis in three cats. J Am Vet Med Assoc 222(10):1380–1384

Vishkautsan P, Papich MG, Thompson GR, Sykes JE (2016) Pharamacokinetics of voriconazole after intravenous and oral administration to healthy cats. Am J Vet Res 77(9):931–939

Whitney BL, Broussard J, Stefanacci JD (2005) Four cats with fungal rhinitis. J Feline Med Surg 7(1):53–58. https://doi.org/10.1016/j.jfms.2004.02.004

Whitney J, Beatty JA, Dhand N, Briscoe K, Barrs VR (2013) Evaluation of serum galactomannan detection for the diagnosis of feline upper respiratory tract aspergillosis. Vet Microbiol 162(1):180–185. https://doi.org/10.1016/j.vetmic.2012.09.002

Wilkinson GT, Sutton RH, Grono LR (1982) *Aspergillus* spp infection associated with orbital cellulitis and sinusitis in a cat. J Small Anim Pract 23(3):127–131

Winstanley EW (1974) Trephining frontal sinuses in the treatment of rhinitis and sinusitis in the cat. Vet Rec 95:289–292

Zonderland JL, Stork CK, Saunders JH, Hamaide AJ, Balligand MH, Clercx CM (2002) Intranasal infusion of enilconazole for treatment of sinonasal aspergillosis in dogs. J Am Vet Med Assoc 221(10):1421–1425

Part IV

Antifungals and Risk of Resistance

Antifungal Use in Veterinary Practice and Emergence of Resistance

16

Seyedmojtaba Seyedmousavi, Nathan P. Wiederhold, Frank Ebel, Mohammad T. Hedayati, Haleh Rafati, and Paul E. Verweij

Abstract

Invasive fungal infections can cause significant morbidity and mortality in humans and different animal species, worldwide. Antifungal therapy remains a central component of protecting human and vertebrate animals against fungal infections. Depending on the strategy chosen, topical and/or systemic drugs can be used based on the clinical picture of the host and mycological identification of

S. Seyedmousavi (✉)
Laboratory of Clinical Immunology and Microbiology (LCIM), National Institute of Allergy and Infectious Diseases (NIAID), National Institutes of Health (NIH), Bethesda, MD, USA
e-mail: Seyedmousavi@nih.gov; S.Seyedmousavi@gmail.com

N. P. Wiederhold
Fungus Testing Laboratory, Department of Pathology and Laboratory Medicine, University of Texas Health Science Center at San Antonio, San Antonio, TX, USA
e-mail: wiederholdn@uthscsa.edu

F. Ebel
Institute for Infectious Diseases and Zoonoses, Munich, Germany
e-mail: Frank.Ebel@lmu.de

M. T. Hedayati
Department of Medical Mycology and Parasitology, Invasive Fungi Research Center, School of Medicine, Mazandaran University of Medical Sciences, Sari, Iran
e-mail: hedayatimt@gmail.com

H. Rafati
Department of Microbiology and Immunology, Center of Expertise in Microbiology, Infection Biology and Antimicrobial Pharmacology, Tehran, Iran
e-mail: halehrafati@gmail.com

P. E. Verweij
Department of Medical Microbiology, Radboud University Medical Center, Nijmegen, The Netherlands

Center of Expertise in Mycology RadboudUMC/CWZ, Nijmegen, The Netherlands
e-mail: Paul.Verweij@radboudumc.nl

S. Seyedmousavi et al. (eds.), *Emerging and Epizootic Fungal Infections in Animals*,
https://doi.org/10.1007/978-3-319-72093-7_16

the etiologic agent. For effective treatment, it is important to correctly identify the causative agents at the species level, which will enable administration of suitable therapeutics and initiation of appropriate therapeutic modalities. In addition, the management of fungal infections in animals usually includes systemic or topical treatment of the animal and environmental decontamination if necessary. Only a few products are licensed for animals, and, as a consequence, off-label use of the drugs approved for use in humans is quite common.

This chapter focuses on the topical and systemic antifungal agents currently in use in veterinary practice. The therapeutic uses that have proved successfully in various animal species are summarized. This chapter also summarizes the currently available evidence for the emergence of resistance against these agents.

16.1 Introduction

Most fungi that cause disease in humans can also cause invasive infections in different animal species and may be associated with significant morbidity and mortality (Fisher et al. 2012; Kohler et al. 2015). In addition, infections of the hair, skin, and nails, allergy, and mycotoxicoses may also occur in animals (Yanong 2003; Seyedmousavi et al. 2013a, 2015a), although often caused by related but nonidentical species.

Antifungal therapy remains a central component of protecting human and animals against fungal infections (Rochette et al. 2003). An ideal antifungal agent is a drug that selectively destroys fungal pathogens with minimal side effects to the host (Seyedmousavi et al. 2017). Depending on the strategy chosen, topical and/or systemic antifungal drugs can be used based on the clinical picture and the species identification of the etiologic agent (Seyedmousavi et al. 2017).

Despite recent advances in antifungal pharmacology, therapeutic options against fungal infections in animals are limited, because (a) very few options are available in the existing chemical drug classes with antifungal activity (Kathiravan et al. 2012) and (b) only a few products are licensed for animals. However, off-label use of antifungals is quite common, and many of the antifungal agents that are used in humans are also used in animals (Foy and Trepanier 2010). These drugs include the polyenes (e.g., amphotericin B and nystatin), the azoles including both the imidazoles and triazoles, the allylamines (e.g., terbinafine), the echinocandins and the nucleoside analogs. Many of the limitations observed for these drugs in humans also occur in some animal species, including variable pharmacokinetics, adverse effects, and drug interactions (Lewis 2011). However, differences for these agents are also observed between humans and animals. Resistance to antifungals can also occur in different animal species that receive these drugs (Beernaert et al. 2009a; Wang et al. 2013; Singer et al. 2014; Ziolkowska et al. 2014; Cafarchia et al. 2015), although the true epidemiology of antifungal resistance in animals is unknown, and options to treat infections by resistant fungi are limited.

The present chapter discusses the currently available classes and representatives of systemic antifungal drugs used in both human and animals. The development and epidemiology of antifungal resistance in animals are also reviewed.

16.2 Systemic Antifungal Drugs

Systemic antifungals that are currently licensed for humans and used in animals for the treatment of invasive fungal infections (IFIs) can be grouped into four classes based on their site of action: polyenes, azoles, echinocandins, and nucleoside analogs (Fig. 16.1) (Groll et al. 1998; Dodds Ashley et al. 2006). Figure 16.2 shows the development timeline for systemic antifungals from the 1950s to present.

A wide variety of topical agents belonging to different classes of antifungals are available as creams, ointments, gels, lotions, powders, shampoos, and other formulations. Topical antifungals can be applied to the skin, hairs, nails, or mucosa to kill or inactivate fungi. Regardless of the actual mechanism of action of the drug or the viscosity, hydrophobicity, and acidity of the formulation, the drug's ability to penetrate or permeate deeper skin layers is an important property impacting the therapeutic efficacy of topical antifungals (Durdu et al. 2017). Table 16.1 summarizes the uses of various antifungals that have proved successfully in various animal species.

16.2.1 Polyenes

Polyenes constitute the oldest class of systemic antifungal drugs. More than 200 polyene macrolides have antifungal activity, and most are produced by the

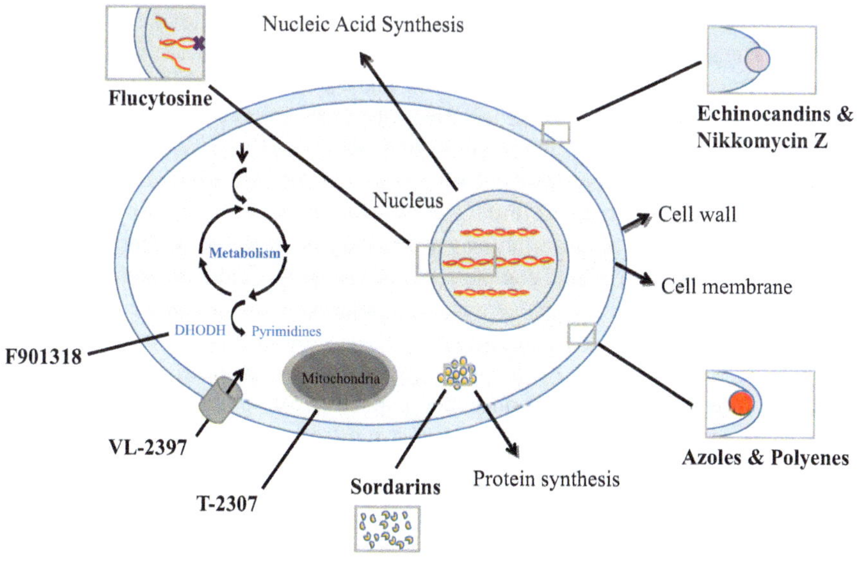

Fig. 16.1 Targets of systemic antifungal agents. F901318 is the leading representative of a novel class of drug, the orotomide, that inhibits dihydroorotate dehydrogenase (DHODH). Nikkomycin Z is an experimental chitin synthase inhibitor currently under development. T-2307 is an investigational arylamidine structurally similar to a class of aromatic diamidines that causes collapse of mitochondrial membrane potential in yeasts. Sordarins derivatives represent a novel class of naturally occurring and semisynthetic products that inhibit fungal protein synthesis through their interaction with the elongation factor 2 (EF2) in the ribosome. VL-2397 (formerly ASP2397) is a novel second generation echinocandin. Active transport of this agent into fungal cells occurs via siderophore iron transporter 1

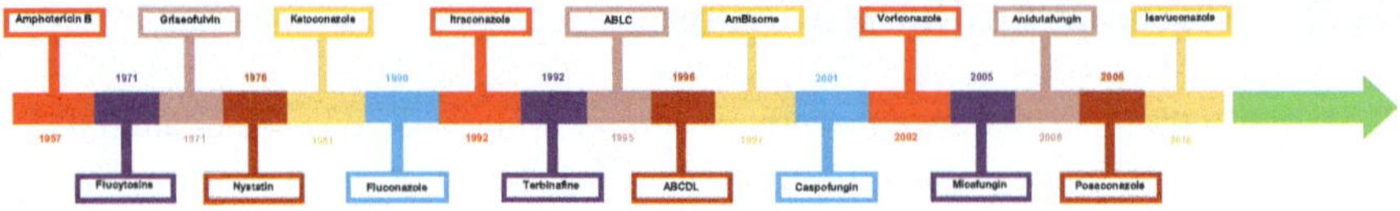

Fig. 16.2 Timeline of development of systemic antifungals from the 1950s to present

soil actinomycete *Streptomyces*. Polyenes bind to ergosterol, the main component of fungal membrane sterols (de Kruijff et al. 1974; Gray et al. 2012). This interaction results in the formation of transmembrane pores, which disrupt cell membrane integrity and results in rapid cellular damage or death (Bolard 1986).

16.2.1.1 Amphotericin B

Amphotericin B is a polyene antifungal that contains a macrolide lactone ring with a series of conjugated double bonds. Its antifungal activity was discovered in 1953, and it was approved for clinical use in the USA in 1957 (Dutcher 1968). The chemical structure of amphotericin B deoxycholate is shown in Fig. 16.3. Amphotericin B is insoluble in water but forms soluble salts under both acidic and basic conditions (Gallis et al. 1990). In the current clinical formulation for humans, it is available for parenteral administration as a colloidal suspension using sodium deoxycholate as a dispersing agent and sodium phosphate as a buffer. The primary drawbacks of amphotericin B deoxycholate use are its dose-limited toxicity and significant side effects in various patients. The significant toxicities of amphotericin B include infusion-related adverse effects (fever, chills, arrhythmia, hypotension, and respiratory distress), nephrotoxicity, neurotoxicity, hematological side effects, and allergic reactions (Hamill 2013). Overall, two approaches have been used to improve the clinical response to amphotericin B in humans: the development of less toxic preparations in the 1990s and direct delivery of amphotericin B to target organs (intranasal, aerosolized, intracavitary, and intraperitoneal administration). There are three available lipid preparations of amphotericin B, which are amphotericin B lipid complex (ABLC), amphotericin B colloidal dispersion (ABCD), and liposomal amphotericin B (AmBisome) (Ostrosky-Zeichner et al. 2003a). However, these products are quite expensive and not available in some regions. In addition, the nephrotoxicity observed with the lipid formulations is reduced compared to amphotericin B deoxycholate, but not eliminated.

In vitro, amphotericin B is active against most common pathogenic yeasts that cause disseminated mycoses in humans and animals. *Cryptococcus* (Barchiesi et al. 1994) and *Candida* spp. (Ostrosky-Zeichner et al. 2003b; Pfaller et al. 2004) are quite sensitive, except *Candida lusitaniae* (Hadfield et al. 1987) and *Trichosporon beigelii* (Walsh et al. 1990), which show decreased susceptibility. *Aspergillus* spp. are sensitive, except *A. terreus* (Sutton et al. 1999). Most members of the order *Mucorales* (formerly indicated as *Zygomycetes*) are susceptible (Eng et al. 1981). Dimorphic fungi such as *Blastomyces dermatitidis*, *Paracoccidioides brasiliensis*, *Histoplasma capsulatum*, and *Coccidioides* spp. (*C. immitis* and *C. posadasii*) are also sensitive (Collins and Pappagianis 1977).

Table 16.1 Recommended indications of antifungals in veterinary practice

Drug	Animal species	Indications	Recommended dosages
Amphotericin B	Birds	Aspergillosis, candidiasis	Conventional AmB: IV, 1.5 mg/kg, q8 h, 3–5 days
			Nebulization: 15 min, 1 mg/kg q24 h, 10–14 days
	Dogs	Aspergillosis, cryptococcosis, blastomycosis, histoplasmosis, coccidioidomycosis, mucormycosis	Conventional AmB: 0.5 mg/kg IV q48 (slow infusion) to a cumulative dose of 4–8 mg/kg
			Liposomal AmB: 3 mg/kg/day IV, at a rate of more than 90–120 mg/kg, 3 times a week, up until 12 treatments
	Cats	Aspergillosis, cryptococcosis, blastomycosis, histoplasmosis, coccidioidomycosis, mucormycosis	Conventional AmB: 0.25 mg/kg IV q48 (slow infusion) to a cumulative dose of 4–8 mg/kg
			Liposomal AmB: 1 mg/kg/day IV, at a rate of more than 90–120 mg/kg, 3 times a week, up until 12 treatments
	Horses	Aspergillosis, candidiasis, histoplasmosis, coccidioidomycosis, sporotrichosis, mucormycosis	Conventional AmB: 0.3 mg/kg IV for 3 consecutive days, and repeat after 24–48 h drug-free interval, long-term treatment needed
Nystatin	Birds	Candidiasis of the gastrointestinal tract	100,000–300,000 IU/kg, q12 h, 7–10 days, with antibiotic therapy
Terbinafine	Dogs	Cryptococcosis, sporotrichosis, dermatophytosis, and *Malassezia* dermatitis	10 mg/kg daily, if resistance to azole
	Cats	Cryptococcosis, sporotrichosis, dermatophytosis	10 mg/kg daily, if resistance to azole
Ketoconazole	Birds	Aspergillosis, candidiasis	30 mg/kg, q 12 h, 14–30 days
	Dogs	Blastomycosis, histoplasmosis, cryptococcosis, coccidioidomycosis, sporotrichosis, *Malassezia* dermatitis, and dermatophytosis	10 mg/kg PO q 12 h for 3–6 months
	Cats	Blastomycosis, histoplasmosis, cryptococcosis, coccidioidomycosis, sporotrichosis, dermatophytosis	10 mg/kg PO q 12 h for 3–6 months
Parconazole	Birds (guinea fowl)	Candidiasis (thrush)	Prophylaxis: 30 mg/kg feed Treatment: 60 mg/kg feed for 7–10 days

(continued)

Table 16.1 (continued)

Drug	Animal species	Indications	Recommended dosages
Fluconazole	Birds	Candidiasis	2–5 mg/kg, q 24 h, 7–10 days
	Dogs	Cryptococcosis, blastomycosis, aspergillosis (nasal)	
	Cats	Aspergillosis (CNS infection), cryptococcosis, blastomycosis, coccidioidomycosis	1.25–10 mg/kg/day q12 h, 50 mg/cat q12 h, for 3–6 months
Itraconazole	Birds	Aspergillosis, candidiasis	Treatment: 5–15 mg/kg, q12 h with food for 7–21 days, or 10 mg/kg q24 h for 3 weeks
			Prevention: 10 mg/kg q24 h for 10 days, 20 mg/kg q24 h, or 15–25 mg/kg/day, for 1 week
	Dogs	Aspergillosis, blastomycosis, histoplasmosis, cryptococcosis, coccidioidomycosis, sporotrichosis, dermatophytosis, and *Malassezia* dermatitis	2.5 mg/kg q12 h or 5 mg/kg q24 h PO (give with food), for 15–30 days
	Cats	Dermatophytosis	5 mg/kg on a week on/week off basis (5 weeks)
		Aspergillosis, sporotrichosis, cryptococcosis, blastomycosis, histoplasmosis, phaeohyphomycosis	2.5 mg/kg q12 h, 5 mg/kg q24 h, or 50–100 mg/cat PO (give with food), for 15–30 days
	Horses	Aspergillosis, coccidioidomycosis, mycotic keratitis, dermatophytosis	2.5 mg/kg q12 or 5 mg/kg q24 h PO
	Rodents, rabbits, and fur animals	Dermatophytosis	5 mg/kg q24 h
Voriconazole	Birds	Aspergillosis	10–18 mg/kg q12 h
	Dogs	Aspergillosis, scedosporiosis	4–5 mg/kg q12 h PO
	Cats	Aspergillosis	4–5 mg/kg q12 h PO
	Horses	Aspergillosis (systemic), *Aspergillus* keratitis	2–4 mg/kg q24 h or 3 mg/kg q24 h PO, topical voriconazole solution for keratitis and intracorneal administration

Posaconazole	Dogs	Aspergillosis, mucormycosis	5–10 mg/kg q12–24 h
	Cats	Aspergillosis, mucormycosis	5 mg/kg q24 h
Flucytosine	Cats	Cryptococcosis	25–50 mg/kg PO QID in combination with amphotericin B. Do not use as single treatment
Griseofulvin	Dogs	Dermatophytosis	25–50 mg/kg, q12–24 h, 4–6 weeks
	Cats	Dermatophytosis	25–50 mg/kg, q12–24 h, 4–6 weeks
	Horses	Dermatophytosis, sporotrichosis	Foal: 15 mg/kg/day for 2–4 weeks Pony: 10 mg/kg/day for 1–3 weeks
	Ruminants	Dermatophytosis	10 mg/kg body weight for duration of 7 days in mild infections and 2–3 weeks in severe cases
	Rodents, rabbits, and fur animals	Dermatophytosis	25–50 mg/kg for 3–4 weeks
Clotrimazole	Birds (raptors)	Aspergillosis	Nebulization: 45 min q24 h
	Dogs	Aspergillosis, dermatophytosis, and *Malassezia* dermatitis	Single or multiple intranasal local instillation, topical treatment
	Cats	Aspergillosis, dermatophytosis	Single or multiple intranasal local instillation, topical treatment
	Rodents, rabbits, and fur animals	Dermatophytosis	Rubbing powder in the fur or treating house/nest and sandbox
Miconazole	Birds	Aspergillosis	45 min/day in raptors
	Dogs	*Malassezia* dermatitis	Topical treatment (in association with chlorhexidine)
	Cats	Dermatophytosis, *Malassezia* dermatitis	Topical treatment (in association with chlorhexidine)
	Rodents, rabbits, and fur animals	Dermatophytosis	Rubbing powder in the fur or treating house/nest and sandbox
Enilconazole	Birds	Aspergillosis	Nebulization: 0.1 ml/kg for 30 min q24 h (5 days on/2 days off)
		Disinfection (*Aspergillus* and other pathogenic fungi)	Disinfection of environment: Flush with solutions as recommended for use in poultry houses
	Dogs	Dermatophytosis, *Malassezia* dermatitis	Topical treatment (0.2% solution twice a week)
		Aspergillosis	10 mg/kg q12 h instilled into nasal sinus for 14 days (10% solution diluted 50/50 with water)

(continued)

Table 16.1 (continued)

Drug	Animal species	Indications	Recommended dosages
	Cats	Dermatophytosis, *Malassezia* dermatitis	Topical treatment (0.2% solution twice a week)
		Aspergillosis	10 mg/kg q12 h instilled into nasal sinus for 14 days (10% solution diluted 50/50 with water)
	Horses	Dermatophytosis	Topical treatment (0.2% solution twice a week)
		Disinfection (dermatophytes and other pathogenic fungi)	Disinfection of environment with spray or smoke generator
	Ruminants	Dermatophytosis	Topical treatment (0.2% solution twice a week)
		Disinfection (dermatophytes and other pathogenic fungi)	Disinfection of environment with spray or smoke generator
	Rodents, rabbits, and fur animals	Dermatophytosis	Topical treatment (0.2% solution twice a week)
		Disinfection (dermatophytes and other pathogenic fungi)	Disinfection of environment with spray or smoke generator
Natamycin	Horses	Dermatophytosis	Wash lesions with topical suspension 2 times with interval of 4–5 days and repeat after 2 weeks
	Ruminants	Dermatophytosis	Wash lesions with topical suspension 2 times with interval of 4–5 days and repeat after 2 weeks
Thiabendazole	Birds	Disinfection	Use smoke generator
	Horses	Dermatophytosis	Wash/spray lesions with topical solution or use ointment for 2 weeks
	Ruminants	Dermatophytosis	Wash/spray lesions with topical solution or use ointment for 2 weeks
	Rodents, rabbits, and fur animals	Dermatophytosis	25–50 mg/kg for 3–4 weeks

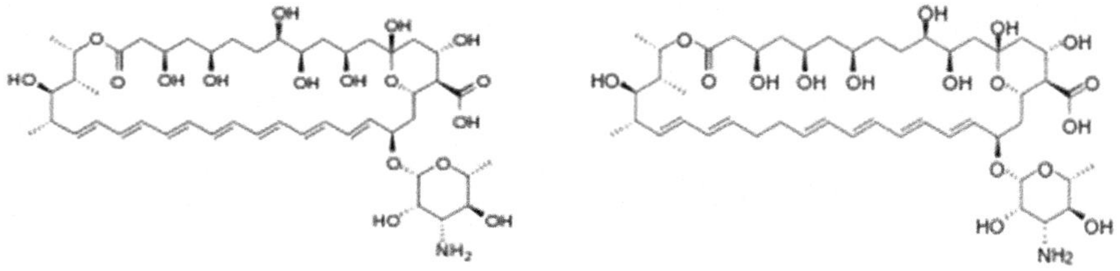

Fig. 16.3 Chemical structures of amphotericin B (left) and nystatin (right)

In humans, amphotericin B in combination with 5-flucytosine is recommended for the treatment of cryptococcal meningitis (Perfect et al. 2010; Day et al. 2013), while monotherapy is effective against a wide range of life-threatening IFIs such as blastomycosis, systemic candidiasis, coccidioidomycosis, histoplasmosis, mucormycosis, and sporotrichosis. The current clinical guidelines recommend lipid formulations of amphotericin B for the treatment of aspergillosis as an alternative to voriconazole or as salvage therapy in patients who are refractory or intolerant to other antifungal therapies (Walsh et al. 2008).

None of the amphotericin B formulations are licensed for veterinary use. However, the off-label use of amphotericin B has been recommended for similar systemic fungal infections in animals (Foy and Trepanier 2010), because many of the newer drugs available for humans are cost-prohibitive in veterinary settings. Successful therapeutic response with amphotericin B has been reported in blastomycosis in dogs (Krawiec et al. 1996), coccidioidomycosis in dogs and cats (Graupmann-Kuzma et al. 2008), and aspergillosis in dogs (Schultz et al. 2008). Equine endometrial candidiasis was also successfully treated with amphotericin B (Brook 1982). In addition, nasopharyngeal conidiobolomycosis can be treated successfully with intralesional injection of amphotericin B in combination with administration of sodium iodide and potassium iodide, but there is a possibility of recrudescence of infection (Zamos et al. 1996).

16.2.1.2 Nystatin

Structurally, nystatin is an amphoteric tetraene originally isolated from *Streptomyces noursei* (Hazen and Brown 1951). It is a polyene antifungal agent, which was first approved by the US Food and Drug Administration (FDA) in 1955 for the treatment of vaginal candidiasis. Nystatin is not absorbed by intact mucosal surfaces and following oral administration, it is passed unchanged in the feces; therefore, it is only active against yeasts present in the gastrointestinal tract (Hofstra et al. 1979). Oral or topical nystatin is well-tolerated; however, patients with renal insufficiency receiving oral therapy with conventional dosages may experience toxicity occasionally. Of note, an investigational lipid formulation of nystatin, liposomal nystatin, showed slightly reduced incidence of toxicity (Semis et al. 2012) with expanded antifungal activity against molds (Oakley et al. 1999; Arikan 2002).

The spectrum of activity of nystatin includes *Candida* spp., *Cryptococcus neoformans* (Bergan and Vangdal 1983), *Trichosporon* spp., and *Rhodotorula* spp. (Hussain Qadri et al. 1986; Pfaller and Diekema 2004). Most dimorphic fungi, such as *B. dermatitidis*, *Paracoccidioides brasiliensis*, *Coccidioides* spp., and

Histoplasma capsulatum, are sensitive to nystatin; however, it is inactive against dermatophytes and most *Aspergillus* spp. (Arikan 2002).

In humans, topical and oral nystatin are recommended for the treatment of superficial *Candida* spp. infections of the skin, oral cavity, and esophagus including diaper dermatitis, angular cheilitis (Rezabek and Friedman 1992), and oral or vaginal candidiasis (Pappas et al. 2009). Oral nystatin has also been used for the prevention of systemic candidiasis in patients who are specifically at risk, such as those with hematologic malignancies and those undergoing induction of chemotherapy (Pappas et al. 2009). However, the response has often been disappointing. Furthermore, it is noteworthy that currently available oral azoles have been found to be more effective (Hope et al. 2012; Ullmann et al. 2012).

In veterinary practice, nystatin is licensed for use in dogs and cats as an ingredient in otic preparations for the treatment of *Malassezia* otitis (Nesbitt and Fox 1981). Nystatin can also be used for treatment of intestinal candidiasis. In birds, nystatin has been used for the treatment of candidiasis of the crop and/or gastrointestinal tract. Regurgitation following nystatin administration may be the result of taste, not toxicity (Orosz and Frazier 1995).

16.2.1.3 Terbinafine

Terbinafine is an allylamine antifungal agent that has largely replaced the use of griseofulvin for the treatment of dermatophytic infections and onychomycosis in humans (Shear et al. 1991). Its antifungal activity is mediated via the noncompetitive inhibition of squalene epoxidase (SE), an enzyme that acts on its substrate squalene, an early intermediate in the fungal ergosterol biosynthesis pathway (Favre and Ryder 1996; Krishnan-Natesan 2009). Notably, terbinafine inhibits the enzymatic activity of fungal SE at a very low concentration (noncompetitive inhibition) than that required to inhibit the mammalian counterpart (4000-fold higher concentration needed; competitive inhibition) (Ryder 1992).

This drug is well absorbed from the gastrointestinal tract and then rapidly diffuses from the bloodstream into several skin tissue compartments, including the dermis and epidermis. In addition, terbinafine can remain in the stratum corneum and nails for several months after terminating the medication, even after very short-term therapy in humans (Jensen 1989; Faergemann et al. 1994). Moreover, terbinafine is highly lipophilic and is highly (>99%) protein-bound in human plasma, which impairs its distribution to the brain and cerebrospinal fluid and leads to high concentration in the hair follicles, skin, nail plate, and adipose tissue (Faergemann et al. 1994).

Overall, terbinafine has a broad spectrum with potent activity against dermatophytes, *Aspergillus* spp., *Sporothrix* spp., and *Malassezia* yeasts, although its potency against *Candida* species may be reduced (Petranyi et al. 1987; Krishnan-Natesan 2009). Terbinafine has been widely reported to elicit a strong clinical response against *Trichophyton* species, with cure rates reaching >80% (Deng et al. 2011; Grover et al. 2012). In some instances, terbinafine is combined with other antifungals for the treatment of infections caused by highly resistance fungi due to reports of in vitro synergy (Gomez-Lopez et al. 2003a; Cuenca-Estrella et al.

2006; Cordoba et al. 2008). Terbinafine has demonstrated a good toxicity profile at the recommended dosage. Most of the reported side effects are generally limited to gastrointestinal upset and, rarely, hepatotoxicity (Hall et al. 1997; Jaiswal et al. 2007).

In humans, its use is limited to the treatment of dermatophyte infections due to high concentrations in the nails and stratum corneum that persist for long periods, while lower concentrations are found in the plasma (Faergemann et al. 1993). In contrast, one pharmacokinetic study reported elevated concentrations in the plasma and deep tissues of raptors, including red-tailed hawks (Bechert et al. 2010). In humans, terbinafine has been reported to accumulate in the lungs following high-dose administration over time (Dolton et al. 2014).

Reports of terbinafine use in dogs and cats are sparse and limited primarily to the treatment of *Malassezia* and dermatophyte infections (Rosales et al. 2005). Terbinafine was shown to be well-tolerated and effective for the treatment of cutaneous or lymphocutaneous sporotrichosis (Chapman et al. 2004). In dogs, terbinafine seems to be equivalent or superior to ketoconazole for the treatment of *Malassezia* dermatitis showing reduction in both yeast counts and pruritus with little evidence of acquired resistance during treatment (Hofbauer et al. 2002; Guillot et al. 2003; Rosales et al. 2005). Higher doses of terbinafine (30–40 mg/kg once daily over 2 weeks period) are required for treatment of dermatophytosis in cats (Kotnik et al. 2001, Kotnik 2002). Recently, an otic formulation containing terbinafine has been approved for use in dogs with *Malassezia* otitis. With this formulation, a weekly administration is sufficient.

16.2.2 Azoles

Overall, the azoles are the most widely used class of antifungal drugs in humans (Lass-Florl 2011) and in animals for the treatment and prophylaxis against deep fungal infections. Azoles are cyclic organic molecules characterized by a core five-member azole ring, which can be divided into two groups based on the number of nitrogen atoms in the azole ring. These are the imidazoles and triazoles, which contain two and three nitrogen atoms, respectively, within the azole ring (Maertens 2004).

Ketoconazole belongs to the imidazoles (Maertens 2004), whereas fluconazole, itraconazole, voriconazole, posaconazole, and isavuconazole are prominent triazoles. The chemical structures of azoles used for systemic antifungal therapy are shown in Fig. 16.4. The azoles inhibit the synthesis of ergosterol from lanosterol in the fungal cell membrane by binding of the free nitrogen atom of the azole ring to the iron atom of the heme group of a fungal enzyme (Groll et al. 2003; Mohr et al. 2008). Their target enzyme is the cytochrome P450 CYP-dependent 14-α-demethylase (CYP51 or Erg11p), which catalyzes the targeted synthetic reaction. The inhibition depletes ergosterol, and methylated sterols accumulate in the cell

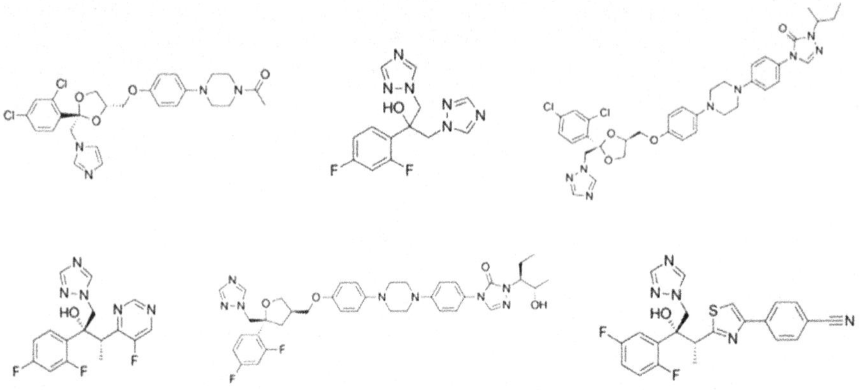

Fig. 16.4 Chemical structures of ketoconazole (top-left), fluconazole (top-middle), itraconazole (top-right), voriconazole (bottom-left), posaconazole (bottom-middle), and isavuconazole (bottom-right)

membrane, which either inhibits growth or induces death of the fungal cells, depending on the species and antifungal compound involved. Of note, the triazoles have different affinities for the CYP-dependent 14-α-demethylase, which results in variable susceptibilities of different fungi, side effects, and drug–drug interactions (Warrilow et al. 2010).

16.2.2.1 Imidazoles

Ketoconazole

Ketoconazole is a member of the class of imidazole antifungals and is structurally similar to clotrimazole and miconazole. Released in the early 1980s, it was the first broad-spectrum oral antifungal treatment for systemic and superficial fungal infections (Maertens 2004). Prior to the introduction of the triazoles, ketoconazole was regarded as the standard and only available oral agent for the treatment of chronic mucocutaneous candidiasis for over a decade, and an effective alternative to amphotericin B in less severe (non-immunocompromised) cases of blastomycosis, histoplasmosis, paracoccidioidomycosis, and coccidioidomycosis in humans (Maertens 2004; Gupta and Lyons 2015). However, within the first few years of its approval, numerous clinically relevant shortcomings of this compound became evident such as hepatotoxicity, several drug interactions, considerable interindividual variation influenced by gastric pH, and poor penetration of the blood–brain barrier, which rendered it ineffective and unsuitable for the treatment of fungal meningitis. In addition, ketoconazole is largely fungistatic and has been proved to be less effective in immunocompromised patients (Maertens 2004; Gupta and Lyons 2015). Due to its significant side effects, the use of ketoconazole in humans is now largely reserved for topical administration for fungal and seborrheic dermatitis. However, systemic ketoconazole may be indicated for the treatment of endemic mycoses (blastomycosis, coccidioidomycosis, histoplasmosis, chromomycosis, and paracoccidioidomycosis) where alternatives are not available or feasible (Gupta and Lyons 2015).

Ketoconazole is licensed for dogs only in some European countries. In dogs and cats, ketoconazole has been used successfully to treat systemic mycoses, including blastomycosis (Arceneaux et al. 1998), histoplasmosis (Clinkenbeard et al. 1988), cryptococcosis (O'Brien et al. 2006), dermatophytosis (De Keyser and Van den Brande 1983), systemic coccidioidomycosis (Hinsch 1988), and sporotrichosis (Nakamura and Saato 1995) and *Malassezia* dermatitis (Godfrey 1998; Foy and Trepanier 2010). However, its successful response rate is lower than itraconazole and fluconazole (Hodges et al. 1994; Legendre et al. 1996). As in humans, the topical ketoconazole formulations (shampoo, cream, ointment) can be used for treatment of dermatophytosis and *Malassezia* infection in animals. A review indicated that ketoconazole was effective to treat *Malassezia* dermatitis at both 5 and 10 mg/kg per day for 3 weeks (Negre et al. 2009). Ketoconazole use is contraindicated in pregnant animals (Frymus et al. 2013). Topical ketoconazole formulations are also used in cats (Willard et al. 1986). However, most of the dermatophytosis cases in cats require a systemic treatment (Rochette et al. 2003).

Parconazole

Parconazole is another systemic fungicide belonging to the imidazole group, with a broad spectrum against yeasts dermatophytes, and other filamentous fungi. It is a registered veterinary drug in poultry and birds breeding in many countries. Parconazole is used in veterinary medicine as an oral fungicide against candidiasis in guinea fowls. As a medicated feed, it is recommended at a concentration of 30 mg/kg for prophylaxis and 60 mg/kg as a therapeutic dose for 7–10 days (EMA 1998).

16.2.2.2 Triazoles

The poor response rates and frequent recurrences of major fungal infections as well as the toxicity associated with ketoconazole therapy led to the development of a second group of azole derivatives, namely, the triazoles (Maertens 2004). In humans, five triazole compounds (fluconazole, itraconazole, voriconazole, posaconazole, and isavuconazole) have been clinically approved and are currently widely used for the prevention and treatment of several life-threatening fungal diseases (EMA 2012a, b). The triazoles have different affinities for the CYP-dependent 14-α-demethylase, which in turn results in variability on the susceptibilities of fungi, side effects, and drug–drug interactions (Warrilow et al. 2010).

Fluconazole

Fluconazole exhibits antifungal activity against most common clinical isolates of *Candida* and *Cryptococcus* spp. and the endemic dimorphic fungi *B. dermatitidis*, *Coccidioides* spp., *H. capsulatum*, and *P. brasiliensis*. However, fluconazole lacks efficacy against molds such as *Aspergillus* spp., and, therefore, it is not suitable for targeted prophylaxis or treatment of aspergillosis (EMA 2012c). Typically, *Candida glabrata*, *Microsporum*, and *Malassezia* species are inherently less susceptible to fluconazole, while *C. krusei* is considered to be intrinsically resistant to this triazole (Odds et al. 1986).

Fluconazole is available both in oral and intravenous formulations, with excellent bioavailability following oral administration (~90%) in humans (Brammer et al. 1990). It is also reported to be completely absorbed following administration to dogs (Humphrey et al. 1985). Fluconazole is cleared primarily by the kidneys, with greater than 70% of the dose eliminated unchanged in the urine in humans and various animal species (Humphrey et al. 1985; Diflucan 2013). The pharmacokinetics of fluconazole is similar in dogs and cats, and this triazole penetrates well into various fluids, including cerebrospinal fluid, aqueous humor, and bronchial epithelial fluid (Humphrey et al. 1985; Vaden et al. 1997; Diflucan 2013). Overall, fluconazole is very well-tolerated with few significant adverse effects, such as nausea, vomiting, and headache (Diflucan 2013). Increases in hepatic enzymes can occur with administration, but this is a class effect of the azoles, and hepatic failure is rare. Drug interactions can also occur with fluconazole due to its potent inhibition of CYP2C9 and moderate inhibition of CYP3A4 (Diflucan 2013).

In humans, fluconazole is approved for the treatment of vaginal, oropharyngeal, and esophageal candidiasis, as well as systemic *Candida* infections, fungal infections of the urinary tract, peritonitis, cryptococcal meningitis, and prophylaxis of patients undergoing bone marrow transplantation (Diaz et al. 1992; Pappas et al. 1995). Previous studies have shown that high doses of fluconazole (≥4–8 mg/kg per week) applied for long durations (12–16 weeks) are required for treating tinea capitis regardless of the fungus causing the infection (Shemer et al. 2013).

Fluconazole shows a predictable pharmacokinetics in animals and does not require therapeutic drug monitoring (Latimer et al. 2001). However, due to its teratogenic potential, fluconazole should be avoided during gestation (Pursley et al. 1996). Fluconazole has been recommended for veterinary patients with systemic mycoses affecting the central nervous system or eyes. Fluconazole pharmacokinetics following intravenous and oral administration suggests that a dosage of 5–10 mg/kg per day in dogs and a dosage of 50 mg per cat per day exceed minimum inhibitory concentrations for most pathogenic fungi (Vaden et al. 1997). It has been used successfully in dogs and cats with cryptococcosis (Malik et al. 1992; O'Brien et al. 2006) and blastomycosis (Arceneaux et al. 1998). Fluconazole at oral doses of 2.5–5 mg/kg appears to be effective for treatment of canine nasal aspergillosis; however the success outcome was much lower compared with topical enilconazole (Sharp et al. 1991).

Itraconazole

Itraconazole is another first-generation triazole antifungal. It is a strongly hydrophobic and water-insoluble compound with a high molecular weight, is highly protein-bound (>99%), and is structurally similar to ketoconazole (Fig. 16.4). The spectrum of activity of itraconazole is increased compared to that of fluconazole, in that it does have activity against certain molds, including *Aspergillus* species and some members of the order *Mucorales*, as well as *B. dermatitidis*, *Coccidioides* spp., and *H. capsulatum*. Itraconazole is also active against *Sporothrix* species, although this may be variable (Kohler et al. 2004). Variable activity has also been reported against some species within the order *Mucorales*. Resistance may develop in *Candida* species, which has also been documented in *Aspergillus* species, including *A. fumigatus* (Seyedmousavi et al. 2014a; Verweij et al. 2016).

In the USA, itraconazole is only available in formulations for oral administration. In humans, this includes capsules and an oral solution. Unfortunately, the bioavailability following administration of these formulations can be variable in both humans and animals (Van Cauteren et al. 1987; Boothe et al. 1997), with various factors influencing the amount that is absorbed. For the capsule, the use of medications that increase the gastric pH, such as antacids, H2 antagonists, and proton pump inhibitors, may significantly reduce the bioavailability due to the need for an acidic pH for solubility (Denning et al. 1994; Glasmacher et al. 1998; Kageyama et al. 1999). Although less affected by gastric pH since itraconazole is already dissolved, the oral solution is associated with significant gastrointestinal adverse effects, including nausea, vomiting, and diarrhea. The osmotic diarrhea that occurs with the oral solution is due to the cyclodextrin component, which is used to keep itraconazole in solution but is not absorbed following oral administration (Sporanox 2017). Compounded formulations of itraconazole have been used in animals but showed variable pharmacokinetics compared to the clinically available capsule and oral solution formulations (Smith et al. 2010). Following administration, itraconazole is eliminated by hepatic metabolism. In humans and in some animals, including dogs, it is metabolized to hydroxy-itraconazole, a bioactive metabolite with in vitro activity similar to that of the parent drug (Odds and Bossche 2000). However, not all animals metabolize itraconazole to hydroxy-itraconazole. Because it accumulates in various tissues, including the stratum corneum, and has a relatively long half-life, alternative dosing strategies have been utilized in cats for the treatment of cutaneous infections (Pinchbeck et al. 2002). However, the half-life may differ considerably between different animal species (Van Cauteren et al. 1987; Reidarson et al. 1998; Pinchbeck et al. 2002; Manire et al. 2003). As with all azoles, increases in hepatic enzymes may be observed with itraconazole use. Interestingly, itraconazole may have a negative inotropic effect. Because of the potential to result in congestive heart failure in patients with impaired ventricular function, itraconazole carries this as a black box warning in humans (Sporanox 2017). Itraconazole is also associated with significant drug interactions due to inhibition of CYP3A4 as well as p-glycoprotein (Sporanox 2017).

Itraconazole was the first extended spectrum triazole to become available for clinical use in humans. It is commonly used for the treatment of chronic and allergic fungal infection and for the empiric therapy of patients with febrile neutropenia who have suspected fungal infections (Lass-Florl 2011). In addition, it is used for the treatment of pulmonary and extrapulmonary blastomycosis, chronic cavitary pulmonary and disseminated non-meningeal histoplasmosis, and pulmonary and extrapulmonary aspergillosis in patients who are intolerant or refractory to amphotericin B therapy (Lass-Florl 2011). Itraconazole also remains the preferred azole for use in human patients to treat non-life-threatening systemic mycoses that do not involve the central nervous system (Lass-Florl 2011). Itraconazole is effective against both *Microsporum* and *Trichophyton* species and offers an alternative to griseofulvin for the treatment of kerion and noninflammatory tinea capitis (Schauder 2002).

Though relatively expensive for veterinary use, itraconazole has been used extensively in different animals with systemic fungal infections (Boothe et al. 1997; Miller et al. 2002; Pinchbeck et al. 2002; Manire et al. 2003; Bunting et al. 2009). In dogs, itraconazole is effective against blastomycosis, histoplasmosis, cryptococcosis, and coccidioidomycosis (Hodges et al. 1994; Legendre et al. 1996; Jacobs et al. 1997; Arceneaux et al. 1998; Graupmann-Kuzma et al. 2008). Oral itraconazole is also effective for treatment of dermatophytosis, sporotrichosis, *Malassezia* dermatitis, and otitis in dogs (Sykes et al. 2001; Pinchbeck et al. 2002).

Itraconazole is currently the preferred drug in feline dermatophytosis and is licensed for this indication in many countries. It is comparable (or superior) in efficacy to ketoconazole or griseofulvin and is much better tolerated by cats (Frymus et al. 2013). In horses, long-term treatment with itraconazole appears to be effective in the treatment of nasal aspergillosis (Korenek et al. 1994) and coccidioidomycosis (Foley and Legendre 1992). Itraconazole has been also used in a variety of raptors, psittacines, and waterfowl as first choice for treatment of aspergillosis, and no toxicity has been reported in birds (Orosz and Frazier 1995).

Voriconazole

Voriconazole is a low molecular weight water-soluble triazole, which is structurally similar to fluconazole. Because of its potent in vitro and in vivo efficacy against *Aspergillus* species (EMA 2012a), voriconazole is considered the drug of choice for the treatment of invasive aspergillosis in humans, and it is also used for this purpose in animals, including dogs, cats, penguins, falcons, and marine mammals, which may be highly susceptible to *Aspergillus* infections (Herbrecht et al. 2002; Di Somma et al. 2007; Taylor et al. 2014; Hyatt et al. 2015; Patterson et al. 2016). Voriconazole also has potent in vitro activity against yeasts, including *C. krusei*, *Cryptococcus* species, endemic fungi (*B. dermatitidis*, *Coccidioides* species, *H. capsulatum*), as well as different molds, including some *Scedosporium* (e.g., *S. apiospermum* and *S. boydii*) and *Fusarium* species (Lackner et al. 2012). However, voriconazole lacks activity against the *Mucorales* (Lewis et al. 2011; Sun and Singh 2011) and has reduced to no activity against certain species of *Scedosporium* (e.g., *S. aurantiacum*, *Lomentospora* (formerly *Scedosporium*) *prolificans*) and *Fusarium* (e.g., *F. solani*) (Alastruey-Izquierdo et al. 2008; Tortorano et al. 2008; Lackner et al. 2012, 2014).

Both oral and intravenous formulations are available for administration, and in humans oral dosing is recommended to be done on an empty stomach, which is not always feasible when administering to different animal species. Once administered, voriconazole is primarily eliminated via hepatic metabolism by the cytochrome P450 isoenzymes 2C19, 2C9, and 3A4. In human adults, the metabolism may be saturable resulting in disproportional increases in concentrations with small increases in the dose due to nonlinear pharmacokinetics (Lazarus et al. 2002; Walsh et al. 2004). In rodents, voriconazole may induce its own metabolism resulting in low or undetectable concentrations (Roffey et al. 2003), and this may also occur in some bird species (Flammer et al. 2008; Beernaert et al. 2009b). The variability in voriconazole concentrations may be due to numerous factors, including drug–drug interactions, drug–disease interactions (e.g., vomiting, diarrhea, mucositis), as well as CYP2C19 polymorphisms (Matsumoto et al. 2009). Because of this, therapeutic drug monitoring is often performed. The therapeutic window for voriconazole is narrow, with trough levels of at least 1 mg/L needed for efficacy (Smith et al. 2006; Pascual et al. 2008), while higher concentrations ($\geq$ 5.5 mg/L) may result in concentration-dependent toxicities, which have been reported in humans and animals (Imhof et al. 2006; Pascual et al. 2008; Troke et al. 2011; Pascual et al. 2012; Hyatt et al. 2015).

Increases in hepatic enzymes can also occur with voriconazole. There is some debate as to whether increases in hepatic enzymes may or may not be concentration-dependent (Denning et al. 2002; Tan et al. 2006). However, it has been shown in humans and some animal species that central nervous system toxicities, including encephalopathy, are indeed concentration-dependent (Imhof et al. 2006; Pascual et al. 2008; Hyatt et al. 2015). Other toxicities that have been reported with voriconazole include visual abnormalities, alopecia, phototoxicity, and periostitis (Malani et al. 2014; Moon et al. 2014; Williams et al. 2014). Voriconazole is also associated with significant drug interactions, as it is a substrate and inhibitor of CYP 2C19, 2C9, and 3A4.

Clinical reports of voriconazole use in animals are sparse and limited primarily to cases with systemic aspergillosis and other difficult-to-treat fungal infections. In dogs, voriconazole undergoes extensive metabolism; it also induces its own metabolism over time (Roffey et al. 2003). Voriconazole is recommended drug of choice for invasive aspergillosis in dogs and cats because of its fungicidal activity and its safety as compared with amphotericin B. Voriconazole also has been used in birds to control aspergillosis. In some species of wild and captive-held falcons and quails, it appears to be effective and safe for the treatment of aspergillosis (Di Somma et al. 2007; Tell et al. 2010; Gentry et al. 2014). Voriconazole has also been used with success to treat disseminated scedosporiosis in dogs (Taylor et al. 2014).

Topical administration of voriconazole can be used successfully to treat dematiaceous fungal keratitis in dogs (Pucket et al. 2012). Treatment with topical 1% voriconazole solution was successful in resolving *Aspergillus flavus* keratomycosis (Labelle et al. 2009). Intracorneal administration of 5% voriconazole solution also resulted in resolution of clinical disease, specifically stromal abscessation and secondary uveitis (Smith et al. 2014).

Posaconazole

Posaconazole is a lipophilic triazole, which is structurally similar to itraconazole. Posaconazole is distinguished from the other azoles by its potent in vitro activity against the *Mucorales*, which show reduced susceptibility to other triazoles, and improved activity against *Aspergillus* spp. compared to itraconazole (Manavathu et al. 2000).

Posaconazole is currently available in oral and intravenous formulations for humans. The oral formulations include a suspension and a tablet. The oral suspension should be administered with a high-fat meal and multiple times per day due to saturable absorption. The bioavailability of this formulation is limited by several factors, including the concomitant use of agents affecting gastric pH and gastric motility, as well as vomiting, diarrhea, and mucositis (Krishna et al. 2009). Due to significant variability in bioavailability with the oral suspension and a higher percentage of patients having low or undetectable concentrations, a delayed-release tablet was developed that prevents the release of the drug in the low pH environment of the stomach but allows its release in the small intestine where absorption is maximized (Krishna et al. 2012; Kersemaekers et al. 2015). Unfortunately, the delayed-release tablet cannot be crushed so that the dose can be adjusted, and there is limited experience with the use of this formulation in animals.

Posaconazole appears to be well-tolerated in humans, although there are some reports of adverse effects associated with elevated posaconazole concentrations (Martino et al. 2015; Parkes et al. 2016). In dogs, neuronal phospholipidosis has been observed in the central and peripheral nervous systems, although no neurologic deficiencies were reported (Cartwright et al. 2009).

In humans, posaconazole is approved only for the following groups of patients: those who are aged 18 years or older (EMA 2012b), receiving remission–induction chemotherapy for acute myelogenous leukemia (AML) or myelodysplastic syndromes (MDS) expected to result in prolonged neutropenia, and those at high risk of developing IFIs. It is also indicated for prophylaxis in recipients of hematopoietic stem cell transplants (HSCT) undergoing high-dose immunosuppressive therapy for graft-versus-host disease and who are at high risk for developing IFIs and salvage therapy of invasive aspergillosis in patients with a disease condition that is refractory to amphotericin B or itraconazole and in patients who are intolerant of the other medicinal products (Herbrecht et al. 2002; Cornely et al. 2007; Ullmann et al. 2007; Walsh et al. 2008).

In animals, although there is little experience with its use in different species, there are emerging data that posaconazole may be effective for the treatment of invasive aspergillosis in dogs (Corrigan et al. 2016; Stewart and Bianco 2017). Posaconazole is primarily metabolized by glucuronidation and thus should be used with caution in cats, which lack major UDP-glucuronosyltransferase enzymes (Court 2013). However, there is a case report in which posaconazole was successfully used with minimal adverse effects to treat a cat with invasive aspergillosis caused by an itraconazole-resistant isolate (McLellan et al. 2006). Posaconazole administered at a dosage of 5 mg/kg PO q12 h appears to be safe for prolonged treatment of disseminated *Aspergillus* infections in dogs. Long-term survival >1 year is possible with prolonged treatment, but relapse is common (Corrigan et al. 2016). Successful treatment with posaconazole has been also reported in two cats with fungal disease in which other antifungal treatment had failed (Mawby et al. 2016). Those cases involved a *Mucor* subcutis infection of the nose and invasive orbital aspergillosis (McLellan et al. 2006; Wray et al. 2008).

A veterinary product containing orbifloxacin and posaconazole in a mineral oil-based system is available for treatment of otitis externa in dogs associated with susceptible strains of yeast (*Malassezia pachydermatis*) and bacteria (coagulase positive *Staphylococci*, *Pseudomonas aeruginosa*, and *Enterococcus faecalis*).

Isavuconazole

Isavuconazole is the newest broad-spectrum triazole available for treatment of severe invasive and life-threatening fungal diseases in humans (Seyedmousavi et al. 2015b). The prodrug isavuconazonium sulfate (BAL8557) (Ohwada et al. 2003) is a highly potent water-soluble triazole suitable for both oral and intravenous administration. BAL8557 consists of a triazolium salt linked to an aminocarboxyl (N-[3-acetoxypropyl]-N-methylamino–carboxymethyl) via an ester moiety. After oral and intravenous administration, it rapidly and almost completely (>99%) undergoes enzymatic activation via plasma esterases followed by spontaneous chemical degradation to release the active drug isavuconazole (BAL4815, formerly RO-094815) and an inactive cleaved product, BAL8728 (Odds 2006; Schmitt-Hoffmann et al. 2006a, b).

Isavuconazole has a broad spectrum of in vitro activity and in vivo efficacy against a wide range of yeasts and molds including *Aspergillus* spp., *Fusarium* spp., *Candida* spp., the *Mucorales*, *Cryptococcus* spp., melanized and dimorphic fungi, dermatophytes that can reproduce in culture (by unilateral budding), and their filamentous co-spp. (Odds 2006; Warn et al. 2006; Guinea et al. 2008; Martin de la Escalera et al. 2008; Perkhofer et al. 2009; Rudramurthy et al. 2011; Shivaprakash et al. 2011; Espinel-Ingroff et al. 2013; Gregson et al. 2013; Seyedmousavi et al. 2013b).

Isavuconazole is available in oral (capsule) and intravenous formulations. In both formulations, the prodrug, isavuconazonium sulfate, is present and is rapidly converted to the active moiety isavuconazole following administration. The half-life is extensive in humans (~130 h) and very short in mice (~3 h) (Warn et al. 2009). Unfortunately, the pharmacokinetics of isavuconazole in other animal species is unknown.

Isavuconazole is effective for the treatment of invasive aspergillosis and mucormycosis (Seyedmousavi et al. 2015b; Maertens et al. 2016; Marty et al. 2016). To date, no study has been conducted using isavuconazole for the treatment of systemic fungal infections in veterinary patients.

16.2.3 Echinocandins

The echinocandins are the only class of antifungal agents that directly target the fungal cell wall (Denning 2002; Mukherjee et al. 2011). They are semisynthetic amphiphilic lipopeptides formed during the fermentation of certain fungi such as *Zalerion arboricola* or *A. nidulans* var. *echinulatus* (Nyfeler and Keller-Schierlein

1974). The echinocandins inhibit β-1,3-D-glucan synthase, which catalyzes the biosynthesis of β-1,3-D-glucan, a key component of the fungal cell wall (Kurtz and Douglas 1997). Of note, mammalian cells do not contain this enzyme, and, therefore, direct human cell toxicity is minimal (Eschenauer et al. 2007).

The echinocandins are highly active (fungicidal) against *Candida* spp. including isolates that are resistant to triazoles (Bachmann et al. 2002a) and spp. that form biofilms (Bachmann et al. 2002b). These agents have modest activity (fungistatic) against *Aspergillus* spp. (Bowman et al. 2006) as well as dimorphic (Kohler et al. 2000) and melanized fungi (Seyedmousavi et al. 2014b) and weak activity against the *Mucorales*, *Fusarium* spp., *Scedosporium* spp., *Cryptococcus* spp., and *Trichosporon* spp. (Eschenauer et al. 2007).

The echinocandins that are currently approved for humans are not orally bioavailable and, therefore, must be administered by slow intravenous infusion (1–2 h). The chemical structures of the three echinocandins, which are in clinical use, are shown in Fig. 16.5.

Caspofungin is derived from pneumocandin B0, a fermentation product of *Glarea lozoyensis*. It was the first echinocandin approved by the FDA and is recommended for adult and pediatric patients for the primary treatment of invasive *Candida* infections, salvage therapy of invasive aspergillosis infections in patients who are refractory or intolerant to other therapies, and empiric antifungal therapy for presumed fungal infections (such as those caused by *Candida* or *Aspergillus*) in persistently febrile patients with neutropenia (Maschmeyer and Glasmacher 2005).

Micafungin was the second marketed echinocandin synthesized by chemical modification of a fermentation product of *Coleophoma empetri*. Micafungin is approved for the treatment of esophageal candidiasis and prophylaxis of invasive *Candida* infections in patients undergoing hematopoietic stem cell transplantation. The indication for its use was later expanded to include candidemia, acute disseminated candidiasis, *Candida* abscesses, and peritonitis (de la Torre and Reboli 2014).

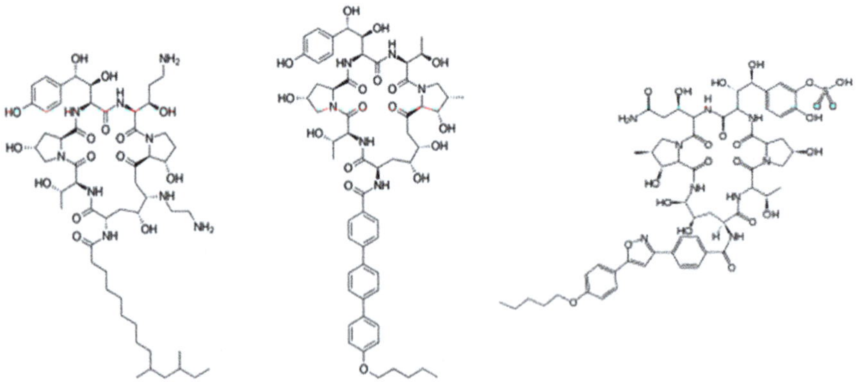

Fig. 16.5 Chemical structures of caspofungin (left), anidulafungin (middle), and micafungin (right)

Anidulafungin is derived from hexapeptides produced by *A. nidulans* and was the third echinocandin antifungal agent to receive approval from the FDA. It is approved for the treatment of candidemia and other *Candida* infections (intra-abdominal abscess and peritonitis) and esophageal candidiasis (Estes et al. 2009).

Although echinocandins hold promise for the treatment of systemic yeast infections in veterinary patients, their potential to treat the common dimorphic fungal infections in dogs and cats is poor (Foy and Trepanier 2010).

16.2.4 Nucleoside Analogs

Flucytosine (5-fluorocytosine or 5-FC) is the only systemic antifungal agent belonging to the class of nucleoside analogs. It was the first agent used for the treatment of invasive mycoses in humans in 1968 (Tassel and Madoff 1968). Flucytosine is the fluorinated analog of cytosine and was discovered in 1957 as an analog of the cytostatic chemotherapeutic agent 5-fluorouracil (5-FU) used for antitumor therapy (Heidelberger et al. 1957). After it penetrates the cell wall, which is controlled by the enzyme cytosine permease, 5-FC is converted to 5-FU by the cytosine deaminase and then further to 5-fluorouridine (Polak and Scholer 1975).

After three phosphorylation steps, it is incorporated into RNA instead of uracil, which results in the blockade of protein synthesis. This inhibition leads to reduced DNA synthesis because of a reduction in the available nucleotide pool (Cutler et al. 1978). Therefore, fungi lacking cytosine deaminase are not susceptible to 5-FC (Polak 1977), and it is noteworthy that this enzyme is absent in mammalian cells. Figure 16.6 shows the chemical structures of cytosine and two fluorinated pyrimidines (5-fluorocytosine and 5-fluorouracil).

Flucytosine can be administered both orally and intravenously and is well distributed in almost all body fluids including the lacrimal fluid, urine, and cerebrospinal fluid (CSF) (Cutler et al. 1978). Flucytosine is well-tolerated, but at concentrations above 100 mg/L, it may induce liver and bone marrow toxicity including bone marrow suppression, myocardial toxicity, and renal failure (Benson and Nahata 1988).

5-FC is active against most clinically important yeast such as those of the *Candida* (Pfaller et al. 2002) and *Cryptococcus* genera (Pfaller et al. 2005), shows limited activity against melanized fungi and *Aspergillus* spp. (Verweij et al. 2008), and is ineffective against dimorphic fungi (Shadomy 1969). The in vitro activity of 5-FC is affected by a variety of test conditions such as incubation time, test medium, medium pH, and endpoint determination criteria (34).

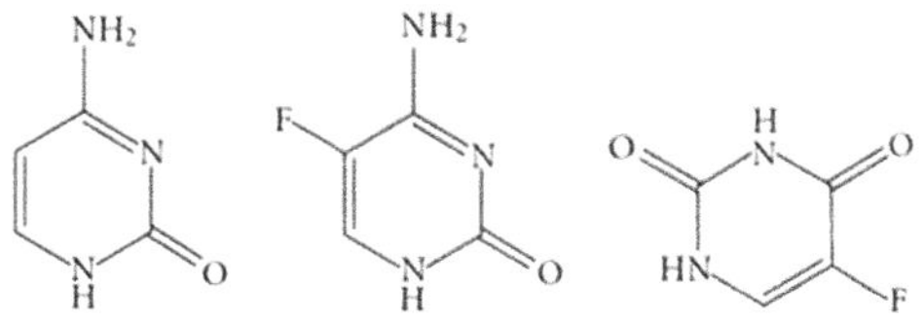

Fig. 16.6 Chemical structures of cytosine (left), 5-fluorocytosine (middle), and 5-fluorouracil (right)

In humans, flucytosine combined with amphotericin B is primarily recommended for the treatment of cryptococcal infections of the central nervous system (Perfect et al. 2010). In addition, flucytosine in combination with an azole (fluconazole) may have a role in the treatment of disseminated *Candida* infections that are refractory to first-line antifungal agents (triazoles and lipid formulations of amphotericin B) (Pappas et al. 2004). Rapid resistance occurs if flucytosine is used as monotherapy (Hospenthal and Bennett 1998), and, therefore, it should never be used alone.

In cats, flucytosine in combination with amphotericin B has been used for treatment of systemic cryptococcosis, whereas this combination therapy is not recommended in dogs due to potential severe toxic reactions (Malik et al. 1996).

16.2.5 Griseofulvin

Griseofulvin is the only antifungal approved for systemic administration by the FDA for veterinary use. Griseofulvin is a metabolite of *P. griseofulvum*, *P. janczewski*, and *P. patulum* that inhibits cell wall synthesis. It binds to tubulin and thereby induces conformational changes in the alpha and beta subunits. Indirectly, it causes an impaired processing of newly synthesized cell wall constituents at the growing tips of hyphae (De Carli and Larizza 1988). Griseofulvin mainly concentrates in keratinocytes; therefore, it is only used for noninvasive dermatophyte infections (Lewis 2011).

Overall, dermatophytes (*Microsporum* spp., *Trichophyton* spp., and *Epidermophyton*) are susceptible to griseofulvin; however it is not active against yeasts (*Malassezia* and *Candida*). Since the late 1950s, griseofulvin has been the gold standard for the systemic therapy of human tinea capitis (Shemer et al. 2013). However, the advent of newer antifungals that exhibit more favorable pharmacokinetic and toxicity profiles has largely relegated griseofulvin to a second-line agent against dermatophytoses in humans. The main disadvantage of griseofulvin is the long duration of treatment that is required (6–12 weeks or longer), which may lead to reduced compliance (Elewski 1999). Another challenge is the high expense of griseofulvin because of the large quantity of drug required for a cure. Moreover, the efficacy of griseofulvin has decreased in recent years owing to decreased susceptibility of the infective fungi due to changes in epidemiology and mutations (Chen and Friedlander 2001), resulting in a need for larger doses and longer treatment durations.

Griseofulvin is licensed for treatment of dermatophytosis in dogs and cats (Medleau and Whiteweithers 1992; Moriello and Deboer 1995) and is still widely used in some countries. It is administered orally for at least 4–6 weeks at 25–50 mg/kg q12–24 h with a fatty meal to increase absorption. Adverse reactions include anorexia, vomiting, diarrhea, and bone marrow suppression, particularly in Siamese, Himalayan, and Abyssinian cats. The use of griseofulvin is contraindicated in kittens younger than 6 weeks of age and in pregnant animals as the compound is teratogenic, particularly during the first weeks of gestation (Frymus et al. 2013).

In ruminants, griseofulvin formulations as powder or as granules mixed into the feed have been used for the treatment of dermatophytosis (Andrews and Edwardson 1981; Power and Malone 1987). Long-term treatment with griseofulvin is shown to be effective against *T. equinum* ringworm in horses (Plumb 1999). Equine sporotrichosis has also been successfully cured with griseofulvin (Fishburn and Kelley 1967; Greydanus-van der Putten et al. 1994). Oral griseofulvin may also be prescribed for rabbits, rodents, and fur animals (Rochette et al. 2003).

16.3 Topical Antifungals

A wide variety of topical agents belonging to different classes of antifungals are available as creams, ointments, gels, lotions, powders, shampoos, and other formulations, which may differ widely in their efficacy. Of note, topical antifungal compounds alone can reduce the transmission of spores (Brammer et al. 1990).

Topical antifungals can be applied to the skin, nails, or mucosa to kill or inactivate fungi. Regardless of the actual mechanism of action of the drug, or the viscosity, hydrophobicity, and acidity of the formulation, the drug's ability to penetrate or permeate deeper skin layers is an important property impacting the therapeutic efficacy of topical antifungals (Trommer and Neubert 2006; Gupta and Cooper 2008; Sugiura et al. 2014).

Natamycin is an allylamine antifungal agent licensed for veterinary use only for cattle and horses. It has a moderate activity against dermatophytes, yeasts, and *Aspergillus*. It can also be used for environmental decontamination (Rochette et al. 2003). For a successful treatment, twice whole-body spray (0.1%) treatment of natamycin in an interval of 3 days is recommended in cattle naturally infected with *T. verrucosum* (Oldenkamp and Spanoghe 1977; Oldenkamp 1979a). Natamycin is used topically in the horse and is reported as being effective in successful elimination of dermatophytosis within 4 weeks. After treatment the recovered animals did not show any evidence of reinfection for up to 6 months (Oldenkamp 1979b).

Clotrimazole, miconazole, and enilconazole are external imidazoles. Because of a high first-pass effect, the topical antifungals show minimal systemic bioavailability and are therefore confined to topical use (Conte et al. 1992). Miconazole, chlorhexidine, and selenium sulfide were components of the treatment protocol for *Malassezia* dermatitis in dogs in several randomized and controlled trials. Miconazole in combination with polymyxin B and prednisolone is licensed for use in dogs and cats and is included in an otic preparation for treatment of otitis externa (Cornelissen and Van den Bossche 1983). A shampoo containing 2% miconazole nitrate + 2% chlorhexidine, when used twice a week for 3 weeks, has been demonstrated to be effective for treatment of *Malassezia* dermatitis in dogs (Negre et al. 2009).

Clotrimazole has been reported to be effective for the treatment of localized dermatophytosis (McCurdy et al. 1981) and sinonasal aspergillosis in dogs and cats (Peeters and Clercx 2007; Furrow and Groman 2009). Like miconazole, it is commonly used against *Malassezia* otitis in dogs and cats in combination with an antibiotic and an anti-inflammatory agent (Bensignor and Grandemange 2006).

Enilconazole is an imidazole with broad-spectrum activity against dermatophytes (*Microsporum* sp., *Trichophyton* sp.). It has activity against *Aspergillus* spp., *Penicillium* spp. and yeasts, including *Candida* spp. In veterinary medicine, enilconazole is used as a topical antimycotic (4 mg/kg bw) against dermatophytes in cattle (Dekeyser 1981), horses (Mayer 1983), dogs, and cats (Hnilica et al. 2000). Enilconazole can be applied topically for treatment of dermatophytosis and aspergillosis. It has been used safely in cats, dogs, cattle, horses, and chickens and is prepared as a 0.2% solution for treatment of fungal skin infection. Treatment consists of four whole-body applications at 3–4-day intervals. Enilconazole can be applied topically for sinus irrigation to treat canine nasal aspergillosis (Benitah 2006). In several countries, it is also used as a fungicide formulation (smoke generator or emulsifiable concentrate) for the disinfecting of farm buildings including poultry houses and rabbit farms, which showed highly effective against spores of dermatophytes and *Aspergillus* spp. (Van Cutsem et al. 1985, 1988).

The anthelmintic thiabendazole is also an imidazole with potent efficacy in the treatment of bovine dermatophytosis caused by *T. verrucosum* and superficial yeast infections in animals. Thiabendazole has been licensed for use in dogs and cats as ingredient of topical antifungals (Gabal 1986).

16.4 Antifungal Resistance in Animals

16.4.1 Mechanisms of Antifungal Resistance

Resistance to antifungal drugs can occur through various mechanisms. These can include (a) non-synonymous point mutations within the gene encoding the target enzyme leading to alterations in the amino acid sequence, (b) increased expression of the target enzyme through increased transcription of the gene encoding it, (c) decreased concentrations of the drug within the fungal cells due to drug efflux, and (d) changes in the biosynthetic pathway resulting in reduced production of the target of the antifungal drugs. For the azoles, each of these mechanisms has been associated with reduced susceptibility in *Candida albicans*, and several are associated with resistance in other *Candida* species. Alterations in the target enzyme (lanosterol 14-α-demethylase) due to point mutations in the encoding gene *ERG11* lead to decreased susceptibilities to the azoles (Vanden Bossche et al. 1990; Parkinson et al. 1995; Lamb et al. 1997; Loffler et al. 1997; White 1997; Franz et al. 1998; Miyazaki et al. 1998; Lopez-Ribot et al. 1999; Sanglard et al. 1999; Lamb et al. 2000; Katiyar and Edlind 2001; Brun et al. 2004; Sanguinetti et al. 2005). Overexpression of the *CDR1*, *CDR2*, and *MDR1* genes that encode for efflux pumps leads to azole resistance (Fling et al. 1991; Marr et al. 1998). Azole resistance has also been documented in *A. fumigatus* and is due to point mutations within the *CYP51A* gene that encodes the enzyme responsible for converting lanosterol to ergosterol (Howard et al. 2009; Vermeulen et al. 2013; Seyedmousavi et al.

2014a). In isolates with environmental exposure to the azoles, tandem repeats in the promoter region along with point mutations in the gene (e.g., TR_{34}/L98H and TR_{46}/Y121F/T289A) have been found and cause increased expression of *CYP51A* (Mellado et al. 2007).

16.4.2 Reports of Antifungal Resistance in Different Animal Species

Several studies have analyzed fungal isolates from different animals for resistance to antimycotic agents, and many of them reported surprisingly high levels of azole resistance in yeasts and molds.

In a retrospective study, Beltaire et al. (2012) analyzed fungal strains isolated from equine uterine collected between 1999 and 2011 and showed resistance rates of 19% and 2% for itraconazole and fluconazole, respectively. Cordeiro Rde et al. (2015) investigated 59 *C. tropicalis* isolates predominantly derived from healthy animals and found resistance to fluconazole and/or itraconazole in 50%, whereas all isolates were susceptible to caspofungin and amphotericin B. Using the same microbroth dilution assay, Brilhante et al. (2016) analyzed *Candida* isolates from the nasolacrimal duct of healthy horses and found that 40% of the *C. tropicalis* isolates were resistant to fluconazole and itraconazole (Brilhante et al. 2016). The same group also found high rates of fluconazole and itraconazole resistance also for *Candida* isolates from rheas and cockatiels (Sidrim et al. 2010; Brilhante et al. 2013), and efflux pumps were a major resistance mechanism (Rocha et al. 2017). Using a commercial kit covering 11 commonly used agents, Lord et al. (2010) tested 144 *Candida*, *Cryptococcus*, *Rhodotorula*, and *Trichosporon* isolates from bird feces for antifungal resistance. They reported that 45.8% of the strains were resistant to at least 4 of the 11 drugs, and 18.1% were resistant to all antifungals tested. A recent study found similar resistant levels for 111 *C. glabrata* isolates from the feces of sea gulls and 79 *C. glabrata* isolates from human patients, while other have reported only moderate azole resistance in *Candida* strains isolated from raptors (Brilhante et al. 2012; Al-Yasiri et al. 2016). Antifungal drug susceptibility of *Cryptococcus neoformans*/*Cryptococcus gattii* species complex isolated from dogs and cats in North America showed that *C. neoformans* strains had higher MICs for flucytosine and itraconazole. However, *C. gattii* isolates exhibited a wider range of MICs than *C. neoformans* (Singer et al. 2014). Fluconazole resistance has also been reported in *Cryptococcus neoformans* var. *grubii* obtained from a case of feline cryptococcosis, due to overexpression of the ERG11 and ABC transporter (Kano et al. 2015).

In another study, Cafarchia et al. determined the MIC distribution and the epidemiological cutoff values (ECVs) of 62 *Malassezia pachydermatis* from dogs with dermatitis and 78 *M. furfur* isolates from humans with bloodstream infections, using Clinical and Laboratory Standards Institute (CLSI) methodology. Overall, MIC data for azoles of *M. pachydermatis* were four twofold dilutions lower than those of *M. furfur*. Itraconazole and posaconazole displayed lower MICs than voriconazole, regardless of the *Malassezia* species. In addition, fluconazole resistance was detected only in *M. pachydermatis* isolates (Cafarchia et al. 2015). These studies indicate that resistance to certain azoles is a common phenomenon in pathogenic yeasts isolated from some animals. Strikingly, the azole resistance rates of *C. albicans* and *C. tropicalis* isolated from healthy animals are higher than those reported in some human studies (Pfaller et al. 2013; Goncalves et al. 2016). This indicates that the elevated resistance levels found in animals may not simply reflect a natural resistance of the respective species. However, differences in the methodology and breakpoints used, as well as the limited number of isolates included in several animal studies, make it difficult to directly compare data obtained for animal and human isolates.

Recent changes in the taxonomy of *Aspergillus* have major implications for our understanding of drug susceptibility profiles (Van Der Linden et al. 2011). New sibling species of *A. fumigatus* exhibit in vitro susceptibility profiles that differ significantly from that of *A. fumigatus*. While acquired resistance is an emerging problem in *A. fumigatus* (Verweij 2007; Mellado et al. 2007; Verweij 2009a), other *Aspergillus* species may be intrinsically resistant to, e.g., amphotericin B and azoles (Van Der Linden et al. 2011). MICs of *A. flavus* clinical isolates to amphotericin B are consistently twofold dilution steps higher than those of *A. fumigatus* (Gomez-Lopez et al. 2003b; Garcia-Effron et al. 2003b). Using CLSI methodology (CLSI 2008), *A. nidulans* was shown to have MIC values of 1–2 mg/L for amphotericin B, which is higher than commonly observed with *A. fumigatus* (Kontoyiannis et al. 2001). Itraconazole and voriconazole cross-resistance and variable susceptibilities for caspofungin were observed in vitro against *A. felis,* another possibly intrinsically resistant sibling of the *A. fumigatus* species complex (Coelho et al. 2011; Barrs et al. 2013; Pelaez et al. 2013). In the section *Usti*, azoles are not active against *A. calidoustus* with MICs of $\geq$8 mg/L, while also other classes of antifungal drugs also appear less active (Varga et al. 2008). Resistance of *A. terreus* to amphotericin B is well known (Lass-Florl et al. 2009). Based on susceptibility to azoles, three different susceptibility patterns were distinguished in the black aspergilli (section *Nigri*). Some isolates showed low azole MICs, others showed high MICs, and a third group showed an uncommon paradoxical effect. However, these groups did not coincide with species boundaries, making it difficult to interpret as an intrinsic or acquired property (Alcazar-Fuoli et al. 2009).

There is no evidence of emerging azole resistance among *A. fumigatus* isolates from dogs and cats, and topical azole therapy should be effective against most isolates (Talbot et al. 2015). However, acquired resistance to itraconazole and voriconazole has been reported for avian *A. fumigatus* strains obtained from domestic and wild birds in Belgium and the Netherlands (Beernaert et al. 2009a), where azole resistance is widespread both in clinical and environmental isolates (Beernaert et al. 2009b). The source of these resistant isolates is unclear. However, two of the four resistant strains were isolated from birds that received itraconazole. This is important, and a fungicide-driven route of resistance selection in *A. fumigatus* may have implications for the management of aspergillosis in animals. Another possibility in these birds can also be considered an indication of the presence of acquired resistance in the surrounding environment (Bromley et al. 2014). Of note, resistance to medical triazoles may be associated with resistance selection to azole fungicides in the environment (Verweij et al. 2009a, b). In humans azole-resistant *Aspergillus* disease can be observed in patients without previous azole therapy, indicating that hosts inhale both azole-susceptible and azole-resistant *A. fumigatus* conidia (ECDC Technical Report 2013).

In another study, Ziółkowska et al. investigated the in vitro susceptibility of *A. fumigatus* strains isolates obtained from the oral cavity, lungs, and air sacs of healthy domestic geese, birds with aspergillosis, and from their environment. All of the strains were susceptible to enilconazole, itraconazole, and voriconazole, but, irrespective of source, showed various degree of resistance to miconazole, clotrimazole (MIC90 = 16 μg/mL), and amphotericin B (MIC90 = 16 μg/mL) (Ziolkowska et al. 2014). To assess the potential risk of azole resistance emergence in avian farms where azole compounds were used for the control of avian mycoses, a drug susceptibility study including *A. fumigatus* isolates from birds and avian farms was also conducted in France and Southern China (Wang et al. 2013). A total number of 175 *A. fumigatus* isolates were analyzed. No resistant isolate was detected, and the distribution of MICs was similar for isolates collected in farms with or without azole chemoprophylaxis. For 61 randomly selected isolates, the full coding sequence of *Cyp51A* gene was determined to detect mutations. Nine amino acid alterations were found in the target enzyme, three of which were new mutations.

Of note, invasive infections caused by azole-resistant *A. fumigatus* are challenging to treat due to the lack of therapeutic options. In humans, combination of an azole with echinocandins or lipid formulations of amphotericin B can be used, and 5-flucytosine has also been recommended to be added to other therapies in patients with central nervous system infections caused by resistant isolates (Verweij et al. 2015). However, both antifungals have limitations, including toxicities, which may prohibit their long-term use in both humans and animals. Depending on the mechanism of resistance, higher doses of certain triazoles may be attempted, and there is a recent report of the successful treatment of invasive aspergillosis caused by an *A. fumigatus* isolate harboring a TR_{46}/Y121F/T289A mutation in a bottlenose dolphin with high-dose posaconazole (Bunskoek et al. 2017). Here, the oral solution of posaconazole was incorporated into gelatin capsules and administered with a goal of achieving trough concentrations of >3 mg/L, which was achieved after prolonged administration and resulted in clinical improvement.

16.5 Conclusions and Future Directions

There are a variety of systemic and topical antifungals available for use in veterinary medicine, and acquisition of antifungal resistance can occur under prolonged therapy or may be associated with resistance selection to fungicides in the environment. There are many similar characteristics between the use of antifungals in humans and animals; however, many unanswered questions remain, including optimal dosage regimens that maximize clinical outcomes while avoiding toxicities. In addition, the true epidemiology of antifungal resistance in different animal species remains unknown, and treatment options for invasive infections caused by resistant fungi are limited.

Although amphotericin B and the echinocandins are useful agents for the treatment of numerous invasive mycoses, these agents are only available for intravenous administration and cannot be administered orally for the treatment of invasive fungal infections. In addition, amphotericin B is associated with significant adverse effects, including nephrotoxicity. As in humans, the triazoles are the main antifungals that are used to treat invasive mycoses in animals, as they can be administered orally for the long periods that are often required to treat these infections, and many veterinarians have experience using the members of this class. In addition, terbinafine is used in combination for difficult-to-treat infections.

Acknowledgement The work of Seyedmojtaba Seyedmousavi was supported by the Intramural Research Program of the NIH, NIAID.

References

Alastruey-Izquierdo A, Cuenca-Estrella M, Monzon A, Mellado E, Rodriguez-Tudela JL (2008) Antifungal susceptibility profile of clinical *Fusarium* spp. isolates identified by molecular methods. J Antimicrob Chemother 61(4):805–809

Alcazar-Fuoli L, Mellado E, Alastruey-Izquierdo A, Cuenca-Estrella M, Rodriguez-Tudela JL (2009) Species identification and antifungal susceptibility patterns of species belonging to *Aspergillus* section *Nigri*. Antimicrob Agents Chemother 53(10):4514–4517

Al-Yasiri MH, Normand AC, L'Ollivier C, Lachaud L, Bourgeois N, Rebaudet S, Piarroux R, Mauffrey JF, Ranque S (2016) Opportunistic fungal pathogen *Candida glabrata* circulates between humans and yellow-legged gulls. Sci Rep 6:36157

Andrews AH, Edwardson J (1981) Treatment of ringworm in calves using griseofulvin. Vet Rec 108:4980500

Arceneaux KA, Taboada J, Hosgood G (1998) Blastomycosis in dogs: 115 cases (1980–1995). J Am Vet Med Assoc 213(5):658–664

Arikan S (2002) Lipid-based antifungal agents: a concise overview. Cell Mol Biol Lett 7(3): 919–922

Bachmann SP, Patterson TF, Lopez-Ribot JL (2002a) In vitro activity of caspofungin (MK-0991) against Candida albicans clinical isolates displaying different mechanisms of azole resistance. J Clin Microbiol 40(6):2228–2230

Bachmann SP, VandeWalle K, Ramage G, Patterson TF, Wickes BL, Graybill JR, Lopez-Ribot JL (2002b) In vitro activity of caspofungin against Candida albicans biofilms. Antimicrob Agents Chemother 46(11):3591–3596

Barchiesi F, Colombo AL, McGough DA, Rinaldi MG (1994) Comparative study of broth macrodilution and microdilution techniques for in vitro antifungal susceptibility testing of yeasts by using the National Committee for Clinical Laboratory Standards' proposed standard. J Clin Microbiol 32(10):2494–2500

Barrs VR, van Doorn TM, Houbraken J, Kidd SE, Martin P, Pinheiro PD, Richardson M, Varga J, Samson RA (2013) *Aspergillus felis* sp. nov., an emerging agent of invasive aspergillosis in humans, cats, and dogs. PLoS One 8(6):e64871

Bechert U, Christensen JM, Poppenga R, Fahmy SA, Redig P (2010) Pharmacokinetics of terbinafine after single oral dose administration in red-tailed hawks (Buteo jamaicensis). J Avian Med Surg 24(2):122–130

Beernaert LA, Baert K, Marin P, Chiers K, De Backer P, Pasmans F, Martel A (2009a) Designing voriconazole treatment for racing pigeons: balancing between hepatic enzyme auto induction and toxicity. Med Mycol 47(3):276–285

Beernaert LA, Pasmans F, Van Waeyenberghe L, Dorrestein GM, Verstappen F, Vercammen F, Haesebrouck F, Martel A (2009b) Avian *Aspergillus fumigatus* strains resistant to both itraconazole and voriconazole. Antimicrob Agents Chemother 53(5):2199–2201

Beltaire KA, Cheong SH, Coutinho da Silva MA (2012) Retrospective study on equine uterine fungal isolates and antifungal susceptibility patterns (1999–2011). Equine Vet J Suppl 43:84–87

Benitah N (2006) Canine nasal aspergillosis. Clin Tech Small Anim Pract 21(2):82–88

Bensignor E, Grandemange E (2006) Comparison of an antifungal agent with a mixture of antifungal, antibiotic and corticosteroid agents for the treatment of *Malassezia* species otitis in dogs. Vet Rec 158(6):193–195

Benson JM, Nahata MC (1988) Clinical use of systemic antifungal agents. Clin Pharm 7(6): 424–438

Bergan T, Vangdal M (1983) In vitro activity of antifungal agents against yeast species. Chemotherapy 29(2):104–110

Bolard J (1986) How do the polyene macrolide antibiotics affect the cellular membrane properties? Biochim Biophys Acta 864(3–4):257–304

Boothe DM, Herring I, Calvin J, Way N, Dvorak J (1997) Itraconazole disposition after single oral and intravenous and multiple oral dosing in healthy cats. Am J Vet Res 58(8):872–877

Bowman JC, Abruzzo GK, Flattery AM, Gill CJ, Hickey EJ, Hsu MJ, Kahn JN, Liberator PA, Misura AS, Pelak BA, Wang TC, Douglas CM (2006) Efficacy of caspofungin against *Aspergillus flavus, Aspergillus terreus*, and *Aspergillus nidulans*. Antimicrob Agents Chemother 50 (12):4202–4205

Brammer KW, Farrow PR, Faulkner JK (1990) Pharmacokinetics and tissue penetration of fluconazole in humans. Rev Infect Dis 12(Suppl 3):S318–S326

Brilhante RS, Castelo Branco DS, Duarte GP, Paiva MA, Teixeira CE, Zeferino JP, Monteiro AJ, Cordeiro RA, Sidrim JJ, Rocha MF (2012) Yeast microbiota of raptors: a possible tool for environmental monitoring. Environ Microbiol Rep 4(2):189–193

Brilhante RS, de Alencar LP, Cordeiro Rde A, Castelo-Branco Dde S, Teixeira CE, Macedo Rde B, Lima DT, Paiva Mde A, Monteiro AJ, Alves ND, Franco de Oliveira M, Sidrim JJ, Rocha MF, Bandeira Tde J, Rodrigues Tde J (2013) Detection of *Candida* species resistant to azoles in the microbiota of rheas (*Rhea americana*): possible implications for human and animal health. J Med Microbiol 62(Pt 6):889–895

Brilhante RS, Bittencourt PV, Castelo-Branco Dde S, de Oliveira JS, Alencar LP, Cordeiro Rde A, Pinheiro M, Nogueira-Filho EF, Pereira-Neto Wde A, Sidrim JJ, Rocha MF (2016) Trends in antifungal susceptibility and virulence of *Candida* spp. from the nasolacrimal duct of horses. Med Mycol 54(2):147–154

Bromley MJ, van Muijlwijk G, Fraczek MG, Robson G, Verweij PE, Denning DW, Bowyer P (2014) Occurrence of azole-resistant species of *Aspergillus* in the UK environment. J Glob Antimicrob Resist 2(4):276–279

Brook D (1982) Diagnosis of equine endometrial candidiasis by direct smear and successful treatment with amphotericin B and oxytetracycline. J S Afr Vet Assoc 53(4):261–263

Brun S, Berges T, Poupard P, Vauzelle-Moreau C, Renier G, Chabasse D, Bouchara JP (2004) Mechanisms of azole resistance in petite mutants of *Candida glabrata*. Antimicrob Agents Chemother 48(5):1788–1796

Bunskoek PE, Seyedmousavi S, Gans SJ, van Vierzen PB, Melchers WJ, van Elk CE, Mouton JW, Verweij PE (2017) Successful treatment of azole-resistant invasive aspergillosis in a bottlenose dolphin with high-dose posaconazole. Med Mycol Case Rep 16:16–19

Bunting EM, Abou-Madi N, Cox S, Martin-Jimenez T, Fox H, Kollias GV (2009) Evaluation of oral itraconazole administration in captive Humboldt penguins (*Spheniscus humboldti*). J Zoo Wildl Med 40(3):508–518

Cafarchia C, Iatta R, Immediato D, Puttilli MR, Otranto D (2015) Azole susceptibility of *Malassezia pachydermatis* and *Malassezia furfur* and tentative epidemiological cut-off values. Med Mycol 53(7):743–748

Cartwright ME, Petruska J, Arezzo J, Frank D, Litwak M, Morrissey RE, MacDonald J, Davis TE (2009) Phospholipidosis in neurons caused by posaconazole, without evidence for functional neurologic effects. Toxicol Pathol 37(7):902–910

Chapman SW, Pappas P, Kauffmann C, Smith EB, Dietze R, Tiraboschi-Foss N, Restrepo A, Bustamante AB, Opper C, Emady-Azar S, Bakshi R (2004) Comparative evaluation of the efficacy and safety of two doses of terbinafine (500 and 1000 mg day(−1)) in the treatment of cutaneous or lymphocutaneous sporotrichosis. Mycoses 47(1–2):62–68

Chen BK, Friedlander SF (2001) Tinea capitis update: a continuing conflict with an old adversary. Curr Opin Pediatr 13(4):331–335

Clinkenbeard KD, Cowell RL, Tyler RD (1988) Disseminated histoplasmosis in dogs: 12 cases (1981–1986). J Am Vet Med Assoc 193(11):1443–1447

CLSI (2008) Reference method for broth dilution antifungal susceptibility testing of filamentous fungi; approved standard-second edition. CLSI document. M38-A2. Clinical and Laboratory Standards Institute, Wayne

Coelho D, Silva S, Vale-Silva L, Gomes H, Pinto E, Sarmento A, Pinheiro MD (2011) *Aspergillus viridinutans*: an agent of adult chronic invasive aspergillosis. Med Mycol 49(7):755–759

Collins MS, Pappagianis D (1977) Uniform susceptibility of various strains of *Coccidioides immitis* to amphotericin B. Antimicrob Agents Chemother 11(6):1049–1055

Conte L, Ramis J, Mis R, Vilageliu J, Basi N, Forn J (1992) Pharmacokinetic study of [C-14] flutrimazole after oral and intravenous administration in dogs – comparison with clotrimazole. Arzneimittelforschung 42-1(6):854–858

Cordeiro Rde A, de Oliveira JS, Castelo-Branco Dde S, Teixeira CE, Marques FJ, Bittencourt PV, Carvalho VL, Bandeira Tde J, Brilhante RS, Moreira JL, Pereira-Neto Wde A, Sidrim JJ, Rocha MF (2015) *Candida tropicalis* isolates obtained from veterinary sources show resistance to azoles and produce virulence factors. Med Mycol 53(2):145–152

Cordoba S, Rodero L, Vivot W, Abrantes R, Davel G, Vitale RG (2008) In vitro interactions of antifungal agents against clinical isolates of *Fusarium* spp. Int J Antimicrob Agents 31(2): 171–174

Cornelissen F, Van den Bossche H (1983) Synergism of the antimicrobial agents miconazole, bacitracin and polymyxin B. Chemotherapy 29(6):419–427

Cornely OA, Maertens J, Winston DJ, Perfect J, Ullmann AJ, Walsh TJ, Helfgott D, Holowiecki J, Stockelberg D, Goh YT, Petrini M, Hardalo C, Suresh R, Angulo-Gonzalez D (2007) Posaconazole vs. fluconazole or itraconazole prophylaxis in patients with neutropenia. N Engl J Med 356(4):348–359

Corrigan VK, Legendre AM, Wheat LJ, Mullis R, Johnson B, Bemis DA, Cepero L (2016) Treatment of disseminated aspergillosis with posaconazole in 10 dogs. J Vet Intern Med 30 (1):167–173

Court MH (2013) Feline drug metabolism and disposition: pharmacokinetic evidence for species differences and molecular mechanisms. Vet Clin North Am Small Anim Pract 43(5):1039–1054

Cuenca-Estrella M, Gomez-Lopez A, Buitrago MJ, Mellado E, Garcia-Effron G, Rodriguez-Tudela JL (2006) In vitro activities of 10 combinations of antifungal agents against the multiresistant pathogen *Scopulariopsis brevicaulis*. Antimicrob Agents Chemother 50(6):2248–2250

Cutler RE, Blair AD, Kelly MR (1978) Flucytosine kinetics in subjects with normal and impaired renal function. Clin Pharmacol Ther 24(3):333–342

Day JN, Chau TT, Lalloo DG (2013) Combination antifungal therapy for cryptococcal meningitis. N Engl J Med 368(26):2522–2523

De Carli L, Larizza L (1988) Griseofulvin. Mutat Res 195(2):91–126
De Keyser H, Van den Brande M (1983) Ketoconazole in the treatment of dermatomycosis in cats and dogs. Vet Q 5(3):142–144
de Kruijff B, Gerritsen WJ, Oerlemans A, Demel RA, van Deenen LL (1974) Polyene antibiotic-sterol interactions in membranes of Acholeplasma laidlawii cells and lecithin liposomes. I. Specificity of the membrane permeability changes induced by the polyene antibiotics. Biochim Biophys Acta 339(1):30–43
de la Torre P, Reboli AC (2014) Micafungin: an evidence-based review of its place in therapy. Core Evid 9:27–39
Dekeyser H (1981) Activity of repeated administration of enilconazole (R-23979) against *Trichophyton verrucosum* in cattle – a clinical-trial. Tijdschr Diergeneeskd 106(16):799–805
Deng S, Hu H, Abliz P, Wan Z, Wang A, Cheng W, Li R (2011) A random comparative study of terbinafine versus griseofulvin in patients with tinea capitis in Western China. Mycopathologia 172(5):365–372
Denning DW (2002) Echinocandins: a new class of antifungal. J Antimicrob Chemother 49(6): 889–891
Denning DW, Lee JY, Hostetler JS, Pappas P, Kauffman CA, Dewsnup DH, Galgiani JN, Graybill JR, Sugar AM, Catanzaro A et al (1994) NIAID mycoses study group multicenter trial of oral itraconazole therapy for invasive aspergillosis. Am J Med 97(2):135–144
Denning DW, Ribaud P, Milpied N, Caillot D, Herbrecht R, Thiel E, Haas A, Ruhnke M, Lode H (2002) Efficacy and safety of voriconazole in the treatment of acute invasive aspergillosis. Clin Infect Dis 34(5):563–571
Di Somma A, Bailey T, Silvanose C, Garcia-Martinez C (2007) The use of voriconazole for the treatment of aspergillosis in falcons (*Falco species*). J Avian Med Surg 21(4):307–316
Diaz M, Negroni R, Montero-Gei F, Castro LG, Sampaio SA, Borelli D, Restrepo A, Franco L, Bran JL, Arathoon EG, Pan-American Study Group et al (1992) A Pan-American 5-year study of fluconazole therapy for deep mycoses in the immunocompetent host. Clin Infect Dis 14(Suppl 1):S68–S76
Diflucan (2013) Product information: DIFLUCAN oral tablets, IV injection, oral suspension, fluconazole oral tablets, IV injection, oral suspension. Pfizer Inc., New York
Dodds Ashley ES, Lewis R, Lewis JS, Martin C, Andes D (2006) Pharmacology of systemic antifungal agents. Clin Infect Dis 43:S28–S39
Dolton MJ, Perera V, Pont LG, McLachlan AJ (2014) Terbinafine in combination with other antifungal agents for treatment of resistant or refractory mycoses: investigating optimal dosing regimens using a physiologically based pharmacokinetic model. Antimicrob Agents Chemother 58(1):48–54
Durdu M, Ilkit M, Tamadon Y, Tolooe A, Rafati H, Seyedmousavi S (2017) Topical and systemic antifungals in dermatology practice. Expert Rev Clin Pharmacol 10(2):225–237
Dutcher JD (1968) The discovery and development of amphotericin B. Dis Chest 54(Suppl 1):296–298
ECDC Technical Report (Kleinkauf N, Verweij PE, Arendrup MC, Donnelly PJ, Cuenca-Estrella M, Fraaije B, Melchers WJG, Adriaenssens N, Kema GHJ, Ullmann A, Bowyer P, Denning DW) (2013) Risk assessment on the impact of environmental usage of triazoles on the development and spread of resistance to medical triazoles in *Aspergillus species*. www.ecdc.europa.eu
Elewski BE (1999) Treatment of tinea capitis: beyond griseofulvin. J Am Acad Dermatol 40(6 Pt 2): S27–S30
EMA (1998) Parconazole. EMEA/MRL, 351/98-Final. The European agency for evalution of meicinal products. Veterinary medicines evluation. Unit. http://www.ema.europa.eu/docs/en_GB/document_library/Maximum_Residue_Limits_-_Report/2009/11/WC500015522.pdf
EMA (2012a) European public assessment report (EPAR) for Diflucan. European Medicines Agency
EMA (2012b) European public assessment report (EPAR) for Noxafil. European Medicines Agency
EMA (2012c) European public assessment report (EPAR) for Vfend. European Medicines Agency

Eng RH, Person A, Mangura C, Chmel H, Corrado M (1981) Susceptibility of *zygomycetes* to amphotericin B, miconazole, and ketoconazole. Antimicrob Agents Chemother 20(5):688–690
Eschenauer G, Depestel DD, Carver PL (2007) Comparison of echinocandin antifungals. Ther Clin Risk Manag 3(1):71–97
Espinel-Ingroff A, Chowdhary A, Gonzalez GM, Lass-Florl C, Martin-Mazuelos E, Meis J, Pelaez T, Pfaller MA, Turnidge J (2013) Multicenter study of isavuconazole MIC distributions and epidemiological cutoff values for *Aspergillus* spp. for the CLSI M38-A2 broth microdilution method. Antimicrob Agents Chemother 57(8):3823–3828
Estes KE, Penzak SR, Calis KA, Walsh TJ (2009) Pharmacology and antifungal properties of anidulafungin, a new echinocandin. Pharmacotherapy 29(1):17–30
Faergemann J, Zehender H, Denouel J, Millerioux L (1993) Levels of terbinafine in plasma, stratum corneum, dermis-epidermis (without stratum corneum), sebum, hair and nails during and after 250 mg terbinafine orally once per day for four weeks. Acta Derm Venereol 73(4):305–309
Faergemann J, Zehender H, Millerioux L (1994) Levels of terbinafine in plasma, stratum corneum, dermis-epidermis (without stratum corneum), sebum, hair and nails during and after 250 mg terbinafine orally once daily for 7 and 14 days. Clin Exp Dermatol 19(2):121–126
Favre B, Ryder NS (1996) Characterization of squalene epoxidase activity from the dermatophyte *Trichophyton rubrum* and its inhibition by terbinafine and other antimycotic agents. Antimicrob Agents Chemother 40(2):443–447
Fishburn F, Kelley DC (1967) Sporotrichosis in a horse. J Am Vet Med Assoc 151(1):45–46
Fisher MC, Henk DA, Briggs CJ, Brownstein JS, Madoff LC, McCraw SL, Gurr SJ (2012) Emerging fungal threats to animal, plant and ecosystem health. Nature 484(7393):186–194
Flammer K, Nettifee Osborne JA, Webb DJ, Foster LE, Dillard SL, Davis JL (2008) Pharmacokinetics of voriconazole after oral administration of single and multiple doses in African grey parrots (Psittacus erithacus timneh). Am J Vet Res 69(1):114–121
Fling ME, Kopf J, Tamarkin A, Gorman JA, Smith HA, Koltin Y (1991) Analysis of a *Candida albicans* gene that encodes a novel mechanism for resistance to benomyl and methotrexate. Mol Gen Genet 227(2):318–329
Foley JP, Legendre AM (1992) Treatment of coccidioidomycosis osteomyelitis with itraconazole in a horse. A brief report. J Vet Intern Med 6(6):333–334
Foy DS, Trepanier LA (2010) Antifungal treatment of small animal veterinary patients. Vet Clin North Am Small Anim Pract 40(6):1171–1188
Franz R, Kelly SL, Lamb DC, Kelly DE, Ruhnke M, Morschhauser J (1998) Multiple molecular mechanisms contribute to a stepwise development of fluconazole resistance in clinical *Candida albicans* strains. Antimicrob Agents Chemother 42(12):3065–3072
Frymus T, Gruffydd-Jones T, Pennisi MG, Addie D, Belak S, Boucraut-Baralon C, Egberink H, Hartmann K, Hosie MJ, Lloret A, Lutz H, Marsilio F, Mostl K, Radford AD, Thiry E, Truyen U, Horzinek MC (2013) Dermatophytosis in cats: ABCD guidelines on prevention and management. J Feline Med Surg 15(7):598–604
Furrow E, Groman RP (2009) Intranasal infusion of clotrimazole for the treatment of nasal aspergillosis in two cats. J Am Vet Med Assoc 235(10):1188–1193
Gabal MA (1986) Study on the evaluation of the use of thiabendazole in the treatment and control of bovine dermatophytosis. Mycopathologia 93(3):163–168
Gallis HA, Drew RH, Pickard WW (1990) Amphotericin B: 30 years of clinical experience. Rev Infect Dis 12(2):308–329
Gentry J, Montgerard C, Crandall E, Cruz-Espindola C, Boothe D, Bellah J (2014) Voriconazole disposition after single and multiple, oral doses in healthy, adult red-tailed hawks (*Buteo jamaicensis*). J Avian Med Surg 28(3):201–208
Glasmacher A, Molitor E, Hahn C, Bomba K, Ewig S, Leutner C, Wardelmann E, Schmidt-Wolf IG, Mezger J, Marklein G, Sauerbruch T (1998) Antifungal prophylaxis with itraconazole in neutropenic patients with acute leukaemia. Leukemia 12(9):1338–1343
Godfrey DR (1998) A case of feline paraneoplastic alopecia with secondary *Malassezia*-associated dermatitis. J Small Anim Pract 39(8):394–396

Gomez-Lopez A, Cuenca-Estrella M, Mellado E, Rodriguez-Tudela JL (2003a) In vitro evaluation of combination of terbinafine with itraconazole or amphotericin B against *Zygomycota*. Diagn Microbiol Infect Dis 45(3):199–202

Gomez-Lopez A, Garcia-Effron G, Mellado E, Monzon A, Rodriguez-Tudela JL, Cuenca-Estrella M (2003b) In vitro activities of three licensed antifungal agents against spanish clinical isolates of *Aspergillus* spp. Antimicrob Agents Chemother 47(10):3085–3088

Goncalves SS, Souza AC, Chowdhary A, Meis JF, Colombo AL (2016) Epidemiology and molecular mechanisms of antifungal resistance in *Candida* and *Aspergillus*. Mycoses 59(4): 198–219

Graupmann-Kuzma A, Valentine BA, Shubitz LF, Dial SM, Watrous B, Tornquist SJ (2008) Coccidioidomycosis in dogs and cats: a review. J Am Anim Hosp Assoc 44(5):226–235

Gray KC, Palacios DS, Dailey I, Endo MM, Uno BE, Wilcock BC, Burke MD (2012) Amphotericin primarily kills yeast by simply binding ergosterol. Proc Natl Acad Sci U S A 109(7):2234–2239

Gregson L, Goodwin J, Johnson A, McEntee L, Moore CB, Richardson M, Hope WW, Howard SJ (2013) In vitro susceptibility of *Aspergillus fumigatus* to isavuconazole: correlation with itraconazole, voriconazole, and posaconazole. Antimicrob Agents Chemother 57(11): 5778–5780

Greydanus-van der Putten SW, Klein WR, Blankenstein B, de Hoog GS, Koeman J (1994) Sporotrichosis in a horse. Tijdschr Diergeneeskd 119(17):500–502

Groll AH, Piscitelli SC, Walsh TJ (1998) Clinical pharmacology of systemic antifungal agents: a comprehensive review of agents in clinical use, current investigational compounds, and putative targets for antifungal drug development. Adv Pharmacol 44:343–500

Groll AH, Gea-Banacloche JC, Glasmacher A, Just-Nuebling G, Maschmeyer G, Walsh TJ (2003) Clinical pharmacology of antifungal compounds. Infect Dis Clin North Am 17(1):159–191, ix

Grover C, Arora P, Manchanda V (2012) Comparative evaluation of griseofulvin, terbinafine and fluconazole in the treatment of tinea capitis. Int J Dermatol 51(4):455–458

Guillot J, Bensignor E, Jankowski F, Seewald W, Chermette R, Steffan J (2003) Comparative efficacies of oral ketoconazole and terbinafine for reducing *Malassezia* population sizes on the skin of Basset Hounds. Vet Dermatol 14(3):153–157

Guinea J, Pelaez T, Recio S, Torres-Narbona M, Bouza E (2008) In vitro antifungal activities of isavuconazole (BAL4815), voriconazole, and fluconazole against 1,007 isolates of *zygomycete, Candida, Aspergillus, Fusarium*, and *Scedosporium* species. Antimicrob Agents Chemother 52(4):1396–1400

Gupta AK, Cooper EA (2008) Update in antifungal therapy of dermatophytosis. Mycopathologia 166(5–6):353–367

Gupta AK, Lyons DC (2015) The rise and fall of oral ketoconazole. J Cutan Med Surg 19(4): 352–357

Hadfield TL, Smith MB, Winn RE, Rinaldi MG, Guerra C (1987) Mycoses caused by Candida lusitaniae. Rev Infect Dis 9(5):1006–1012

Hall M, Monka C, Krupp P, O'Sullivan D (1997) Safety of oral terbinafine: results of a postmarketing surveillance study in 25,884 patients. Arch Dermatol 133(10):1213–1219

Hamill RJ (2013) Amphotericin B formulations: a comparative review of efficacy and toxicity. Drugs 73(9):919–934

Hazen EL, Brown R (1951) Fungicidin, an antibiotic produced by a soil *actinomycete*. Proc Soc Exp Biol Med 76(1):93–97

Heidelberger C, Chaudhuri NK, Danneberg P, Mooren D, Griesbach L, Duschinsky R, Schnitzer RJ, Pleven E, Scheiner J (1957) Fluorinated pyrimidines, a new class of tumour-inhibitory compounds. Nature 179(4561):663–666

Herbrecht R, Denning DW, Patterson TF, Bennett JE, Greene RE, Oestmann JW, Kern WV, Marr KA, Ribaud P, Lortholary O, Sylvester R, Rubin RH, Wingard JR, Stark P, Durand C, Caillot D, Thiel E, Chandrasekar PH, Hodges MR, Schlamm HT, Troke PF, de Pauw B, Invasive Fungal Infections Group of the European Organisation for Research and Treatment of Cancer and the

Global Aspergillus Study Group (2002) Voriconazole versus amphotericin B for primary therapy of invasive aspergillosis. N Engl J Med 347(6):408–415
Hinsch B (1988) Ketoconazole treatment of disseminated coccidioidomycosis in a dog. Mod Vet Pract 69:161–162
Hnilica K, Medleau L, Cornelius L (2000) Evaluation of the toxicity of topical enilconazole in cats. In: 4th World Congress of Veterinary Dermatology, San Francisco, CA, USA
Hodges RD, Legendre AM, Adams LG, Willard MD, Pitts RP, Monce K, Needels CC, Ward H (1994) Itraconazole for the treatment of histoplasmosis in cats. J Vet Intern Med 8(6):409–413
Hofbauer B, Leitner I, Ryder NS (2002) In vitro susceptibility of *Microsporum canis* and other dermatophyte isolates from veterinary infections during therapy with terbinafine or griseofulvin. Med Mycol 40(2):179–183
Hofstra W, de Vries-Hospers HG, van der Waaij D (1979) Concentrations of nystatin in faeces after oral administration of various doses of nystatin. Infection 7(4):166–170
Hope WW, Castagnola E, Groll AH, Roilides E, Akova M, Arendrup MC, Arikan-Akdagli S, Bassetti M, Bille J, Cornely OA, Cuenca-Estrella M, Donnelly JP, Garbino J, Herbrecht R, Jensen HE, Kullberg BJ, Lass-Florl C, Lortholary O, Meersseman W, Petrikkos G, Richardson MD, Verweij PE, Viscoli C, Ullmann AJ, ESCMID Fungal Infection Study Group (2012) ESCMID* guideline for the diagnosis and management of *Candida diseases* 2012: prevention and management of invasive infections in neonates and children caused by *Candida* spp. Clin Microbiol Infect 18(Suppl 7):38–52
Hospenthal DR, Bennett JE (1998) Flucytosine monotherapy for cryptococcosis. Clin Infect Dis 27(2):260–264
Howard SJ, Cerar D, Anderson MJ, Albarrag A, Fisher MC, Pasqualotto AC, Laverdiere M, Arendrup MC, Perlin DS, Denning DW (2009) Frequency and evolution of azole resistance in *Aspergillus fumigatus* associated with treatment failure. Emerg Infect Dis 15(7):1068–1076
Humphrey MJ, Jevons S, Tarbit MH (1985) Pharmacokinetic evaluation of UK-49,858, a metabolically stable triazole antifungal drug, in animals and humans. Antimicrob Agents Chemother 28 (5):648–653
Hussain Qadri SM, Flournoy DJ, Qadri SG, Ramirez EG (1986) Susceptibility of clinical isolates of yeasts to anti-fungal agents. Mycopathologia 95(3):183–187
Hyatt MW, Georoff TA, Nollens HH, Wells RL, Clauss TM, Ialeggio DM, Harms CA, Wack AN (2015) Voriconazole toxicity in multiple penguin species. J Zoo Wildl Med 46(4):880–888
Imhof A, Schaer DJ, Schanz U, Schwarz U (2006) Neurological adverse events to voriconazole: evidence for therapeutic drug monitoring. Swiss Med Wkly 136(45–46):739–742
Jacobs GJ, Medleau L, Calvert C, Brown J (1997) Cryptococcal infection in cats: factors influencing treatment outcome, and results of sequential serum antigen titers in 35 cats. J Vet Intern Med 11(1):1–4
Jaiswal A, Sharma RP, Garg AP (2007) An open randomized comparative study to test the efficacy and safety of oral terbinafine pulse as a monotherapy and in combination with topical ciclopirox olamine 8% or topical amorolfine hydrochloride 5% in the treatment of onychomycosis. Indian J Dermatol Venereol Leprol 73(6):393–396
Jensen JC (1989) Clinical pharmacokinetics of terbinafine (Lamisil). Clin Exp Dermatol 14(2): 110–113
Kageyama S, Masuya M, Tanaka I, Oka K, Morita K, Tamaki S, Tsuji K, Katayama N, Sugimoto H, Kagawa Y, Kojima M, Shiku H (1999) Plasma concentration of itraconazole and its antifungal prophylactic efficacy in patients with neutropenia after chemotherapy for acute leukemia. J Infect Chemother 5(4):213–216
Kano R, Okubo M, Yanai T, Hasegawa A, Kamata H (2015) First isolation of azole-resistant *Cryptococcus neoformans* from feline cryptococcosis. Mycopathologia 180(5–6):427–433
Kathiravan MK, Salake AB, Chothe AS, Dudhe PB, Watode RP, Mukta MS, Gadhwe S (2012) The biology and chemistry of antifungal agents: a review. Bioorg Med Chem 20(19):5678–5698
Katiyar SK, Edlind TD (2001) Identification and expression of multidrug resistance-related ABC transporter genes in *Candida krusei*. Med Mycol 39(1):109–116

Kersemaekers WM, Dogterom P, Xu J, Marcantonio EE, de Greef R, Waskin H, van Iersel ML (2015) Effect of a high-fat meal on the pharmacokinetics of 300-milligram posaconazole in a solid oral tablet formulation. Antimicrob Agents Chemother 59(6):3385–3389

Kohler S, Wheat LJ, Connolly P, Schnizlein-Bick C, Durkin M, Smedema M, Goldberg J, Brizendine E (2000) Comparison of the echinocandin caspofungin with amphotericin B for treatment of histoplasmosis following pulmonary challenge in a murine model. Antimicrob Agents Chemother 44(7):1850–1854

Kohler LM, Monteiro PC, Hahn RC, Hamdan JS (2004) In vitro susceptibilities of isolates of *Sporothrix schenckii* to itraconazole and terbinafine. J Clin Microbiol 42(9):4319–4320

Kohler JR, Casadevall A, Perfect J (2015) The spectrum of fungi that infects humans. Cold Spring Harb Perspect Med 5(1):a019273

Kontoyiannis DP, Lewis RE, May GS, Osherov N, Rinaldi MG (2001) *Aspergillus nidulans* is frequently resistant to amphotericin B. Mycoses 45:406–407

Korenek NL, Legendre AM, Andrews FM, Blackford PYW, Breider MA, Rinaldi MG (1994) Treatment of mycotic rhinitis with itraconazole in three horses. J Vet Intern Med 8:224–227

Kotnik T (2002) Drug efficacy of terbinafine hydrochloride (Lamisil (R)) during oral treatment of cats, experimentally infected with *Microsporum canis*. J Vet Med B Infect Dis Vet Public Health 49(3):120–122

Kotnik T, Erzen NK, Kuzner J, Drobnic-Kosorok M (2001) Terbinafine hydrochloride treatment of Microsporum canis experimentally-induced ringworm in cats. Vet Microbiol 83(2):161–168

Krawiec DR, McKiernan BC, Twardock AR, Swenson CE, Itkin RJ, Johnson LR, Kurowsky LK, Marks CA (1996) Use of an amphotericin B lipid complex for treatment of blastomycosis in dogs. J Am Vet Med Assoc 209(12):2073–2075

Krishna G, Moton A, Ma L, Medlock MM, McLeod J (2009) Pharmacokinetics and absorption of posaconazole oral suspension under various gastric conditions in healthy volunteers. Antimicrob Agents Chemother 53(3):958–966

Krishna G, Ma L, Martinho M, O'Mara E (2012) Single-dose phase I study to evaluate the pharmacokinetics of posaconazole in new tablet and capsule formulations relative to oral suspension. Antimicrob Agents Chemother 56(8):4196–4201

Krishnan-Natesan S (2009) Terbinafine: a pharmacological and clinical review. Expert Opin Pharmacother 10(16):2723–2733

Kurtz MB, Douglas CM (1997) Lipopeptide inhibitors of fungal glucan synthase. J Med Vet Mycol 35(2):79–86

Labelle AL, Hamor RE, Barger AM, Maddox CW, Breaux CB (2009) *Aspergillus flavus* keratomycosis in a cat treated with topical 1% voriconazole solution. Vet Ophthalmol 12(1): 48–52

Lackner M, de Hoog GS, Verweij PE, Najafzadeh MJ, Curfs-Breuker I, Klaassen CH, Meis JF (2012) Species-specific antifungal susceptibility patterns of *Scedosporium* and *Pseudallescheria* species. Antimicrob Agents Chemother 56(5):2635–2642

Lackner M, de Hoog GS, Yang L, Moreno LF, Ahmed SA, Andreas F, Kaltseis J, Nagl M, Lass-Florl C (2014) Proposed nomenclature for *Pseudallescheria, Scedosporium* and related genera. Fungal Divers 67(1):1–10

Lamb DC, Kelly DE, Schunck WH, Shyadehi AZ, Akhtar M, Lowe DJ, Baldwin BC, Kelly SL (1997) The mutation T315A in *Candida albicans* sterol 14alpha-demethylase causes reduced enzyme activity and fluconazole resistance through reduced affinity. J Biol Chem 272(9): 5682–5688

Lamb DC, Kelly DE, White TC, Kelly SL (2000) The R467K amino acid substitution in *Candida albicans* sterol 14 alpha-demethylase causes drug resistance through reduced affinity. Antimicrob Agents Chemother 44(1):63–67

Lass-Florl C (2011) Triazole antifungal agents in invasive fungal infections: a comparative review. Drugs 71(18):2405–2419

Lass-Florl C, Alastruey-Izquierdo A, Cuenca-Estrella M, Perkhofer S, Rodriguez-Tudela JL (2009) In vitro activities of various antifungal drugs against *Aspergillus terreus*: global assessment using the methodology of the European committee on antimicrobial susceptibility testing. Antimicrob Agents Chemother 53(2):794–795

Latimer FG, Colitz CM, Campbell NB, Papich MG (2001) Pharmacokinetics of fluconazole following intravenous and oral administration and body fluid concentrations of fluconazole following repeated oral dosing in horses. Am J Vet Res 62(10):1606–1611

Lazarus HM, Blumer JL, Yanovich S, Schlamm H, Romero A (2002) Safety and pharmacokinetics of oral voriconazole in patients at risk of fungal infection: a dose escalation study. J Clin Pharmacol 42(4):395–402

Legendre AM, Rohrbach BW, Toal RL, Rinaldi MG, Grace LL, Jones JB (1996) Treatment of blastomycosis with itraconazole in 112 dogs. J Vet Intern Med 10(6):365–371

Lewis RE (2011) Current concepts in antifungal pharmacology. Mayo Clin Proc 86(8):805–817

Lewis RE, Liao G, Wang W, Prince RA, Kontoyiannis DP (2011) Voriconazole pre-exposure selects for breakthrough mucormycosis in a mixed model of *Aspergillus fumigatus-Rhizopus oryzae* pulmonary infection. Virulence 2(4):348–355

Loffler J, Kelly SL, Hebart H, Schumacher U, Lass-Florl C, Einsele H (1997) Molecular analysis of cyp51 from fluconazole-resistant *Candida albicans* strains. FEMS Microbiol Lett 151(2): 263–268

Lopez-Ribot JL, McAtee RK, Perea S, Kirkpatrick WR, Rinaldi MG, Patterson TF (1999) Multiple resistant phenotypes of *Candida albicans* coexist during episodes of oropharyngeal candidiasis in human immunodeficiency virus-infected patients. Antimicrob Agents Chemother 43(7): 1621–1630

Lord AT, Mohandas K, Somanath S, Ambu S (2010) Multidrug resistant yeasts in synanthropic wild birds. Ann Clin Microbiol Antimicrob 9:11

Maertens JA (2004) History of the development of azole derivatives. Clin Microbiol Infect 10(Suppl 1):1–10

Maertens JA, Raad II, Marr KA, Patterson TF, Kontoyiannis DP, Cornely OA, Bow EJ, Rahav G, Neofytos D, Aoun M, Baddley JW, Giladi M, Heinz WJ, Herbrecht R, Hope W, Karthaus M, Lee DG, Lortholary O, Morrison VA, Oren I, Selleslag D, Shoham S, Thompson GR, Lee M, Maher RM, Schmitt-Hoffmann AH, Zeiher B, Ullmann AJ (2016) Isavuconazole versus voriconazole for primary treatment of invasive mould disease caused by *Aspergillus* and other filamentous fungi (SECURE): a phase 3, randomised-controlled, non-inferiority trial. Lancet 387(10020):760–769

Malani AN, Kerr L, Obear J, Singal B, Kauffman CA (2014) Alopecia and nail changes associated with voriconazole therapy. Clin Infect Dis 59(3):e61–e65

Malik R, Wigney DI, Muir DB, Gregory DJ, Love DN (1992) Cryptococcosis in cats – clinical and mycological assessment of 29 cases and evaluation of treatment using orally-administered fluconazole. J Med Vet Mycol 30(2):133–144

Malik R, Craig AJ, Wigney DI, Martin P, Love DN (1996) Combination chemotherapy of canine and feline cryptococcosis using subcutaneously administered amphotericin B. Aust Vet J 73(4):124–128

Manavathu EK, Cutright JL, Loebenberg D, Chandrasekar PH (2000) A comparative study of the in vitro susceptibilities of clinical and laboratory-selected resistant isolates of *Aspergillus* spp. to amphotericin B, itraconazole, voriconazole and posaconazole (SCH 56592). J Antimicrob Chemother 46(2):229–234

Manire CA, Rhinehart HL, Pennick GJ, Sutton DA, Hunter RP, Rinaldi MG (2003) Steady-state plasma concentrations of itraconazole after oral administration in Kemp's ridley sea turtles, Lepidochelys kempi. J Zoo Wildl Med 34(2):171–178

Marr KA, Lyons CN, Rustad TR, Bowden RA, White TC (1998) Rapid, transient fluconazole resistance in Candida albicans is associated with increased mRNA levels of CDR. Antimicrob Agents Chemother 42(10):2584–2589

Martin de la Escalera C, Aller AI, Lopez-Oviedo E, Romero A, Martos AI, Canton E, Peman J, Garcia Martos P, Martin-Mazuelos E (2008) Activity of BAL 4815 against filamentous fungi. J Antimicrob Chemother 61(5):1083–1086

Martino J, Fisher BT, Bosse KR, Bagatell R (2015) Suspected posaconazole toxicity in a pediatric oncology patient. Pediatr Blood Cancer 62(9):1682

Marty FM, Ostrosky-Zeichner L, Cornely OA, Mullane KM, Perfect JR, Thompson GR, Alangaden GJ, Brown JM, Fredricks DN, Heinz WJ, Herbrecht R, Klimko N, Klyasova G, Maertens JA, Melinkeri SR, Oren I, Pappas PG, Racil Z, Rahav G, Santos R, Schwartz S, Vehreschild JJ, Young JA, Chetchotisakd P, Jaruratanasirikul S, Kanj SS, Engelhardt M, Kaufhold A, Ito M, Lee M, Sasse C, Maher RM, Zeiher B, Vehreschild MJ, VITAL and FungiScope Mucormycosis Investigators (2016) Isavuconazole treatment for mucormycosis: a single-arm open-label trial and case-control analysis. Lancet Infect Dis 16(7):828–837

Maschmeyer G, Glasmacher A (2005) Pharmacological properties and clinical efficacy of a recently licensed systemic antifungal, caspofungin. Mycoses 48(4):227–234

Matsumoto K, Ikawa K, Abematsu K, Fukunaga N, Nishida K, Fukamizu T, Shimodozono Y, Morikawa N, Takeda Y, Yamada K (2009) Correlation between voriconazole trough plasma concentration and hepatotoxicity in patients with different CYP2C19 genotypes. Int J Antimicrob Agents 34(1):91–94

Mawby DI, Whittemore JC, Fowler LE, Papich MG (2016) Posaconazole pharmacokinetics in healthy cats after oral and intravenous administration. J Vet Intern Med 30(5):1703–1707

Mayer H (1983) Beitrag zur Therapie der Dermatomykosen beim Pferd. Berliner und Müncherner Tierärztliche Wochenschrift 96:458–459

McCurdy HD, Hepler DI, Larson KA (1981) Effectiveness of a topical antifungal agent (clotrimazole) in dogs. J Am Vet Med Assoc 179(2):163–165

McLellan GJ, Aquino SM, Mason DR, Kinyon JM, Myers RK (2006) Use of posaconazole in the management of invasive orbital aspergillosis in a cat. J Am Anim Hosp Assoc 42(4):302–307

Medleau L, Whiteweithers NE (1992) Treating and preventing the various forms of dermatophytosis. Vet Med 87(11):1096–1100

Mellado E, Garcia-Effron G, Alcazar-Fuoli L, Melchers WJ, Verweij PE, Cuenca-Estrella M, Rodriguez-Tudela JL (2007) A new *Aspergillus fumigatus* resistance mechanism conferring in vitro cross-resistance to azole antifungals involves a combination of cyp51A alterations. Antimicrob Agents Chemother 51(6):1897–1904

Miller WG, Padhye AA, van Bonn W, Jensen E, Brandt ME, Ridgway SH (2002) Cryptococcosis in a bottlenose dolphin (*Tursiops truncatus*) caused by *Cryptococcus neoformans* var. *gattii*. J Clin Microbiol 40(2):721–724

Miyazaki H, Miyazaki Y, Geber A, Parkinson T, Hitchcock C, Falconer DJ, Ward DJ, Marsden K, Bennett JE (1998) Fluconazole resistance associated with drug efflux and increased transcription of a drug transporter gene, PDH1, in *Candida glabrata*. Antimicrob Agents Chemother 42:1695–1701

Mohr J, Johnson M, Cooper T, Lewis JS, Ostrosky-Zeichner L (2008) Current options in antifungal pharmacotherapy. Pharmacotherapy 28(5):614–645

Moon WJ, Scheller EL, Suneja A, Livermore JA, Malani AN, Moudgal V, Kerr LE, Ferguson E, Vandenberg DM (2014) Plasma fluoride level as a predictor of voriconazole-induced periostitis in patients with skeletal pain. Clin Infect Dis 59(9):1237–1245

Moriello KA, Deboer DJ (1995) Efficacy of griseofulvin and itraconazole in the treatment of experimentally-induced dermatophytosis in cats. J Am Vet Med Assoc 207(4):439–444

Mukherjee PK, Sheehan D, Puzniak L, Schlamm H, Ghannoum MA (2011) Echinocandins: are they all the same? J Chemother 23(6):319–325

Nakamura Y, Saato H (1995) Sporothrix schenkii isolated from a cat in Japan. Mycoses 39:125–128

Negre A, Bensignor E, Guillot J (2009) Evidence-based veterinary dermatology: a systematic review of interventions for *Malassezia dermatitis* in dogs. Vet Dermatol 20(1):1–12

Nesbitt G, Fox P (1981) Clinical evaluation of Panolog Cream used to treat canine and feline dermatoses. Vet Med Small Anim Clin 76:535–538

Nyfeler R, Keller-Schierlein W (1974) Metabolites of microorganisms. 143. Echinocandin B, a novel polypeptide-antibiotic from *Aspergillus nidulans* var. *echinulatus*: isolation and structural components. Helv Chim Acta 57(8):2459–2477

O'Brien CR, Krockenberger MB, Martin P, Wigney DI, Malik R (2006) Long-term outcome of therapy for 59 cats and 11 dogs with cryptococcosis. Aust Vet J 84(11):384–392

Oakley KL, Moore CB, Denning DW (1999) Comparison of in vitro activity of liposomal nystatin against *Aspergillus* species with those of nystatin, amphotericin B (AB) deoxycholate, AB colloidal dispersion, liposomal AB, AB lipid complex, and itraconazole. Antimicrob Agents Chemother 43(5):1264–1266

Odds FC (2006) Drug evaluation: BAL-8557--a novel broad-spectrum triazole antifungal. Curr Opin Investig Drugs 7(8):766–772

Odds FC, Bossche HV (2000) Antifungal activity of itraconazole compared with hydroxy-itraconazole in vitro. J Antimicrob Chemother 45(3):371–373

Odds FC, Cheesman SL, Abbott AB (1986) Antifungal effects of fluconazole (UK 49858), a new triazole antifungal, in vitro. J Antimicrob Chemother 18(4):473–478

Ohwada J, Tsukazaki M, Hayase T, Oikawa N, Isshiki Y, Fukuda H, Mizuguchi E, Sakaitani M, Shiratori Y, Yamazaki T, Ichihara S, Umeda I, Shimma N (2003) Design, synthesis and antifungal activity of a novel water soluble prodrug of antifungal triazole. Bioorg Med Chem Lett 13(2):191–196

Oldenkamp EP (1979a) Natamycin treatment of ringworm in cattle in the United Kingdom. Vet Rec 105(24):5554–5556

Oldenkamp EP (1979b) Treatment of ringworm in horses with natamycin. Equine Vet J 11:36–38

Oldenkamp EP, Spanoghe L (1977) Natamycin-S treatment of ringworm in cattle. Epidemiological aspects. Tijdschr Diergeneeskd 102(2):124–125

Orosz S, Frazier DL (1995) Antifungal agents: a review of their pharmacology and therapeutic indications. J Avian Med Surg 9:8–18

Ostrosky-Zeichner L, Marr KA, Rex JH, Cohen SH (2003a) Amphotericin B: time for a new "gold standard". Clin Infect Dis 37(3):415–425

Ostrosky-Zeichner L, Rex JH, Pappas PG, Hamill RJ, Larsen RA, Horowitz HW, Powderly WG, Hyslop N, Kauffman CA, Cleary J, Mangino JE, Lee J (2003b) Antifungal susceptibility survey of 2,000 bloodstream *Candida* isolates in the United States. Antimicrob Agents Chemother 47 (10):3149–3154

Pappas PG, Bradsher RW, Chapman SW, Kauffman CA, Dine A, Cloud GA, Dismukes WE (1995) Treatment of blastomycosis with fluconazole: a pilot study. The National Institute of Allergy and Infectious Diseases Mycoses Study Group. Clin Infect Dis 20(2):267–271

Pappas PG, Rex JH, Sobel JD, Filler SG, Dismukes WE, Walsh TJ, Edwards JE, Infectious Diseases Society of America (2004) Guidelines for treatment of candidiasis. Clin Infect Dis 38(2):161–189

Pappas PG, Kauffman CA, Andes D, Benjamin DK Jr, Calandra TF, Edwards JE Jr, Filler SG, Fisher JF, Kullberg BJ, Ostrosky-Zeichner L, Reboli AC, Rex JH, Walsh TJ, Sobel JD, Infectious Diseases Society of America (2009) Clinical practice guidelines for the management of candidiasis: 2009 update by the Infectious Diseases Society of America. Clin Infect Dis 48(5):503–535

Parkes LO, Cheng MP, Sheppard DC (2016) Visual hallucinations associated with high posaconazole concentrations in serum. Antimicrob Agents Chemother 60(2):1170–1171

Parkinson T, Falconer DJ, Hitchcock CA (1995) Fluconazole resistance due to energy-dependent drug efflux in *Candida glabrata*. Antimicrob Agents Chemother 39:1696–1699

Pascual A, Calandra T, Bolay S, Buclin T, Bille J, Marchetti O (2008) Voriconazole therapeutic drug monitoring in patients with invasive mycoses improves efficacy and safety outcomes. Clin Infect Dis 46(2):201–211

Pascual A, Csajka C, Buclin T, Bolay S, Bille J, Calandra T, Marchetti O (2012) Challenging recommended oral and intravenous voriconazole doses for improved efficacy and safety: population pharmacokinetics-based analysis of adult patients with invasive fungal infections. Clin Infect Dis 55(3):381–390

Patterson TF, Thompson GR, Denning DW, Fishman JA, Hadley S, Herbrecht R, Kontoyiannis DP, Marr KA, Morrison VA, Nguyen MH, Segal BH, Steinbach WJ, Stevens DA, Walsh TJ,

Wingard JR, Young JA, Bennett JE (2016) Practice guidelines for the diagnosis and management of aspergillosis: 2016 update by the Infectious Diseases Society of America. Clin Infect Dis 63(4):e1–e60

Peeters D, Clercx C (2007) Update on canine sinonasal aspergillosis. Vet Clin North Am Small Anim Pract 37(5):901–916, vi

Pelaez T, Alvarez-Perez S, Mellado E, Serrano D, Valerio M, Blanco JL, Garcia ME, Munoz P, Cuenca-Estrella M, Bouza E (2013) Invasive aspergillosis caused by cryptic *Aspergillus* species: a report of two consecutive episodes in a patient with leukaemia. J Med Microbiol 62(Pt 3):474–478

Perfect JR, Dismukes WE, Dromer F, Goldman DL, Graybill JR, Hamill RJ, Harrison TS, Larsen RA, Lortholary O, Nguyen MH, Pappas PG, Powderly WG, Singh N, Sobel JD, Sorrell TC (2010) Clinical practice guidelines for the management of cryptococcal disease: 2010 update by the infectious diseases society of America. Clin Infect Dis 50(3):291–322

Perkhofer S, Lechner V, Lass-Florl C (2009) In vitro activity of isavuconazole against aspergillus species and *zygomycetes* according to the methodology of the European Committee on Antimicrobial Susceptibility Testing. Antimicrob Agents Chemother 53(4):1645–1647

Petranyi G, Meingassner JG, Mieth H (1987) Antifungal activity of the allylamine derivative terbinafine in vitro. Antimicrob Agents Chemother 31(9):1365–1368

Pfaller MA, Diekema DJ (2004) Rare and emerging opportunistic fungal pathogens: concern for resistance beyond *Candida albicans* and *Aspergillus fumigatus*. J Clin Microbiol 42(10):4419–4431

Pfaller MA, Messer SA, Boyken L, Huynh H, Hollis RJ, Diekema DJ (2002) In vitro activities of 5-fluorocytosine against 8,803 clinical isolates of Candida spp.: global assessment of primary resistance using National Committee for Clinical Laboratory Standards susceptibility testing methods. Antimicrob Agents Chemother 46(11):3518–3521

Pfaller MA, Diekema DJ, Messer SA, Boyken L, Hollis RJ, Jones RN (2004) In vitro susceptibilities of rare Candida bloodstream isolates to ravuconazole and three comparative antifungal agents. Diagn Microbiol Infect Dis 48(2):101–105

Pfaller MA, Messer SA, Boyken L, Rice C, Tendolkar S, Hollis RJ, Doern GV, Diekema DJ (2005) Global trends in the antifungal susceptibility of *Cryptococcus neoformans* (1990 to 2004). J Clin Microbiol 43(5):2163–2167

Pfaller MA, Messer SA, Woosley LN, Jones RN, Castanheira M (2013) Echinocandin and triazole antifungal susceptibility profiles for clinical opportunistic yeast and mold isolates collected from 2010 to 2011: application of new CLSI clinical breakpoints and epidemiological cutoff values for characterization of geographic and temporal trends of antifungal resistance. J Clin Microbiol 51(8):2571–2581

Pinchbeck LR, Hillier A, Kowalski JJ, Kwochka KW (2002) Comparison of pulse administration versus once daily administration of itraconazole for the treatment of *Malassezia pachydermatis* dermatitis and otitis in dogs. J Am Vet Med Assoc 220(12):1807–1812

Plumb DC (1999) Veterinary drug handbook. PharmaVet Publishing, White Bear Lake

Polak A (1977) 5-Fluorocytosine--current status with special references to mode of action and drug resistance. Contrib Microbiol Immunol 4:158–167

Polak A, Scholer HJ (1975) Mode of action of 5-fluorocytosine and mechanisms of resistance. Chemotherapy 21(3–4):113–130

Power SB, Malone A (1987) An outbreak of ringworm in sheep in Ireland caused by *Trichophyton verrucosum*. Vet Rec 121:218–220

Pucket JD, Allbaugh RA, Rankin AJ (2012) Treatment of dematiaceous fungal keratitis in a dog. J Am Vet Med Assoc 240(9):1104–1108

Pursley TJ, Blomquist IK, Abraham J, Andersen HF, Bartley JA (1996) Fluconazole-induced congenital anomalies in three infants. Clin Infect Dis 22(2):336–340

Reidarson TH, Harrell JH, Rinaldi MG, McBain J (1998) Bronchoscopic and serologic diagnosis of *Aspergillus fumigatus* pulmonary infection in a bottlenose dolphin (Tursiops truncatus). J Zoo Wildl Med 29(4):451–455

Rezabek GH, Friedman AD (1992) Superficial fungal infections of the skin. Diagnosis and current treatment recommendations. Drugs 43(5):674–682

Rocha MF, Bandeira SP, de Alencar LP, Melo LM, Sales JA, Paiva MA, Teixeira CE, Castelo-Branco DS, Pereira-Neto WA, Cordeiro RA, Sidrim JJ, Brilhante RS (2017) Azole resistance in *Candida albicans* from animals: highlights on efflux pump activity and gene overexpression. Mycoses 60(7):462–468

Rochette F, Engelen M, Vanden Bossche H (2003) Antifungal agents of use in animal health – practical applications. J Vet Pharmacol Ther 26(1):31–53

Roffey SJ, Cole S, Comby P, Gibson D, Jezequel SG, Nedderman AN, Smith DA, Walker DK, Wood N (2003) The disposition of voriconazole in mouse, rat, rabbit, guinea pig, dog, and human. Drug Metab Dispos 31(6):731–741

Rosales MS, Marsella R, Kunkle G, Harris BL, Nicklin CF, Lopez J (2005) Comparison of the clinical efficacy of oral terbinafine and ketoconazole combined with cephalexin in the treatment of *Malassezia* dermatitis in dogs--a pilot study. Vet Dermatol 16(3):171–176

Rudramurthy SM, Chakrabarti A, Geertsen E, Mouton JW, Meis JF (2011) In vitro activity of isavuconazole against 208 *Aspergillus flavus* isolates in comparison with 7 other antifungal agents: assessment according to the methodology of the European Committee on Antimicrobial Susceptibility Testing. Diagn Microbiol Infect Dis 71(4):370–377

Ryder NS (1992) Terbinafine: mode of action and properties of the squalene epoxidase inhibition. Br J Dermatol 126(Suppl 39):2–7

Sanglard D, Ischer F, Calabrese D, Majcherczyk PA, Bille J (1999) The ATP binding cassette transporter gene CgCDR1 from *Candida glabrata* is involved in the resistance of clinical isolates to azole antifungal agents. Antimicrob Agents Chemother 43(11):2753–2765

Sanguinetti M, Posteraro B, Fiori B, Ranno S, Torelli R, Fadda G (2005) Mechanisms of azole resistance in clinical isolates of *Candida glabrata* collected during a hospital survey of antifungal resistance. Antimicrob Agents Chemother 49(2):668–679

Schauder S (2002) Itraconazole in the treatment of tinea capitis in children. Case reports with long-term follow-up evaluation. Review of the literature. Mycoses 45(1–2):1–9

Schmitt-Hoffmann A, Roos B, Heep M, Schleimer M, Weidekamm E, Brown T, Roehrle M, Beglinger C (2006a) Single-ascending-dose pharmacokinetics and safety of the novel broad-spectrum antifungal triazole BAL4815 after intravenous infusions (50, 100, and 200 milligrams) and oral administrations (100, 200, and 400 milligrams) of its prodrug, BAL8557, in healthy volunteers. Antimicrob Agents Chemother 50(1):279–285

Schmitt-Hoffmann A, Roos B, Maares J, Heep M, Spickerman J, Weidekamm E, Brown T, Roehrle M (2006b) Multiple-dose pharmacokinetics and safety of the new antifungal triazole BAL4815 after intravenous infusion and oral administration of its prodrug, BAL8557, in healthy volunteers. Antimicrob Agents Chemother 50(1):286–293

Schultz RM, Johnson EG, Wisner ER, Brown NA, Byrne BA, Sykes JE (2008) Clinicopathologic and diagnostic imaging characteristics of systemic aspergillosis in 30 dogs. J Vet Intern Med 22(4):851–859

Semis R, Nili SS, Munitz A, Zaslavsky Z, Polacheck I, Segal E (2012) Pharmacokinetics, tissue distribution and immunomodulatory effect of intralipid formulation of nystatin in mice. J Antimicrob Chemother 67(7):1716–1721

Seyedmousavi S, Guillot J, de Hoog GS (2013a) Phaeohyphomycoses, emerging opportunistic diseases in animals. Clin Microbiol Rev 26(1):19–35

Seyedmousavi S, Rijs AJ, Melchers WJ, Mouton JW, Verweij PE (2013b) In vitro activity of isavuconazole compared with itraconazole, voriconazole, and posaconazole in azole-resistant *Aspergillus fumigatus*. Proceeding of the 53rd Interscience Conference on Antimicrobial Agents and Chemotherapy (ICAAC), Denver, CO

Seyedmousavi S, Mouton JW, Melchers WJ, Bruggemann RJ, Verweij PE (2014a) The role of azoles in the management of azole-resistant aspergillosis: from the bench to the bedside. Drug Resist Updat 17(3):37–50

Seyedmousavi S, Samerpitak K, Rijs AJ, Melchers WJ, Mouton JW, Verweij PE, de Hoog GS (2014b) Antifungal susceptibility patterns of opportunistic fungi in the genera *Verruconis* and *Ochroconis*. Antimicrob Agents Chemother 58(6):3285–3292

Seyedmousavi S, Guillot J, Arne P, de Hoog GS, Mouton JW, Melchers WJ, Verweij PE (2015a) *Aspergillus* and aspergilloses in wild and domestic animals: a global health concern with parallels to human disease. Med Mycol 53(8):765–797

Seyedmousavi S, Verweij PE, Mouton JW (2015b) Isavuconazole, a broad-spectrum triazole for the treatment of systemic fungal diseases. Expert Rev Anti-Infect Ther 13(1):9–27

Seyedmousavi S, Rafati H, Ilkit M, Tolooe A, Hedayati MT, Verweij P (2017) Systemic antifungal agents: current status and projected future developments. Methods Mol Biol 1508:107–139

Shadomy S (1969) In vitro studies with 5-fluorocytosine. Appl Microbiol 17(6):871–877

Sharp NJH, Harvey CE, Obrien JA (1991) Treatment of canine nasal aspergillosis penicilliosis with fluconazole (UK-49,858). J Small Anim Pract 32(10):513–516

Shear NH, Villars VV, Marsolais C (1991) Terbinafine: an oral and topical antifungal agent. Clin Dermatol 9(4):487–495

Shemer A, Plotnik IB, Davidovici B, Grunwald MH, Magun R, Amichai B (2013) Treatment of tinea capitis – griseofulvin versus fluconazole – a comparative study. J Dtsch Dermatol Ges 11(8):737–741. 737–742

Shivaprakash MR, Geertsen E, Chakrabarti A, Mouton JW, Meis JF (2011) In vitro susceptibility of 188 clinical and environmental isolates of *Aspergillus flavus* for the new triazole isavuconazole and seven other antifungal drugs. Mycoses 54(5):e583–e589

Sidrim JJ, Maia DC, Brilhante RS, Soares GD, Cordeiro RA, Monteiro AJ, Rocha MF (2010) Candida species isolated from the gastrointestinal tract of cockatiels (*Nymphicus hollandicus*): in vitro antifungal susceptibility profile and phospholipase activity. Vet Microbiol 145(3–4): 324–328

Singer LM, Meyer W, Firacative C, Thompson GR, Samitz E, Sykes JE (2014) Antifungal drug susceptibility and phylogenetic diversity among *Cryptococcus* isolates from dogs and cats in North America. J Clin Microbiol 52(6):2061–2070

Smith J, Safdar N, Knasinski V, Simmons W, Bhavnani SM, Ambrose PG, Andes D (2006) Voriconazole therapeutic drug monitoring. Antimicrob Agents Chemother 50(4):1570–1572

Smith JA, Papich MG, Russell G, Mitchell MA (2010) Effects of compounding on pharmacokinetics of itraconazole in black-footed penguins (Spheniscus demersus). J Zoo Wildl Med 41(3):487–495

Smith KM, Pucket JD, Gilmour MA (2014) Treatment of six cases of equine corneal stromal abscessation with intracorneal injection of 5% voriconazole solution. Vet Ophthalmol 17(Suppl 1):179–185

Sporanox (2017) Product information: SPORANOX oral capsules. Janssen Pharmaceuticals Inc., Titusville

Stewart J, Bianco D (2017) Treatment of refractory sino-nasal aspergillosis with posaconazole and terbinafine in 10 dogs. J Small Anim Pract 58(9):504–509

Sugiura K, Sugimoto N, Hosaka S, Katafuchi-Nagashima M, Arakawa Y, Tatsumi Y, Jo Siu W, Pillai R (2014) The low keratin affinity of efinaconazole contributes to its nail penetration and fungicidal activity in topical onychomycosis treatment. Antimicrob Agents Chemother 58(7): 3837–3842

Sun HY, Singh N (2011) Mucormycosis: its contemporary face and management strategies. Lancet Infect Dis 11(4):301–311

Sutton DA, Sanche SE, Revankar SG, Fothergill AW, Rinaldi MG (1999) In vitro amphotericin B resistance in clinical isolates of *Aspergillus terreus*, with a head-to-head comparison to voriconazole. J Clin Microbiol 37(7):2343–2345

Sykes JE, Torres SM, Armstrong PJ, Lindeman CJ (2001) Itraconazole for treatment of sporotrichosis in a dog residing on a Christmas tree farm. J Am Vet Med Assoc 218(9):1440

Talbot JJ, Kidd SE, Martin P, Beatty JA, Barrs VR (2015) Azole resistance in canine and feline isolates of *Aspergillus fumigatus*. Comp Immunol Microbiol Infect Dis 42:37–41
Tan K, Brayshaw N, Tomaszewski K, Troke P, Wood N (2006) Investigation of the potential relationships between plasma voriconazole concentrations and visual adverse events or liver function test abnormalities. J Clin Pharmacol 46(2):235–243
Tassel D, Madoff MA (1968) Treatment of *Candida* sepsis and *Cryptococcus* meningitis with 5-fluorocytosine. A new antifungal agent. JAMA 206(4):830–832
Taylor A, Talbot J, Bennett P, Martin P, Makara M, Barrs VR (2014) Disseminated *Scedosporium prolificans* infection in a Labrador retriever with immune mediated haemolytic anaemia. Med Mycol Case Rep 6:66–69
Tell LA, Clemons KV, Kline Y, Woods L, Kass PH, Martinez M, Stevens DA (2010) Efficacy of voriconazole in Japanese quail (Coturnix japonica) experimentally infected with *Aspergillus fumigatus*. Med Mycol 48(2):234–244
Tortorano AM, Prigitano A, Dho G, Piccinini R, Dapra V, Viviani MA (2008) In vitro activity of conventional antifungal drugs and natural essences against the yeast-like alga *Prototheca*. J Antimicrob Chemother 61(6):1312–1314
Troke PF, Hockey HP, Hope WW (2011) Observational study of the clinical efficacy of voriconazole and its relationship to plasma concentrations in patients. Antimicrob Agents Chemother 55(10):4782–4788
Trommer H, Neubert RH (2006) Overcoming the stratum corneum: the modulation of skin penetration. A review. Skin Pharmacol Physiol 19(2):106–121
Ullmann AJ, Lipton JH, Vesole DH, Chandrasekar P, Langston A, Tarantolo SR, Greinix H, Morais de Azevedo W, Reddy V, Boparai N, Pedicone L, Patino H, Durrant S (2007) Posaconazole or fluconazole for prophylaxis in severe graft-versus-host disease. N Engl J Med 356(4):335–347
Ullmann AJ, Akova M, Herbrecht R, Viscoli C, Arendrup MC, Arikan-Akdagli S, Bassetti M, Bille J, Calandra T, Castagnola E, Cornely OA, Donnelly JP, Garbino J, Groll AH, Hope WW, Jensen HE, Kullberg BJ, Lass-Florl C, Lortholary O, Meersseman W, Petrikkos G, Richardson MD, Roilides E, Verweij PE, Cuenca-Estrella M, ESCMID Fungal Infection Study Group (2012) ESCMID* guideline for the diagnosis and management of *Candida* diseases 2012: adults with haematological malignancies and after haematopoietic stem cell transplantation (HCT). Clin Microbiol Infect 18(Suppl 7):53–67
Vaden SL, Heit MC, Hawkins EC, Manaugh C, Riviere JE (1997) Fluconazole in cats: pharmacokinetics following intravenous and oral administration and penetration into cerebrospinal fluid, aqueous humour and pulmonary epithelial lining fluid. J Vet Pharmacol Ther 20(3):181–186
Van Cauteren H, Heykants J, De Coster R, Cauwenbergh G (1987) Itraconazole: pharmacologic studies in animals and humans. Rev Infect Dis 9(Suppl 1):S43–S46
Van Cutsem J, Van Gerven F, Geerts H, Rochette F (1985) Treatment with enilconazole spray of dermatophytosis in rabbit farms. Mykosen 28(8):400–407
Van Cutsem J, Van Gerven F, Janssen PA (1988) In vitro activity of enilconazole against *Aspergillus* spp. and its fungicidal efficacy in a smoke generator against *Aspergillus fumigatus*. Mycoses 31(3):143–147
Van Der Linden JW, Warris A, Verweij PE (2011) *Aspergillus* species intrinsically resistant to antifungal agents. Med Mycol 49(Suppl 1):S82–S89
Vanden Bossche H, Marichal P, Gorrens J, Bellens D, Moereels H, Janssen AJ (1990) Mutation in cytochrome P-450-dependent 14a-demethylase results in decreased affinity for azole antifungals. Biochem Soc Trans 18:56–59
Varga J, Houbraken J, Van Der Lee HA, Verweij PE, Samson RA (2008) *Aspergillus calidoustus* sp. nov., causative agent of human infections previously assigned to Aspergillus ustus. Eukaryot Cell 7(4):630–638
Vermeulen E, Lagrou K, Verweij PE (2013) Azole resistance in *Aspergillus fumigatus*: a growing public health concern. Curr Opin Infect Dis 26(6):493–500

Verweij PE, Mellado E, Melchers WJ (2007) Multiple-triazole-resistant aspergillosis. N Engl J Med 356(14):1481–1483

Verweij PE, Te Dorsthorst DT, Janssen WH, Meis JF, Mouton JW (2008) In vitro activities at pH 5.0 and pH 7.0 and in vivo efficacy of flucytosine against *Aspergillus fumigatus*. Antimicrob Agents Chemother 52(12):4483–4485

Verweij PE, Howard SJ, Melchers WJ, Denning DW (2009a) Azole-resistance in *Aspergillus*: proposed nomenclature and breakpoints. Drug Resist Updat 12(6):141–147

Verweij PE, Snelders E, Kema GH, Mellado E, Melchers WJ (2009b) Azole resistance in *Aspergillus fumigatus*: a side-effect of environmental fungicide use? Lancet Infect Dis 9(12):789–795

Verweij PE, Ananda-Rajah M, Andes D, Arendrup MC, Bruggemann RJ, Chowdhary A, Cornely OA, Denning DW, Groll AH, Izumikawa K, Kullberg BJ, Lagrou K, Maertens J, Meis JF, Newton P, Page I, Seyedmousavi S, Sheppard DC, Viscoli C, Warris A, Donnelly JP (2015) International expert opinion on the management of infection caused by azole-resistant *Aspergillus fumigatus*. Drug Resist Updat 21–22:30–40

Verweij PE, Chowdhary A, Melchers WJ, Meis JF (2016) Azole resistance in *Aspergillus fumigatus*: can we retain the clinical use of mold-active antifungal azoles? Clin Infect Dis 62 (3):362–368

Walsh TJ, Melcher GP, Rinaldi MG, Lecciones J, McGough DA, Kelly P, Lee J, Callender D, Rubin M, Pizzo PA (1990) *Trichosporon beigelii*, an emerging pathogen resistant to amphotericin B. J Clin Microbiol 28(7):1616–1622

Walsh TJ, Karlsson MO, Driscoll T, Arguedas AG, Adamson P, Saez-Llorens X, Vora AJ, Arrieta AC, Blumer J, Lutsar I, Milligan P, Wood N (2004) Pharmacokinetics and safety of intravenous voriconazole in children after single- or multiple-dose administration. Antimicrob Agents Chemother 48(6):2166–2172

Walsh TJ, Anaissie EJ, Denning DW, Herbrecht R, Kontoyiannis DP, Marr KA, Morrison VA, Segal BH, Steinbach WJ, Stevens DA, van Burik JA, Wingard JR, Patterson TF, Infectious Diseases Society of America (2008) Treatment of aspergillosis: clinical practice guidelines of the Infectious Diseases Society of America. Clin Infect Dis 46(3):327–360

Wang DY, Gricourt M, Thierry S, Arné P, Seguin D, Deville M, Dannaoui E, Chermette R, Botterel F, Guillot J (2013) Mutations in the cyp51A gene and susceptibility to itraconazole in *Aspergillus fumigatus* isolated from avian farms in France and China. Med Mycol 93 (1):12–15

Warn PA, Sharp A, Denning DW (2006) In vitro activity of a new triazole BAL4815, the active component of BAL8557 (the water-soluble prodrug), against *Aspergillus* spp. J Antimicrob Chemother 57(1):135–138

Warn PA, Sharp A, Parmar A, Majithiya J, Denning DW, Hope WW (2009) Pharmacokinetics and pharmacodynamics of a novel triazole, isavuconazole: mathematical modeling, importance of tissue concentrations, and impact of immune status on antifungal effect. Antimicrob Agents Chemother 53(8):3453–3461

Warrilow AG, Martel CM, Parker JE, Melo N, Lamb DC, Nes WD, Kelly DE, Kelly SL (2010) Azole binding properties of *Candida albicans* sterol 14-alpha demethylase (CaCYP51). Antimicrob Agents Chemother 54(10):4235–4245

White TC (1997) The presence of an R467K amino acid substitution and loss of allelic variation correlate with an azole-resistant lanosterol 14alpha demethylase in *Candida albicans*. Antimicrob Agents Chemother 41(7):1488–1494

Willard MD, Nachreiner RF, Howard VC, Fooshee SK (1986) Effect of long-term administration of ketoconazole in cats. Am J Vet Res 47(12):2510–2513

Williams K, Mansh M, Chin-Hong P, Singer J, Arron ST (2014) Voriconazole-associated cutaneous malignancy: a literature review on photocarcinogenesis in organ transplant recipients. Clin Infect Dis 58(7):997–1002

Wray JD, Sparkes AH, Johnson EM (2008) Infection of the subcutis of the nose in a cat caused by Mucor species: successful treatment using posaconazole. J Feline Med Surg 10(5):523–527
Yanong RP (2003) Fungal diseases of fish. Vet Clin North Am Exot Anim Pract 6(2):377–400
Zamos DT, Schumacher J, Loy JK (1996) Nasopharyngeal conidiobolomycosis in a horse. J Am Vet Med Assoc 208(1):100–101
Ziolkowska G, Tokarzewski S, Nowakiewicz A (2014) Drug resistance of *Aspergillus fumigatus* strains isolated from flocks of domestic geese in Poland. Poult Sci 93(5):1106–1112

Index

S. Seyedmousavi et al. (eds.), *Emerging and Epizootic Fungal Infections in Animals*,
https://doi.org/10.1007/978-3-319-72093-7

The manufacturer's authorised representative in the EU is Springer Nature Customer Service Centre GmbH, Europaplatz 3, 69115 Heidelberg, Germany. If you have any concerns regarding our products, please contact ProductSafety@springernature.com

Printed and bound by CPI Group (UK) Ltd, Croydon, CR0 4YY
01/10/2026
02225972-0002